Composite Materials

Composite Materials
Mechanics, Manufacturing and Modeling

Sumit Sharma

CRC Press
Taylor & Francis Group
Boca Raton London New York

CRC Press is an imprint of the
Taylor & Francis Group, an **informa** business

First edition published 2021
by CRC Press
6000 Broken Sound Parkway NW, Suite 300, Boca Raton, FL 33487-2742

and by CRC Press
2 Park Square, Milton Park, Abingdon, Oxon, OX14 4RN

First edition published by CRC Press 2021

CRC Press is an imprint of Taylor & Francis Group, LLC

ISBN: 9780367687557 (hbk)
ISBN: 9781003147756 (ebk)

Typeset in Times
by codeMantra

My Parents

Mr. Ashok Sharma & Mrs. Nirmal Sharma

My Daughter & Wife

Dhriti Sharma & Mrs. Rajni Sharma

My Supervisors

Late Prof. Rakesh Chandra

Dr. Pramod Kumar

Contents

Preface

This book has resulted due to the guidance of my supervisors **Dr. Rakesh Chandra** and **Dr. Pramod Kumar** at Dr. B.R. Ambedkar National Institute of Technology, Jalandhar. I am very grateful to my parents, my wife, and my little daughter, who have always motivated me during my life. There was an urgent need for such a book because it caters to the need of students and researchers working in the field of composite materials. This book will provide the readers with an overview of the basic terminology associated with composites. The author is working in the area of composite materials for the last 12 years. He has also worked in the area of molecular dynamics (MD) simulation of composites for many years. Before that, the author has worked in the area of "finite element modeling" of composites using NISA and MATLAB.

After reading Chapter 1, the readers will be able to define and identify various types of composite materials. Why composites are necessary and what are their applications? These topics have been discussed at length in this chapter. A comparative view has been presented with respect to the properties of composites in comparison with other materials. This will enable the readers to gauge the importance of composite materials in various areas such as aerospace, missiles, automobiles, and medical and renewable filed. A detailed classification of composite materials has been made in this chapter.

Chapter 2 focuses on the basic ingredients of a fiber-reinforced composite, namely, the fiber and the matrix and, to some extent, the interface. Though this chapter relies largely on a materials science and chemistry perspective, it is nevertheless important for researchers to be aware of this terminology and these basic ideas, particularly if the mechanician is to work in an interdisciplinary environment. Various types of fibers such as plant/vegetable fibers, animal fibers, and advanced fibers have been discussed in detail. Also, different types of matrices, viz., metallic, polymeric, and ceramic, have been explained with examples and their applications.

In Chapter 3, different techniques of manufacturing composites have been discussed in detail. The methods discussed include the basics of hand layup, filament winding, resin transfer molding (RTM), and pultrusion for polymer matrix composites. For metal matrix composites, the techniques discussed are powder metallurgy, diffusion bonding, spark plasma sintering, compocasting, etc. Also, techniques such as melt infiltration and polymer infiltration and pyrolysis (PIP) have been explained in detail for the benefit of the readers. The basics of curing have also been discussed at length. After reading this chapter, the readers will be able to make their own composites.

In Chapter 4, the basics of mechanics of composite materials have been discussed in detail. The concepts related to lamina, laminate, tensors, continuity, and compatibility equations have been explained in order to give the readers an insight into the mechanics of composites. Minimum principles, which have their foundation in the calculus of variations and have proven to be very effective for obtaining approximate solutions to problems in solid mechanics, have also been explained. The constitutive equations of stress and strain have also been dealt with at length. The basics of anisotropy, monoclinic material, orthotropic material, transversely isotropic material, and isotropic material have been explained with the help of several equations and figures.

In Chapter 5, the fundamental equations of the three-dimensional stress–strain relations for fiber-reinforced composites have been developed. These equations form the base for developing the two-dimensional stress–strain relations in further chapters. The stress–strain relations developed in this chapter have been obtained by smearing the properties of the fiber and the matrix in an equivalent homogeneous orthotropic material. The compliance and stiffness matrices have been developed for various types of composites. In addition, the effects of free thermal strain and free moisture strain have also been explained in detail.

Chapter 6 discusses several kinds of micromechanical models. Results from the various models are presented, and comparisons among the models are made. The primary interest with the models is the prediction of composite material properties. However, to provide insight into failure, the stresses in the fiber and the matrix have also been discussed. The approach will be to introduce some of the more complex models first. In considering fibers and the surrounding matrix, it has been assumed that the fibers are spaced periodically in square-packed or hexagonal-packed arrays. It has been assumed that the fibers are infinitely long. Three approaches for determining the four elastic moduli have been discussed in detail, namely, the strength-of-materials approach, semi-empirical modeling approach, and elasticity approach.

One of the most frequently used key assumptions in the analysis of the mechanical behavior of materials, the plane stress assumption, is the topic of Chapter 7. The three-dimensional stress–strain behavior of Chapter 5 is simplified to account for the plane stress assumption, including thermal expansion effects. The consequences of these simplifications are emphasized with numerical examples. In Chapter 8, the response of an off-axis element of fiber-reinforced material in a state of plane stress has been explained in detail. The concept of coupling of various stress and strain components has been well developed. The response of a fiber-reinforced material under the effect of free thermal strains as well as free moisture strains has been explained at length. Also, the coupling of in-plane thermal effects through Poisson's ratios and through the thickness thermal effects has been explained in a lucid manner. Understanding of these concepts will help the readers in finding the response of a laminate, which is the subject of Chapter 9.

Chapter 9 starts with the nomenclature of different types of laminates. This will enable the readers to analyze the orientation of each lamina in a laminate. The Kirchhoff hypothesis has been used to derive the laminate strains and the corresponding displacements. The implications of using the Kirchhoff hypothesis on the in-plane and out-of-plane displacements have been discussed in detail. In Chapter 10, the missing link in the interpretation of the concepts developed so far has been discussed in detail. The procedure to find the force and moment resultants from the given reference surface strains and curvatures, and vice versa, has been explained with the help of suitable examples. The laminate stiffness matrix, also called the ABD matrix, has been developed for several laminates. In the end, a brief classification of laminates has been done. The effect of the type of laminate on the ABD matrix has also been discussed.

In Chapter 11, the failure theories popularly used for lamina failure analysis of a composite material have been discussed in brief. The aim of this chapter is to give the readers an introduction to the various failure criteria, viz., Hill's theory of failure, Tsai–Hill theory of failure, Hoffman theory of failure, maximum stress failure theory, maximum strain theory, the Tsai–Wu failure criterion, and Hashin theory. After reading this chapter, the readers will be able to apply the concepts for the failure analysis of a lamina of any type of composite material. In Chapter 12, several approaches and methods of the analysis of the interrelations between the microstructures, and the mechanical behavior and strength of materials in particular have been discussed. This chapter reviews and evaluates models that predict the stiffness of short-fiber composites. These include the dilute model based on Eshelby's equivalent inclusion, the self-consistent model for finite-length fibers, Mori–Tanaka-type models, bounding models, the Halpin–Tsai equation, and shear-lag models.

In Chapter 13, the basics of fracture mechanics have been explained with special focus on the interfacial damage. The sources of energy absorption have been dealt with in detail. The slow crack growth in the composites is examined for the conditions where the fast fracture is not favored energetically. In Chapter 14, the equilibrium considerations have been made to demonstrate the conditions under which the interlaminar shear and normal stresses must be nonzero over some portion of any plane $z=z^*$. It has been shown that the existence of these interlaminar stresses can often be determined using lamination theory. There are some cases where interlaminar stresses are nonzero but self-equilibrating, in which case lamination theory cannot prove existence. Interlaminar shear forces and bending moments under uniform strain and curvature loadings have been derived. These will help the readers in calculating the interlaminar stresses for any laminate.

In Chapter 15, the tools necessary to study flat laminated plates have been developed and their application to several problems that illustrate the unique response characteristics of fiber-reinforced structures in general and plates in particular has been discussed. The governing equations and boundary conditions for laminated plates have been discussed in detail. The Kirchhoff free-edge condition has been explained in a lucid manner. The governing equations for laminated plates have also been derived in displacement form. Lastly, the governing equations have been simplified by making certain assumptions. In Chapter 16, the viscoelastic and dynamic behavior of composites has been explained with the help of suitable equations and figures. The viscoelastic analysis includes the Boltzmann superposition integral, spring–dashpot models, the quasi-elastic approach, complex modulus, and the elastic–viscoelastic correspondence principle. The dynamic behavior includes the longitudinal wave propagation, flexural vibration, and damping analysis of composites. Chapter 17 explains the basic terminology associated with mechanical testing of composites. First, various societies for testing standards have been listed, viz., ASTM International, Composites Research Advisory Group, and Society of Automobile Engineers. Second, various primary and physical properties have been discussed in detail. These include tensile strength, fiber volume fraction, and void content. Lastly, the standards used for testing of composites have been discussed. The testing methods include the tensile testing, compression test, and shear test.

This chapter will enable the reader to practice any of the above-stated tests for predicting the properties of a composite material.

An attempt has been made here to cover thoroughly all the topics related to manufacturing, mechanics, and modeling of composites so that the users working in this area can use this book as the text for as per their requirement. The author will be highly grateful to the potential readers for sending their valuable suggestions, if any, so that this book can be improved further.

Sumit Sharma
Dr. B.R. Ambedkar National Institute of Technology
Jalandhar-144011, Punjab
INDIA
August 2020

MATLAB® is a registered trademark of The MathWorks, Inc. For product information, please contact:

The MathWorks, Inc.
3 Apple Hill Drive
Natick, MA 01760-2098 USA
Tel: 508-647-7000
Fax: 508-647-7001
E-mail: info@mathworks.com
Web: www.mathworks.com

Author

Dr. Sumit Sharma is working as an Assistant Professor in the Department of Mechanical Engineering in Dr. B.R. Ambedkar National Institute of Technology (NIT) Jalandhar, Punjab. Before joining this institute, he worked as an Assistant Professor in the School of Mechanical Engineering at Lovely Professional University, Phagwara. He completed his Ph.D. in Composite Materials from NIT Jalandhar in 2015. He did his M.Tech. in Mechanical Engineering (Gold Medalist), also from NIT Jalandhar in 2010. He graduated in Mechanical Engineering (with honors) from Kurukshetra University in 2007.

Dr. Sumit Sharma is working in the area of composites for the last 12 years. He has more than 40 research articles in reputed journals such as *Computational Materials Science*, *Composites Part B*, *Composite Science and Technology*, *Journal of Composite Materials*, *Journal of Molecular Modeling*, *JOM*, and *IEEE*. He is also the reviewer of various journals such as *Computational Materials Science*, *Composites Part B*, *Composite Science and Technology*, *Computational Condensed Matter*, and *Carbon*. He is a member of ASTM International, MRS, and ISME societies. He has published the following books:

1. **Sharma S.** (2020), *An Introduction to Molecular Dynamics Simulation of Polymer Composites*, NOVA Publishers, ISBN: 9781536174083.
2. Han B., **Sharma S.**, Nguyen T.A., Longbiao L., Bhat K.S. (2020), *Fiber-Reinforced Nanocomposites: Fundamentals and Applications*, Elsevier Publishers, ISBN:9780128199046.
3. **Sharma S.** (2019), *Molecular Dynamics Simulation of NanoComposites Using Biovia Materials Studio, LAMMPS and Gromacs*, Elsevier Publishers, ISBN: 9780128169544.
4. **Sharma S.** (2019), *Metallic Glass Based Nanocomposites: Molecular Dynamics Study Of Properties*, Boca Raton, CRC Press, ISBN: 9780367076702.

His research interests include molecular dynamics simulations, mechanics of composite materials, metallic glasses, multiscale modeling of composites, strength of materials, materials science and engineering, fracture mechanics, mechanical vibrations, and finite element modeling.

1 Introduction

An article was published by Professor A. Kelly [1] in Composites Science and Technology titled "Composites in Context." In that article, large-scale social changes which have influenced the development of new materials were reviewed, and new materials and processing methods were described and contrasted with some recent advances in composite materials science. Emerging technologies at the time included in situ metal matrix composites, carbon fiber-reinforced thermoplastic composites, SiC-reinforced aluminum as well as toughening of ceramics through the use of fiber reinforcement. Tremendous developments have been made in many aspects of composites research and technology during the two decades since the publication of Kelly's paper. Recent advances in producing nanostructured materials with novel material properties have stimulated research to create multifunctional macroscopic engineering materials by designing structures at the nanometer scale. Motivated by the recent enthusiasm in nanotechnology, development of nanocomposites is one of the rapidly evolving areas of composites research.

According to American Ceramic Society, nanotechnology can be broadly defined as "The creation, processing, characterization, and utilization of materials, devices, and systems with dimensions of the order of 1–100 nm, exhibiting novel and significantly enhanced physical, chemical, and biological properties, functions, phenomena, and processes due to their nanoscale size." By US National Nanotechnology Initiative (NNI) standards, nanotechnology involves the following according to Hunt [2]:

a. Research and technology development at the atomic, molecular, or macromolecular levels, approximately 1–100 nm in length (approximately 80,000 smaller than a human hair)
b. Creation and use of structures, devices, and systems that have novel properties and functions because of their small and/or intermediate size
c. Ability to control or manipulate on the atomic scale.

Current interests in nanotechnology encompass nano-biotechnology, nanosystems, nanoelectronics, and nanostructured materials, of which nanocomposites are a significant part. Through nanotechnology, it is envisioned that nanostructured materials will be developed using a bottom-up approach. "More materials and products will be made from the bottom-up, that is, by building them from atoms, molecules, and the nanoscale powders, fibers, and other small structural components made from them. This differs from all previous manufacturing, in which raw materials get pressed, cut, molded, and otherwise coerced into parts and products." Scientists and engineers working with fiber-reinforced composites have practiced this bottom-up approach in processing and manufacturing for decades. When designing a composite, the material properties are tailored for the desired performance across various length scales. From selection and processing of matrix and fiber materials, and design and optimization of the fiber–matrix interface/interphase at the submicron scale to the manipulation of yarn bundles in 2-D and 3-D textiles to the layup of lamina in laminated composites and finally the net-shape forming of the macroscopic composite part, the integrated approach used in composites processing is a remarkable example in the successful use of the "bottom-up" approach.

Expansion of length scales from meters (finished woven composite parts), micrometers (fiber diameter), and sub-micrometers (fiber–matrix interphase) to nanometers (nanotube diameter) presents tremendous opportunities for innovative approaches in the processing, characterization, and analysis/modeling of the new generation of composite materials. As scientists and engineers seek to make practical materials and devices from nanostructures, understanding material behavior across length scales from the atomistic to macroscopic levels is required. Knowledge of how the nanoscale structure influences the bulk properties will enable the design of the nanostructure to create multifunctional composites.

A morphological characteristic that is of fundamental importance in understanding of structure–property relationship of nanocomposites is the surface area/volume ratio of the reinforcement materials. The change in particle diameter, layer thickness, or fibrous material diameter from micrometer to nanometer changes the ratio by three orders in magnitude. At this scale, there is often distinct size dependence of material properties. In addition, with the drastic increase in interfacial area, properties of the composite become dominated more by properties of the interface or interphase.

1.1 WHAT IS A COMPOSITE?

Many materials are effectively composites. This is particularly true of natural biological materials, which are often made up of at least two constituents. In many cases, a strong and stiff component is present, often in elongated form, embedded in a softer constituent forming the *matrix*. For example, wood is made up of fibrous chains of cellulose molecules in a matrix of lignin, while bone and teeth are both essentially composed of hard inorganic crystals (hydroxyapatite or osteones) in a matrix of a tough organic constituent called collagen. Commonly, such composite materials show a marked *anisotropy* – that is to say, their properties vary significantly when measured in different directions. This usually arises because the harder constituent is in fibrous form with the fiber axes preferentially aligned in particular directions. In addition, one or more of the constituents may exhibit inherent anisotropy as a result of their crystal structure. In natural materials, such anisotropy of mechanical properties is often exploited within the structure. For example, wood is much stronger in the direction of the fiber tracheids, which are usually aligned parallel to the axis of the trunk or branch, than it is in the transverse directions. High strength is required in the axial direction since a branch becomes loaded like a cantilevered beam by its own weight and the trunk is stressed in a similar way by the action of the wind.

Such beam bending causes high stresses along its length, but not through the thickness. Now, a formal definition of a composite may be given as follows:

A composite is a structural material that consists of two or more constituents that are combined at a macroscopic level and are not soluble in each other. One constituent is called the reinforcing phase, and the one in which it is embedded is called the matrix. The reinforcing phase material may be in the form of fibers, particles, or flakes. The matrix phase materials are generally continuous. Examples of composite systems include concrete reinforced with steel and epoxy reinforced with graphite fibers, etc.

Thus, a composite material is heterogeneous at a microscopic scale but statistically homogeneous at a macroscopic scale. The materials that form the composite are also called as *constituents* or *constituent materials*. The constituent materials of a composite have significantly different properties. Further, it should be noted that the properties of the composite formed may not be obtained from these constituents. However, a combination of two or more materials with significant properties will not suffice to be called as a composite material. In general, the following conditions must be satisfied to be called a composite material:

i. The combination of materials should result in significant property changes. One can see significant changes when one of the constituent materials is in platelet or fibrous from.
ii. The content of the constituents is generally more than 10% (by volume).
iii. In general, the property of one constituent is much greater (approx. 5 times) than the corresponding property of the other constituent.

The composite materials can be natural or artificially made materials. In the following sections, we will see the examples of these materials.

1.2 WHY COMPOSITES?

There is unabated thirst for new materials with improved desired properties. All the desired properties are difficult to find in a single material. For example, a material that needs high fatigue life may not be cost-effective. The list of the desired properties, depending upon the requirement of the application, is given below:

i. Strength
ii. Stiffness
Iii. Toughness
iv. High corrosion resistance
v. High wear resistance
vi. High chemical resistance
vii. High environmental degradation resistance
viii. Reduced weight
ix. High fatigue life

TABLE 1.1
Specific Modulus and Specific Strength of Different Materials

Material	Specific Gravity	Young's Modulus (GPa)	Ultimate Strength (MPa)	Specific Modulus (E/ρ), GPa-m^{3}kg^{-1}	Specific Strength (S/ρ), MPa-m^{3}kg^{-1})
Unidirectional graphite/epoxy	1.6	181	1500	0.1131	0.9377
Unidirectional glass/epoxy	1.8	38.60	1062	0.02144	0.5900
Cross-ply graphite/epoxy	1.6	95.98	373	0.06000	0.2331
Cross-ply glass/epoxy	1.8	23.58	88.25	0.01310	0.0490
Quasi-isotropic graphite/epoxy	1.6	69.64	276.48	0.04353	0.1728
Quasi-isotropic glass/epoxy	1.8	18.96	73.08	0.01053	0.0406
Steel	7.8	206.84	648.1	0.02652	0.08309
Aluminum	2.6	68.95	275.8	0.02652	0.1061

x. Thermal insulation or conductivity
xi. Electrical insulation or conductivity
xii. Acoustic insulation
xiii. Radar transparency
xiv. Energy dissipation
xv. Reduced cost
xvi. Attractiveness.

The list of desired properties is in-exhaustive. It should be noted that the most important characteristics of composite materials is that their properties are *tailorable*; that is, one can design the required properties. Table 1.1 shows the specific modulus (E/ρ) and the specific strength (S/ρ) of various materials. From Table 1.1, it can be clearly seen that composites have higher values of E/ρ and S/ρ in comparison with other materials. In considering the formulation of a composite material for a particular type of application, it is important to consider the properties exhibited by the potential constituents. The properties of particular interest are the stiffness (Young's modulus), strength, and toughness. Density is of great significance in many situations since the mass of the component may be of critical importance. Thermal properties, such as expansivity and conductivity, must also be taken into account. In particular, because composite materials are subject to temperature changes (during manufacture and/or in service), a mismatch between the thermal expansivities of the constituents leads to internal residual stresses. These can have a strong effect on the mechanical behavior. Some representative property data are shown in Table 1.2 for various types of matrix and reinforcement, as well as for some typical engineering materials and a few representative composites. Inspection of the data shows that some attractive property combinations, for example, high stiffness/strength and low density, can be obtained with composites. An outline of how such properties can be predicted from those of the individual constituents forms an important part of the contents of this book.

TABLE 1.2
Comparison of Properties Exhibited by Different Class of Materials

Material	Density (ρ), Mg m^{-3}	Young's Modulus (E), GPa	Tensile Strength (S), MPa	Fracture Toughness (K_c), MPa m$^{1/2}$	Thermal Conductivity (K), W (m K)$^{-1}$	Thermal Expansivity (α), 10^{-6} K^{-1}
Thermosetting resin (epoxy)	1.25	3.5	50	0.5	0.3	60
Engineering thermoplastic (nylon)	1.1	2.5	80	4	0.2	80
Rubber (polyurethane)	1.2	0.01	20	0.1	0.2	200
Metal (mild steel)	7.8	208	400	140	60	17
Construction ceramic (concrete)	2.4	40	20	0.2	2	12
Engineering ceramic (alumina)	3.9	380	500	4	25	8
Wood (load parallel to grain)	0.6	16	80	6	0.5	3
Wood (load perpendicular to grain)	0.6	1	2	0.5	0.3	10
General PMC (in-plane)	1.8	20	300	40	8	20
Advanced PMC (load parallel to fibers)	1.6	200	1500	40	200	0
Advanced PMC (load perpendicular to fibers)	1.6	3	50	5	40	30
MMC (Al-20% SiC)	2.8	90	500	15	140	18

1.3 HISTORY OF COMPOSITES

The existence of composite is not new. The word "composite" has become very popular in recent four-five decades due to the use of modern composite materials in various applications. The composites have existed from 10000 BC. The evolution of materials and their relative importance over the years are depicted in Figure 1.1. The common composite was straw bricks, used as a construction material. Then, the next composite material can be seen from Egypt around 4000 BC where fibrous composite materials were used for preparing the writing material. These were the laminated writing materials fabricated from the papyrus plant. Further, Egyptians made containers from coarse fibers that were drawn from heat-softened glass. One more

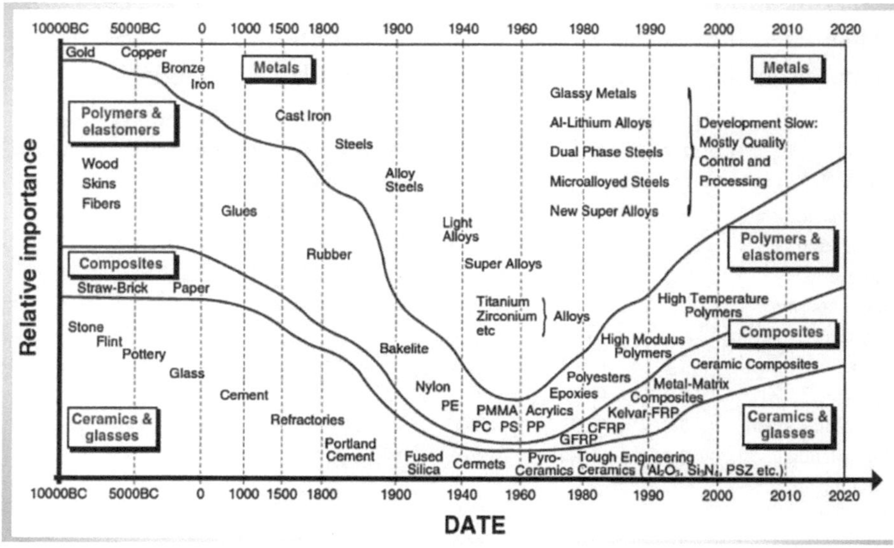

FIGURE 1.1 Evolution of materials and their relative importance over the years.

important application of composites can be seen around 1200 BC from Mongols. Mongols invented the so-called modern composite bow. The history shows that the earliest proof of existence of composite bows dates back to 3000 BC – as predicted by *Angara Dating*. The bow used various materials like wood, horn, sinew (tendon), leather, bamboo, and antler. The horn and antler were used to make the main body of the bow as it is very flexible and resilient.

Sinews were used to join and cover the horn and antler together. Glue was prepared from the bladder of fish which is used to glue all the things in place. The string of the bow was made from sinew, horse hair, and silk. The composite bow so prepared used to take almost a year for fabrication. The bows were so powerful that one could shoot the arrows almost 1.5 km away. Until the discovery of gun powder, the composite bow used to be a very lethal weapon as it was a short and handy weapon.

As said, "Need is the mother of all inventions," the modern composites – that is, polymer composites – came into existence during World War II. During World War II due to constraint impositions on various nations for crossing boundaries as well as importing and exporting the materials, there was scarcity of materials, especially in the military applications. During this period, the fighter planes were the most advanced instruments of war. The lightweight yet strong materials were in high demand. Further, applications like housing of electronic radar equipment require nonmetallic materials. Hence, the glass fiber-reinforced plastics (GFRPs) were first used in these applications. Phenolic resins were used as the matrix material. The first use of composite laminates can be seen in the *Havilland Mosquito Bomber* of the British Royal Air Force. The composites exist in day-to-day life applications as well. The most common existence is in the form of concrete. Concrete is a composite made from gravel, sand, and cement. Further, when it is used along with steel to form structural components in construction, it forms one further form of composite.

The other material is wood, which is a composite made from cellulose and lignin. The advanced forms of wood composites can be plywoods. These can be particle-bonded composites or a mixture of wooden planks/blocks with some binding agent. Nowadays, these are widely used to make furniture and as construction materials.

An excellent example of natural composite is muscles of human body. The muscles are present in a layered system consisting of fibers at different orientations and in different concentrations. These result in a very strong, efficient, versatile, and adaptable structure. The muscles impart strength to bones, and vice versa. These two together form a structure that is unique. The bone itself is a composite structure. The bone contains mineral matrix material which binds the collagen fibers together. The other examples include wings of a bird, fins of a fish, trees, and grass. A leaf of a tree is also an excellent example of composite structure. The veins in the leaf not only transport food and water, but also impart the strength to the leaf so that the leaf remains stretched with the maximum surface area. This helps the plant to extract more energy from sun during photosynthesis.

1.4 CLASSIFICATION OF COMPOSITES

The composite materials can be classified into four categories:

i. Fibrous composite materials
ii. Laminated composite materials
iii. Particulate composite materials
iv. Combination of some or all of the above three types.

1.4.1 Fiber-Reinforced Composites

Long fibers are generally more stiff and strong in comparison with the same material in bulk form. For example, ordinary plate glass fractures at stresses approaching a few thousand pounds per square inch (20 MPa), whereas the glass fibers have strengths of 400,000–700,000 psi (2800–4800 MPa) in commercially available forms and about 7000 MPa in laboratory-prepared forms. This difference in strength is attributable to less number of defects in the fiber than in a bulk material. A fiber is characterized geometrically not only by its very high length-to-diameter ratio but also by its near-crystal-sized diameter.

A fiber is an individual filament of the material. A filament with a length-to-diameter ratio above 1000 is called a fiber. The fibrous form of the reinforcement is widely used. The fibers can be in the following two forms:

i. *Continuous fibers:* If the fibers used in a composite are very long and unbroken or cut, then it forms a continuous fiber composite. A composite thus formed using continuous fibers is called as *fibrous composite*. The fibrous composite is the most widely used form of composite.
Ii. *Short/chopped fibers:* The fibers are chopped into small pieces when used in fabricating a composite. A composite with short fibers as reinforcements is called as *short fiber composite.*

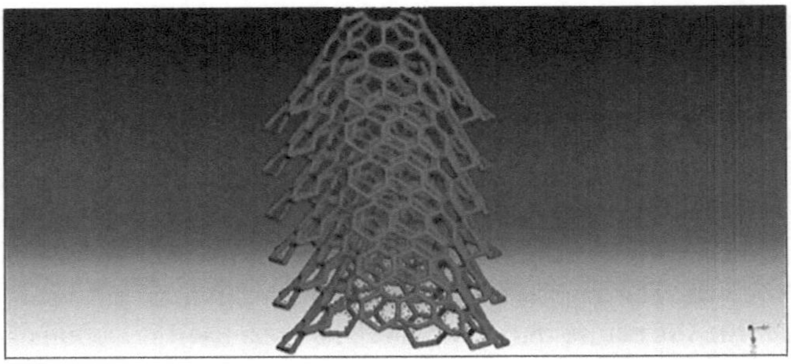

FIGURE 1.2 Molecular model of stacked cones.

In the fiber-reinforced composites, the fiber is the major load-carrying constituent.

Vapor-grown carbon nanofibers (CNFs) have been used to reinforce a variety of polymers, including polypropylene, polycarbonate, nylon, poly(ether sulfone), poly(ethylene terephthalate), poly(phenylene sulfide), acrylonitrile butadiene styrene (ABS), and epoxy. CNFs are known to have wide-ranging morphologies from structures with a disordered bamboo-like structure to highly graphitized "cup-stacked" structures as shown in Figure 1.2, where the conical shells of nanofiber are nested within each other. CNFs typically have diameters of the order of 50–200 nm. CNFs have transport and mechanical properties that approach the theoretical values of single-crystal graphite, similar to the fullerenes, but they can be made in high volumes at low cost, ultimately lower than that of conventional carbon fibers. In equivalent production volumes, CNFs are projected to have a cost comparable to E-glass on a per-pound basis, yet they possess properties that far exceed those of glass and are equal to, or exceed, those of much more costly commercial carbon fiber.

1.4.2 Laminated Composites

Laminated composite materials consist of layers of at least two different materials bonded together. Lamination is used to combine the best aspects of the constituent layers and bonding material in order to achieve a more useful material. The properties that can be emphasized by lamination are strength, stiffness, low weight, corrosion resistance etc. The laminated composites can be further classified as:

i. Bimetals
ii. Clad metals
iii. Laminated glass
iv. Plastic-based laminates.

1.4.2.1 Bimetals

Bimetals are laminates of two different metals that have significantly different coefficients of thermal expansion. When the temperature changes, bimetals warp or deflect, making these well suited for temperature-measuring devices. Figure 1.3 shows a simple thermostat made from a cantilever strip of two metals bonded together.

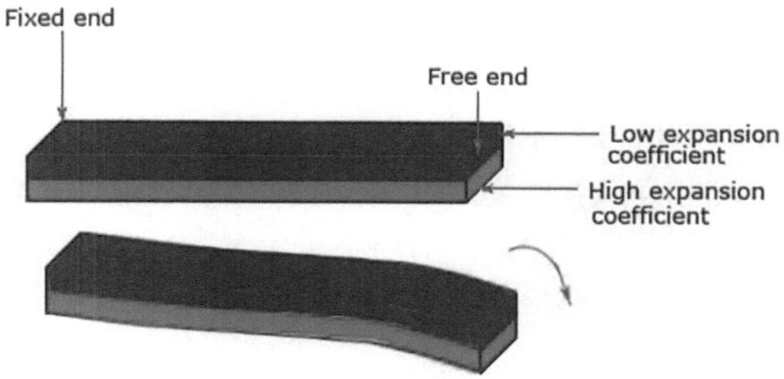

FIGURE 1.3 A simple thermostat made from a cantilever strip of two metals bonded together.

Metal A (upper layer) has a coefficient of thermal expansion α_A, and metal B (lower layer) has a coefficient of thermal expansion α_B with $\alpha_B > \alpha_A$. When the temperature is increased, strip B wants to expand more than strip A but since they are bonded together, so strip B causes the bimetallic strip to bend.

1.4.2.2 Clad Metals

Clad metals are the metals bonded to at least one layer of a different metal. Cladding can be achieved by a number of different processes, including extrusion, pressing, electroplating, and a variety of chemical techniques. The advantages of cladding range from enhanced appearance and corrosion resistance to superior thermal and electrical performance, and the process is often used to protect less wear-resistant metals. Virtually, any metal can be clad, including alloys, whether as an overlay, full clad, or inlay. The process is regularly used in the production of electrical components, decorative products, currency, machine parts, aerospace components, and shielding solutions; even cookware and automotive parts use clad metals. Clad metals are considered composites, and they often display the beneficial characteristics of both the metals involved.

Roll bonding is frequently used to produce clad metals. In roll bonding, multiple strips of cleaned and prepared, dissimilar metals are simultaneously passed through a high-pressure rolling mill. Due to the pressure exerted by the rolls, the metals coalesce into a single material that is bonded on an atomic level. Generally, the resulting material is heat-treated to enhance the strength of the bond. Explosive bonding, which uses the energy produced from explosive charges, is also used to produce clad metals. Both the thickness and distribution of the cladding can be controlled during the production process. Manufacturers can also apply specialized coatings to specific portions of the metal to prevent it from bonding. Aluminum is one of the most frequently used metals in cladding as it provides added wear resistance and strength. Aluminum clad parts are used in catalytic converters and aerospace components. Stainless steel, nickel, and copper are also widely used cladding materials. In many cases, cladding provides additional cost benefits because it allows the effective use of less-expensive materials in place of solid alloys.

FIGURE 1.4 A copper clad aluminum wire.

High-strength aluminum alloys do not resist corrosion but pure aluminum and some aluminum alloys are highly corrosion resistant but relatively weak. Thus, a high-strength aluminum alloy covered with a corrosion-resistant aluminum alloy is a composite material having both high strength and corrosion resistance. Another example of clad metals is the aluminum wire cladded with about 10% copper. It was introduced in the 1960s as a replacement for copper wire in the electrical wiring market. Aluminum wire is economical and lightweight but gets overheated and is difficult to connect to terminals at wall switches and outlets. Aluminum wire connections expand and contract when the current is turned on or off so that the fatigue breaks the wire causing shorts and potential fires. On the other hand, copper wire is costly and heavy but it stays cool and can be easily connected to wall switches and outlets. The copper clad aluminum wire is lightweight and connectable, stays cool, and is comparatively less expensive than the copper wire. Copper clad aluminum wire is also insusceptible to the construction site problem of theft because of far lower salvage value than copper wire. Figure 1.4 shows a copper clad aluminum wire.

1.4.2.3 Laminated Glass

Laminated glass is a safety and security glass that is made by sandwiching a laminated sheet between two pieces of glass. The laminated sheet is usually polyvinyl butyryl (PVB) sheet. The PVB sheet in the middle of the glass helps in sticking the glass pieces to it when the glass is broken. The laminated glass is designed to prevent it from shattering into pieces and thus ensures safety. Laminated glass is made by sandwiching PVB sheet between two panes of glass. They are sealed by a series of pressure rollers and then autoclaved. During the manufacturing process of laminated glass, mechanical and chemical bonding is developed between the PVB sheet and the glass. The adhesive nature of PVB creates the mechanical bond, and the hydrogen bonding between PVB and glass is the reason for chemical bonding. Because of the chemical bonding of PVB with glass, even when there is any breakage, the laminated glass remains intact, protecting people from injury. Laminated glass is very safe in overhead glazing. The PVB interlayer material has viscoelastic property, which in turn helps in the reduction of sound acoustic insulation. Laminated glass reduces the transmission of UV rays and hence protects the furniture from fading.

It is durable and maintains color and strength for a much longer time. The installation of laminated glass is similar to any other type of glass. Laminated glass has the following applications:

i. It is widely used in building and housing products.
ii. It is used in automotive and transport industries.
iii. It is mostly used in building facades and car windscreens.
iv. It plays a major role in overhead glazing like skylights, glass ceilings, and roofs.
v. It can also be used in greenhouse.

Laminated glass has the same strength as ordinary glass but it consists of two pieces of glass containing a sandwich of plastic interlayer. If the glass does get broken, this interlayer holds the whole piece in place so there is no hole left in the window for an intruder to get in through for example or large free shards that can cut.

Figure 1.5 shows a laminated glass composite. Figure 1.6 shows the benefits of laminated glass.

1.4.2.4 Plastic-Based Laminates

Many materials can be saturated with various plastics for a variety of purposes. Formica is one of the plastic-based laminates consisting of layers of heavy kraft paper impregnated with a phenolic resin overlaid by a plastic-saturated decorative thin sheet which in turn is overlaid with a plastic-saturated cellulose mat. Heat and pressure are used to bond the layers together. Figure 1.7 shows different designs of Formica used as decorative sheets for kitchen countertop. Another variation of Formica is obtained by placing an aluminum layer between the decorative layer and the kraft paper layer. This helps in quick dissipation of heat when a hot utensil is placed on the kitchen counter, thus preventing the formation of a burned spot. Layers of glass or asbestos fabrics can be impregnated with silicones to yield a composite

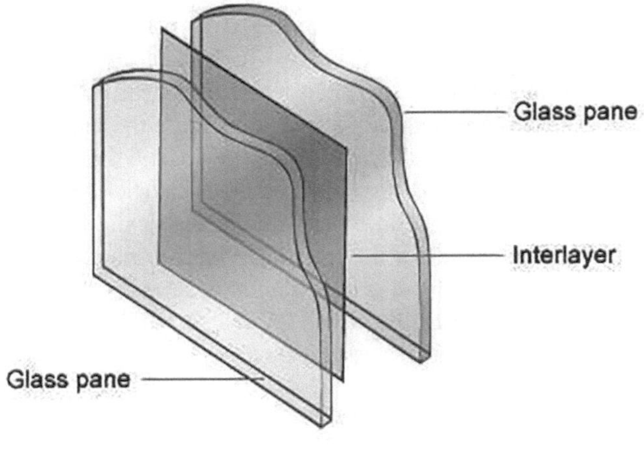

FIGURE 1.5 A laminated glass composite.

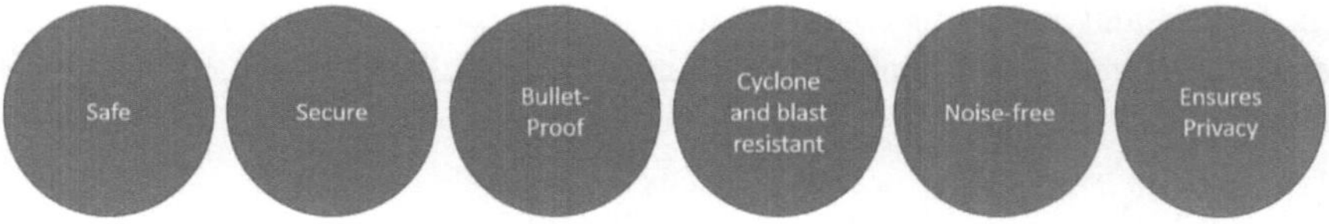

FIGURE 1.6 Benefits of laminated glass.

FIGURE 1.7 Different designs of Formica used as decorative sheets for kitchen countertop.

material with significant high-temperature properties. Glass, Kevlar, or nylon fabrics can be laminated with various resins to yield an impact and penetration-resistant composite material that can be used a lightweight personnel armor.

1.4.3 Particulate Composites

The reinforcement is in the form of particles which are of the order of a few microns in diameter. The particles are generally added to increase the modulus and decrease the ductility of the matrix materials. In this case, the load is shared by both particles and matrix materials. However, the load shared by the particles is much larger than the matrix material. For example, in an automobile application, carbon black (as a particulate reinforcement) is added in rubber (as matrix material). The composite with reinforcement in particle form is called a *particulate composite*. The particles can be either metallic or nonmetallic as can the matrix. The four possible combinations of these constituents are as discussed below.

1.4.3.1 Nonmetallic Particles in Nonmetallic Matrix

The most common example in this category is concrete, which is a combination of particles of sand and gravel that are bonded with a mixture of cement and water that has chemically reacted and hardened. The strength of concrete is normally that of the gravel because the cement matrix is stronger than the gravel.

Flakes of nonmetallic materials such as mica or glass can form an effective composite when suspended in a glass or plastic, respectively. Flakes are primarily two dimensional with stiffness and strength in two directions as opposed to only one for fibers. Flakes are packed parallel to each other, resulting in a higher density than the other fiber packing geometries. A flake composite material is much more impervious to liquids than an ordinary composite because of the overlapping of flakes. Mica in glass composite is extensively used in electrical applications because of good insulating and machining qualities.

1.4.3.2 Metallic Particles in Nonmetallic Matrix

Solid rocket propellants consist of inorganic particles such as aluminum powder and perchlorate oxidizers in a flexible organic binder such as polyurethane or polysulfide rubber. The particles comprise nearly 75% of the propellant leaving only 25% for the binder. A steadily burning reaction is obtained which provides a controlled thrust. Another example is aluminum paint, which is actually aluminum flakes suspended in paint. Upon application, the flakes orient themselves parallel to the surface and give very good coverage. Silver flakes can also be applied to give good electrical conductivity.

Cold solder is a metal powder suspended in a thermosetting resin. The composite material is strong and hard, and conducts heat and electricity. Inclusion of copper in an epoxy resin significantly increases the conductivity. Many metallic additives to plastic increase the thermal conductivity, lower the coefficient of thermal expansion, and decrease wear.

1.4.3.3 Metallic Particles in Metallic Matrix

Unlike an alloy, a metallic particle in a metallic matrix does not dissolve. Lead particles are commonly used in copper alloys and steel to improve the machinability. In addition, lead is a natural lubricant in bearings made from copper alloys.

Many metals are naturally brittle at room temperature, so must be machined when hot. However, particles of these metals, such as tungsten, chromium, and molybdenum, can be suspended in a ductile matrix. The resulting composite material is ductile, yet has the elevated temperature properties of the brittle constituents. The actual process used to suspend the brittle particles is called liquid sintering and involves infiltration of the matrix material around the brittle particles. Fortunately, in the liquid sintering process, the brittle particles become rounded and therefore naturally more ductile.

1.4.3.4 Nonmetallic Particles in Metallic Matrix

Nonmetallic particles such as ceramics can be suspended in a metal matrix. The resulting composite material is called a cermet. Two common classes of cermets are oxide-based and carbide-based composite materials. As a slight departure from the present classification scheme, oxide-based cermets can be either oxide particles in a metal matrix or metal particles in an oxide matrix. Such cermets are used in tool making and high-temperature applications where erosion resistance is needed. Carbide-based cermets have particles of carbides of tungsten, chromium, and titanium. Tungsten carbide in a cobalt matrix is used in machine parts requiring very high hardness, such as wire-drawing dies and valves. Chromium carbide in a cobalt matrix has high corrosion and abrasion resistance; it also has a coefficient of thermal expansion close to that of steel, so is well suited for use in valves. Titanium carbide in either a nickel or a cobalt matrix is often used in high-temperature applications such as turbine parts. Cermets are also used as nuclear reactor fuel elements and control rods. Fuel elements can be uranium oxide particles in stainless steel ceramic, whereas boron carbide in stainless steel is used for control rods.

1.4.4 Combination of Composites

Numerous multiphase composite materials exhibit more than one characteristic of the various classes – fibrous, laminated, or particulate composite materials – as discussed above. For example, reinforced concrete is both particulate (because the concrete is composed of gravel in a cement-paste binder) and fibrous (because of the steel reinforcement). Also, laminated fiber-reinforced composite materials are obviously both laminated and fibrous composite materials. Thus, any classification system is arbitrary and imperfect. Nevertheless, the system should serve to acquaint the reader with the broad possibilities of composite materials. Laminated fiber-reinforced composite materials are a hybrid class of composite materials involving both fibrous composite materials and lamination techniques. Here, layers of fiber-reinforced material are bonded together with the fiber directions of each layer typically oriented in different directions to give different strengths and stiffness of the laminate in various directions. Thus, the strengths and stiffness of the laminated fiber-reinforced composite material can be tailored to the specific design requirements of the structural element being built. Examples of laminated fiber-reinforced composite materials include rocket motor cases, boat hulls, aircraft wing panels and body sections, tennis rackets, and golf club shafts.

1.5 NANOMATERIALS

Nanostructured materials are single-phase or multiphase polycrystalline solids with a typical average grain size of a few nanometers (1 nm = 10^{-9}m), typically less than 100nm. Such materials exhibit properties that are substantially different from and often superior to those of conventional coarse-grained materials, due to their unique microstructure. Since the grain sizes are so small, a significant volume fraction of the atoms resides in grain boundaries. Thus, the material is characterized by a large number of interfaces in which the atomic arrangements are different from those of the crystal lattice. At such small grain sizes, the surfaces start to dominate the bulk behavior of the material, and consequently, nanostructured materials exhibit special and completely new and unexpected properties, often not observed in coarse-grained materials. Accordingly, nanostructured materials have been shown to exhibit very high strength, increased diffusivity, reduced sintering temperatures, useful catalytic properties, and attractive physical properties. Due to their novel and improved properties and varied potential applications in different fields of technology, these materials are attracting increasing attention from researchers all over the world.

A theoretical interpretation of a nanomaterial must be based on some primary concept. All materials are composed of granules, which in turn consist of atoms. These granules may be visible or invisible to the naked eye, depending on their size, which can range from hundreds of microns to centimeters. The term "material" usually refers to a substance in a solid state, which can be in either crystalline or glassy phase. Since nanoparticles can form both nanomaterials and nanopowders, it is worth pointing out that in mechanics, materials are assumed to have a stability of shape, while powders are regarded as granular media. Usually, powders are made into materials by the application of special techniques such as compression, sintering, and irradiation. As already mentioned, the size of granules in nanopowders and nanomaterials ranges from 10 to 100 nm in at least one dimension (usually in all three). Many authors believe that nanomaterials, that is, materials with an internal structure of nanoscale dimension, are hardly something new to science, just it has been discovered only recently that some formations of oxides, metals, ceramics, and other substances are nanomaterials. For example, black carbon was discovered at the beginning of the 19th century. Fumed silica powder, a component of silicon rubber, came into commercial use in 1940. However, it has only recently become clear that these two substances are nanomaterials. It should be noted that the size of a particle is not the only characteristic of nanoparticle, nanocrystal, or nanomaterial. In opinion of Guz et al. [3], one quite important and specific property of many nanomaterials is that the majority of their atoms are located on the surface of a particle as opposed to conventional materials, in which atoms are distributed over the volume of a particle.

Carbon particles are a well-studied class of nanoparticles. Science has long been aware of three forms of carbon: diamond, graphite, and amorphous carbon. Highly symmetric molecule of carbon C_{60} was discovered in 1985. It has a spherical form, resembling a football, with carbon atoms on the surface. It contains 60 atoms in five-atom rings separated by six-atom rings. These molecules were named "fullerenes" and became the subject of extensive and fruitful studies. Scientists who studied fullerenes were awarded the Nobel Prize in Chemistry in 1996. Since then, the number of different varieties of fullerenes has increased considerably, reaching many thousands to date. The molecule C_{60} has got its name from R.B. Fuller (1895–1983), an architect who built a house from pentagons and hexagons. The molecule C_{70} resembles a rugby ball and includes only six-atom rings, and also bears the name "fullerene." Figure 1.8a and b depicts C_{60} and C_{70} molecules. Fullerenes can form crystals called fullerites. These crystals have a face-centered cubic lattice with cavities of two types: tetrahedral and octahedral. Placing ions of potassium, rubidium, or cesium into these cavities produces new nanoparticles with unusual properties. Fullerenes may be deposited onto a surface, forming a monolayer. There are also rope-like fullerene formations. But the most important thing is that fullerene molecules can form carbon nanotubes, which may be considered as related to graphite.

The molecular structure of graphite looks like a sheet of chicken wire, a tessellation of hexagonal rings of carbon. In graphite, these sheets are stacked one over another and can slide past each other. This explains why graphite is soft and greasy, and is often used as a lubricant. When graphite lattices are rolled up into a tube, they form nanotubes, molecules with a very large number of atoms, $C_{10,000}-C_{1,000,000}$. Nanotubes differ in diameter, length, and the way they are rolled. The internal

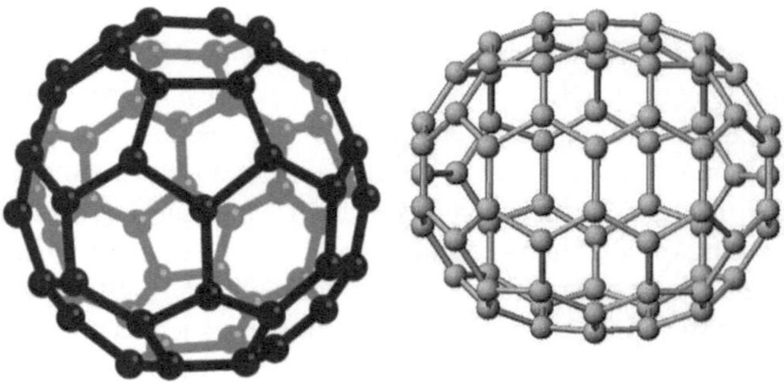

(a) (b)

FIGURE 1.8 (a) C_{60} and (b) C_{70} molecules.

cavities may also be different, and tubes may have more than one sheet. Atoms form "hemispherical caps" at ends of fullerene molecule. Sheets may be rolled differently forming zigzag, armchair, and chiral structures. Any cylindrical nanotube can be considered as being a rolled graphite sheet. A sheet is schematically depicted as a periodic hexagonal structure or lattice consisting of regular hexagons with carbon atoms at vertices as shown in Figure 1.9. The procedure of rolling a sheet is as follows: Choose two atoms A and B that belong to hexagons spaced far enough from each other (it is clear that a tube cannot be made from three hexagons); connect the points with a straight line; cut out a strip from the sheet by two cuts perpendicular to the AB line; and roll the strip so as to superpose the points A and B. Length of AB will be the width of the rolled-up strip and simultaneously the length of a parallel circle of the resultant circular cylindrical tube. Such an arbitrary tube is called a chiral tube. Chirality, which can be left or right, is a concept borrowed from stereochemistry. It refers to the property of nonidentity of an object with its mirror image. Three types of chirality are central, axial, and planar, corresponding to three chiral elements, namely, center, axis, and plane.

Chirality can be measured with the help of a vector:

$$\vec{r} = \vec{AB} = n\vec{e}_1 + m\vec{e}_2 \quad (n,m \in N), \tag{1.1}$$

where $\vec{e}_1$ and $\vec{e}_2$ are the vectors, defining two directions intersecting at A. In general, at any vertex of a hexagon, there are three directions, important to the lattice. One of them (type 1) is the direction towards the opposite vertex of the hexagon. The straight line crosses the hexagon and then passes along the common side of neighboring hexagons, with this sequence repeating periodically. The side of a hexagon with two adjacent sides resembles an armchair. The other two directions (type 2) pass through A, and the third (counting from A) vertex either on the left or on the right. These two directions define two straight lines that are specific to the lattice; they are differently directed but otherwise identical. The sides connecting all the three vertices (A, the second and the third ones) resemble a zigzag, and this sequence repeats periodically. The vector $\vec{e}_1$ always lies on a type 2 line (horizontal line in Figure 1.9), while the vector $\vec{e}_2$ can be directed arbitrarily.

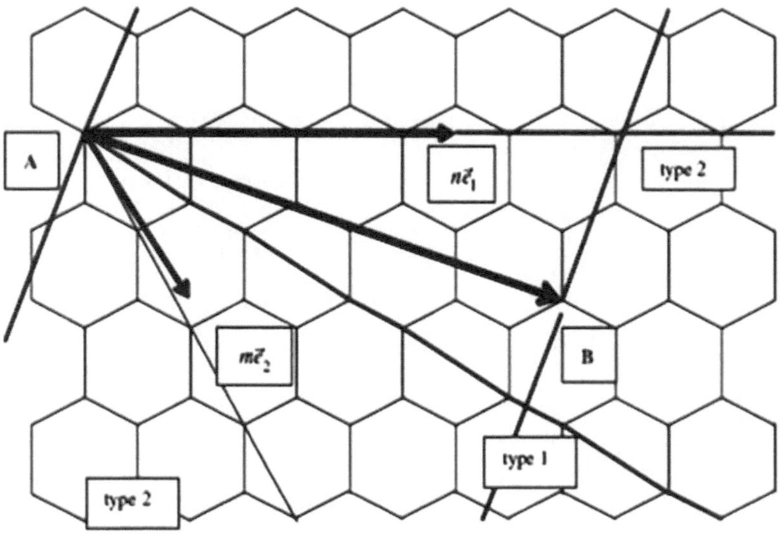

FIGURE 1.9 A cylindrical nanotube represented as a rolled graphite sheet with a periodic hexagonal structure consisting of regular hexagons with carbon atoms at vertices.

Chiral structures are distinguished by the numbers n and m, and denoted by (n, m). If $\vec{e}_1$ and $\vec{e}_2$ are codirectional – that is, B and $\vec{e}_1$ lie on the same straight line – then a zigzag chiral structure is obtained. Its typical feature is that m in the notation (n, m) is always equal to zero. If the point B lies on a straight line of type 1, then $\vec{e}_1$ lies on the second straight line of type 2 and the type 1 line is the bisector of the angle formed by the type 2 lines. It is an armchair structure and its numbers n and m are always equal. Any other structure is called a chiral structure. Applications of nanotechnology are shown in Figure 1.10. Figure 1.11 shows SWCNTs (single-walled carbon nanotubes) with different chirality for the particular values of (n, m). The length of nanotubes can be 1000 times (and more) greater than their diameter. They can be single-walled and multiwalled. Multiwalled nanotubes, between 4 and 30 nm in diameter, were discovered by Iijima in 1991. The thinnest single-walled nanotube has a wall thickness of 0.4 nm. By comparison, the diameter of the carbon atom is equal to 0.15 nm. An assembly of nanotubes may have a diameter of 30–50 nm and a length of more than 50 μm. Single-walled nanotubes may form rough structures or ropes. Ropes consist of nanotubes bundled in an ordered manner to form a lattice. Nanotube ropes may have a diameter of 10–20 nm and a length of 100 μm and more.

Nanotubes are technologically advantageous over conventional carbon fibers since they are produced from colloidal solutions at room temperatures, whereas carbon fibers require high-temperature environment. Research of nanotubes follows mainly three directions: (1) mechanical and electrical properties of nanotubes in polymeric films, (2) arrangement of nanotubes in polymeric composites, and (3) properties of the nanotube matrix transition layer.

Major constituents in a fiber-reinforced composite material are the reinforcing fibers and a matrix, which acts as a binder for the fibers. Other constituents that

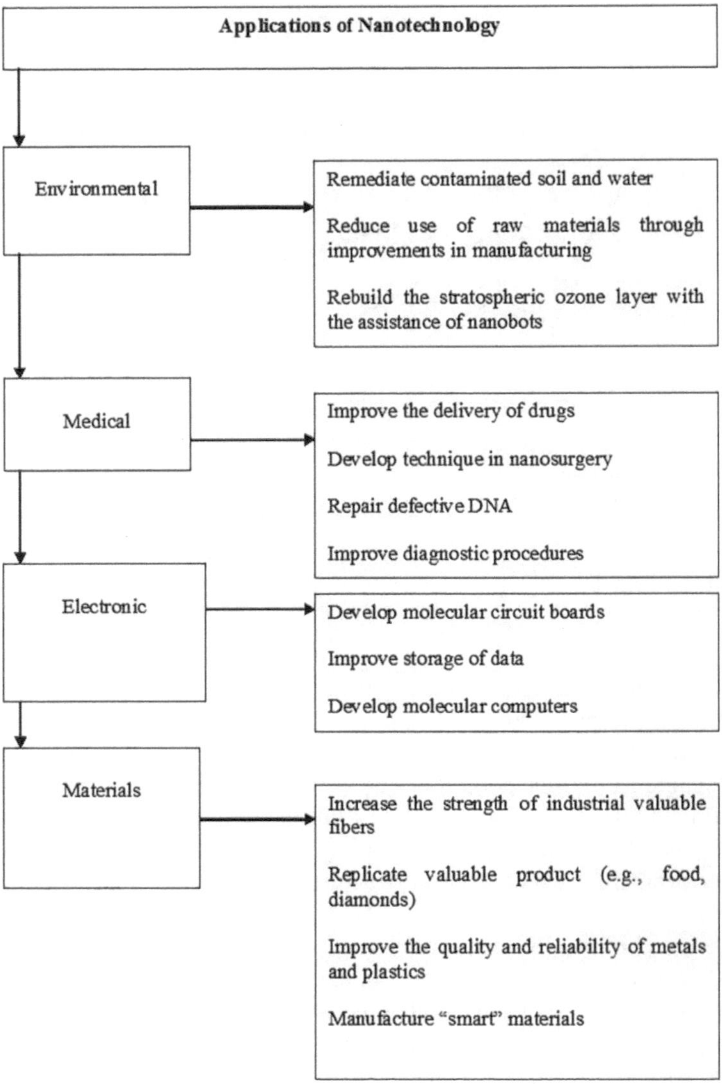

FIGURE 1.10 Applications of nanotechnology.

may also be found are coupling agents, coatings, and fillers. Coupling agents and coatings are applied on the fibers to improve their wetting with the matrix as well as to promote bonding across the fiber–matrix interface. Both in turn promote a better load transfer between the fibers and the matrix. Fillers are used with some polymeric matrices primarily to reduce cost and improve their dimensional stability. Manufacturing of a composite structure starts with the incorporation of a large number of fibers into a thin layer of matrix to form a lamina. The thickness of a lamina is usually in the range of 0.1–1 mm. If continuous (long) fibers are used in making the lamina, they may be arranged either in a unidirectional orientation (i.e., all fibers

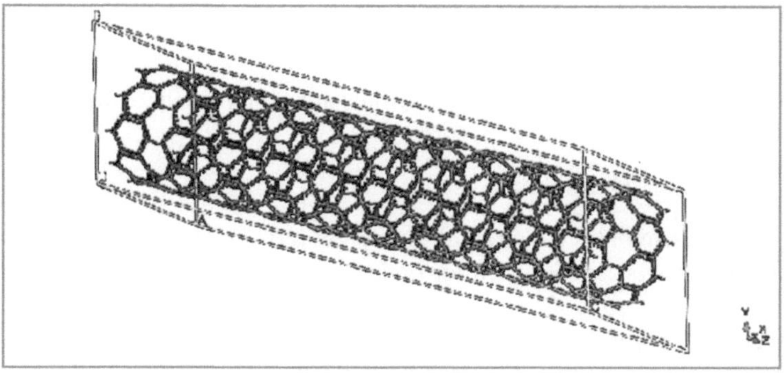

A zigzag (10,0) SWCNT.

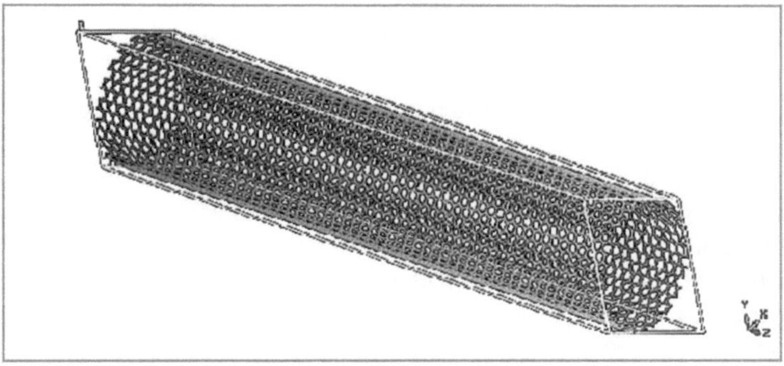

An armchair (20,20) SWCNT.

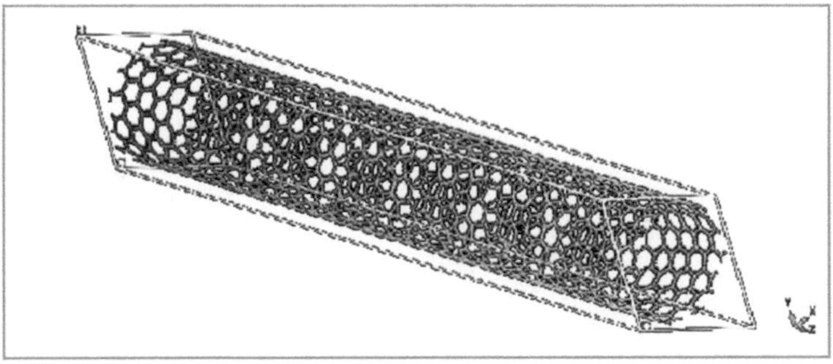

A chiral (12,6) SWCNT.

FIGURE 1.11 Examples of SWCNTs with different chirality for particular values of (n, m). From top to bottom: (a) Zigzag, (b) armchair, and (c) chiral nanotubes.

in one direction, Figure 1.11a), in a bidirectional orientation (i.e., fibers in two directions, usually normal to each other, Figure 1.11b), or in a multidirectional orientation (i.e., fibers in more than two directions, Figure 1.11c). The bi- or multidirectional orientation of fibers is obtained by weaving or other processes used in the textile industry.

1.6 APPLICATIONS OF COMPOSITE MATERIALS

Though, a complete list of applications of composite materials cannot be prepared but some of the major areas where composites are used have been discussed below.

1.6.1 Aerospace Applications

In order to make use of higher strength-to-weight ratio and higher stiffness-to-weight ratio, composite materials are used for making various parts such as

i. Aircraft, spacecraft, satellites, space telescopes, space shuttle, space station, missiles, boosters, rockets, helicopters (due to high specific strength, stiffness, fatigue life, and dimensional stability).
ii. All composite voyager aircraft flew nonstop around the world with refueling.
iii. Carbon/carbon composite is used on the leading edges nose cone of the shuttle.
iv. B2 bomber – both fiber glass and graphite fibers are used with epoxy matrix and polyimide matrix.
v. The indigenous light combat aircraft (LCA – Tejas) has Kevlar composite in nose cone; glass composites in tail fin and carbon composites form almost all part of the fuselage and wings, except the control surfaces of the wing.
vi. Further, the indigenous light combat helicopter (LCH – Dhruv) has carbon composites for its main rotor blades. The other composites are used in tail rotor, vertical fin, stabilizer, cowling, radome, doors, cockpit, side shells etc.

Figures 1.12–1.14 show the use of composite materials in modern aircrafts. The Airbus 320 uses a range of components made from composites, including the fin and tailplane. This has allowed a weight saving of 800 kg over its equivalent in

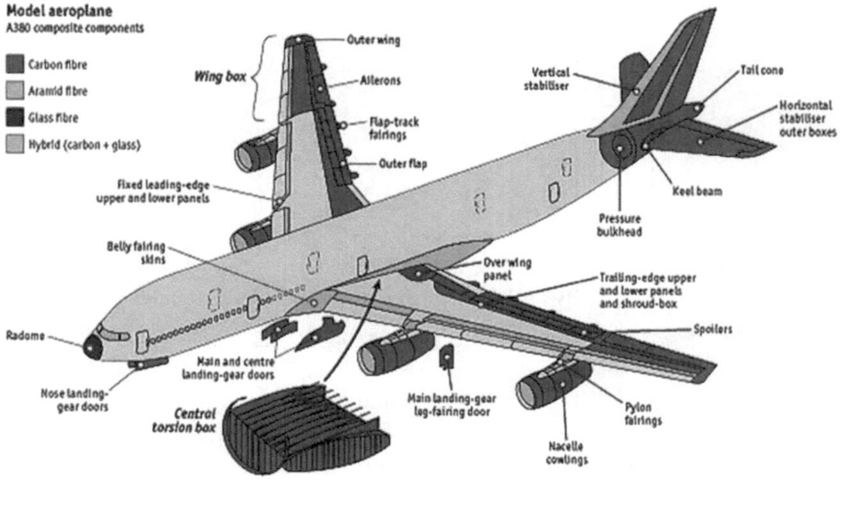

FIGURE 1.12 Use of composite materials in commercial aircraft.

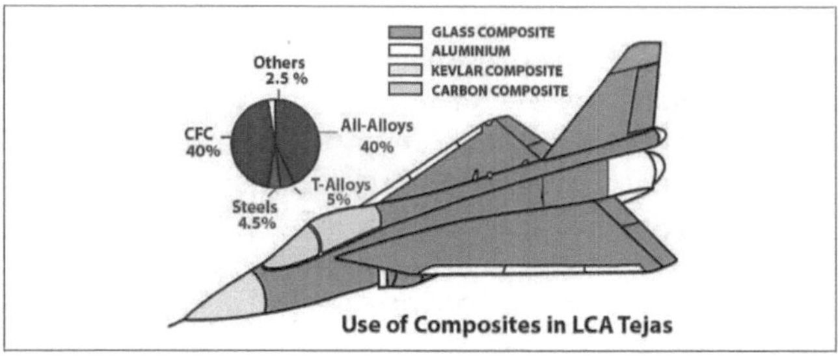

FIGURE 1.13 Use of composite materials in light combat aircraft (LCA-Tejas) developed by India.

FIGURE 1.14 Use of composite materials in Eurofighter Typhoon. Here, the different parts are (1) Radar Transparent Radome: Epoxy or BMI prepreg or RTM resins and woven preforms made of glass or quartz; (2) Foreplane Canard Wings: Epoxy carbon prepregs; (3) Fuselage Panel Sections: Epoxy carbon prepregs, nonmetallic honeycomb core and HexBond adhesives; (4) Leading Edge Devices: Epoxy carbon and glass prepregs; (5) Fin Fairings: Epoxy glass and carbon prepregs; (6) Wing Skins and Ribs: Epoxy carbon and glass prepregs; (7) Fin Tip: Epoxy/quartz prepregs; (8) Rudder: Epoxy carbon prepreg; (9) Fin: Epoxy carbon/glass prepreg; (10) Flying Control Surfaces: Epoxy carbon and glass prepregs. Honeycomb core material and HexBond adhesives.

aluminum alloy. Composite materials comprise more than 20% of the A380's airframe. Carbon-fiber reinforced polymer (CFRP) and GFRP are used extensively in wings, fuselage sections (such as the undercarriage and rear end of fuselage), tail surfaces, and doors. The successful application of composites in missiles has led to the development of primary structures for space vehicles. In fact, space applications lend themselves in many ways to the utilization of new materials. For satellites, for example, the timescales from concept to manufacture can be as little as 2 years and the short product runs normally involved; the materials element in the final cost is often relatively low. Also in many applications, no other material is suitable for technical reasons.

Once in orbit, mechanical loads are comparatively low. Environmental conditions can be extreme and severe thermal cycling can occur, as well as the effects of high vacuum and erosion through atomic oxygen or micro-meteoroid impacts. Glass fiber composite (GRP) is used in applications where thermal insulation is important, for example, in local bracketry. The material is also used in some antenna reflectors. Carbon fiber composite (CFRP), however, is most often associated with space applications. The potential for very high stiffness and excellent thermal stability over a wide temperature range makes CFRPs ideal. Examples of their application include fairings, manipulator arms, antennae reflectors, solar array panels, and optical platforms and benches. They have also recently been used for primary structure applications. In the past, the need for a combination of stiffness and strength, and for thermal and electrical conductivities, has favored metals. However, the constant pressure for weight reduction means that now some satellites have been built with a predominantly composite structure subsystem.

1.6.2 Missile Applications

Composite materials are used in missiles for making the following components:

i. Rocket motor cases
ii. Nozzles
iii. Igniter
iv. Interstage structure
v. Equipment section
vi. Aerodynamic fairings.

The material commonly used as matrix is epoxy along with carbon fiber, Kevlar fiber, and glass fibers as reinforcements.

1.6.3 Launch Vehicle Applications

Composite materials are used in launch vehicle applications for making the following components:

i. Rocket motor case
ii. Interstage structure
iii. Payload fairings and dispensers
iv. High-temperature nozzle
v. Nose cone
vi. Control surfaces.

1.6.4 Railways

Composite materials are used in railways for making:

i. Bodies of railway bogeys
ii. Seats
iii. Driver's cabin

iv. Stabilization of ballasted rail tracks
v. Doors
vi. Sleepers for railway girder bridges (Figure 1.15a)
vii. Gear case (Figure 1.15b)
viii. Pantographs (Figure 1.15c).

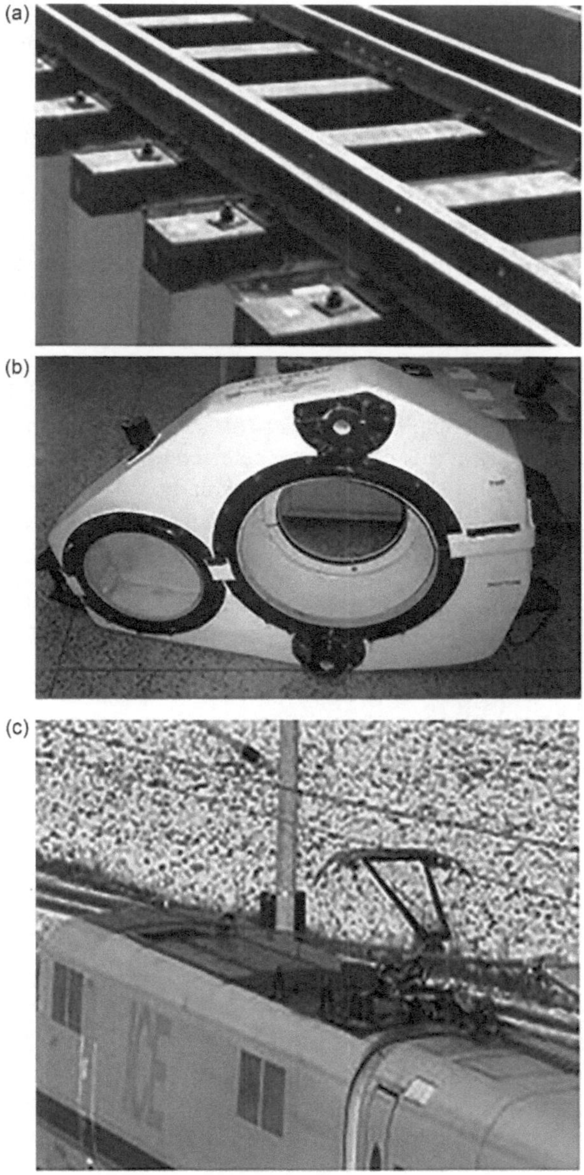

FIGURE 1.15 (a) Sleepers for railway girder bridges, (b) gear case made of carbon fiber polymer composite, and (c) pantograph.

1.6.5 Sports Equipment

Modern-day sports equipment pieces are widely made of composite materials. The main areas include the following:

i. Tennis rackets, golf clubs, baseball bats, helmets, skis, hockey sticks, fishing rods, boat hulls, wind surfing boards, water skis, sails, canoes and racing shells, paddles, yachting rope, speed boat, scuba diving tanks, race cars etc. Figure 1.16 shows the use of composite materials in skis used for ice-skiing.
ii. Modern-day tennis rackets are made from a high-modulus graphite and/or carbon fiber, which is used to keep the frame lightweight and stiff for increased racket head stability and performance.
iii. Advantages include reduced weight, maintenance, and corrosion resistance.
iv. Club heads for drivers and other woods may be made from stainless steel, titanium, or graphite fiber-reinforced epoxy. Face inserts may be made from zirconia ceramic or a titanium metal matrix ceramic composite. Oversize metal woods are usually filled with synthetic polymer foam.
v. Composite baseball bats, opposed to aluminum or wood baseball bats, incorporate a reinforced carbon fiber polymer, or composite, into the bat's construction. This composite material can make up all or part of the bat. Bats made entirely of this polymer are referred to as composite bats.
vi. *Helmets*: CFRP shell reinforced with tow carbon fiber fabric. Integrating the ultrathin carbon fiber fabric into the shell has reduced its weight by 21%.

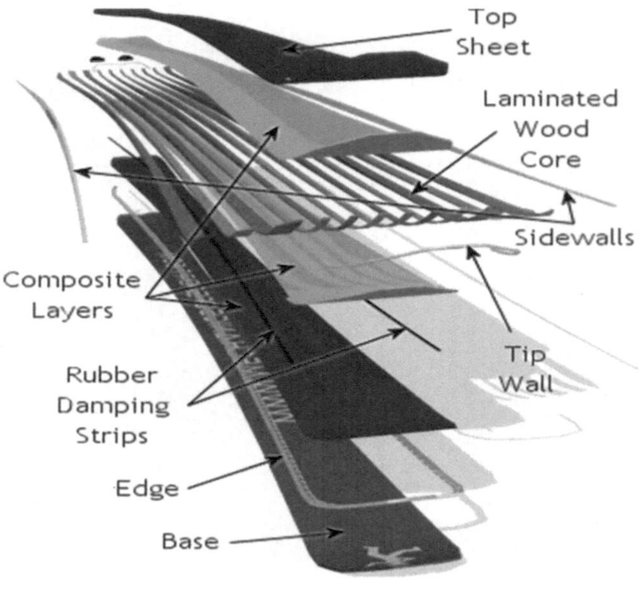

FIGURE 1.16 Schematic of ski board used in ice-skiing showing the use of composite materials.

1.6.6 Automotives

Composite materials when used for making automotive parts provide the advantages such as lower weight, greater durability, corrosion resistance, fatigue life, wear, and impact resistance. Some of the automotive parts that are made of composites are as follows:

i. Drive shafts, fan blades and shrouds, springs, bumpers, interior panels, tires, brake shoes, clutch plates, gaskets, hoses, belts, and engine parts
ii. Carbon and glass fiber pultruded over aluminum cylinder to create drive shaft
iii. Fuel saving – braking energy can be stored in carbon fiber super flywheels
iv. Other applications include mirror housings, radiator end caps, air filter housing, accelerating pedals, rear-view mirrors, headlamp housings, intake manifolds, and fuel tanks.

Various applications of composite materials include structural (chassis and body-in-white), powertrain (engine, suspension, and others), interiors (dashboard, door panel, and others), exteriors (door module, hood, bumper, and others), and other applications (switches and modules and other electrical components). Of all the different composite materials used to manufacture automobile components, polymer matrix composite (PMC) is the most widely used as polymer composites can directly reduce the weight of a vehicle body and chassis by up to 50%, which subsequently reduces the vehicle's fuel consumption. The advancements in manufacturing technologies of these materials have increased their penetration in the growing markets, including India and China, which are also the largest vehicle-producing countries. The growth in the metal matrix composite (MMC) materials is mainly due to the high strength and lightweight properties of this material. Apart from these properties, the low cost as compared to PMC is one of the major drivers for the increasing MMC market.

There are two classification systems of composite materials: One is based on the matrix material (metal matrix composites, MMCs; ceramic matrix composites, CMCs; polymer matrix composites, PMCs), and the second is based on the material structure: particulate (random orientation of particles; preferred orientation of particles), fibrous (short-fiber-reinforced composites; long-fiber-reinforced composites), and laminate composites. The various PMCs are classified as thermoplastic and thermosets, and can be reinforced with various types of fibers depending upon the applications. PMCs are used in various automotive applications like crashworthiness, body panels, and bumpers. CMCs are used in elevated temperatures of various engine components and braking systems. The MMCs use magnesium, copper, and aluminum as their matrix with fibers to be used in various engine and crash-absorbing components. Moreover, MMCs with aluminum matrix and ceramic-based composites find some automotive applications with supercar brake disks. PMCs have heavily been used in the automotive industry. Polymers used in automotive applications are divided into thermoplastics and thermosets. Thermoplastics are high molecular weight materials that soften or melt on the application of heat. Thermoset processing requires the nonreversible conversion of a low molecular weight base resin to a polymerized structure. The resultant material cannot be re-melted or re-formed. In automotive applications, reinforced plastics are the major composite material. For polymer composites, common fillers used include calcium carbonate ($CaCO_3$), talc, wollastonite, glass, and carbon fiber. Some of the common processing techniques for polymer composites are injection molding, sheet molding compound (SMC), glass-mat thermoplastic (GMT) compression molding, resin transfer molding (RTM), and reaction injection molding (RIM).

1.6.7 Infrastructure

Corrosion is a major design consideration such as in the chemical and on offshore oil plate forms. Fiber-reinforced polymer (FRP) composites have found extensive applications in the oil and gas industry for the last three decades in areas such as modules, protection, equipment, spoolable pipes, and pressure vessels. Significant advances have been made in the areas of composite pipe work and fluid handling. The high cost to replace steel piping in retrofit applications and increased longevity in new construction are driving the use of composites, which withstand the severe conditions experienced in offshore environment. In the offshore oil and gas industry, composites offer several potential advantages such as the cost of manufacturing and erecting oil rigs, which could be significantly reduced if heavy metal pipelines could be replaced with lighter ones made of composites.

Composite pipes could be used for fire-water piping, seawater cooling, draining systems, and sewerage. The cost advantages of composite products are much greater when they replace expensive corrosion-resistant metals, such as copper-nickel alloys, duplex/superduplex stainless steel, and titanium, used in offshore platforms for various applications. Their resistance to corrosion helps in improving reliability and safety, and also leads to lower life cycle costs.

Piping systems using GFRP composites offer complete solutions for the offshore environment against highly corrosive fluids at various pressures, temperatures, and adverse soil and weather conditions (especially in oil exploration, desalination, chemical plants, fire mains, dredging, portable water etc.). Composite pipes are commonly used in oil transportation where resistance to crude oil, paraffin buildup as well as ability to withstand relatively high pressures is required. The system is also being used on offshore rigs for seawater cooling lines, air vent systems, drilling fluids, fire-fighting, ballasts, and drinking water lines in offshore application. The lightweight properties of GFRP help reduce heavy and expensive construction costs. Established oil fields use composite pipes for high-pressure and steam-injection lines for the recovery of oil preserves. Composites can withstand the detrimental effect of brackish water when expelled under pressure from fire mains, making rupture less likely and the system more reliable. The chemical resistance and service temperature of such composites in a particular fluid depend on resin formulations and additives used.

Composite grids/gratings, hand rails, cable trays, ladders, decking, and flooring have been used on fixed and floating offshore platforms world over for more than 30 years. In topside applications, the inherent corrosion resistance of composite materials reduces life cycle costs by minimizing their maintenance. Conventionally, grids/gratings are made of mild steel/cast iron. Due to the limitations such as corrosion resistance, weight, durability, and life cycle costs for the metallic gratings, composite grids/gratings perform much better due to their superior properties under aggressive environments as in chemical process industry.

Composite ladders are stronger than wood or aluminum, and do not absorb water, rot, or corrode. The products can be pigmented with a suitable color along with the resin during the pultrusion process. With the color throughout the part, there is no chipping or peeling. Unlike aluminum, composites have excellent insulation properties which substantially reduce the hazard of electrocution by contacting high-voltage power lines. For rough jobs where a ladder takes a beating, composite provides the ultimate ruggedness and long-term durability. Composite coil tube replaces the existing steel coil tubing for high-pressure downhole applications in offshore platforms. The tube can be coiled or uncoiled on a drum and can easily be transported to the desired location of the wells.

Composite riser is the pipeline that connects the rig of the water surface to the well bore at the seabed. It must separate the oil, gas, and drilling fluids from seawater. The weight of riser can drastically come down with the use of composite material as alternative to heavy metallic risers. Composite risers can be designed to withstand highly corrosive chemicals, salts, and fluids under different environmental conditions. The durability and life cycle costs in offshore platforms can be improved.

To accommodate the relative motions between the platform and the riser, in case of tension leg platforms, a telescopic joint is used at the upper extremity of each riser. These joints require a tensioning system capable of storing and releasing large amounts of energy as movement takes place. Tension is applied through gas-pressurized tensioners with accumulator bottles. In older designs, steel accumulator bottles were used but recently considerable success has been achieved with composite bottles. The composite bottles offer significant weight and cost saving being less than a third of the weight of equivalent steel bottles. These bottles can withstand very high internal pressures.

Caissons are attractive applications for composites as an offshoot of composite piping technology. In general, caissons are used to provide the service fluids to enter or leave the sea. These are located at splash zones in the seawater. Caissons are designed to withstand flexural fatigue loads created by waving loads and corrosion to aqueous fluids in the sea.

1.6.8 Medical Applications

FRP composites are finding more applications in the medical sector because of their lightweight, high-stiffness characteristics, and biocompatibility. The use of external components such as artificial limbs offer advantages such as the high specific properties, fatigue resistance, and flexibility of manufacture of composites. The chemical inertness of carbon fiber has led to a number of surgical applications where the material is used in conjunction or instead of metallic or polymeric materials.

Early designs of artificial limbs employed press-molded carbon fiber composites as the main load-bearing structure. More recent developments incorporate a wider use of carbon fiber to include load-carrying links in the joint mechanisms, foot keels with sprung energy return, and hybrid designs incorporating elastomers to dampen shock loads.

The behavior of artificial joints, which conventionally consist of an ultra-high molecular weight polyethylene (UHMWPE) component articulating against a polished steel part, can be improved by enhancing the wear characteristics of the polymer. UHMWPE can be reinforced by a random distribution of short (~3 mm) carbon fibers to provide the desired tribological properties.

Carbon composites are being implanted into cartilage to promote biological resurfacing of damaged areas. The open weave structure of the material promotes cell growth along and between individual fibers ultimately resulting in a suitable repair. Carbon fiber tows, either used individually or in plaits, are also being employed in the repair of damaged ligament. Loops of material are passed through holes drilled in the adjacent bone structures, and then their length is adjusted to achieve the correct tension for the particular patient.

The mechanical properties of bone repair materials, often a self-curing acrylic, can be enhanced by the addition of carbon fibers. Tensile, compressive, and shear strengths as well as creep and fatigue performance are all improved, and this could lead to a wider clinical use of the material.

1.6.9 Renewables

The use of FRPs in wind turbines is an important technical element. Components must exhibit excellent fatigue strength, resist random loading and corrosion, require minimal maintenance, and serve for 30+ years. Uncertainties over the performance of initial experiments with steel and aluminum have been overcome with the use of composite members, on which production is now almost entirely based. The blades are the main components, predominantly manufactured from glass fiber, and the performance of the turbine is ultimately dictated by their efficiency. The use of lighter-weight FRP materials means that the turbines can produce more power per unit volume, minimizing the impact on the landscape.

To reduce the cost of energy from wind turbines to levels competitive with coal- and gas-fired electricity production, producers have raised tower heights to place turbines in stronger winds and lengthened blades to capture more wind. As this strategy reaches the upper limits of practicality, designers are now exploring the use of carbon fiber as a means to further push the design envelope and decrease the cost of energy. Compared to conventional all-glass-fiber designs, composites that replace some of the glass with carbon fiber reinforcements can produce the same blade using less fiber and resin, while increasing blade stiffness, improving aerodynamics, and decreasing the loads imposed by the blades on the tower and hub. A design that incorporates carbon fiber also can make power input from the blades more predictable.

Offshore wind farms are a relatively recent, high impact development. FRP materials will be instrumental in the success of offshore programs due to their proven performance in corrosive and hostile environments, which will maintain efficiency of the structures under increased locational loads.

Due to their inherent properties, FRP composites work well as marine turbine blades as they do not rust. Epoxy resin is less susceptible to water penetration and is generally, but not always, used as the matrix in an underwater composite. Excellent fatigue performance means that the turbine does not need to be constantly maintained.

As natural insulators with high dielectric strength, glass fiber composites revolutionized the handling of electricity when they first replaced wood and metal in 1959. Today, utility companies are working with composite suppliers to take advantage of composites for power transmission towers and distribution poles, cables, crossarms – traditionally the province of wood and steel – and the aluminum conductor cables they support. Pultruded and filament wound composite utility poles and crossarms have begun to overcome buyer resistance as electric power companies employ them primarily as replacements for aging wood poles in remote and/or extremely humid locations. Composite-reinforced aluminum conductor cables (CRAC) replace traditional steel-strength members in cables with a pultruded continuous fiber core, which is expected to reduce weight and to increase power transmission efficiency by an estimated 200%. If successful in upcoming tests and demonstration projects, CRAC technologies may find application in infrastructure modernization projects. CRAC cable developers claim that power needs will actually increase, by as much as 19%, in the next 10 years, making CRAC cabling an attractive alternative for upgrading powerlines on the existing grid, without erecting new towers or obtaining additional rights of way.

Composite materials are likely candidates for the eventual materials of choice used to make the bipolar plates, end plates, fuel tanks, and other components in fuel cell systems. Fuel cell technologies of several types offer a "clean" (near-zero Volatile Organic Compounds (VOC)) means to convert hydrogen to electrical power in automotive and stationary power systems. Due to their conductivity, corrosion resistance, dimensional stability, and flame retardancy, vinyl ester-based bulk molding compounds (BMCs) with carbon fiber reinforcement have already been selected in a least one commercially available stationary unit.

REFERENCES

1. Kelly A. (1985), "Composites in context", *Composite Science and Technology*, Vol. 23(3), pp. 171–199.
2. Hunt Warren H. (2004), "Nano-materials: Nomenclature, Novelty and Necessity", *Journal of Material Science*, Vol. 56, pp. 13–18.
3. Guz I.A., Rodger A.A., Guz A.N., Rushchitsky J.J. (2007), "Developing the mechanical models for nano-materials", *Composites Part A: Applied Science and Manufacturing*, Vol. 38, pp. 1234–1250.

2 Materials

In a composite, typically, there are two constituents: One acts as a reinforcement and the other acts as a matrix. Sometimes, the constituents are also referred as phases. Both of these constituents will be discussed in the following sections. The functions of reinforcing agents are as follows:

i. These agents are the main load-carrying constituents.
ii. The reinforcing materials, in general, have significantly higher desired properties. Hence, they contribute the desired properties to the composite.
iii. They transfer the strength and stiffness to the matrix material.

The matrix also performs various functions, which are as follows:

i. The matrix material holds the fibers together.
ii. It plays an important role to keep the fibers at desired positions. The desired distribution of the fibers is very important from micromechanical point of view.
iii. It keeps the fibers separate from each other so that the mechanical abrasion between them does not occur.
iv. It transfers the load uniformly between fibers. Further, in case a fiber is broken or fiber is discontinuous, then it helps to redistribute the load in the vicinity of the break site.
v. It provides protection to fibers from environmental effects.
vi. It provides better finish to the final product.
vii. It enhances some of the properties of the resulting material and structural component (that fiber alone is not able to impart), such as transverse strength of a lamina and impact resistance.

2.1 FIBERS

Fiber is an individual filament of the material. A filament with a length-to-diameter ratio above 1000 is called a fiber. Fibers are the principal constituents in a fiber-reinforced composite material. They occupy the largest volume fraction in a composite laminate and share the major portion of the load acting on a composite structure. Proper selection of the fiber type, fiber volume fraction, fiber length, and fiber orientation is very important since it influences the following characteristics of a composite laminate:

i. Density
ii. Tensile strength and modulus
iii. Compressive strength and modulus

iv. Fatigue strength as well as fatigue failure mechanisms
v. Electrical and thermal conductivities
vi. Cost.

Before discussing the types of fibers, various terms associated with the fibers should be understood first. The main terms associated with the fibers are given below.

1. *Filament*: An individual element.
2. *Strand:* Bundles of 204 filaments or multiple of these.
3. *Roving*: Combination of strands to form thicker parallel bundles.
4. *Yarns*: Strands are twisted to form yarns.
5. *Aspect ratio:* The ratio of length to diameter of a fiber.
6. *Bicomponent fibers:* A fiber made by spinning two compositions concurrently in each capillary of the spinneret.
7. *Blend:* A mix of natural staple fiber (such as cotton or wool) and synthetic staple fibers (such as nylon and polyester). Blends are made to take advantages of the natural and synthetic fibers.
8. *Braiding:* Two or more yarns are intertwined to form an elongated structure. The long direction is called the bias direction or machine direction.
9. *Carding:* Process of making fibers parallel by using rollers covered with needles.
10. *Chopped strands:* Fibers are chopped to various lengths, 3–50 mm, for mixing with resins.
11. *Continuous fibers:* Continuous strands of fibers, generally, available as wound fiber spools.
12. *Cord:* A relatively thick fibrous product made by twisting together two or more plies of yarn.
13. *Covering power:* The ability of fiber to occupy space. Noncircular fibers have greater covering power than circular fibers.
14. *Crimp:* Waviness along the fiber length. Some natural fibers, for example, wool, have a natural crimp. In synthetic polymeric fibers, crimp can be introduced by passing the filament between rollers having teeth. Crimp can also be introduced by chemical means. This is done by controlling the coagulation of the filament to produce an asymmetrical cross section.
15. *Denier:* A unit of linear density. It is the weight in grams of 9000-m-long yarn. This unit is commonly used in the US textile industry.
16. *Fabric:* A kind of planar fibrous assembly. It allows the high degree of anisotropic characteristic of yarn to be minimized, although not completely eliminated.
17. *Felt:* Homogeneous fibrous structure made by interlocking fibers via the application of heat, moisture, and pressure.
18. *Filament:* Continuous fiber, that is, fiber with aspect ratio approaching infinity.
19. *Fill:* It is similar to weft.
20. *Handle:* Also known as softness of handle. It is a function of denier (or tex), compliance, cross section, crimp, moisture absorption, and surface roughness of the fiber.

21. *Knitted fabric:* One set of yarn is looped and interlocked to form a planar structure.
22. *Knitting:* This involves drawing loops of yarns over previous loops, also called interlooping.
23. *Mat:* Randomly dispersed chopped fibers or continuous fiber strands, held together with a binder. The binder can be resin compatible if the mat is to be used to make a polymeric composite.
24. *Microfibers:* Also known as micro-denier fibers. These are fibers having less than 1 denier per filament (or less than 0.11 tex per filament). Fabrics made of such microfibers have superior silk-like handle and dense construction. They find applications in stretch fabrics, lingerie, rain wear, etc.
25. *Monofilament:* A large-diameter continuous fiber, generally, with a diameter greater than 100 m.
26. *Nonwovens:* Randomly arranged fibers without making fiber yarns. Nonwovens can be formed by spun bonding, resin bonding, or needle punching. A planar sheet-like fabric is produced from fibers without going through the yarns spinning step. Chemical bonding and/or mechanical interlocking is achieved. Fibers (continuous or staple) are dispersed in a fluid (i.e., a liquid or air) and laid in a sheet-like planar form on a support, and then chemically bonded or mechanically interlocked. Paper is perhaps the best example of a wet-laid nonwoven fabric where we generally use wood or cellulosic fibers. In spun-bonded nonwovens, continuous fibers are extruded and collected in a random planar network and bonded.
27. *Particle:* Extreme case of a fibrous form – it has a more or less equiaxial form; that is, the aspect ratio is about 1.
28. *Plaiting:* It is similar to braiding.
29. *Rayon:* Term used to designate any of the regenerated fibers made by the viscose, cuprammonium, or acetate processes. They are considered to be natural fibers because they are made from regenerated, natural cellulose.
30. *Retting:* A biological process of degrading pectin and lignin associated with vegetable fibers, loosening the stem and fibers, followed by their separation.
31. *Ribbon:* Fiber of rectangular cross section with a width-to-thickness ratio greater than 4.
32. *Rope:* Linear flexible structure with a minimum diameter of 4 mm. It generally has three strands twisted together in a helix. The rope characteristics are defined by two parameters, namely, unit mass and break length. Unit mass is simply g m^{-1} or *ktex*, while breaking length is the length of rope that will break under the force of its own weight when freely suspended. Thus, break length equals mass at break/unit mass.
33. *Roving:* A bundle of yarns or tows of continuous filaments (twisted or untwisted).
34. *Spinneret:* A vessel with numerous shaped holes at the bottom through which a material in molten state is forced out in the form of fine filaments or threads.
35. *Spun bonding:* Process of producing a bond between nonwoven fibers by heating the fibers to near their melting point.

36. *Staple fiber:* Fibers having short, discrete lengths (10–400 mm long) that can be spun into a yarn are called staple fibers. This spinning quality can be improved if the fiber imparted a waviness or crimp. Staple fibers are excellent for providing bulkiness for filling, filtration etc. Frequently, staple natural fibers, for example, cotton or wool, are blended with staple synthetic fibers, for example, nylon or polyester, to obtain the best of both types.
37. *Tenacity:* A measure of fiber strength that is commonly used in the textile industry. Commonly, the units are gram-force per denier, gram-force per tex, or Newton per tex. It is a specific strength unit; that is, there is a factor of density involved. Thus, although the tensile strength of glass fiber is more than double that of nylon fiber, both glass and nylon fibers have a tenacity of about 6 g den^{-1}. This is because the density of glass is about twice that of nylon.
38. *Tex:* A unit of linear density. It is the weight in grams of 1000 m of yarn. Tex is commonly used in Europe.
39. *Tow:* Bundle of twisted or untwisted continuous fibers. A tow may contain tens or hundreds of thousands of individual filaments.
40. *Twist:* The angle of twist that individual filaments may have about the yarn axis. Most yarns have filaments twisted because it is easier to handle a twisted yarn than an untwisted one.
41. *Wire:* Metallic filament.
42. *Warp:* Lengthwise yarn in a woven fabric.
43. *Weft:* Transverse yarn in a woven fabric. Also called fill.
44. *Whisker:* Tiny, whisker-like fiber (a few mm long, a few m in diameter) that is a single crystal and almost free of dislocations. Note that this term involves a material requirement. The small size and crystalline perfection make whiskers extremely strong, approaching the theoretical strength.
45. *Woven fabric:* Flat, drapeable sheet made by interlacing yarns or tows.
46. *Woven roving:* Heavy, drapeable fabric woven from continuous rovings.
47. *Yarn:* A generic term for a bundle of untwisted or twisted fibers (short or continuous). A yarn can be produced from staple fibers by yarn spinning. The yarn spinning process consists of some fiber alignment, followed by locking together by twisting. Continuous synthetic fibers are also used to make yarns. Continuous fibers are easy to align parallel to the yarn axis. Generally, the degree of twist is low, just enough to give some interfilament cohesion.

2.2 TYPES OF FIBERS

The fibers that are used in the fabrication of a composite can be divided into two broad categories as follows:

1. Natural fibers
2. Advanced fibers.

1. *Natural fibers*: The natural fibers are divided into the following three subcategories:
    i. *Animal fibers*: Silk, wool, spider silk, sinew, camel hair etc.
    ii. *Plant/vegetable fibers*: Cotton (seed), jute (stem), hemp (stem), sisal (leaf), ramie, bamboo, maze, sugarcane, banana, kapok, coir, abaca, kenaf, flax, raffia palm etc.
    iii. *Mineral fibers*: Asbestos, basalt, mineral wool, glass wool etc.
2. *Advanced fibers*: An advanced fiber is defined as a fiber which has a high specific stiffness (i.e., the ratio of Young's modulus to the density of the material, E/ρ) and a high specific strength (i.e., the ratio of ultimate strength to the density of the material, S/ρ). The fibers made from the following materials are the advanced fibers.
    i. Carbon and/or graphite
    ii. Glass fibers
    iii. Alumina
    iv. Aramid
    v. Silicon carbide
    vi. Sapphire.

The materials of the advanced fibers are lighter than the conventional metals. These materials occupy higher position as compared to metals in the periodic table. Thus, these materials have higher specific properties (property per unit weight) than metals.

2.3 NATURAL FIBERS

Some of the principle natural fibers as listed in the previous section have been discussed below.

2.3.1 Silk Fiber

Silk is a natural protein fiber, some forms of which can be woven into textiles. The protein fiber of silk is mainly composed of fibroin and is produced by certain insect larvae to form cocoons. The best-known silk is obtained from the cocoons of the larvae of the mulberry silkworm *Bombyx mori* reared in captivity (sericulture). The life cycle of the silkworm is shown in Figure 2.1. The shimmering appearance of silk is due to the triangular prism-like structure of the silk fiber, which allows silk cloth to refract incoming light at different angles, thus producing different colors.

Silk fibers from the *Bombyx mori* silkworm have a triangular cross section with rounded corners, 5–10 μm wide. The fibroin heavy chain is mostly composed of beta sheets, due to a 59-mer amino acid repeat sequence with some variations. Silk has a smooth, soft texture that is not slippery, unlike many synthetic fibers. It is one of the strongest natural fibers, but loses up to 20% of its strength when wet. It has a good moisture regain of 11%. Its elasticity is moderate to poor: If elongated even a small amount, it remains stretched. It can be weakened if exposed to too much sunlight. It may also be attacked by insects, especially if left dirty.

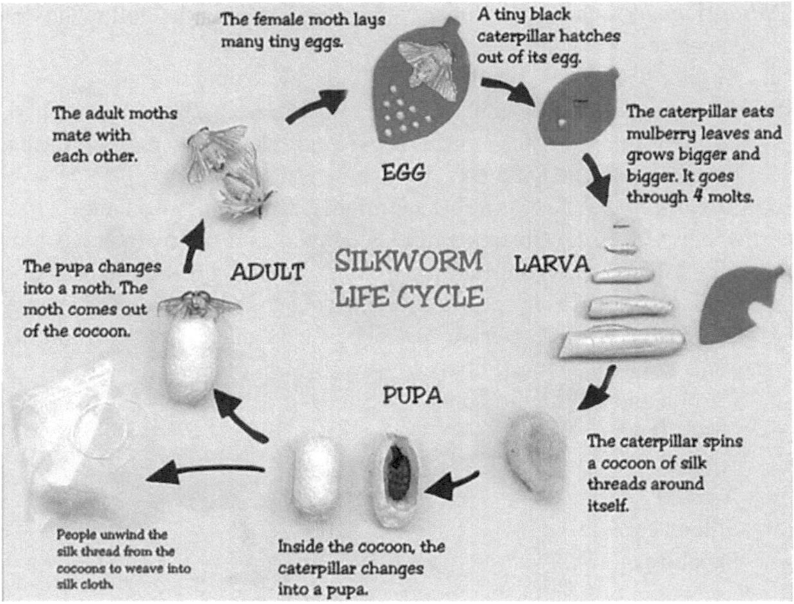

FIGURE 2.1 Life cycle of a silkworm.

FIGURE 2.2 Sources of wool.

2.3.2 Wool Fiber

Wool is the textile fiber obtained from sheep and certain other animals, including cashmere from goats, mohair from goats, qiviut from muskoxen, angora from rabbits, and other types of wool from camelids. These sources of wool are shown in Figure 2.2. It has several qualities that distinguish it from hair or fur: It is crimped, it is elastic, and it grows in staples (clusters). Wool fibers readily absorb moisture, but are not hollow. Wool can absorb almost one-third of its own weight in water. It absorbs sound like many other fabrics. It is generally a creamy white in color, although some breeds of sheep produce natural colors, such as black, brown, silver, and random mixes. It ignites at a higher temperature than cotton and some syntheticfibers. It has a lower rate of flame spread, a lower rate of heat release, and a lower heat of combustion, and does not melt or drip. It forms a char which is insulating and self-extinguishing, and it contributes less to toxic gases and smoke than other flooring products when used in carpets. Wool carpets are specified for high safety environments, such as trains and aircraft. Wool is usually specified for garments for firefighters, soldiers, and others in occupations where they are exposed to the likelihood of fire. It is also separated into grades based on the measurement of the wool's diameter in microns and also its style. These grades may vary depending on the breed or purpose of the wool. For example:

<15.5: Ultrafine Merino
15.6–18.5: Superfine Merino
18.6–20: Fine Merino
20.1–23: Medium Merino.

2.3.3 Spider Silk

Spider silk is a protein fiber spun by spiders. Spiders use their silk to make webs or other structures, which function as nets to catch other animals, or as nests or cocoons to protect their offspring. They can also use their silk to suspend themselves. Most silks, in particular dragline silk, have exceptional mechanical properties. They exhibit a unique combination of high tensile strength and extensibility (ductility). This enables a silk fiber to absorb a lot of energy before breaking (toughness, the area under a stress–strain curve). Weight for weight, silk is stronger than steel, but not as strong as Kevlar. As can be seen from Figure 2.3, silk is, however, tougher than both.

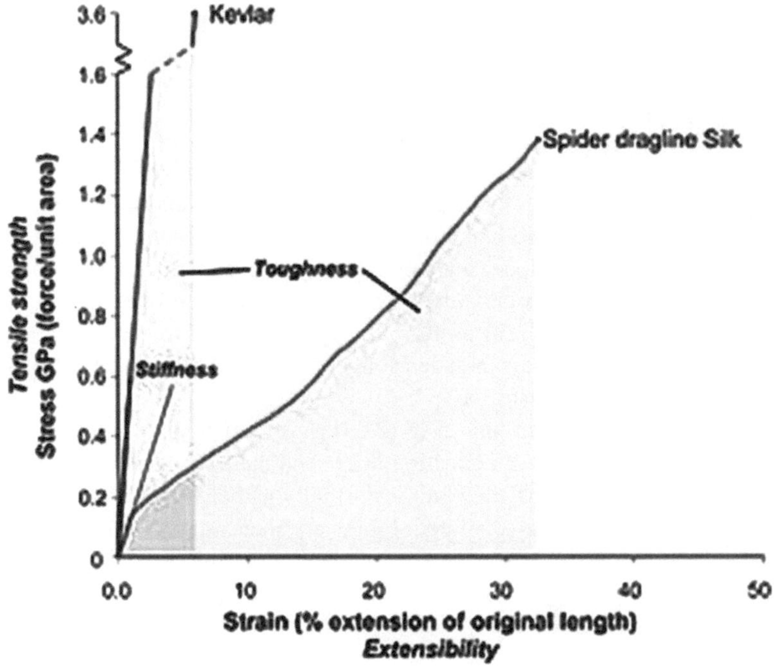

FIGURE 2.3 Comparison of stress–strain curve of silk and Kevlar.

A dragline silk's tensile strength is comparable to that of high-grade alloy steel (450–1970 MPa) and about half as strong as aramid filaments, such as Kevlar (3000 MPa). The combination of strength and ductility gives dragline silks a very high toughness (or work to fracture), which "equals that of commercial polyaramid (aromatic nylon) filaments, which themselves are benchmarks of modern polymer fiber technology." Silks are about a sixth of the density of steel (1.31 g cm^{-3}). As a result, a strand long enough to circle the Earth would weigh less than 500 g. Spider dragline silk has a tensile strength of roughly 1.3 GPa. The tensile strength listed for steel might be slightly higher, for example, 1.65 GPa, but spider silk is a much less dense material, so that a given weight of spider silk is five times as strong as the same weight of steel.

2.3.4 Sinew Fiber

A piece of tough fibrous tissue uniting muscle to bone through a tendon or ligament is known as SINEW. The process for breaking tendons to obtain sinew fibers is very straightforward. A leg or backstrap tendon is taken and is beaten with a stone, a hammer, or a wooden mallet until the fibers start separating. At least on deer leg tendons, the color of the tendon changes from brown to white when this is happening. Make sure that whatever is used to separate the fibers does not have any sharp corners: We do not want to cut the sinew fibers. Also, for the same reason, use only just enough force to separate the fibers. Once the tendon has been well beaten, we can take the tendon and start ripping it into pieces. A good strategy is to rip the entire tendon into two parts, split those, and then split until the desired fiber thickness is reached. In practice, it doesn't seem to matter much whether we start splitting from the ends or from the middle. If we have trouble getting splits to start, we can use a thick needle or a sharpened piece of hardened, round steel (~3 mm thick) to get us started. The whole process is shown in Figure 2.4.

2.3.5 Camel Hair

Camel hair is a type of cloth made from pure camel hair or a blend of camel hair and another fiber. The outer protective fur (guard hair) is coarse and inflexible, and can be woven into haircloth. Guard hair can be made soft and plush by blending it, especially with wool. The camel's pure undercoat is very soft, gathered when camels molt, and is frequently used for coats. Camel hair is collected from the Bactrian camel, which is found across Asia from eastern Turkey and China to Siberia. Significant supplier countries of camel hair include Mongolia, Tibet, Afghanistan, Iran, Russia, China, New Zealand, and Australia.

Each camel can produce around 5 lb (2.25 kg) of hair a year. Hair may be collected by shearing, combing, or collecting fiber shed naturally during the 6–8-week molting season in late spring. Fallen hair is normally gathered by hand. After collection, coarse and fine hairs are separated. Fibers are then washed to remove any dirt or debris and then spun into yarn suitable for weaving or knitting.

2.3.6 Cotton Fiber

Cotton is a soft, fluffy staple fiber that grows in a boll, or protective case, around the seeds of cotton plants of the genus Gossypium in the family of Malvaceae. The fiber

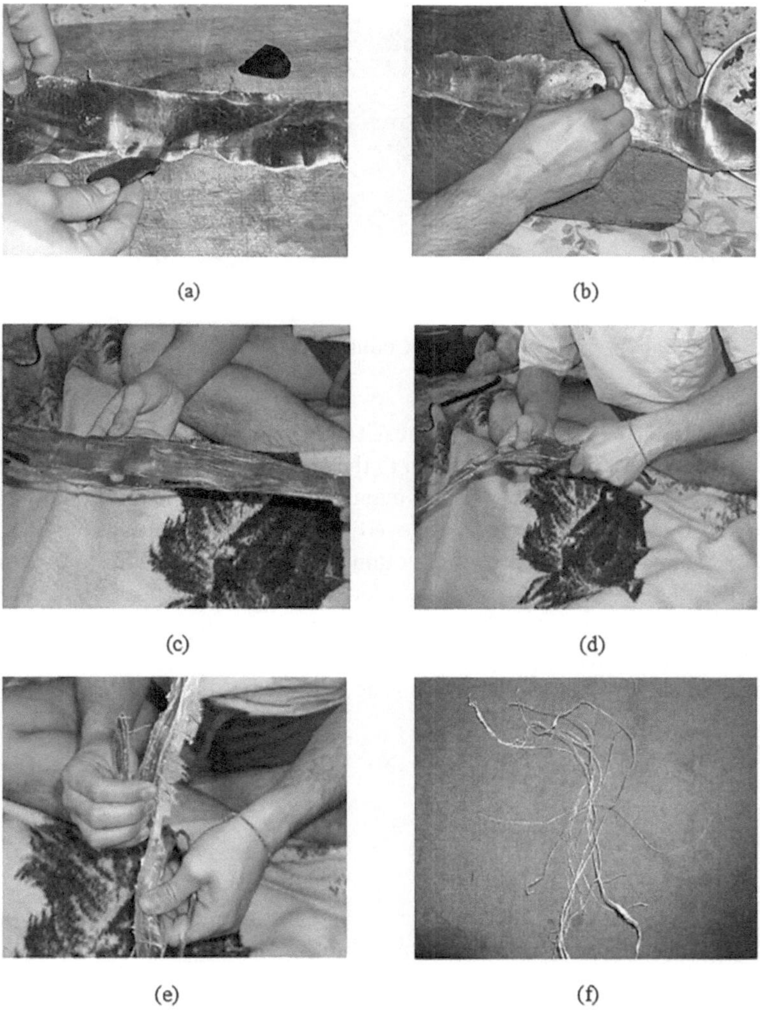

FIGURE 2.4 (a–f) Steps of obtaining a sinew fiber.

is almost pure cellulose. The plant is a shrub native to tropical and subtropical regions around the world, including the Americas, Africa, and India. The fiber is most often spun into yarn or thread, and used to make a soft, breathable textile. Current estimates for the world production are about 25 million tons or 110 million bales annually, accounting for 2.5% of the world's arable land. China is the world's largest producer of cotton, but most of this is used domestically. The United States has been the largest exporter for many years.

Cotton is used to make a number of textile products. These products include terrycloth for highly absorbent bath towels and robes; denim for blue jeans; cambric, popularly used in the manufacture of blue work shirts (from which we get the term "blue-collar"); and corduroy, seersucker, and cotton twill. These products are shown

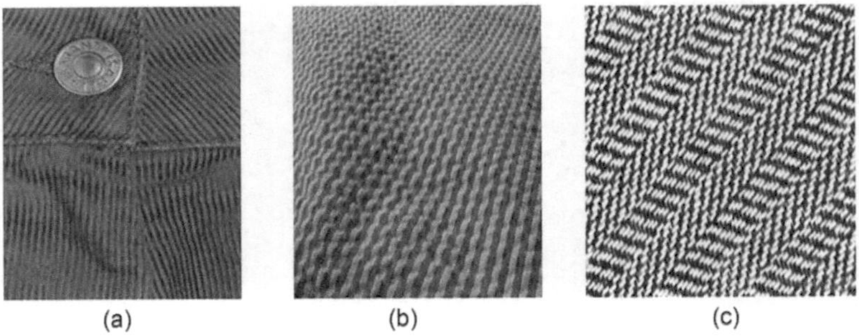

FIGURE 2.5 Clothes made of cotton. (a) Corduroy, (b) seersucker, (c) cotton twill.

in Figure 2.5. Socks, underwear, and most T-shirts are made from cotton. Bedsheets often are made from cotton. In addition to the textile industry, cotton is used in fishing nets, coffee filters, tents, explosives manufacture, cotton paper, and bookbinding. Cotton is also used to make yarn employed in crochet and knitting. Fabric can also be made from recycled or recovered cotton that otherwise would be thrown away during the spinning, weaving, or cutting process. While many fabrics are made completely of cotton, some materials blend cotton with other fibers, including rayon and synthetic fibers such as polyester.

The chemical composition of cotton is as follows:

i. Cellulose = 91.00%
ii. Water = 7.85%
iii. Protoplasm, pectins = 0.55%
iv. Waxes, fatty substances = 0.40%
v. Mineral salts = 0.20%.

The properties of cotton fiber are shown in Table 2.1.

2.3.7 Jute Fiber

Jute is a long, soft, shiny vegetable fiber that can be spun into coarse, strong threads. It is produced from plants in the category called as genus Corchorus, as shown in Figure 2.6. It is one of the most affordable natural fibers and is second only to cotton in amount produced and variety of uses of vegetable fibers. Jute fibers are primarily composed of the plant materials such as cellulose and lignin. Jute falls into the bast fiber category (fiber collected from bast, the phloem of the plant, sometimes called the "skin"). **Lignin** is a class of complex organic polymers that form important structural materials in the support tissues of vascular plants and some algae. Lignins are particularly important in the formation of cell walls, especially in wood and bark, because they lend rigidity and do not rot easily. Chemically lignins are cross-linked phenol polymers.

The jute fiber comes from the stem and ribbon (outer skin) of the jute plant. The fibers are first extracted by retting. The retting process consists of bundling jute

TABLE 2.1
Properties of Cotton Fiber

Property	Evaluation
Shape	Fairly uniform in width, 12–20 μm; Length varies from 1 to 6 cm (½ to 2½ in); Typical length is 2.2 to 3.3 cm (⅞ to 1¼ in).
Luster	High
Tenacity (strength)	
Dry	3.0–5.0 g d^{-1}
Wet	3.3–6.0 g d^{-1}
Resiliency	Low
Density	1.54–1.56 g cm^{-3}
Moisture absorption	8.5%
raw: conditioned	15%–25%
saturation	8.5%–10.3%
mercerized: conditioned	15%–27%+
saturation	
Dimensional stability	Good
Resistance to	Damage, weaken fibers
acids	Resistant; no harmful effects
alkali	High resistance to most
organic solvents	Prolonged exposure weakens fibers.
sunlight	Mildew and rot-producing bacteria damage fibers.
microorganisms	Silverfish damage fibers.
insects	
Thermal reactions	Decomposes after prolonged exposure to temperatures of 150°C or over.
to heat	
to flame	Burns readily.

FIGURE 2.6 Plants in the category called as genus Corchorus.

stems together and immersing them in slow running water. There are two types of retting: stem and ribbon. After the retting process, stripping begins. In the stripping process, nonfibrous matter is scraped off, then the workers dig in and grab the fibers from within the jute stem.

Advantages of jute include the following:

i. Jute has good insulating and antistatic properties, as well as having low thermal conductivity and a moderate moisture regain.
ii. It has acoustic insulating properties and can be used for manufacturing with no skin irritations.
iii. Its fiber is 100% biodegradable and recyclable, and thus environmentally friendly.
iv. It has low pesticide and fertilizer needs.
v. It is a natural fiber with golden and silky shine, and hence called the golden fiber.
vi. It is the cheapest vegetable fiber procured from the bast or skin of the plant's stem.
vii. It is the second most important vegetable fiber after cotton, in terms of usage, global consumption, production, and availability.
viii. It has high tensile strength and low extensibility, and ensures better breathability of fabrics. Therefore, it is very suitable in agricultural commodity bulk packaging.

2.3.8 Kenaf Fiber

Kenaf – *Hibiscus cannabinus* – is a plant in the Malvaceae family. *Hibiscus cannabinus* is in the genus Hibiscus and is probably native to southern Asia, though its exact natural origin is unknown. The name also applies to the fiber obtained from this plant. Kenaf is one of the allied fibers of jute and shows the similar characteristics. It is an annual or biennial herbaceous plant (rarely a short-lived perennial) growing to 1.5–3.5 m tall with a woody base. The stems are 1–2 cm diameter, often but not always branched, which are shown in Figure 2.7. The leaves are 10–15 cm long, variable in shape, with leaves near the base of the stems being deeply lobed with 3–7 lobes. The flowers are 8–15 cm in diameter, and white, yellow, or purple; when white or yellow, the center is still dark purple. The fruit capsule is 2 cm in diameter and contains several seeds. The fibers in kenaf are found in the bast (bark) and the core (wood). The bast constitutes 40% of the plant. "Crude fiber" separated from the bast is multicellular, consisting of several individual cells stuck together. The individual fiber cells are about 2–6 mm long and slender. The cell wall is thick (6.3 μm). The core is about 60% of the plant, and has thick (≈38 μm) but short (0.5 mm) and thin-walled (3 μm) fiber cells. Paper pulp is produced from the whole stem

and therefore contains two types of fibers, from the bast and from the core. The pulp quality is similar to hardwood. Kenaf fiber is mainly used as rope, twine, coarse cloth (similar to that made from jute), and paper. Uses of kenaf fiber include engineered wood, insulation, clothing-grade cloth, soilless potting mixes, animal bedding, packing material, and material that absorbs oil and liquids.

FIGURE 2.7 Kenaf – *Hibiscus cannabinus.*

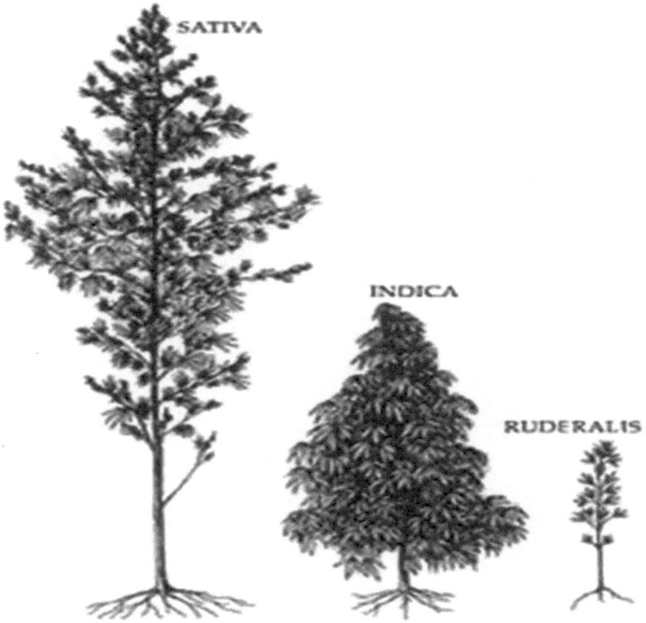

FIGURE 2.8 Varieties of the cannabis plant.

2.3.9 Hemp Fiber

Hemp is a commonly used term for high-growing industrial varieties of the cannabis plant (shown in Figure 2.8) and its products, which include fiber, oil, and seed. Hemp is refined into products such as hemp foods, hemp oil, wax, resin, rope, cloth, pulp, paper, and fuel. Items ranging from rope, to fabrics, to industrial materials were made from hemp fiber. Hemp was often used to make sail canvas, and the word "canvas" is derived from cannabis. Today, a modest hemp fabric industry exists, and hemp fibers can be used in clothing. Pure hemp has a texture similar to linen.

2.3.10 Flax Fiber

Flax fiber is extracted from the bast beneath the surface of the stem of the flax plant (as shown in Figure 2.9). It is soft, lustrous, and flexible; bundles of fiber have the appearance of blonde hair, hence the description "flaxen." It is stronger than cotton fiber, but less elastic. The best grades are used for linen fabrics such as damasks, lace,

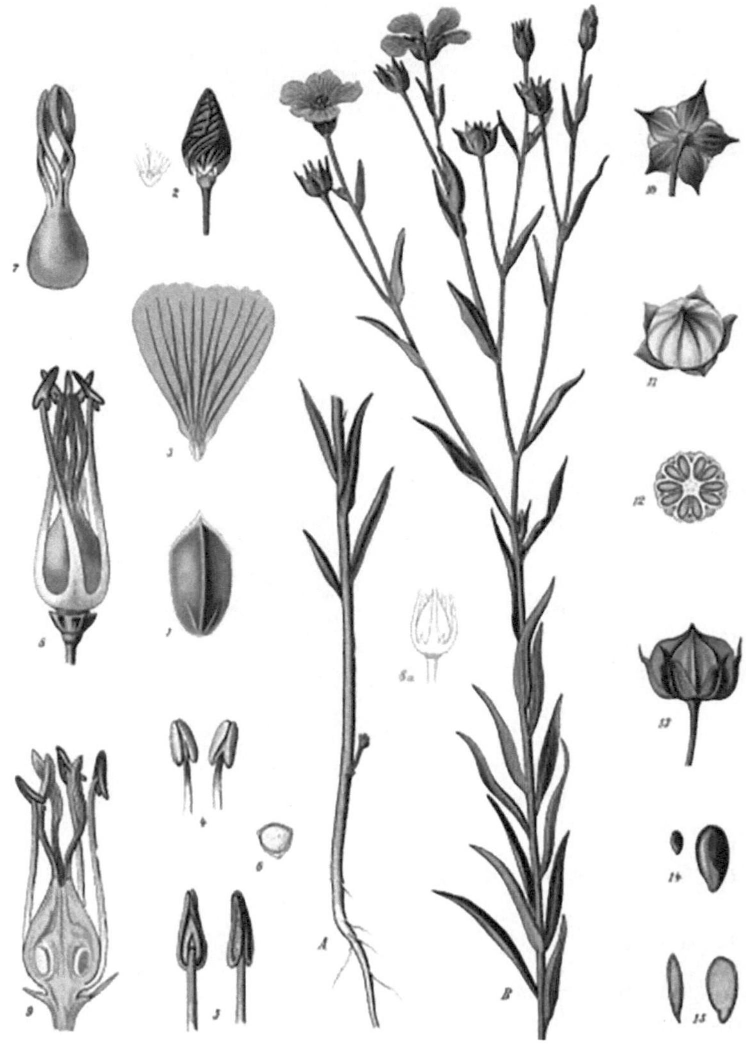

FIGURE 2.9 Source of flax fiber.

and sheeting. Coarser grades are used for the manufacturing of twine and rope, and historically for canvas and webbing equipment. Flax fiber is a raw material used in the high-quality paper industry for the use of printed banknotes and rolling paper for cigarettes and tea bags. Flax mills for spinning flaxen yarn were invented by John Kendrew and Thomas Porthouse of Darlington, England, in 1787.

2.3.11 Ramie Fiber

Ramie is a flowering plant in the nettle family Urticaceae, native to eastern Asia. It is an herbaceous perennial growing to 1–2.5 m tall; the leaves are heart-shaped, 7–15 cm long and 6–12 cm broad, and white on the underside with dense small hairs – this gives it a silvery appearance; unlike stinging nettles, the hairs do not sting. The true ramie or China grass is also called Chinese plant or white ramie. A second type, known as green ramie or rhea, is believed to have originated in the Malay Peninsula. It has smaller leaves which are green on the underside, and it appears to be better suited to tropical conditions. The word "ramie" is derived from the Malay word "rami."

Ramie is one of the strongest natural fibers. It exhibits even greater strength when wet. Ramie fiber is known especially for its ability to hold shape, reduce wrinkling, and introduce a silky luster to the fabric appearance. It is not as durable as other fibers, and so is usually used as a blend with other fibers such as cotton or wool. It is similar to linen in absorbency, density, and microscopic appearance. However, it will not dye as well as cotton. Because of its high molecular crystallinity, ramie is stiff and brittle, and will break if folded repeatedly in the same place; it lacks resiliency and is low in elasticity and elongation potential. The properties of Ramie fiber are shown in Tables 2.2 and 2.3.

Ramie is used to make such products as industrial sewing thread, packing materials, fishing nets, and filter cloths. It is also made into fabrics for household furnishings (upholstery, canvas) and clothing, frequently in blends with other textile

TABLE 2.2
Physical and Chemical Properties of Ramie Fiber

Cellulose (wt.%)	Lignin (wt.%)	Hemicellulose (wt.%)	Pectin (wt.%)	Wax (wt.%)	Microfibrillar Angle (°)	Moisture Content (wt.%)	Density (g cm^{-3})
68.6–76.2	0.6–0.7	13.1–16.7	1.9	0.3	7.5	8.0	1.50

TABLE 2.3
Mechanical Properties of Untreated Ramie Fibers

Fiber Diameter (mm)	Fracture Load (N)	Tensile Strength (MPa)	Fracture Strain (%)
0.034	0.467	560	0.025

fibers (for instance, when used in admixture with wool, shrinkage is reported to be greatly reduced when compared with pure wool). Shorter fibers and waste are used in paper manufacture. Ramie ribbon is used in fine bookbinding as a substitute for traditional linen tape.

2.3.12 Sisal Fiber

The sisal fiber is traditionally used for rope and twine, and has many other uses, including paper, cloth, wall coverings, carpets, and dartboards. Fiber is extracted by a process known as decortication, where leaves are crushed and beaten by a rotating wheel set with blunt knives, so that only fibers remain. The fiber is then dried, brushed, and baled for export. Proper drying is important as fiber quality largely depends on moisture content. The procedure of obtaining sisal fibers from the sisal plant is shown in Figure 2.10. Apart from ropes, twines, and general cordage, sisal is used in low-cost and specialty paper, dartboards, buffing cloth, filters, geotextiles, mattresses, carpets, handicrafts, wire rope cores etc. Sisal has been utilized as an environment-friendly strengthening agent to replace asbestos and fiberglass in composite materials in various uses, including the automobile industry.

2.3.13 Bamboo Fiber

The bamboos are a subfamily of flowering perennial evergreen plants in the grass family Poaceae (shown in Figure 2.11). Bamboo fiber is a cellulose fiber extracted or fabricated from natural bamboo, and possibly other additives, and is made from

FIGURE 2.10 (a–d) Procedure of obtaining sisal fibers from the sisal plant.

FIGURE 2.11 Perennial evergreen plants in the grass family Poaceae.

(or in the case of material fabrication, is) the pulp of bamboo plants. It is usually not made from the fibers of the plant, but is a synthetic viscose made from bamboo cellulose. Bamboo is extremely resilient and durable as a fiber. In studies comparing it to cotton and polyester, it is found to have a high breaking tenacity, better moisture-wicking properties, and better moisture absorption. Bamboo fibers are extracted through mechanical needling and scraping or through a steam explosion process where bamboo is injected with steam and placed under pressure and then exposed to the atmosphere where small explosions within the bamboo due to steam release allow for the collection of bamboo fiber.

Bamboo fiber can be in a pulped form in which the material is extremely fine and in a powdered state. In bamboo, the internodal regions of the stem are usually hollow and the vascular bundles in the cross section are scattered throughout the stem instead of in a cylindrical arrangement. Due to its versatile properties, bamboo fibers are mainly used in textile industry for making attires, towels, and bathrobes. Due to its antibacterial nature, its fiber is used for making bandages, masks, nurse wears, and sanitary napkins.

2.3.14 Maize (Corn) Fiber

Corn fiber is a man-made fiber entirely derived from annually renewable resources. These fibers have the performance advantages often associated with synthetic materials, and complementing properties of natural products such as cotton and wool. The trade name of this fiber is Ingeo. This fiber combines the qualities of natural and synthetic fibers in a unique way. Strength and resilience are balanced with comfort, softness, and drape in textiles. In addition, it has good moisture management characteristics. The process for manufacturing the polymer used to make corn fiber on an industrial scale centers on the fermentation, distillation, and polymerization of a simple plant sugar, maize dextrose. The sugars are fermented in a process similar to making yogurt. After fermentation, the products are transformed into a high-performance polymer called polylactide, which can then be spun or otherwise processed into corn fiber for use in a wide range of textile applications.

The production and use of corn fiber means that less greenhouse gases are added to the atmosphere. Greenhouse gases are the chief contributor to global climate change. Compostability and chemical recyclability mean that under the right conditions and with the right handling, the complete life cycle of production, consumption, disposal, and reuse is neatly closed. Corn is available in both spun and filament forms in a wide variety of counts from micro-denier for the finest lightest fabrics to high counts for more robust applications. It is derived from naturally occurring plant sugars. When products come to the end of their useful life, they can be returned to the Earth, unlike petroleum-based products, which can only be disposed of through thermal recycling, physical recycling, or landfill. Corn also uses no chemical additives or surface treatments, and amazingly, is naturally flame retardant. It is reported to have outstanding moisture management properties and low odor retention, giving the wearer optimum comfort and confidence. Corn fiber filament is said to have a subtle luster and fluid drape with a natural hand offering a new material to stimulate creativity. Corn fiberfill allows outerwear garment makers to offer a complete story and a more environmentally friendly alternative to polyester and nylon combinations in padded garments. It reportedly outperforms other synthetics in resistance to UV light, retaining strength color and properties over time. The garments made of corn fiber are generally wrinkle-free.

2.3.15 Coir Fiber

Coir fibers are found between the hard, internal shell and the outer coat of a coconut. The individual fiber cells are narrow and hollow, with thick walls made of cellulose. They are pale when immature, but later become hardened and yellowed as a layer of lignin is deposited on their walls. Each cell is about 1 mm (0.04 in) long and 10–20 μm (0.0004–0.0008 in) in diameter. Fibers are typically 10–30 cm (4–12 in) long. The two varieties of coir are brown and white. *Brown coir* harvested from fully ripened coconuts is thick and strong, and has high abrasion resistance. It is typically used in mats, brushes, and sacking. Mature brown coir fibers contain more lignin and less cellulose than fibers such as flax and cotton, so are stronger but less flexible. *White coir* fibers harvested from coconuts before they are ripe are white or light brown in color and are smoother and finer, but also weaker. They are generally spun to make yarn used in mats or rope. It is also widely used for making fishing nets.

2.3.16 Banana Fiber

Banana fiber is extracted from the pseudo stem sheath of the plant. The extraction can be done mainly in three ways: manual, chemical, and mechanical. Of these, mechanical extraction is the best way to obtain fiber of both good quality and quantity in an ecofriendly way. In this process, the fiber is extracted by inserting the pseudo stem sheaths one by one into a raspador machine. The raspador machine removes nonfibrous tissues and the coherent material from the fiber bundle present in the sheath, and gives the fine fiber as output. After extraction, the fiber is shade-dried for a day and packed in high-density polyethylene (HDPE) bags. Then, it is stored away from moisture and light to keep it in good condition until it is used. It is used to make fancy items like bags, table mats, and purses, and its latest use is in weaving of banana fiber fabric.

2.3.17 Kapok Fiber

The scientific name of Kapok fiber is *Ceiba pentandra*. The tree from which Kapok fiber is obtained grows to 70 m (230 ft) with a trunk up to 3 m (9.8 ft) in diameter with buttresses. The trunk and many of the larger branches are often crowded with large simple thorns. The palmate leaves are composed of 5–9 leaflets, each up to 20 cm (7.9 in) long. The trees produce several hundred 15-cm (5.9 in) pods containing seeds surrounded by a fluffy, yellowish fiber that is a mix of lignin and cellulose. Kapok fiber is light, very buoyant, resilient, and resistant to water, but it is very flammable. The process of harvesting and separating the fiber is labor-intensive and manual. It is difficult to spin, but is used as an alternative to down as filling in mattresses, pillows, upholstery, and stuffed toys such as teddy bears, and for insulation. It was previously much used in life jackets and similar devices until synthetic materials largely replaced the fiber. The seeds produce an oil, used locally in soap, and can be used as fertilizer.

2.3.18 Abaca Fiber

Abacá, binomial name *Musa textilis*, is a species of banana native to the Philippines. It is grown as a commercial crop in the Philippines, Ecuador, and Costa Rica. The plant, also known as Manila hemp, has great economic importance, being harvested for its fiber, also called Manila hemp, extracted from the leaf and stems. The plant grows to 13–22 ft (4.0–6.7 m) and averages about 12 ft (3.7 m). The plant from which abaca fiber is obtained is shown in Figure 2.12. The fiber was originally used for

FIGURE 2.12 Source of abaca fiber.

FIGURE 2.13 Source of raffia palm fiber.

making twines and ropes; now it is mostly pulped and used in a variety of specialized paper products, including tea bags, filter paper, and banknotes. It is classified as a hard fiber, along with coir and sisal. Abacá rope is very durable, flexible, and resistant to saltwater damage, allowing its use in hawsers, ship's lines, and fishing nets. A 1-in (2.5 cm) rope can require 4 metric tons (8800 lb) to break.

2.3.19 Raffia Palm Fiber

Raffia fiber is widely used throughout the world. It is used in twine, rope, baskets, placemats, hats, shoes, and textile. As shown in Figure 2.13, the fiber is produced from the membrane on the underside of each individual frond leaf. The membrane is taken off to create a long thin fiber which can be dyed and woven as a textile into products ranging from hats to shoes to decorative mats. Plain raffia fibers are exported and used as garden ties or as a "natural" string in many countries. Especially when one wishes to graft trees, raffia is used to hold plant parts together as this natural rope has many benefits for this purpose.

2.3.20 Sugarcane Fiber

The samples are subjected to juice extractor wherein juice is separated from the sugarcane. The residue left after the extraction of juice called bagasse is collected for the extraction of fibers. The soft core part pith is removed from the bagasse manually to get the outer hard rind. The rind is then cut across the length so that the cut portions are free from nodes. The samples are then subjected to hot water treatment (material:liquor ratio 1:50). In this process, the samples are kept in hot water at around 90°C for 1 hour for the removal of coloring matter and sugar traces. The samples are then dried under the sunlight. Finally, the samples are subjected to chemical extraction. During this process, the samples are treated with 0.1N NaOH solution, at boiling water temperature for 4 hours under atmospheric pressure. The material:liquor ratio taken for this process is 1:100. During this treatment, the samples are subjected to vigorous stirring for an effective separation of fibers. The well-separated fibers are then dried.

Currently, bagasse sugarcane, a waste product of the sugar industry, is mainly burned as fuel in sugar mill boilers. The low cost, low density, and acceptable mechanical properties of bagasse fiber make it an ideal candidate to be considered for value-added applications such as reinforcement in plastic composites. Bagasse also has commercial potential as reinforcement in cement composites. The advantages of incorporating natural fiber as reinforcement in cement composites are related to their mechanical and thermal properties and reasonable cost. Different treatment techniques enhance the adhesion and compatibility between fibers and matrix, hence improving the mechanical properties of the composite.

2.3.21 Asbestos Fiber

Asbestos is the name given to a group of fibrous minerals, chiefly composed of silicates that occur naturally in many parts of the world. Six types of asbestos are commercially exploited. The three main types are as follows:

i. Crocidolite-blue asbestos
ii. Amosite-brown asbestos
iii. Chrysotile-white asbestos.

Asbestos minerals are crystalline and split longitudinally to form very fine fibers. Crocidolite and amosite are amphiboles with straight and relatively brittle fibers, while chrysotile is a serpentine mineral with curled flexible fibers. Asbestos is extremely durable and stable with a high resistance to heat. Some forms also have resistance to acids and alkalis. Because of its fibrous nature, asbestos can be spun and woven into yarns and fabrics, and used to reinforce cement and plastics.

Man-made mineral fibers (MMMF) is a term given to all synthetic mineral fibers. Those mainly encountered in the environment are the inorganic vitreous fibers which have been produced since the early 1950s. The main types of MMMF are as follows:

i. Mineral wools – glasswool, rockwool, slagwool
ii. Continuous filament – glass or textile fiber
iii. Refractory fibers – ceramic and special-purpose fibers.

MMMF are generally coarser than asbestos fibers and also exhibit good resistance to heat and chemicals. These can be woven and formed into insulation blankets, and as with asbestos, these can be used as a reinforcing agent.

The most common use of asbestos fibers has been in corrugated asbestos cement roof sheets typically used for outbuildings, warehouses, and garages. It may also be found in sheets or panels used for ceilings and sometimes for walls and floors. Chrysotile has been a component in joint compound and some plasters. Numerous other items have been made containing chrysotile, including brake linings, fire barriers in fuse boxes, pipe insulation, floor tiles, and gaskets for high-temperature equipment.

2.3.22 Glass Wool

Glass wool is an insulating material made from fibers of glass arranged using a binder into a texture similar to wool. The process traps many small pockets of air between the glass, and these small air pockets result in high thermal insulation properties. Glass wool is produced in rolls or in slabs, with different thermal and mechanical properties. It may also be produced as a material that can be sprayed or applied in place, on the surface to be insulated. After the mixture of natural sand and recycled glass at 1450°C, the glass that is produced is converted into fibers.

It is typically produced in a method similar to making cotton candy, which is forced through a fine mesh by centripetal force and cooled on contact with the air. The cohesion and mechanical strength of the product is obtained by the presence of a binder that "cements" the fibers together. Ideally, a drop of bonder is placed at each fiber intersection. This fiber mat is then heated to around 200°C to polymerize the resin and is calendered to give it strength and stability.

The final stage involves cutting the wool and packing it in rolls or panels under very high pressure before palletizing the finished product in order to facilitate transport and storage. Glass wool is a thermal insulation that consists of intertwined and flexible glass fibers, which causes it to "package" air, resulting in a low density that can be varied through compression and binder content. Glass wool can be a loose fill material, blown into attics, or – together with an active binder – sprayed on the underside of structures, sheets, and panels that can be used to insulate flat surfaces such as cavity wall insulation, ceiling tiles, curtain walls as well as ducting. It is also used to insulate piping and for soundproofing.

2.3.23 Rock Wool

Stone wool is a furnace product of molten rock at a temperature of about 1600°C, through which a stream of air or steam is blown. More advanced production techniques are based on spinning molten rock in high-speed spinning heads somewhat like the process used to produce cotton candy. The final product is a mass of fine, intertwined fibers with a typical diameter of 2–6 μm. Mineral wool may contain a binder, often a Ter-polymer, and an oil to reduce dusting. Though not immune to the effects of a sufficiently hot fire, the fire resistance of fiberglass, stone wool, and ceramic fibers makes them common building materials when passive fire protection is required, being used as spray fireproofing, in stud cavities in drywall assemblies, and as packing materials in fire-stops. Other uses are in resin-bonded panels, as filler in compounds for gaskets, in brake pads, in plastics in the automotive industry, as a filtering medium, and as a growth medium in hydroponics.

2.3.24 Ceramic Wool

High-temperature insulation wool (HTIW), known as ceramic fiber wool until the 1990s, is one of several types of synthetic mineral wool, generally defined as those resistant to temperatures above 1000°C. The first variety, aluminum silicate fiber, developed in the 1950s, was referred to as refractory ceramic fiber. Silicon carbide

TABLE 2.4
Heat Resistance Exhibited by Different Types of Mineral Wool

Material	Temperature
Glass wool	230°C–260°C
Stone wool	700°C–850°C
Ceramic fiber wool	1200°C

TABLE 2.5
Summary of Properties of Different Types of Fibers

Property	Glass	Flax	Hemp	Jute	Ramie	Coir	Sisal	Cotton
Density (g cm^{-3})	2.55	1.4	1.48	1.46	1.5	1.25	1.33	1.51
Tensile strength (N mm^{-2})	2400	800–1500	550–900	400–800	500	220	600–700	400
Stiffness (kN mm^{-2})	73	60–80	70	10-30	44	6	38	12
Elongation at break (%)	3	1.2–1.6	1.6	1.8	2	15–25	2–3	3–10
Moisture absorption (%)	-	7	8	12	12-17	10	11	8-25
Price of raw fiber ($ kg^{-1})	1.3	0.5–1.5	0.6–1.8	0.35	1.5–2.5	0.25–0.5	0.6–0.7	1.5–2.2

(SiC) and aluminum oxide (Al_2O_3) fibers are examples of ceramic fibers notable for their high-temperature applications in metal and ceramic matrix composites. Their melting points are 2830°C and 2045°C, respectively. Silicon carbide retains its strength well above 650°C, and aluminum oxide has excellent strength retention up to about 1370°C. Both fibers are suitable for reinforcing metal matrices in which carbon and boron fibers exhibit adverse reactivity. Aluminum oxide fibers have lower thermal and electrical conductivities and have higher coefficient of thermal expansion than silicon carbide fibers.

Table 2.4 shows the heat resistance property of different types of mineral wool. Table 2.5 summarizes the properties of different types of fibers.

2.4 ADVANCED FIBERS

Advanced fibers are man-made fibers having superior properties in comparison with the natural fibers. Some of these fibers have been discussed in the sections that follow.

2.4.1 Boron Fiber

This fiber was first introduced by Talley in 1959 [1]. In commercial production of boron fibers, the method of chemical vapor deposition (CVD) is used. CVD is a process in which one material is deposited onto a substrate to produce near theoretical density and small grain size for the deposited material. In CVD, the material is deposited on a thin filament. The material grows on this substrate and produces a thicker filament. The size of the final filament is such that it could not be produced by drawing or other conventional methods of producing fibers. It is the fine and dense structure of the deposited material, which determines the strength and modulus of the fiber. In the fabrication of boron fiber by CVD, the boron trichloride is mixed with hydrogen and boron is deposited according to the reaction:

$$2BCl_3(g) + 3H_2(g) \rightarrow 2B(s) + 6HCl(g) \quad (2.1)$$

In the process, the passage takes place for couple of minutes. During this process, the atoms diffuse into tungsten core to produce the complete boridization and the production of WB_4 and W_2B_5. In the beginning, the tungsten fiber of 12 μm diameter is used, which increases to 12 μm. This step induces significant residual stresses in the fiber. The core is subjected to compression, and the neighboring boron mantle is subjected to tension. CVD method for boron fiber fabrication is shown in Figure 2.14.

2.4.2 Carbon Fiber

The first carbon fiber for commercial use was fabricated by Thomas Edison. There are two types of precursor materials used for C fiber fabrication:

i. Polyacrylonitrile (PAN)
ii. Rayon pitch, that is, the residue of petroleum refining.

2.4.2.1 Fabrication of C Fiber Using PAN

A common method of manufacture involves heating the spun PAN filaments to approximately 300°C in air, which breaks many of the hydrogen bonds and oxidizes the material. The oxidized PAN is then placed into a furnace having an inert atmosphere of a gas such as argon, and heated to approximately 2000°C, which induces graphitization of the material, changing the molecular bond structure. When heated

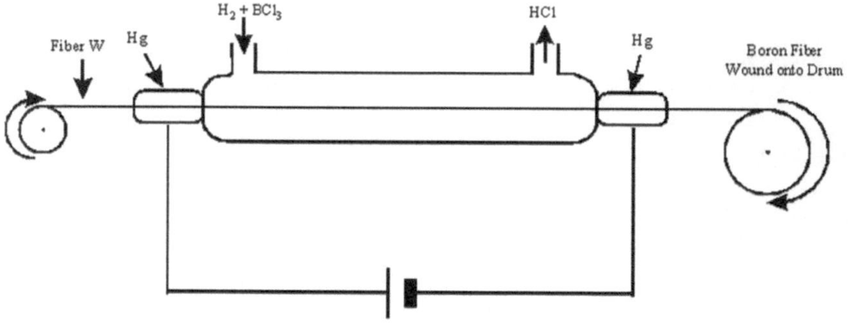

FIGURE 2.14 Schematic of chemical vapor deposition (CVD) process for fabrication of boron fiber.

in the correct conditions, these chains bond side-to-side (ladder polymers), forming narrow graphene sheets which eventually merge to form a single, columnar filament. The result is usually 93%–95% carbon. Lower-quality fiber can be manufactured using pitch or rayon as the precursor instead of PAN. The carbon can become further enhanced, as high modulus, or high-strength carbon, by heat treatment processes. Carbon heated in the range of 1500°C–2000°C (carbonization) exhibits the highest tensile strength (820,000 psi, 5650 MPa or N mm^{-2}), while carbon fiber heated from 2500 to 3000°C (graphitizing) exhibits a higher modulus of elasticity (77,000,000 psi or 531 GPa or 531 kN mm^{-2}). There are a number of types of PAN precursor fibers. They are all acrylic-based fibers and contain at least 85% of acrylonitrile. The balance may include secondary polymers or residual spin bath chemicals left over from the initial PAN precursor fiber processing. There are three stages of C fiber fabrication, which are as follows:

i. Stabilization
ii. Carbonization
iii. Graphitization.

PAN precursor, if heated to a high temperature to promote carbonization, will not yield fibers of high strength and stiffness unless a pre-oxidation, or stabilization, step is used during processing. A thermally stable structure is obtained during this step, so upon further heating, the original fiber architecture is retained. Stabilization generally calls for a heat treatment in the range 200°C–300°C in an oxygen-containing atmosphere. This leads to the polymer backbone of the precursor undergoing a series of chemical reactions that ultimately result in the formation of poly-naphthyridine, a substance with the preferred structural form for the formation of graphite. Cross-linking also occurs, which is induced either by oxidizing agents or by other catalysts. Upon further heating, this precursor structure gives rise to graphitic nuclei whose basal planes of the carbon atoms are oriented parallel to the direction of the polymer chains. Significant shrinkage (e.g., up to 40%) occurs during stabilization; this can be reduced by stretching the fibers during the heat treatment or by infiltrating PAN fibers with fine silica aggregates. The silica particles lodge in the interstices between and around the fibers, and essentially lock the fiber in place. Using such methods can reduce shrinkage to about 20%. Several chemical treatments can be used to speed up the oxidation process. For instance, treating PAN fibers with a diethanol amine or triethanol amine solution prior to heat treatment has been found to substantially reduce the time required for oxidation.

The carbonization step involves heating the stabilized precursor fiber to temperatures up to 1000°C in an inert or mildly oxidative atmosphere. Carbonization can take anywhere from a few minutes to several hours. In one type of process, fibers possessing high modulus and high strength are obtained by first heating PAN fibers in an oxidizing atmosphere until they are permeated with oxygen. The fibers are then heated further, to initiate carbonization, while being held under tension in a nonoxidizing atmosphere. Finally, to increase the ultimate tensile strength, the fibers are heat-treated in an inert atmosphere between 1300°C and 1800°C. Research has shown that if carbonization is done in an oxidizing atmosphere, then the optimal atmosphere should contain between 50 and 170 ppm oxygen. In this study, argon was used as the carrier gas, with the oxygen content varying between 2.8 and 1500 ppm.

Graphitization occurs by heating the carbonized fiber to high temperature (up to 3000°C) in an inert atmosphere. The process can last anywhere from 1 to 20 minutes. Tensioning the fibers during graphitization improves ultimate mechanical properties and reduces any residual shrinkage. Precursor fiber begins the process, and tension is applied to the precursor while heating it to an elevated temperature. The tension is reduced before the pre-oxidation phase and is increased during pre-oxidation. The fiber is then continuously heated and graphitized under high tension. The graphitized fiber is wound onto a take-up reel and dried. Processing speeds can reach as high as 45 m hour^{-1}, but are typically between 6 and 12 m hour^{-1}. All the stages along with temperature and tension profiles are shown in Figure 2.15.

Figure 2.15 also shows other important steps that can take place after the fibers are dried. First, after graphitization, special coating materials, called sizings, are applied to the fiber surface. Sizings, or fiber surface treatments, are used to provide lubrication and to protect the fibers during subsequent processing and handling. Other chemicals are also applied during the sizing operation to assist in bonding the fibers to the matrix. Second, the fibers must eventually be combined with the matrix material. If the matrix can be made into a liquid or semiliquid form, then after the fibers are dried and surface treatment has been applied, the fibers can be combined with the matrix. In one type of process, the fibers are unwound from the take-up reel, the sizings are applied, the liquid matrix material is applied to a fiber, and the wetted fibers are rewound onto a large drum. The drum is usually covered with a sheet of paper coated to prevent the fibers from sticking to the drum, and to act as a backing paper for holding the fiber–matrix system together. Generally, the winding of the fibers as they are impregnated with the matrix in liquid form and the concurrent winding of the backing paper are a continuous process; the result is layer-after-layer impregnated fibers that are wound on the drum and separated by

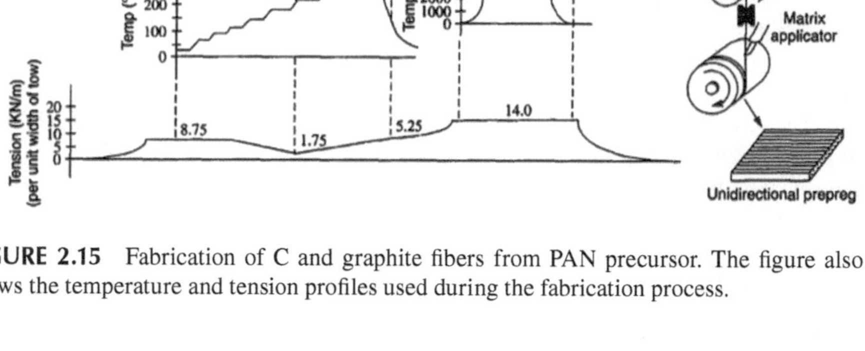

FIGURE 2.15 Fabrication of C and graphite fibers from PAN precursor. The figure also shows the temperature and tension profiles used during the fabrication process.

sheets of backing paper. This matrix-impregnated material is popularly known as unidirectional "prepreg," and it is generally the form in which material is received from the supplier. Thus, prepreg is short for *preimpregnated*, meaning impregnated before the user receives the material. The preimpregnated material on the drum is generally cut to some standard width, ranging from, say, 3 mm wide to 300 mm or wider. The narrower forms are often referred to as tapes and are used in machines that automatically fabricate composite structural components.

2.4.2.2 Fabrication of C Fiber Using Pitch

Two particular factors have led to the use of pitch as a precursor: higher yields and faster production rates. However, pitch-derived fibers are more brittle than those derived from PAN, and they have a higher density, leading to lower specific properties. In addition, the steps leading to pitch fibers are slightly different from the steps leading to PAN-derived fibers. The process of producing specific types of pitch from petroleum products is critically important to the successful production of high-modulus and high-strength carbon-based fibers. The basic process is a distillation of residual oils left over from the thermal or catalytic cracking of crude oil. The residuals from asphalt production, natural asphalt, shale oil, or coal tar can also be used. These heavy oil-based products are introduced into a reactor or series of reactors, where they are heated to temperatures between 350°C and 500°C. Thermal cracking, polymerization, and condensation occur, and the gases and light oils that are released are taken out of the reactor through a condenser. The resulting material is even heavier, has higher carbon content, and is the basis for pitch fiber precursor. The preferred raw material for pitch fiber production is liquid crystal, or mesophase, pitch. The mesophase is a highly anisotropic substance in the form of crystals, called spherulites, mixed in an isotropic pitch medium. The spherulites consist of a collection of relatively long molecules with their long axes normal relative to the boundary of the sphere. Under the influence of heat, the spherulites continue to grow and expand at the expense of the isotropic pitch surrounding them. An interesting feature of mesophase pitch is that it softens above 350°C and it can be mechanically deformed in this state. When the mesophase content reaches about 75%, the carbon substance can be subjected to fiber-forming techniques such as melt spinning.

The melt spinning of pitch fibers from the raw material can occur after the carbon content reaches the range 91%–96.5% and the mean molecular weight is at least 400. After pitch fibers are spun, they are subjected to an oxidizing gas at a temperature below the spinning temperature, or they are subjected to another chemical treatment that renders them infusible. In one method, for example, pitch fibers are treated for 7 hours at a temperature of 100°C with air containing ozone; thereafter, the temperature is raised at 1°C minute^{-1} up to 300°C. This stage of processing is critically important to guarantee that pitch fibers will retain their shape under heat treatment during carbonization and graphitization. However, if they are exposed to oxidization for too long, the fibers become brittle. The carbonization of pitch fibers occurs at somewhat higher temperatures. The heating rate between 100°C and 500°C is critical to prevent fiber rupture from the released volatiles. A typical heating schedule would call for heating from 100° to 500°C at 5°C hour^{-1} and from 500° to 1100°C at 10°C hour^{-1}. The cool down from 1100°C is usually controlled to be less than

30°C hour^{-1}. If desired, the carbonized pitch fibers can be further heated in an inert atmosphere to produce a graphitic microstructure. Graphitization temperatures are typically between 2500°C and 3300°C. Total graphitization times are generally very short, on the order of a few minutes. The properties of the three types of PAN fibers and the rayon fiber are given in Table 2.6 for comparison.

2.4.3 Glass Fiber

Silica, SiO_2, forms the basis of nearly all commercial glasses. It exists in the form of a polymer $(SiO_2)_n$. It does not melt, but gradually softens until reaching a temperature of 2000°C, after which it begins to decompose. When silica is heated until fluid-like and then cooled, it forms a random glassy structure. Only prolonged heating above 1200°C will induce crystallization (i.e., a quartz-type structure). Using silica as a glass is perfectly suitable for many industrial applications. However, its drawback is the high processing temperatures needed to form the glass and work it into useful shapes. Other types of glasses were developed to decrease the complexity of processing and increase the commercialization of glass in fiber form. The main types of glass are as follows:

i. Type A, a soda-lime glass, was the first used, and it is still retained for a few minor applications.
ii. Type E, a borosilicate glass, was developed for better resistance to attack by water and mild chemical concentrations.
iii. Relative to Type E, Type C glass has a much improved durability when exposed to acids and alkalis.

TABLE 2.6
Properties of C-Based Fibers

	PAN			Rayon
Property	**Intermediate Modulus**	**High Modulus**	**Ultra-High Modulus**	
Diameter (μm)	8–9	7–10	7–10	6.5
Density (kg m^{-3})	1780–1820	1670–1900	1860	1530–1660
Tensile modulus (GPa)	228–276	331–400	517	41–393
Tensile strength (MPa)	2410–2930	2070–2900	1720	620–2200
Elongation (%)	1.0	0.5	0.3–0.4	1.5–2.5
Coefficient of thermal expansion (× 10^{-6}/°C)				
Along fiber direction	−0.1 to −0.5	−0.5 to −1.2	−1.0	-
Perpendicular to fiber direction	7–12	7–12	-	-
Thermal conductivity (W (m °C)$^{-1}$)	20	70–105	140	38
Specific heat (J (kg °K)$^{-1}$)	950	925	-	-

iv. The increased strength and stiffness of Type S glass makes it a natural choice for use in high-performance applications, where higher specific strength and specific stiffness are important.

Type E: **Borosilicate glass** is a type of glass with silica and boron trioxide as the main glass-forming constituents. Borosilicate glasses are known for having very low coefficients of thermal expansion (~3×10^{-6}/°C at 20°C), making them resistant to thermal shock, more so than any other common glass. Such glass-made is less subject to thermal stress and is commonly used for the construction of reagent bottles. Borosilicate glass is sold under such trade names as Simax, Borcam, Borosil, Suprax, Kimax, Heatex, Pyrex, Endural, Schott, or Refmex, Kimble.

Type C: **Alkali-lime glass** with high boron oxide content.

Type R-glass: Aluminosilicate glass without MgO and CaO with high mechanical requirements as reinforcement.

S-glass: Aluminosilicate glass without CaO but with high MgO content with high tensile strength.

Table 2.7 presents the four predominant glass compositions used to form continuous glass fibers. The basic glass fiber fabrication process consists of five main components:

i. A fiber-drawing furnace consisting of a heat source, tank, and platinum alloy bushing. Raw material is fed into the furnace tank through an inlet, and fibers are drawn out through tiny nozzles in the bushing. By convention, the number of nozzles is usually 200 or a multiple thereof. Special cooling fins are located immediately under the bushing to stabilize the fiber-drawing process. This drawing process can produce a material with aligned properties. Glass fibers tend to be isotropic.
ii. A light water spray below the bushing to cool the fibers.
iii. A sizing applicator.
iv. A gathering shoe to collect the individual fibers and combine them into tows.

TABLE 2.7
Composition of Different Types of Glass

Constituent	Type A: Soda-Lime Glass (%)	Type C: Chemical Glass (%)	Type E: Electrical Glass (%)	Type S: Structural Glass (%)
SiO_2	72.0	65.0	55.2	65.0
Al_2O_3	2.5	4.0	14.8	25.0
B_2O_3	0.5	5.0	7.3	-
MgO	0.9	3.0	3.3	10.0
CaO	9.0	14.0	18.7	-
Na_2O	12.5	8.5	0.3	-
K_2O	1.5	-	0.2	-
Fe_2O_3	0.5	0.5	0.3	-
F_2	-	-	0.3	-

v. A collet or winding mandrel to collect the tows. The winding of the collet produces tensioning of the tows, which draws the fibers through the bushing nozzles.

The wound fiber tows at this point are referred to as glass fiber cakes. The sizings applied to the fibers are almost all aqueous based, and thus, the cake has a typical water content of about 10%. Before the fiber is shipped for end use, this water content has to be substantially reduced. This is accomplished by oven drying. There are two main objectives of oven drying: (1) Water content must be reduced to less than 0.1%, and (2) the dry fiber must be subjected to heat treatment to allow conglomerates of sizing particles to flow within the tow to impart certain handling characteristics. A nominal drying schedule calls for a temperature between 115° and 125°C for 4–10 hours. If the fibers are to be preimpregnated, the dried fibers are unwound, the matrix material is applied, and the wetted fibers are rewound on a drum. Figure 2.16 shows the glass fiber fabrication procedure. Table 2.8 shows the properties of different types of glass fibers.

2.4.4 Aramid (Kevlar) Fiber

Kevlar is perhaps the most common polymer fiber. It was developed by the DuPont Co. in 1968 and is an aromatic polyamide called poly(paraphenylene terephthalamide). The aromatic rings make the fiber fairly rigid. Kevlar fibers are manufactured by the extrusion and spinning processes. A solution of the polymer and a solvent (hexamethylphosphoramide) is held at a low temperature, between −50°C and −80°C, before

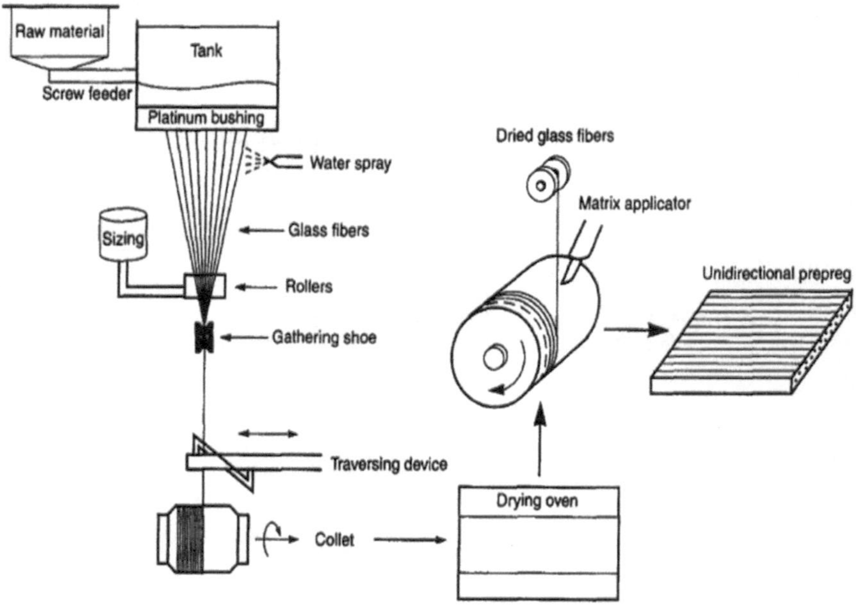

FIGURE 2.16 Glass fiber fabrication procedure.

TABLE 2.8
Properties of Glass Fibers

	Glass Type		
Property	**E**	**C**	**S**
Diameter (μm)	8–14	-	10
Density (kg m^{-3})	2540	2490	2490
Tensile modulus (GPa)	72.4	68.9	85.5
Tensile strength (MPa)	3450	3160	4590
Elongation (%)	1.8–3.2	4.8	5.7
Coefficient of thermal expansion (× 10^{-6}/°C)	5.0	7.2	5.6
Thermal conductivity (W (m °C)$^{-1}$)	1.3	-	-
Specific heat (J (kg °K)$^{-1}$)	840	780	940

being extruded into a hot-walled cylinder at 200°C. The solvent then evaporates, and the fibers are wound onto a drum. The fibers at this stage have low strength and stiffness. The fibers are subsequently subjected to hot stretching to align the polymer chains along the axis of the fiber. Afterward, the aligned fibers show a significant increase in strength and stiffness. Figure 2.17 shows a schematic of its structure. The polymer molecules form rigid planar sheets, and the sheets are stacked on top of each other with only weak hydrogen bonding between them. These sheets are folded in the axial direction and are oriented radially, with the fold generally occurring along the hydrogen-bonding line.

Aramid fibers are highly crystalline aromatic polyamide fibers that have the lowest density and the highest tensile strength-to-weight ratio among the current reinforcing fibers. Kevlar 49 is the trade name of one of the aramid fibers available in the market. As a reinforcement, aramid fibers are used in many marine and aerospace applications where lightweight, high tensile strength, and resistance to impact damage (e.g., caused by accidentally dropping a hand tool) are important. Like carbon fibers, they also have a negative coefficient of thermal expansion in the longitudinal direction, which is used in designing low thermal expansion composite panels.

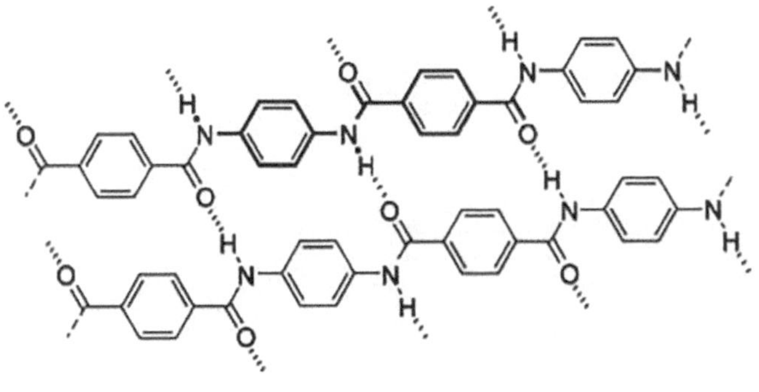

FIGURE 2.17 Microstructure of Kevlar fiber.

The major disadvantages of aramid fiber-reinforced composites are their low compressive strengths and difficulty in cutting or machining.

2.5 WOVEN FABRIC

Basic woven fabrics consist of two sets of yarns interlaced at right angles to create a single layer. Such biaxial or 0/90 fabrics are characterized by the following nomenclature:

a. *Yarn construction*: May include the strand count as well as the number of strands twisted and plied together to make up the yarn. In case of glass fibers, the strand count is given by the yield expressed in yards per pound or in TEX, which is the mass in grams per 1000 m. For example, if the yarn is designated as 150 2/3, its yield is 150 × 100 or 15,000 yd lb^{-1}. The 2/ after 150 indicates that the strands are first twisted in groups of two, and the /3 indicates that three of these groups are plied together to make up the final yarn. The yarns for carbon fiber fabrics are called tows. They have little or no twist and are designated by the number of filaments in thousands in the tow. Denier (abbreviated as de) is used for designating Kevlar yarns, where 1 denier is equivalent to 1 g/9000 m of yarn.
b. *Count*: Number of yarns (ends) per unit width in the warp (lengthwise) and fill (crosswise) directions (Figure 2.18). For example, a fabric count of 60 × 52 means 60 ends per inch in the warp direction and 52 ends per inch in the fill direction.
c. *Weight:* Areal weight of the fabric in ounces per square yard or grams per square meter.
d. *Thickness*: Measured in thousandths of an inch (mil) or in millimeters.
e. *Weave style*: Specifies the repetitive manner in which the warp and fill yarns are interlaced in the fabric. Common weave styles are shown in Figure 2.19.

a. Plain weave, in which warp and fill yarns are interlaced over and under each other in an alternating fashion.

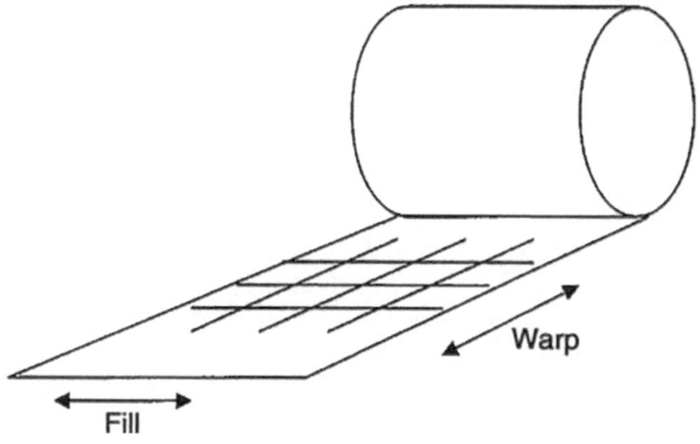

FIGURE 2.18 Warp and fill directions of fabrics.

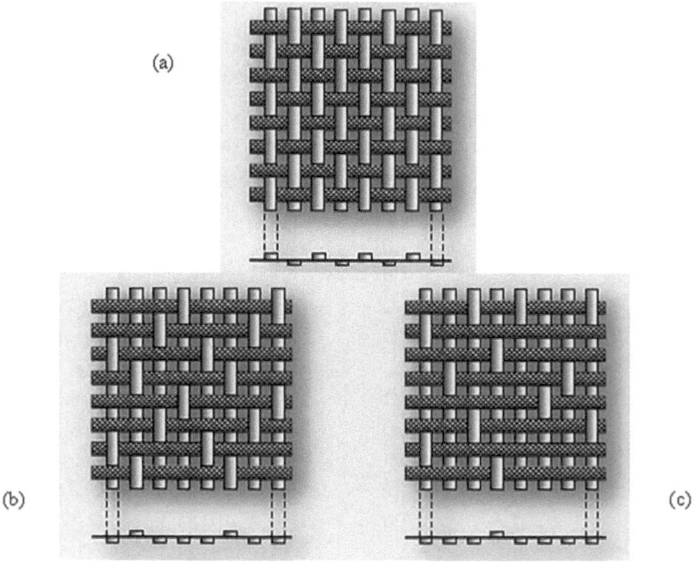

FIGURE 2.19 Common weave styles: (a) Plain weave, (b) crowfoot satin weave, (c) 5 harness satin weave.

b. Basket weave, in which a group of two or more warp yarns are interlaced with a group of two or more fill yarns in an alternating fashion.
c. Satin weave, in which each warp yarn weaves over several fill yarns and under one fill yarn.

Common satin weaves are crowfoot satin or four-harness satin, in which each warp yarn weaves over three and under one fill yarn, five-harness satin (over four, under one), and eight-harness satin (over seven, under one). For a lamina containing unidirectional fibers, the composite màterial has the highest strength and modulus in the longitudinal direction of the fibers. However, in the transverse direction, its strength and modulus are very low. For a lamina containing bidirectional fibers, the strength and modulus can be varied using different amounts of fibers in the longitudinal and transverse directions. For a balanced lamina, these properties are the same in both directions.

In order to give the readers an overview of the properties of different types of fibers, their properties are summarized in Table 2.9.

2.6 MATRICES

The role of a matrix has been discussed in an earlier section. In this section, different types of matrix materials have been discussed. The matrix materials can be classified as follows:

A. Polymeric matrices
   i. Thermoset polymers
      a. *Epoxies*: Principally used in aerospace and aircraft applications

TABLE 2.9
Summary of Properties of Different Types of Fibers

Fiber	Density (Mg m^{-3})	Young's Modulus (GPa)	Poisson's Ratio	Tensile Strength (GPa)	Failure Strain (%)	Thermal Expansivity (10^{-6} K^{-1})	Thermal Conductivity (W (m K)$^{-1}$)
SiC	3	400	0.20	2.4	0.6	4	10
Boron	2.6	400	0.20	4	1	5	38
High-modulus carbon	1.95	Axial 380 Radial 12	0.20	2.4	0.6	Axial −0.7 Radial 10	Axial 105
High-strength carbon	1.75	Axial 230 Radial 20	0.20	3.4	1.1	Axial −0.4 Radial 10	Axial 24
E-glass	2.56	76	0.22	2	2.6	4.9	13
Nicalon	2.6	190	0.20	2	1	6.5	10
Kevlar 49	1.45	Axial 130 Radial 10	0.35	3	2.3	Axial −6 Radial 54	Axial 0.04
FP fiber	3.9	380	0.26	2	0.5	8.5	8
Saffil	3.4	300	0.26	2	0.7	7	5
SiC whisker	3.2	450	0.17	5.5	1.2	4	100
Cellulose (flax)	1.0	80	0.3	2	3	-	-

b. *Polyesters, vinyl esters*: Commonly used in automotive, marine, chemical, and electrical applications
c. *Phenolics*: Used in bulk molding compounds
d. *Polyimides, polybenzimidazoles (PBI), polyphenylquinoxaline (PPQ)*: For high-temperature aerospace applications (temperature range: 250°C–400°C)
e. Cyanate ester.

ii. Thermoplastic polymers
  a. Nylons (such as nylon 6, nylon 6,6), thermoplastic polyesters, such as polyethylene terephthalate (PET).
  b. *Polycarbonate (PC), polyacetals*: Used with discontinuous fibers in injection-molded articles.
  c. *Polyamide-imide (PAI), polyether ether ketone (PEEK), polysulfone (PSUL), polyphenylene sulfide (PPS), polyetherimide (PEI)*: Suitable for moderately high-temperature applications with continuous fibers.

B. Metallic matrices

Aluminum and its alloys, titanium alloys, magnesium alloys, copper-based alloys, nickel-based superalloys, stainless steel: Suitable for high-temperature applications (temperature range: 300°C–500°C).

C. Ceramic matrices

Aluminum oxide (Al_2O_3), carbon, silicon carbide (SiC), silicon nitride (Si_3N_4): Suitable for high-temperature applications.

2.6.1 Polymer Matrix Composite

A polymer is defined as a long-chain molecule containing one or more repeating units of atoms, joined together by strong covalent bonds. A polymeric material (commonly called a plastic) is a collection of a large number of polymer molecules of similar chemical structure (but not of equal length). In the solid state, these molecules are frozen in space, either in a random fashion in amorphous polymers or in a mixture of random fashion and orderly fashion (folded chains) in semicrystalline polymers. In a thermoplastic polymer, individual molecules are not chemically joined together. They are held in place by weak secondary bonds or intermolecular forces, such as van der Waals bonds and hydrogen bonds. With the application of heat, these secondary bonds in a solid thermoplastic polymer can be temporarily broken and the molecules can now be moved relative to each other or flow to a new configuration if pressure is applied on them. On cooling, the molecules can be frozen in their new configuration and the secondary bonds are restored, resulting in a new solid shape. Thus, a thermoplastic polymer can be heat-softened, melted, and reshaped (or postformed) as many times as desired.

In a thermoset polymer, on the other hand, the molecules are chemically joined together by cross-links, forming a rigid, three-dimensional network structure. Once these cross-links are formed during the polymerization reaction (also called the curing reaction), the thermoset polymer cannot be melted by the application of heat. However, if the number of cross-links is low, it may still be possible to soften them at elevated temperatures. There are two unique characteristics of polymeric solids that are not observed in metals under ordinary conditions – namely, that their mechanical properties strongly depend on both the ambient temperature and the loading rate.

Applications of polymer matrix composites range from tennis racquets to the space shuttle. Rather than enumerating only the areas in which polymer-based composites are used, a few examples have been taken from each industry. Emphasis has been placed on why a composite material is the material of choice.

Aircraft: The military aircraft industry has mainly used polymer composites. The percentage of structural weight of composites that was less than 2% in F-15s in the 1970s has increased to about 30% on the AV-8B in the 1990s. In both cases, the weight reduction over metal parts was more than 20%. In commercial airlines, the use of composites has been conservative because of safety concerns. Use of composites is limited to secondary structures such as rudders and elevators made of graphite/epoxy for the Boeing 767 and landing gear doors made of Kevlar–graphite/epoxy. Composites are also used in panels and floorings of airplanes. Some examples of using composites in the primary structure are the all-composite Lear Fan 2100 plane and the tail fin of the Airbus A310-300. In the latter case, the tail fin consists of graphite/epoxy and aramid honeycomb. It not only reduced the weight of the tail fin by 662 lb (300 kg) but also reduced the number of parts from 2000 to 100. Helicopters and tiltrotors (Figure 2.20) use graphite/epoxy and glass/epoxy rotor blades that not only increase the life of blades by more than 100% over metals but also increase the top speeds.

Space: Two factors make composites the material of choice in space applications: high specific modulus and strength, and dimensional stability during large changes

FIGURE 2.20 Tiltrotor using graphite/epoxy and glass/epoxy rotor blades.

in temperature in space. Examples include the graphite/epoxy honeycomb payload bay doors in the space shuttle. Weight savings over conventional metal alloys translate to higher payloads that cost as much as \$1000 lb^{-1} (\$2208 kg^{-1}). Also, for the space shuttles, graphite/epoxy was primarily chosen for weight savings and for small mechanical and thermal deflections concerning the remote manipulator arm, which deploys and retrieves payloads. A high-gain antenna for the space station uses sandwiches made of graphite/epoxy facings with an aluminum honeycomb core. Antenna ribs and struts in satellite systems use graphite/epoxy for their high specific stiffness and their ability to meet the dimensional stability requirements due to large temperature excursions in space.

Sporting goods: Graphite/epoxy is replacing metals in golf club shafts mainly to decrease the weight and use the saved weight in the head. This increase in the head weight has improved driving distances by more than 25 yd (23 m).

Bicycles use a hybrid construction of graphite/epoxy composites wound on an aluminum tubing or chopped S-glass-reinforced urethane foam. The graphite/epoxy composite increases the specific modulus of the tube and decreases the mass of the frame by 25%. Composites also allow frames to consist of one piece, which improves fatigue life and avoids stress concentration found in metallic frames at their joints. Bicycle wheels made of carbon–polymide composites offer low weight and better impact resistance than those made of aluminum. Tennis and racquetball rackets with graphite/epoxy frames are now commonplace. The primary reasons for using composites are that they improve the torsional rigidity of the racquet and reduce risk of elbow injury due to vibration damping. Ice hockey sticks are now manufactured out of hybrids such as Kevlar–glass/epoxy. Kevlar is added for durability and stiffness. Ski poles made of glass/polyester composites have higher strength, flexibility, and lower weight than conventional ski poles. This reduces stress and impact on upper body joints as the skier plants his poles.

Medical devices: Applications here include the use of glass–Kevlar/epoxy lightweight face masks for epileptic patients. Artificial portable lungs are made of graphite–glass/epoxy so that a patient can be mobile. X-ray tables made of graphite/epoxy facing sandwiches are used for their high stiffness, lightweight, and transparency to radiation. The latter feature allows the patient to stay on one bed for an operation as well as X-rays and be subjected to a lower dosage of radiation.

Marine: The application of fiberglass in boats is well known. Hybrids of Kevlar–glass/epoxy are now replacing fiberglass for improved weight savings, vibration damping, and impact resistance. Kevlar–epoxy by itself would have poor compression properties. Housings made of metals such as titanium to protect expensive oceanographic research instruments during explorations of sea wrecks are cost-prohibitive. These housings are now made out of glass/epoxy and sustain pressures as high as 10 ksi (69 MPa) and extremely corrosive conditions. Bridges made of polymer composite materials are gaining wide acceptance due to their low weight, corrosion resistance, longer life cycle, and limited earthquake damage. Although bridge components made of composites may cost \$5 lb^{-1} as opposed to components made of steel, reinforced concrete may only cost \$0.30–\$1.00 lb^{-1}; the former weighs 80% less than the latter. Also, by lifetime costs, fewer composite bridges need to be built than traditional bridges.

Automotive: The fiberglass body of the Corvette® comes to mind when considering automotive applications of polymer matrix composites. In addition, the Corvette has glass/epoxy composite leaf springs with a fatigue life of more than five times that of steel. Composite leaf springs also give a smoother ride than steel leaf springs and give more rapid response to stresses caused by road shock. Moreover, composite leaf springs offer less chance of catastrophic failure and excellent corrosion resistance. By weight, about 8% of today's automobile parts are made of composites, including bumpers, body panels, and doors. However, since 1981, the average engine horsepower has increased by 84%, while average vehicle weight has increased by more than 20%. To overcome the increasing weight but also maintain the safety of modern vehicles, some estimate that carbon composite bodies will reduce the weight by 50%.

Commercial: Fiber-reinforced polymers have many other commercial applications too. Examples include mops with pultruded fiberglass handles.

Some brooms used in pharmaceutical factories have handles that have no joints or seams; the surfaces are smooth and sealed. This keeps the bacteria from staying and growing. To have a handle that is also strong, rigid, and chemically and heat resistant, the material of choice is glass fiber-reinforced polypropylene. Other applications include pressure vessels for purposes such as chemical plants. Garden tools can be made lighter than traditional metal tools and thus are suitable for children and people with physically challenged hands.

2.6.2 Metal Matrix Composites

Metal matrix has the advantage over the polymeric matrix in applications requiring a long-term resistance to severe environments, such as high temperature. The yield strength and modulus of most metals are higher than those for polymers, and this is an important consideration for applications requiring high transverse strength and modulus as well as compressive strength for the composite. Another advantage of using metals is that they can be plastically deformed and strengthened by a variety of thermal and mechanical treatments. However, metals have a number of disadvantages; namely, they have high densities, high melting points (therefore, high process temperatures), and a tendency toward corrosion at the fiber–matrix interface.

The two most commonly used metal matrices are based on aluminum and titanium. Both of these metals have comparatively low densities and are available in a variety of alloy forms. Although magnesium is even lighter, its great affinity toward oxygen promotes atmospheric corrosion and makes it less suitable for many applications. Beryllium is the lightest of all structural metals and has a tensile modulus higher than that of steel. However, it suffers from extreme brittleness, which is the reason for its exclusion as a potential matrix material. Nickel- and cobalt-based superalloys have also been used as matrix; however, the alloying elements in these materials tend to accentuate the oxidation of fibers at elevated temperatures. Aluminum and its alloys have attracted the most attention as matrix material in metal matrix composites. Commercially, pure aluminum has been used for its good corrosion resistance. Aluminum alloys, such as 201, 6061, and 1100, have been used for their higher tensile strength-to-weight ratios. Carbon fiber is used with aluminum alloys; however, at typical fabrication temperatures of 500°C or higher, carbon reacts with aluminum to form aluminum carbide (Al_4C_3), which severely degrades the mechanical properties of the composite. Protective coatings of either titanium boride (TiB_2) or sodium have been used on carbon fibers to reduce the problem of fiber degradation as well as to improve their wetting with the aluminum alloy matrix. Carbon fiber-reinforced aluminum composites are inherently prone to galvanic corrosion, in which carbon fibers act as a cathode owing to a corrosion potential of 1 V higher than that of aluminum. A more common reinforcement for aluminum alloys is SiC.

Titanium alloys that are most useful in metal matrix composites are α, β alloys (e.g., Ti-6Al-9V) and metastable β-alloys (e.g., Ti-10V-2Fe-3Al). These titanium alloys have higher tensile strength-to-weight ratios as well as better strength retentions at 400°C–500°C in comparison with aluminum alloys. The thermal expansion coefficient of titanium alloys is closer to that of reinforcing fibers, which reduces the thermal mismatch between them. One of the problems with titanium alloys is their high reactivity with boron and Al_2O_3 fibers at normal fabrication temperatures. Borsic (boron fibers coated with silicon carbide) and silicon carbide (SiC) fibers show less reactivity with titanium. Improved tensile strength retention is obtained by coating boron and SiC fibers with carbon-rich layers.

Metal matrix composites applications are as follows:

Space: The space shuttle uses boron/aluminum tubes to support its fuselage frame. In addition to decreasing the mass of the space shuttle by more than 320 lb (145 kg), boron/aluminum also reduced the thermal insulation requirements because of its low thermal conductivity. The mast of the Hubble Telescope uses carbon-reinforced aluminum.

Military: Precision components of missile guidance systems demand dimensional stability – that is, the geometries of the components cannot change during use. Metal matrix composites such as SiC/aluminum composites satisfy this requirement because they have high microyield strength. In addition, the volume fraction of SiC can be varied to have a coefficient of thermal expansion compatible with other parts of the system assembly.

Transportation: Metal matrix composites are finding use now in automotive engines that are lighter than their metal counterparts. Also, because of their high strength and low weight, metal matrix composites are the material of choice for gas turbine engines.

2.6.3 Ceramic Matrix Composites

Ceramics are known for their high-temperature stability, high thermal shock resistance, high modulus, high hardness, high corrosion resistance, and low density. However, they are brittle materials and possess low resistance to crack propagation, which is manifested in their low fracture toughness. The primary reason for reinforcing a ceramic matrix is to increase its fracture toughness.

Structural ceramics used as matrix materials can be categorized as either oxides or nonoxides. Alumina (Al_2O_3) and mullite (Al_2O_3–SiO_2) are the two most commonly used oxide ceramics. They are known for their thermal and chemical stabilities. The common nonoxide ceramics are silicon carbide (SiC), silicon nitride (Si_3N_4), boron carbide (B_4C), and aluminum nitride (AlN). Of these, SiC has found wider applications, particularly where high modulus is desired. It also has an excellent high temperature resistance. Si_3N_4 is considered for applications requiring high strength, and AlN is of interest because of its high thermal conductivity. The reinforcements used in ceramic matrix composites are SiC, Si_3N_4, AlN, and other ceramic fibers. Of these, SiC has been the most commonly used reinforcement because of its thermal stability and compatibility with a broad range of both oxide and nonoxide ceramic matrices. The forms in which the reinforcement is used in ceramic matrix composites include whiskers (with length-to-diameter ratio as high as 500), platelets, particulates, and both monofilament and multifilament continuous fibers.

Ceramic matrix composites are finding an increased application in high-temperature areas in which metal and polymer matrix composites cannot be used. This is not to say that Ceramic Matrix Composites (CMCs) are not attractive otherwise, especially considering their high strength and modulus, and low density. Typical applications include cutting tool inserts in oxidizing and high-temperature environments. Textron Systems Corporation has developed fiber-reinforced ceramics with SCS™ monofilaments for future aircraft engines.

Typical properties of the above three types of matrix material are shown in Table 2.10.

2.6.4 Carbon–Carbon Composites

Carbon–carbon composites use carbon fibers in a carbon matrix. These composites are used in very high-temperature environments of up to 6000°F (3315°C), and are 20 times stronger and 30% lighter than graphite fibers. Carbon is brittle and flaw-sensitive like ceramics. Reinforcement of a carbon matrix allows the composite to fail gradually and also gives advantages such as ability to withstand high temperatures, low creep at high temperatures, low density, good tensile and compressive strengths, high fatigue resistance, high thermal conductivity, and high coefficient of friction. Drawbacks include high cost, low shear strength, and susceptibility to oxidations at high temperatures. Typical properties of carbon–carbon composites are given in Table 2.11.

TABLE 2.10
Summary of Properties of Different Types of Matrix

Matrix	Density (Mg m^{-3})	Young's Modulus (GPa)	Poisson's Ratio	Tensile Strength (GPa)	Failure Strain (%)	Thermal Expansivity (10^{-6} K^{-1})	Thermal Conductivity (W (m K)$^{-1}$)
			Thermosets				
Epoxy resin	1.1–1.4	3–6	0.38–0.40	0.035-0.1	1–6	60	0.1
Polyesters	1.2–1.5	2–4.5	0.37–0.39	0.04–0.09	2	100–200	0.2
			Thermoplastics				
Nylon 6,6	1.14	1.4–2.8	0.3	0.06–0.07	40–80	90	0.2
Polypropylene	0.90	1–1.4	0.3	0.02–0.04	300	110	0.2
PEEK	1.26–1.32	3.6	0.3	0.17	50	47	0.2
			Metals				
Al	2.70	70	0.33	0.2–0.6	6–20	24	130–230
Mg	1.80	45	0.35	0.1–0.3	3–10	27	100
Ti	4.5	110	0.36	0.3–1.0	4–12	9	6–22
			Ceramics				
Borosilicate glass	2.3	64	0.21	0.10	0.2	3	12
SiC	3.4	400	0.20	0.4	0.1	4	50
Al_2O_3	3.8	380	0.25	0.5	0.1	8	30

TABLE 2.11
Comparison of Properties of C-C Composites with Other Materials

Property	Units	C-C composite	Steel	Aluminum
Specific gravity	-	1.68	7.8	2.6
Young's modulus	GPa	13.5	206.8	68.95
Ultimate strength	MPa	35.7	648.1	234.4
Coefficient of thermal expansion	μm (m °C)$^{-1}$	2.0	11.7	23

The main uses of carbon–carbon composites are the following:

Space shuttle nose cones: As the shuttle enters Earth's atmosphere, temperatures as high as 3092°F (1700°C) are experienced. Carbon–carbon composite is a material of choice for the nose cone because it has the lowest overall weight of all ablative materials; high thermal conductivity to prevent surface cracking; high specific heat to absorb large heat flux; and high thermal shock resistance to low temperatures in space of –238°F (–150°C) to 3092°F (1700°C) due to reentry. Also, the carbon–carbon nose remains undamaged and can be reused many times.

Aircraft brakes: The carbon–carbon brakes cost \$440 lb^{-1} (\$970 kg^{-1}), which is several times more than their metallic counterpart; however, the high durability (two to four times that of steel), high specific heat (2.5 times that of steel), low braking distances and braking times (three quarters that of beryllium), and large weight savings of up to 990 lb (450 kg) on a commercial aircraft such as Airbus A300-B2K and A300-B4 are attractive. 1 lb (0.453 kg) weight savings on a full-service commercial aircraft can translate to fuel savings of about 360 gal year^{-1} (1360 L year^{-1}). Other advantages include reduced inventory due to longer endurance of carbon brakes.

Mechanical fasteners: Fasteners needed for high-temperature applications are made of carbon–carbon composites because they lose little strength at high temperatures.

2.7 FIBER SURFACE TREATMENT

Though a composite may be made from a strong fiber and a well-suited matrix, the result may not necessarily be a strong material. The reason is that the strength of the fiber–matrix interface is equally important in determining the mechanical performance of a composite. The surface area of the fiber–matrix interfaces for a single layer of a typical graphite–epoxy composite is about 50 times the total surface area of that layer. Ultimately, the successful development of composite materials is determined by the quality of the fiber–matrix interface. To enhance the qualities of the interface, the surface of the fiber is treated by a number of agents or processes, referred to collectively as interfacial treatments and sizings that produce chemical change of the surface. Many different interfacial treatments are used in the composites industry. Lubricants and protectants are used immediately after fiber formation to protect the fibers from damage as they pass over guide rollers and winders. Coupling agents are used to increase adhesion between the fiber and the matrix. Special coatings are sometimes used to protect the fibers from environmental attack, such as corrosion from saltwater. Most fiber sizes are formulated so that several different objectives are met with the same compound. For instance, the same compound that protects the fibers during drawing and winding may later be used as an adhesion promoter between the fiber and the matrix. Other interfacial treatments are also used, including plasma treatment, acid etching, irradiation, and oxidation. For a number of reasons, the technology of fiber sizing is very complicated. Usually, fiber sizes are aqueous dispersions or solutions. The adhesives used in sizes are particulates in dilute suspensions. Most of the technologies of colloid stabilization and surfactant chemistry are closely guarded trade secrets developed through empirical observation. In fact, many of the sizes used in industry were developed for water-based paints and adhesives.

With the multiplicity of objectives for sizing compounds, it is not uncommon to find that improper performance can be traced to problems with sizing, either in its application or in its chemistry. For polymer composites, perhaps the most important issue is the promotion of adhesion between the fiber and the matrix. Currently, there are efforts to have some of the sizing migrate a short distance from the surface of the fiber and combine with the matrix. This provides a gradual transition from matrix to fiber and sizing properties in the vicinity of the fiber. This region of transition is sometimes referred to as the interphase region, and its presence is thought to enhance the fiber–matrix interaction.

2.7.1 Graphite Fiber Treatment

Graphite fibers are generally fragile and subject to abrasion during handling. To protect the fibers from abrasion, epoxy sizings are applied to the fiber surface. In some cases, a vinyl addition polymer may be incorporated into the epoxy sizing to improve handling characteristics. The sizing compounds are generally in the form of a solution, and may contain lubricants and film-formers. Pyrolytic coatings have been shown to improve the tensile strength and increase the oxidation resistance of graphite fibers. These coatings are applied by decomposing the source gas (hydrocarbons, elemental halides) onto the heated surface of the fibers. Coating uniformity is difficult to control, but can be improved if the pyrolysis is carried out in a vacuum. Improvements in tensile strength are also found when a bromine treatment is given to an untreated fiber. In this technique, carbon fibers are immersed in liquid bromine or bromine dissolved in a solvent. The bromine is subsequently removed, but some remains within the fiber. Boron nitride coatings have also been found to improve the oxidative resistance of graphite fibers. Coatings are applied by mixing boric acid with urea, passing the fibers through the solution, drying the fibers to drive off the water, and firing in a nitrogen atmosphere for about one minute at 1000°C. During firing, the urea–boric acid complex is reduced to boron nitride. The coating concentration is about 4% after firing, and the boron nitride is molecularly bonded to the carbon surface. This type of bonding assures a permanent joining of the boron nitride and the fiber.

Boron nitride-coated fibers are extremely stable at high temperatures in oxidizing atmospheres. Metal carbides have also been used instead of boron nitride to protect fibers from oxidation. The carbide impregnates the fiber surface and lodges in the crevices of surface irregularities. When the carbide surface is exposed to an oxidizing atmosphere above 400°C, the carbide is converted to a refractory oxide. This oxide protects the fiber from further oxidation. The refractory oxide acts not only as a chemical barrier but also as a thermal shield, and it improves the overall durability of the fiber.

First attempts at improving the fiber–matrix bonding concentrated on chemical modifications to the resin system. The resin's wetting ability has been used as a criterion to judge the suitability of using a particular resin system with graphite fiber reinforcement. However, some resins having poor wetting ability, such as aromatic polyphenylene resins, possess other desirable characteristics that dictate their use in composite structures. Thus, there has been a need to develop techniques to alter the fiber surface to overcome the poor wetting ability and promote increased fiber–matrix adhesion. As a result, several surface treatment methods exist. The most well-developed methods are acid treatments, oxidation treatments, plasma treatments, carbon coatings, resin coatings, ammonia treatments, and electrolytic treatments.

Acids are used to promote a strong interfacial bond between graphite fibers and resins with poor wetting ability. For instance, fibers are wetted with a sulfonic acid solution, dried to drive off the solvent, heated to allow the acid to react with the surface or with itself, washed to remove any unreacted acid, and then dried to remove the washing solvent. Hypochlorous acid has also been used successfully; the resulting composites show significantly increased shear strengths. Heat-treated graphite fibers in an ammonia atmosphere at 1000°C prior to impregnation with matrix resin increase the shear strength of the composite material. Heating is usually accomplished by passing a current through the fibers as they pass through a controlled atmosphere containing 10%–100% ammonia. The balance can be any nonoxidizing gas like nitrogen, argon, hydrogen, or helium. The exposure time is usually very short, from 1 to 60 seconds. As the concentration of ammonia is increased, the shear strength of the composite material increases; however, there may also be a decrease in tensile strength of fiber and composite. For optimal properties, the competing influences of the ammonia content of the atmosphere, exposure time, and fiber temperature must be balanced. Resins that bond well with fibers treated with ammonia are those that bond to amine functions during cure. This class of resins includes epoxies, polyimides, polyethylene, and polypropylene.

Composite materials made from low-modulus carbon fibers, typically, have a high shear strength. Compared to high-modulus fibers, these types of fibers show superior bonding ability with resins. It is believed that this behavior is a direct result of the presence of an isotropic surface layer of carbon on the low-modulus carbon fibers. The fundamental structural units of these fibers are small and have a layered structure, but are randomly oriented. Therefore, a large portion of the fiber surface will consist of exposed layer edges uniformly distributed over the surface. The exposed edges are believed to be highly reactive and may even bond chemically with epoxy resins. In contrast, high-modulus carbon fibers are more ordered due to graphitization and orientation of the crystallites during manufacture. The surface is anisotropic, consisting of relatively large areas of exposed crystallite basal planes and little exposure of edges.

Crystallite basal planes have low reactivity, and they show poor bonding ability with most resins. One technique for improving interfacial bonding for high-modulus graphite fibers is to deposit a coating of isotropic carbon on the surface. Two methods exist to accomplish this. In the first method, graphite fibers are electrically heated to about 1200°C and then exposed to an atmosphere containing methane and nitrogen. The methane decomposes on the fiber surface, creating a uniform carbon coating. Composites made using these fibers show a twofold increase in shear strength compared to untreated fibers. There is a small loss in tensile strength for these composites. In the second method, the fibers are impregnated with a thermally carbonizable organic precursor such as phenylated polyquinoxaline. The precursor is then pyrolyzed at a high temperature, and it carbonizes on the surface of the fibers. Carbon and graphite fibers can be electrolytically treated to improve their surface characteristics for improved bonding to matrices. Electrolysis is used either (1) to change the reactivity of the fiber surface or (2) to deposit chemical groups on the surface of the fiber that will bond to the matrix.

Electrolysis is accomplished, as in Figure 2.21, by pulling the fibers through a series of rollers that are electrically charged in positive/negative pairs. The negatively charged rollers are immersed in the electrolytic solution. After passing through the electrolysis, the fibers are dried before being wound onto a take-up reel. Electrolytic solutions are usually an aqueous-based caustic mixture. If the fibers are used as the cathode and vinyl monomer is added to the electrolysis solution, then the fibers will be covered with the vinyl polymerization product, greatly improving the bond strength between the fiber and some matrix resins.

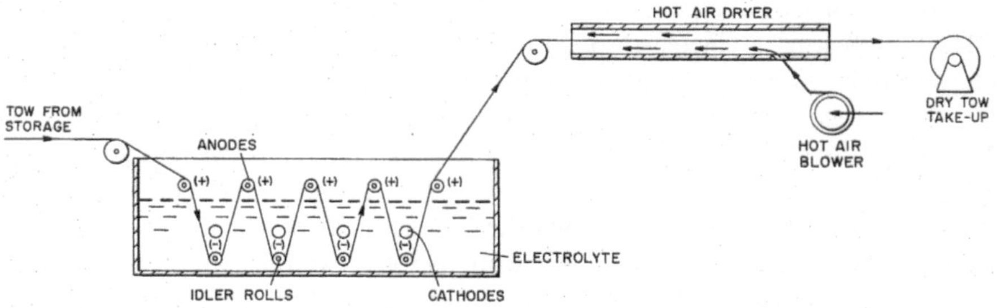

FIGURE 2.21 Method of electrolysis for improving bonding between carbon/graphite fiber and matrix.

Oxidation of carbon and graphite fibers begins at the fiber surface in regions that are irregularly shaped. By heating the fiber to 1000°C in an oxidizing atmosphere, a pitted fiber surface is obtained. The increased surface area leads to greater bond strength compared to untreated fibers. The exposure time must be controlled so that weight loss is less than 1%. In another type of treatment, fibers are treated by exposure to formates, acetates, and nitrate salts of copper, lead, cobalt, cadmium, and vanadium pentoxide. The fibers are subsequently oxidized by exposing them to air or oxygen in the temperature range of 200°C–600°C. This method uniformly roughens the fiber surface.

Significant improvements in interfacial bonding can be realized by plasma treating the fiber surface. Fibers are drawn through a plasma chamber in which a thermal plasma of argon or oxygen, or mixtures of hydrogen and nitrogen or carbon fluoride and oxygen, is generated. Argon plasma generally introduces active sites that are able to react subsequently with atmospheric oxygen. Oxygen plasma introduces oxygen both by direct reaction and by active sites. Carbon fluoride/oxygen plasma essentially results in an etching and oxidation of the fiber surface. Nitrogen/oxygen plasma introduces amine-like groups on the fiber surface that are able to participate in the cross-linking reaction for epoxies. The temperature within the plasma can reach 8000°C. Carbon fiber is continuously drawn through the plasma, and the residence time within the plasma must be carefully controlled so that the fiber surface temperature does not get too high.

A polymer coating on the surface is sometimes incorporated to enhance interfacial bonding. Thermoplastic polymers like PSUL or PC are coated onto the surface of fibers before they are impregnated with a thermosetting matrix. The result is an interphase region between the fiber and the matrix. The resulting composite material shows a significant increase in shear strength. Elastomers like urethane polymers are also used to provide a compatible interphase region between the fiber surface and the matrix. The elastomer can be applied to the fiber in the form of a sizing, or it can be blended with the matrix material.

The development of an interphase region provides for a more efficient distribution of stresses and reduces the tendency for cracks to develop at the interface. In addition, compatible elastomer and matrix material combinations, a result of using a common curing agent in both the elastomer and the matrix, will result in a gradual transition of properties within the interphase. These composites have enhanced toughness and high impact strength. Tailoring of interphase properties may provide for significant improvements in composite mechanical properties. Many other techniques have been used to some degree of success to improve fiber–matrix bond strengths. For instance, whiskerization is a technique in which silicon carbide single crystals are grown on the surface of carbon fibers to roughen the surface and to change the surface chemistry. Typical silicon carbide crystals are 0.01 to 1×10^{-6}m in diameter. They provide a mechanical interlock with the matrix material surrounding the fiber. There is also some evidence that irradiation by neutrons during cure improves fiber–resin bonding. Oxides, organometallics, isocyanates, and metal halides have all been used with some success by suitably coating the carbon fiber surface and then impregnating the fibers to form the composite. With the multiplicity of surface treatment techniques available, much of the development work in characterizing new composite materials revolves around empirical analysis of optimal combinations of surface treatments to produce composites with high tensile and shear strengths, good toughness, and reasonable costs.

The illustrations in Figure 2.22 present a microscopic view of various effects of applying – or not applying – surface treatments to graphite fibers. Figure 2.22a shows a failed graphite fiber-reinforced composite that was not treated. Here, we see inadequate fiber–matrix bonding; the fibers have very little matrix material attached to them, which is evidence of poor interfacial bonding. Figure 2.22b also shows a failed, untreated composite, but this example exhibits much better interfacial bonding. Here, the failure has occurred in the matrix itself, some of which remains attached to the fiber. Figure 2.22c and d provides an alternative point of view and dramatically illustrates the influence of graphite fiber surface treatment on the tendency of a matrix to bond to the fibers. Figure 2.22c illustrates the effects of a poor surface treatment; the thermoplastic matrix, because it is forming small spheres as a consequence of the treatment, is making a minimal contact with the fiber. By contrast, in Figure 2.22d, the contact area between the fiber and the matrix material has been maximized as a result of an improved surface treatment.

2.7.2 Glass Fiber Treatment

A fiber sizing for glass fibers to be used in polymer matrix composites must accomplish several objectives simultaneously. The fiber sizing should (1) promote good adhesion between the glass fiber and the polymer matrix; (2) promote good cohesion between the fibers that make up the tow; (3) impart certain handling characteristics, like hardness or choppability (for shorter fiber composites); (4) provide adequate protection to the fiber during processing; and (5) impart antistatic properties so that static charges do not build up on the fiber surface. For glass fibers to act efficiently as reinforcement, some method of coupling the hydrophilic fibers to the hydrophobic polymer matrix must be used. (A material is said to be *hydrophobic* if it does not absorb water; it is said to be *hydrophilic* if it can absorb water.) Coupling agents are a class of chemicals that are organometallic and, in most cases, organo-silicon, possessing dual functionality. Each metal or silicon atom has attached to its one or more

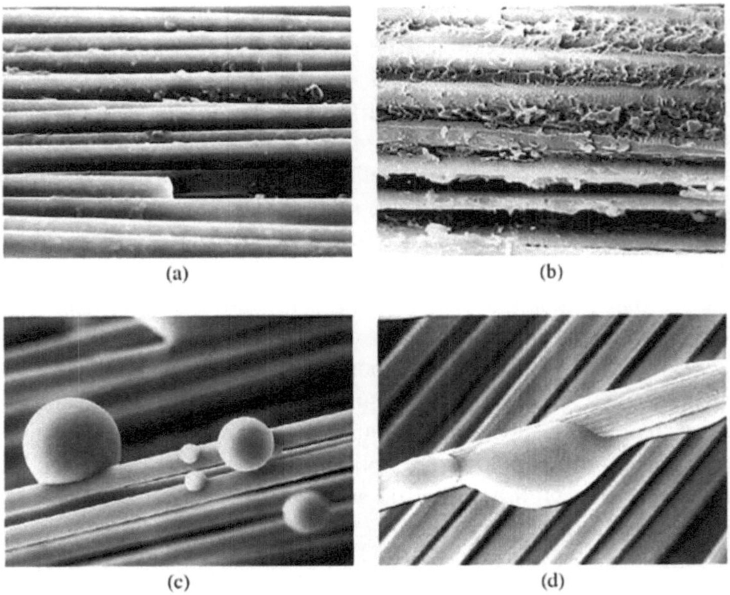

FIGURE 2.22 (a–d) A microscopic view of various effects of applying – or not applying – surface treatments to graphite fibers.

groups that can react with the glass surface, and one or more groups that can co-react with the resin during its polymerization.

With a coupling agent, a chemical bridge is formed between the glass surface and the polymer. In actual practice, the function of coupling agents is a little more complex. There is a significant evidence to suggest that several layers (about eight monomolecular layers) react with the glass surface. This gives rise to an interphase layer within the glass that possesses mechanical properties different from those of the glass fibers and the polymer matrix. There is also evidence to suggest that the flexibility of the interphase is sufficient to permit the breaking and reforming of bonds to the glass surface when the composite is under stress. Table 2.12 shows the effects

TABLE 2.12
Effect of Various Coupling Agents on the Flexural Strength of Otherwise Identical Glass Cloth–Polyester Laminates

Trade Name of Silane Coupling Agent	Flexural Strength at 25°C (MPa)		Wet Strength Retention (%)
	Dry	After 2-hour Boil	
None	42.3	24.7	58
Volan A (DuPont)	50.8	43.7	86
A 172 (Union Carbide)	50.8	48.0	94
A 174 (Union Carbide)	60.0	56.4	94

of coupling agents on the strength of glass fiber-reinforced composite and gives the results for a glass–polyester system that was tested for flexural strength under dry and wet conditions. The dry strength is increased by 42% using the A 174 silane coupling agent (*g*-methacryloxypropyltrimethoxy-silane). The strength enhancement by silane coupling is even more dramatic considering the test results for specimens immersed in boiling water for 2 hours. For this case, the flexural strength is increased by 128% over the untreated specimens. In addition, the wet strength retention (the ratio of the dry strength to the wet strength) is improved. To a large degree, the pathways to strengthen enhancement through the use of coupling agents are empirically driven as the surface chemistry of adhesion is not fully developed for many composite systems.

Another sizing component, film-formers, are materials used to bind the individual fibers together as a tow. The vast bulk of fiber sizes employ polyvinyl acetate (PVA) as a film-former. PVA is in the form of a suspension of particles in an aqueous medium. Photomicrographs of a tow show that PVA is deposited in globules on and between the fibers and that the tow is held together by these globules forming bridges from one fiber to the next. One drawback to the use of PVA as a film-former is that it is an unwelcome addition to the composite material. PVA remains after the fiber tows are impregnated with resin, leading to a possible reduction in mechanical properties of the composite material. Other film-formers more compatible with the intended matrix system are being developed. Certain polyesters have been shown to be suitable for epoxy or polyester matrices, and acrylic polymers have been introduced in fiber sizes as a film-former for thermoplastic composites. Plasticizers are added to PVA emulsions (8%–20%) to increase flexibility and reduce the softening temperature of the fiber tow. By increasing the amount of plasticizer added to fiber sizing, the flexibility of the fiber tow can be increased. The most common plasticizers used are phthalates, phosphates, and polyesters. The use of polyesters as plasticizing agents is of particular interest to the glass fiber-reinforced composites industry since they are much more compatible with epoxy and polyester matrices.

Lubricants are added to fiber sizes in concentrations from 0.2% to 2%. Most lubricants are cationic surface-active agents as they will be attracted to the negative charges normally present on the surface of a glass fiber. The cationic group is usually an amine to which a fatty acid or other lubricating group is attached. Static electricity on fibers is created by friction as they are drawn over rollers or as they slide relative to each other. Static charging increases until the losses from conduction are balanced by the rate at which charges are generated. Conduction along the surface of the fiber is possible if it is moist enough. Thus, one technique used to control static electricity is to carry out the processing of fibers in a humid environment. A relative humidity level of about 70% is usually sufficient. If this is not practical, then antistatic agents must be used to conduct electricity along the fiber surface. The problem that glass fiber manufacturers face is that the amount of antistatic material that must be supplied to ensure an adequate conduction path may be so large that other properties of the tow are sacrificed. For example, the impregnation of resin into the tows may be slowed if the concentration of antistatic agents is too large. Chemicals suitable as antistatic agents must be able to ionize to conduct electricity. For these agents to ionize, they must be hydrophilic. Both lithium chloride and magnesium chloride have been successfully used as antistatic agents.

TABLE 2.13
Formulation for Glass Fiber Sizing

Component	%
Coupling agent	0.3–0.6
Film-former (including plasticizer)	3.5–15.0
Lubricants	0.1–0.3
Surfactants	0–0.5
Antistatic agents	0–0.3
Distilled water	83.3–96.1

The exact formulation for a particular sizing is dictated by a number of factors, such as intended matrix material, cost of components, compatibility among different components, handling characteristics of the fiber tow, and stability of the size in the diluted form. Most of the common fiber sizings are formulated with the concentration ranges listed in Table 2.13.

2.7.3 Polymer Fiber Treatment

Relatively little data have been assembled concerning the surface treatment of polymer fibers. The data that exist can be grouped into two classes: (1) protective coatings and (2) adhesion promoters. Kevlar fibers are susceptible to surface damage during processing operations such as weaving. To minimize this damage, they are coated with a polyvinyl alcohol sizing, which serves as a protective layer covering the fiber surface. Conventional coupling agents used in glass and carbon fiber sizings to improve adhesion do not work well with Kevlar fibers. However, the fiber surface of Kevlar shows a good affinity for some epoxy resins. Thus, a light pretreatment with an epoxy resin has been shown to give an improved adhesion with other polymer matrices. Spectra fibers can be plasma-treated to increase the strength of the interfacial bond with epoxy matrices. Flexural strength has been shown to increase by a factor of three over untreated Spectra–epoxy composites.

2.8 FIBER CONTENT, DENSITY, AND VOID CONTENT

Theoretical calculations for strength, modulus, and other properties of a fiber-reinforced composite are based on the fiber volume fraction in the material. Experimentally, it is easier to determine the fiber weight fraction w_f, from which the fiber volume fraction v_f and composite density ρ_c can be calculated:

$$v_f = \frac{\dfrac{w_f}{\rho_f}}{\left(\dfrac{w_f}{\rho_f}\right) + \left(\dfrac{w_m}{\rho_m}\right)} \tag{2.2}$$

$$\rho_c = \frac{1}{\left(\frac{w_f}{\rho_f}\right) + \left(\frac{w_m}{\rho_m}\right)}, \tag{2.3}$$

where

w_f = fiber weight fraction (=fiber mass fraction)
w_m = matrix weight fraction (= matrix mass fraction) $= (1 - w_f)$
ρ_f = fiber density
ρ_m = matrix density.

In terms of volume fractions, the composite density ρ_c can be written as:

$$\rho_c = \rho_f v_f + \rho_m v_m, \tag{2.4}$$

where v_f is the fiber volume fraction and v_m is the matrix volume fraction. Also, $v_m = 1 - v_f$.

The fiber weight fraction can be experimentally determined by either the ignition loss method (ASTM D2854) or the matrix digestion method (ASTM D3171). The ignition loss method is used for PMC-containing fibers that do not lose weight at high temperatures, such as glass fibers. In this method, the cured resin is burned off from a small test sample at 500°C–600°C in a muffle furnace. In the matrix digestion method, the matrix (either polymeric or metallic) is dissolved away in a suitable liquid medium, such as concentrated nitric acid. In both cases, the fiber weight fraction is determined by comparing the weights of the test sample before and after the removal of the matrix. For unidirectional composites containing electrically conductive fibers (such as carbon) in a nonconductive matrix, the fiber volume fraction can be determined directly by comparing the electrical resistivity of the composite with that of fibers (ASTM D3355). During the incorporation of fibers into the matrix or during the manufacturing of laminates, air or other volatiles may be trapped in the material. The trapped air or volatiles exist in the laminate as micro-voids, which may significantly affect some of its mechanical properties. A high void content (over 2% by volume) usually leads to lower fatigue resistance, greater susceptibility to water diffusion, and increased variation (scatter) in mechanical properties. The void content in a composite laminate can be estimated by comparing the theoretical density with its actual density:

$$v_v = \frac{\rho_c - \rho}{\rho_c}, \tag{2.5}$$

where

v_v = volume fraction of voids
ρ_c = theoretical density, calculated from Eq. (2.3) or (2.4)
ρ = actual density, measured experimentally on composite specimens (which is less than ρ_c due to the presence of voids).

2.9 LOAD TRANSFER MECHANISM

Once you can produce strong, stiff material in the form of fibers, there immediately comes the challenge of how to make use of the material: The fibers need to be aligned with the load, the load needs to be transferred into the fibers, and the fibers need to remain aligned under the load. Equally important, the fibers need to be in a format that makes them readily available and easy to use. Figure 2.23 illustrates the basic mechanism used to transfer a tensile load, F, into a fiber tow. Essentially, the fiber tow is embedded in, surrounded by, and bonded to another material (see Figure 2.23a). The material, which is usually softer and weaker, not only surrounds the tow, but also penetrates the tow and surrounds every fiber in the tow. The embedding material is referred to as the matrix material, or matrix. The matrix transmits the load to the fiber through a shear stress, τ. This can be seen in the section view (Figure 2.23b), along the length of the fiber. Due to F, a shear stress acts on the outer surface of the fiber. This stress, in turn, causes a tensile stress, σ, within the fiber. Near the ends of the fiber, the shear stress on the surface of the fiber is high and the tensile stress within the fiber is low. As indicated in Figure 2.23c, as the distance from the end of the fiber increases, the shear stress decreases in magnitude and the tensile stress increases. After some length, sometimes referred to as the characteristic distance, the shear stress becomes very small and the tensile stress reaches a maximum value.

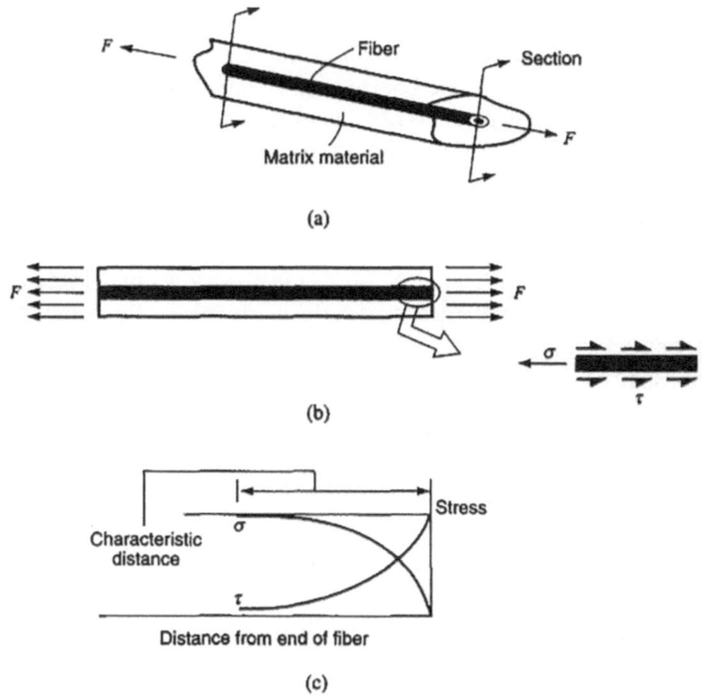

FIGURE 2.23 Basic mechanism used to transfer a tensile load, F, into a fiber tow. (a) Fiber surrounded by matrix, (b) section view, (c) characteristics of fiber stress distribution.

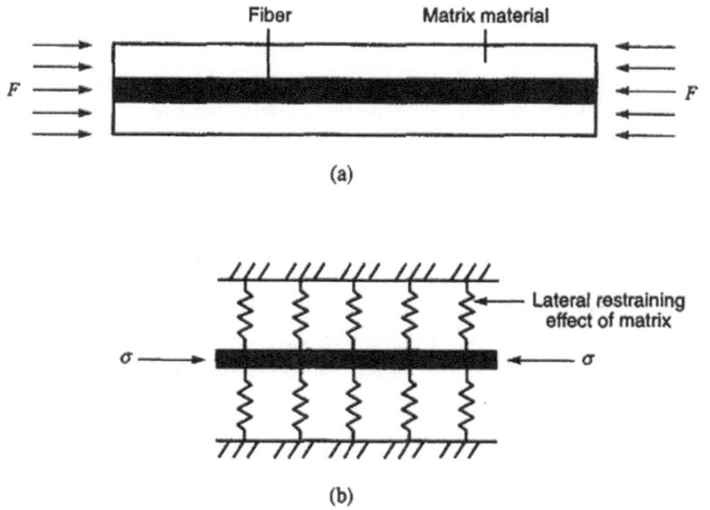

FIGURE 2.24 Basic mechanism used to transfer a compressive load into a fiber tow. (a) Section view, (b) supporting effect of matrix.

This tensile condition continues along the length of the fiber. Generally, this characteristic distance is many times smaller than the length of the fiber.

For loading the fiber in compression (see Figure 2.24), the issue of fiber buckling must be addressed. If the shear stress on the end of the fiber in Figure 2.23c is reversed, then the stress within the fiber becomes compressive and attains a maximum value at some distance from the end. This is exactly like the tension case except that the fiber responds quite differently to a compressive load; specifically, the fiber tends to buckle. The compressive resistance of some types of fibers is so poor that they will kink and fold, much like a string loaded in compression. Other fibers are quite stiff and act like very thin columns; they fail by what might be considered classic column buckling. To prevent the fiber from kinking, folding, or buckling due to a compressive load, it must be restrained laterally, and the matrix provides this restraint. To use a rough analogy, the fiber and the matrix in compression are like a beam column on an elastic foundation. As you might expect, in the presence of a compressive loading, any slight crookedness or waviness in the fiber can be quite detrimental.

Up to this point, we have focused on the idea of a single fiber or fiber tow, how loads are transmitted into it, and how it is prevented from buckling. The matrix serves both these roles. In addition, the matrix keeps the fibers aligned and in a parallel array. The embedding of strong, stiff fibers in a parallel array in a softer material results in a *fiber-reinforced composite material* with superior properties *in the fiber direction.* Clearly, the material properties perpendicular to the fiber direction are not as good. Therefore, to load a composite material perpendicularly to the fiber direction is to load the fiber in the soft and weak diametral direction of the fiber. In addition, if a composite material is loaded perpendicularly to the fiber direction, commonly referred to as the transverse direction, not all of the load is transmitted through the fiber. A portion of the load goes around the fiber and is entirely in the matrix material. The fact that the fibers do not touch means some of the load must be transferred through the matrix. The poorer transverse properties of the fiber, coupled with the softer and weaker properties of the matrix, lead to poor properties of the composite in the direction perpendicular to the fibers. In addition, and more importantly, the transverse properties of the composite depend to a large degree on the integrity of the interface bond between the fibers and the matrix. If this bond is weak, the transverse properties of the composite material are poor, and a poor interface leads to poor transverse strength.

Progressive failure of the interfaces leads to what can be interpreted as low stiffness in the transverse direction. A poor interface results in high resistance to thermal and electrical conduction. Considerable research is directed toward improving the bond at the interface between the fiber and the matrix by treating the surface of the fiber before it is combined with the matrix material to form a composite. Though the use of fibers leads to large gains in the properties in one direction, the properties in the two perpendicular directions are greatly reduced. In addition, the strength and stiffness properties of fiber-reinforced materials are poor in another important aspect.

In Figure 2.25, the three basic components of shear stress are being applied to a small volume of fiber-reinforced material, but in neither case is the inherent strength of the fiber being utilized. In all three cases, the strength of the composite critically depends upon the strength of the fiber–matrix interface, either in shear, as in Figure 2.25a and c, or in tension, as in Figure 2.25b. In addition, the strength of the matrix material is being utilized to a large degree. This lack of good shear properties is as serious as the lack of good transverse properties. Because of their poor transverse and shear properties, and because of the way fiber-reinforced material is supplied, components made of fiber-reinforced composite are usually laminated by using a number of layers of fiber-reinforced material. The number of layers can vary from just a few to several hundred. In a single layer, sometimes referred to as a lamina, all the fibers are oriented in a specific direction. While the majority of the layers in a laminate have their fibers in the direction of the load, some layers have their fibers oriented specifically to counter the poor transverse and shear properties

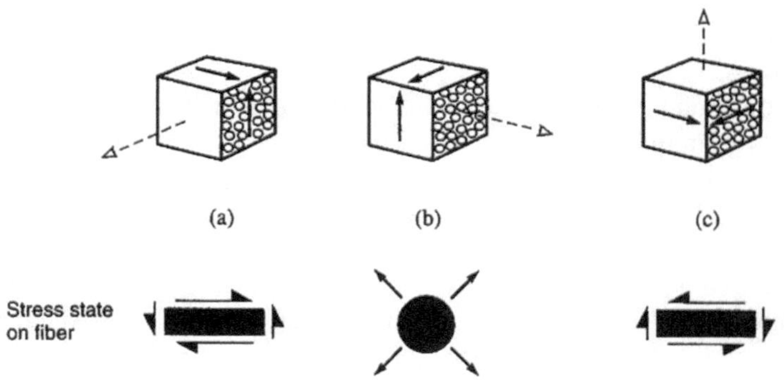

FIGURE 2.25 (a–c) Poor shear properties.

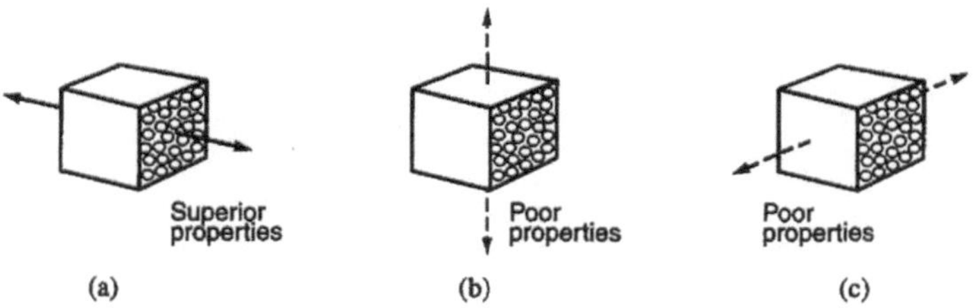

FIGURE 2.26 Poor transverse properties. (a) Fiber direction, (b) transverse direction, (c) transverse direction.

of fiber-reinforced materials. Despite these poor transverse properties, however, the *specific* strength – namely, strength normalized by density – and the *specific* stiffness – stiffness normalized by density – of composite materials are much greater than those of a single homogeneous material. Consequently, the weight of a structure utilizing fiber reinforcement to meet strength and stiffness requirements is reduced.

As shown in Figure 2.26, though the use of fibers leads to large gains in the properties in one direction, the properties in the two perpendicular directions are greatly reduced.

REFERENCE

1. Talley C.P. (1959), "Mechanical properties of glassy boron", *Journal of Applied Physics*, Vol. 30, p. 1114.

3 Manufacturing Techniques

A key ingredient in the successful production application of a material or a component is a cost-effective and reliable manufacturing method. Cost-effectiveness largely depends on the rate of production, and reliability requires a uniform quality from part to part. The early manufacturing method for fiber-reinforced composite structural parts used a hand layup technique. Although hand layup is a reliable process, it is by nature very slow and labor-intensive. In recent years, particularly due to the interest generated in the automotive industry, there is more emphasis on the development of manufacturing methods that can support mass production rates. Compression molding, pultrusion, and filament winding represent three such manufacturing processes. Although they have existed for many years, investigations on their basic characteristics and process optimization started mostly in the mid-1970s. Resin transfer molding (RTM) is another manufacturing process that has received a significant attention in both aerospace and automotive industries for its ability to produce composite parts with complex shapes at relatively high production rates. With the introduction of automation, fast-curing resins, new fiber forms, high-resolution quality control tools, and so on, the manufacturing technology for fiber-reinforced polymer composites has advanced at a remarkably rapid pace. This chapter describes the basic characteristics of major manufacturing methods used in the fiber -reinforced polymer industry.

3.1 POLYMER MATRIX COMPOSITES

Many techniques, originally developed for making glass fiber-reinforced polymer matrix composites, can also be used with other fibers. Glass fiber-reinforced polymer composites represent the largest class of polymer matrix composites (PMCs). Polymeric matrix materials can be conveniently classified as thermosets and thermoplastics. Recall that thermosets harden on curing. Curing or cross-linking occurs in thermosets by appropriate chemical agents and/or application of heat and pressure. Conventionally, thermal energy (heating to 200°C or above) is provided for this purpose. This process, however, brings in the problems of thermal gradients, residual stresses, and long curing times. Residual stresses can cause serious problems in nonsymmetric or very thick PMC laminates, where they may be relieved by warping of the laminate, fiber waviness, matrix micro-cracking, and ply delamination. Electron beam curing offers an alternative that avoids these problems. It is a nonthermal curing process that requires much shorter cure time cycles. Curing by electron beam occurs by electron-initiated reactions at a selectable cure temperature. Different methods of fabrication of PMC-first thermoset-based composites and then thermoplastic-based composites have been discussed below.

3.1.1 Thermoset Matrix Composites

There are many processing methods for composites with thermoset matrix materials, including epoxy, unsaturated polyester, and vinyl ester.

3.1.1.1 Hand LayUp and Spray Techniques

Hand layup and spray techniques are perhaps the simplest polymer processing techniques. Fibers can be laid onto a mold by hand and the resin (unsaturated polyester is one of the most common) is sprayed or brushed on. Frequently, resin and fibers (chopped) are sprayed together onto the mold surface. In both cases, the deposited layers are densified with rollers. Figure 3.1 shows schematics of these processes. Accelerators and catalysts are frequently used. Curing may be done at room temperature or at a moderately high temperature in an oven. Figures 3.1 and 3.2 show the hand layup and the spray techniques, respectively. The advantages of the hand layup process are as follows:

a. It results in low-cost tooling with the use of room-temperature cure resins.
b. The process is simple to use.
c. Any combination of fibers and matrix materials can be used.
d. Higher fiber contents and longer fibers can be used as compared to other processes.

The disadvantages of the hand layup process are as follows:

a. Since the process is worked by hands, there are safety and hazard considerations.
b. The resin needs to be less viscous so that it can be easily worked by hands.
c. The quality of the final product is highly skill dependent of the laborers.

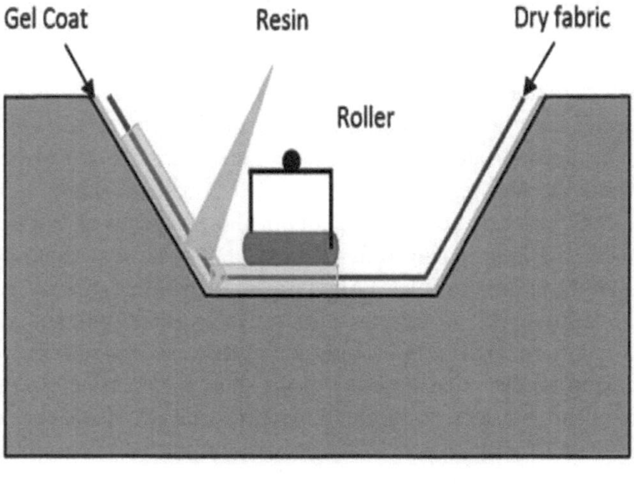

FIGURE 3.1 The hand layup process.

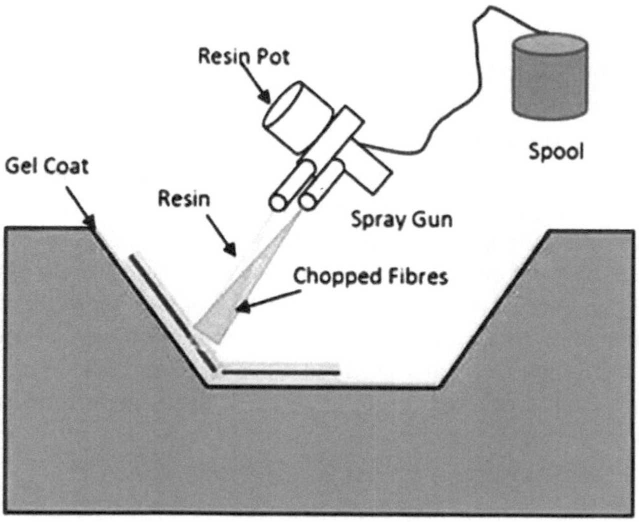

FIGURE 3.2 The spray-up process.

d. Uniform distribution of resin inside the fabric is not possible. It leads to voids in the laminates.
e. Possibility of diluting the contents.

The hand layup process is suitable for the fabrication of wind-turbine blades, boats, and architectural moldings.

The **spray-up process** offers the following advantages:

a. It is suitable for small- to medium-volume parts.
b. It is a very economical process for making small to large parts.
c. It utilizes low-cost tooling as well as low-cost material systems.

The following are some of the limitations of the spray-up process:

a. It is not suitable for making parts that have high structural requirements.
b. It is difficult to control the fiber volume fraction as well as the thickness. These parameters highly depend on operator skill.
c. Because of its open mold nature, styrene emission is a concern.
d. The process offers a good surface finish on the one side and a rough surface finish on the other side.
e. The process is not suitable for parts where dimensional accuracy and process repeatability are prime concerns.
f. Cores, when needed, have to be inserted manually.
g. Only short fibers can be used in this process.
h. Similar to wet/hand layup process, the resins need to be of low viscosity so that it can be sprayed.

The spray-up process has the following applications: making simple enclosures, lightly loaded structural panels, for example, caravan bodies, truck fairings, bathtubs, shower trays, and some small dinghies.

3.1.1.2 Filament Winding

This process is an automated one. This process is used in the fabrication of components or structures made with flexible fibers. This process is primarily used for hollow, generally circular- or oval-sectioned components. Fiber tows are passed through a resin bath before being wound onto a mandrel in a variety of orientations, controlled by the fiber feeding mechanism and rate of rotation of the mandrel. The wound component is then cured in an oven or autoclave. One can use resins like epoxy, polyester, vinyl ester, and phenolic along with any fiber. The fiber can be directly drawn from creel, nonwoven, or stitched into a fabric form. The filament winding process is shown in Figure 3.3.

The advantages of the filament winding process are as follows:

a. Resin content is controlled by nips or dies.
b. The process can be very fast and is economic.
c. Complex fiber patterns can be attained for better load bearing of the structure.

The disadvantages of this process are as follows:

a. Resins with low viscosity are needed.
b. The process is limited to convex-shaped components.
c. Fiber cannot easily be laid exactly along the length of a component.
d. Mandrel costs for large components can be high.
e. The external surface of the component is not smoothly finished.

The filament winding process is used for making pressure bottles, rocket motor casing, chemical storage tanks, pipelines, gas cylinders, firefighters, breathing tanks etc.

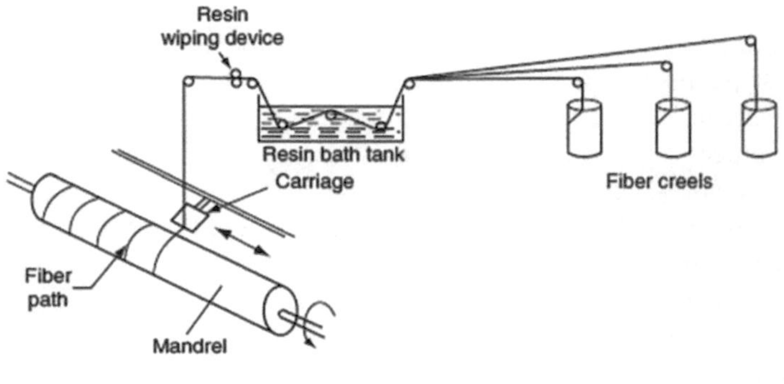

FIGURE 3.3 Filament winding process.

3.1.1.3 Autoclave Curing

An autoclave is a closed vessel for controlling temperature and pressure and is used for curing polymeric matrix composites. Composites to be cured are prepared either through hand layup or through machine placement of individual laminae in the form of fibers tape which has been impregnated with resin. Components are then placed in an autoclave and subjected to a controlled cycle of temperature and pressure. After curing, the composite is "solidified." One can use the fibers like carbon, glass, and aramid along with any resin.

Autoclave-based processing of PMCs results in a very high-quality product. That is the reason it is used to make components in the aerospace field. An autoclave is a closed vessel (round or cylindrical) in which processes (physical and/or chemical) occur under simultaneous application of high temperature and pressure. Heat and pressure are applied to appropriately stacked prepregs. The combined action of heat and pressure consolidates the laminae, removes the entrapped air, and helps cure the polymeric matrix. Autoclave processing of composites thus involves a number of phenomena: chemical reaction (curing of the thermoset matrix), flow of the resin, and heat transfer.

Figure 3.4 shows schematically the setup in an autoclave to make a laminated composite. Figure 3.5 shows the autoclave molding machine employing the vacuum bagging process discussed in the next section.

The advantages of autoclave curing process are highlighted as follows:

a. Large components can be fabricated.
b. Since the curing of matrix material is carried out under a controlled environment, the resin distribution is better as compared to hand or spray layup processes.
c. Less possibility of dilution with foreign particles.
d. Better surface finish.

The disadvantages of this process include the following:

a. Initial cost of tooling is high.
b. Running and maintenance cost is high.
c. Not suitable for small products.

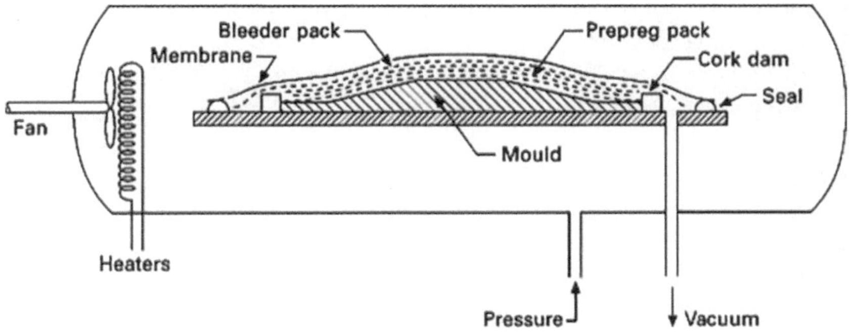

FIGURE 3.4 The setup inside an autoclave molding machine.

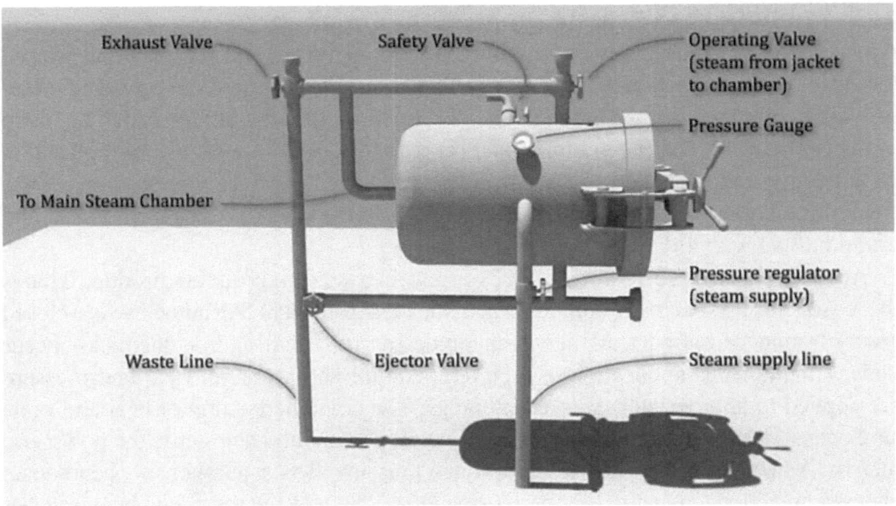

FIGURE 3.5 An autoclave molding machine.

The process is suitable for aerospace, and automobile parts such as wing box, chassis, and bumpers.

3.1.1.4 Vacuum Bagging Process

This is basically an extension of the wet layup process described above where pressure is applied to the laminate once laid up in order to improve its consolidation. This is achieved by sealing a plastic film over the wet laid-up laminate and onto the tool. The air under the bag is extracted by a vacuum pump, and thus, up to one atmosphere of pressure can be applied to the laminate to consolidate it. The process has the following advantages:

a. Higher fiber content laminates can usually be achieved than with the standard wet layup techniques.
b. Lower void contents are achieved than with wet layup.
c. Better fiber wet-out due to pressure and resin flow throughout structural fibers, with excess into bagging materials.
d. *Health and safety*: The vacuum bag reduces the amount of volatiles emitted during cure.

The process has some limitations which are highlighted below.

a. The extra process adds cost both in labor and in disposable bagging materials.
b. A higher level of skill is required by the operators.
c. Mixing and control of resin content are still largely determined by the operator skill.

The application of a vacuum to assist in compressing the plies together (debulking) has proven to be valuable in wet layups and necessary in prepreg layups. The vacuum provides the dual advantage of pressing the layers together and simultaneously withdrawing the excess volatiles. These volatiles could be residual solvent, low molecular weight resin components, absorbed moisture, or trapped air. The method of applying the vacuum that has been developed for composites allows the volatiles to escape freely and also permits good debulking. The vacuum bagging process is used for making large one-off cruising boats, race car components, core bonding in production boats etc.

The procedure for building up such an assembly is as follows (Figure 3.6):

a. Prepare the mold by coating it with an appropriate mold release.
b. Remove prepreg materials from the freezer, and bring them to room temperature before opening the protective bag to prevent contamination from water condensation.
c. Build up the part by placing layers of prepreg on top of each other in the prescribed pattern and to the prescribed thickness. This buildup can either be done manually or be automated. The buildup may also involve the inclusion of any inserts, ribs, or other structural members and the additional prepreg layers necessary to serve as anchoring and support for these inclusions.

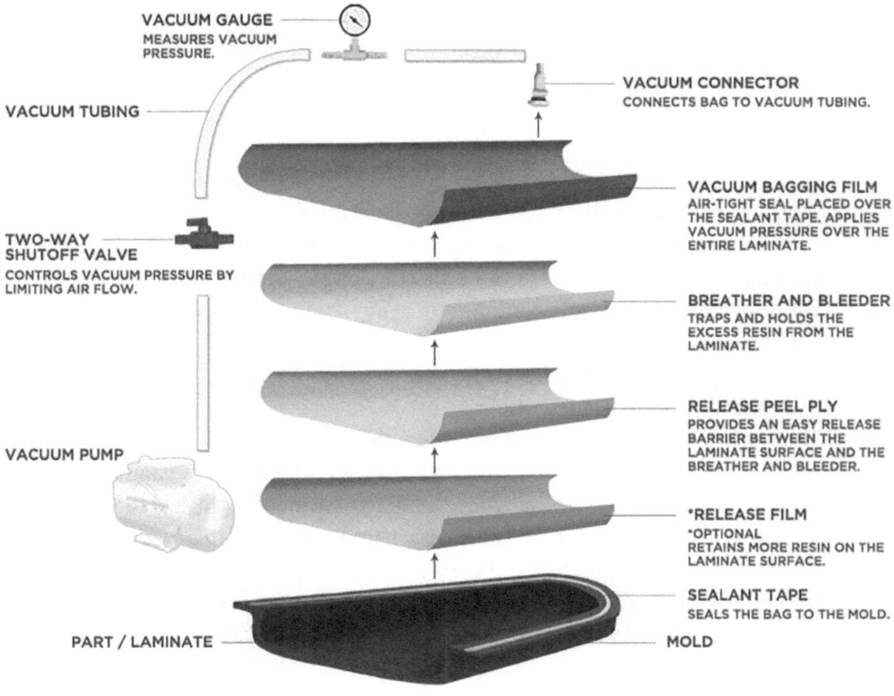

FIGURE 3.6 Components of vacuum bagging.

d. Place the release film or peel ply material over the part. The release material is generally porous to permit excess resin to flow through, while leaving an impression on the part suitable for secondary bonding without further surface preparation.
e. The bleeder material is a mat that absorbs the excess resin. Common bleeder materials are polyester felt or mat, fiberglass (which may be coated with Teflon or mold release), and cotton. The important characteristic is that the bleeder material should have good absorption qualities and not compact under the pressure of an autoclave. The resin content of the final part is dependent on the ability of the bleeder material to absorb a measured amount of resin, as well as on temperature, pressure, viscosity, and amount of bleeder. The bleeder should be conformable so that it does not cause wrinkles in the assembly when it is placed under vacuum.
f. The barrier is a layer that limits the upward movement of the resin and prevents it from reaching or clogging the breather and vacuum lines. The barrier material must not, therefore, allow the passage of resin, but must allow air to pass. In many bagging assemblies, this material is omitted. In autoclave processing, such an omission may lead to resin-plugged vacuum lines and pumps. When the barrier is omitted in wet layup applications, resin traps are often used.
g. The breather material acts as a distributor for air, escaping volatiles, and gases, as well as a buffer between bag wrinkles and the part surfaces. Breather layers should be highly porous and must not collapse under vacuum, temperature, or pressure. Typical materials are fiberglass, polyester felt, and cotton, although the bleeder materials are sometimes used. For near-net or phenolic resin systems, the use of breather is sometimes omitted when the thickness of the bleeder is great and experience has shown that a separate breather system is not required.
h. Vacuum bag sealing tape or sealant is applied around the mold. This sticky material makes an airtight seal with the bagging film.
i. Thermocouples or other monitoring devices are inserted into the assembly. To ensure that the places where the thermocouple wires enter the assembly will be airtight, the outer insulation is often scraped off the thermocouple wire (especially if it has cloth insulation). The wires are then embedded in the sealant material. Thermocouples are usually placed in the outer area of a part that will be trimmed off.
j. The vacuum bag is then laid over the assembly, and a vacuum port is attached through the bag, providing a fitting for attaching the vacuum hose. The bag is then pressed against the sealant to give an airtight seal all around. Vacuum bags can be made of any plastic film material that is strong enough to hold a vacuum, fit and conform to the assembly, and withstand the cure temperature without degrading. In practice, the most common material is nylon or a coextruded nylon that has been heat-stabilized.

The most common problems associated with vacuum bag processing are material quality and bag leaks. Vacuum bagging materials are often procured under vague material, construction, and performance requirements. Poor industry standards often lead to products with varying performance. For instance, uncontrolled or unregulated vacuum bag quality is the leading cause of bag failure, typified by film degradation or decomposition above 82°C. Bag leaks most often occur at the sealant vacuum bag interface. The second most common cause of failure is handling damage to the nylon film before cure. Nylon film is hygroscopic and subject to moisture change in relation to the relative humidity of its environment. Film becomes dry and "brittle" when its moisture content falls below 2%. Dry film is susceptible to damage/cracking when excessively handled or abused (wadded).

Bridging is also a common problem in vacuum bag molding. Bridging occurs when, because of the shape of the part, the vacuum bagging materials are not pressed against all of the part surface. Some areas, therefore, are not properly pressed. (For instance, for a part with a narrow, deep channel, bagging materials may form a material bridge across the top of the channel, preventing the transfer of pressure to the bottom of the channel.) If the nylon bag itself bridges over a narrow gap, the film may be stretched beyond its limits during cure and may burst. Good bagging techniques, such as intentionally putting pleats into the bag during assembly, allow enough excess film to be present to provide bag conformity against all surfaces. Another technique is the placement of preformed rubber pads under the bag in corners and channels to fill areas where bridging is possible. This technique is particularly useful with vacuum bags made of rubber.

3.1.1.5 Pultrusion

It is a continuous process in which composites in the form of fibers and fabrics are pulled through a bath of liquid resin. Then, the fibers wetted with resin are pulled through a heated die. The die plays important roles like completing the impregnation and controlling the resin. Further, the material is cured to its final shape. The die shape used in this process is nothing the replica of the final product. Finally, the finished product is cut to length. In this process, the fabrics may also be introduced into the die. The fabrics provide a fiber direction other than 0°. Further, a variant of this method to produce a profile with some variation in the cross section is available. This is known as pulforming. The pultrusion process is shown in Figure 3.7. The process has the following advantages:

a. The process is suitable for mass production.
b. The process is fast and economic.
c. Resin content can be accurately controlled.
d. Fiber cost is minimized as it can be taken directly from a creel.
e. The surface finish of the product is good.
f. Structural properties of product can be very good as the profiles have very straight fibers.

The process has the following disadvantages:

a. Limited to constant or near-constant cross-sectional components.
b. Heated die costs can be high.
c. Products with small cross sections alone can be fabricated.

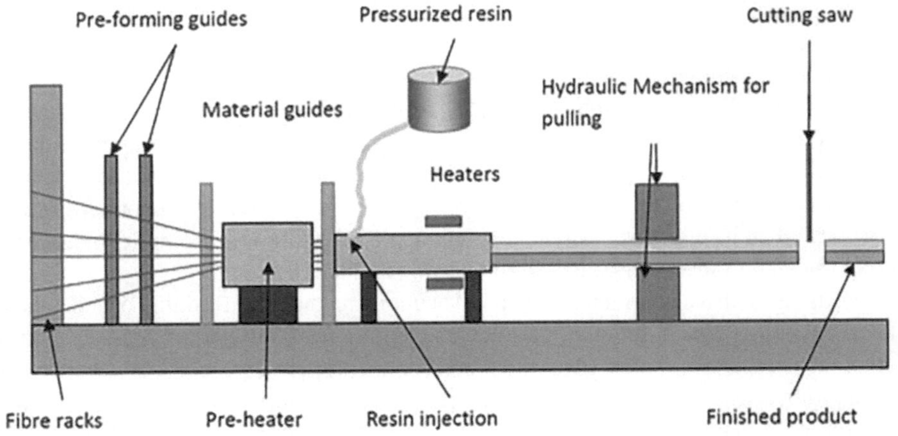

FIGURE 3.7 Pultrusion process.

The pultrusion process is used for making beams and girders used in roof structures, bridges, ladders, frameworks etc.

3.1.1.6 Resin Transfer Molding (RTM)

In RTM, several layers of dry continuous strand mat, woven roving, or cloth are placed in the bottom half of a two-part mold, the mold is closed, and a catalyzed liquid resin is injected into the mold via a centrally located sprue. The resin injection point is usually at the lowest point of the mold cavity. The injection pressure is in the range of 69–690 kPa (10–100 psi). As the resin flows and spreads throughout the mold, it fills the space between the fiber yarns in the dry fiber preform, displaces the entrapped air through the air vents in the mold, and coats the fibers. Depending on the type of the resin–catalyst system used, curing is performed either at room temperature or at an elevated temperature in an air-circulating oven. After the cured part is pulled out of the mold, it is often necessary to trim the part at the outer edges to conform to the exact dimensions. Instead of using flat-reinforcing layers, such as a continuous strand mat, the starting material in an RTM process can be a preform that already has the shape of the desired product (Figure 3.8). The advantages of using a preform are good moldability with complicated shapes (particularly with deep draws) and the elimination of the trimming operation, which is often the most labor-intensive step in an RTM process. In this process, the resins like epoxy, polyester, vinyl ester, and phenolic can be used. Further, the high-temperature resins such as bismaleimides can be used at elevated process temperatures. The fibers of any type can be used. The stitched materials work well in this process since the gaps allow a rapid resin transport. Some specially developed fabrics can assist with resin flow.

The RTM process has been successfully used in molding such parts as cabinet walls, chair or bench seats, hoppers, water tanks, bathtubs, and boat hulls. It also offers a cost-saving alternative to the labor-intensive bag-molding process or the capital-intensive compression molding process. It is particularly suitable for producing low- to mid-volume parts, say 5000–50,000 parts a year. There are several

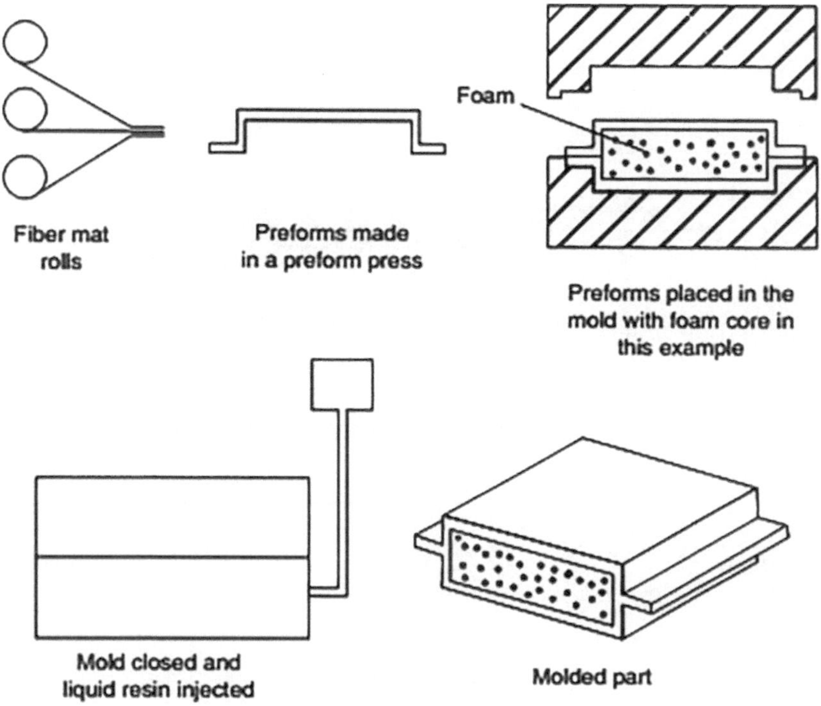

FIGURE 3.8 Resin transfer molding (RTM) process.

variations of the basic RTM process. In one of these variations, known as vacuum-assisted RTM (VARTM), as shown in Figure 3.9, vacuum is used in addition to the resin injection system to pull the liquid resin into the preform. Another manufacturing process very similar to RTM is called structural reaction injection molding (SRIM). It also uses dry fiber preform that is placed in the mold before resin injection.

The difference in RTM and SRIM is mainly in the resin reactivity, which is much higher for SRIM resins than for RTM resins. SRIM is based on the reaction injection molding (RIM) technology in which two highly reactive, low-viscosity liquid streams are impinged on each other at high speeds in a mixing chamber immediately before injecting the liquid mix into a closed mold cavity (Figure 3.10).

The RTM process has the following advantages:

a. The process is very efficient.
b. Suitable for complex shapes.
c. High fiber volume laminates can be obtained with very low void contents.
d. Good health and safety, and environmental control due to enclosure of resin.
e. Possible labor reductions.
f. Both sides of the component have a molded surface. Hence, the final product gets a superior surface finish.
g. Better reproducibility.
h. Relatively low clamping pressure and ability to induce inserts.

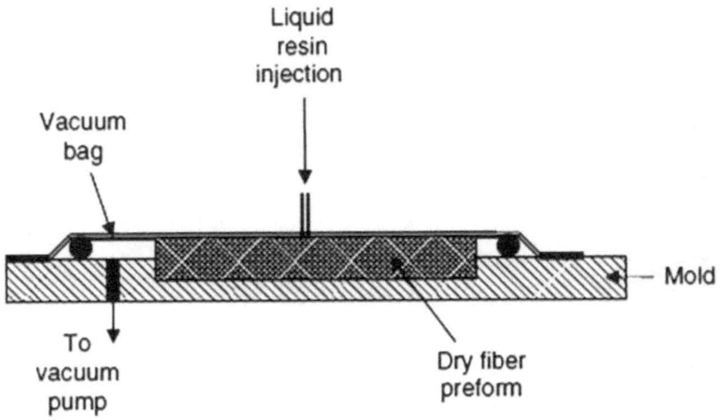

FIGURE 3.9 Vacuum-assisted resin transfer molding process (VARTM).

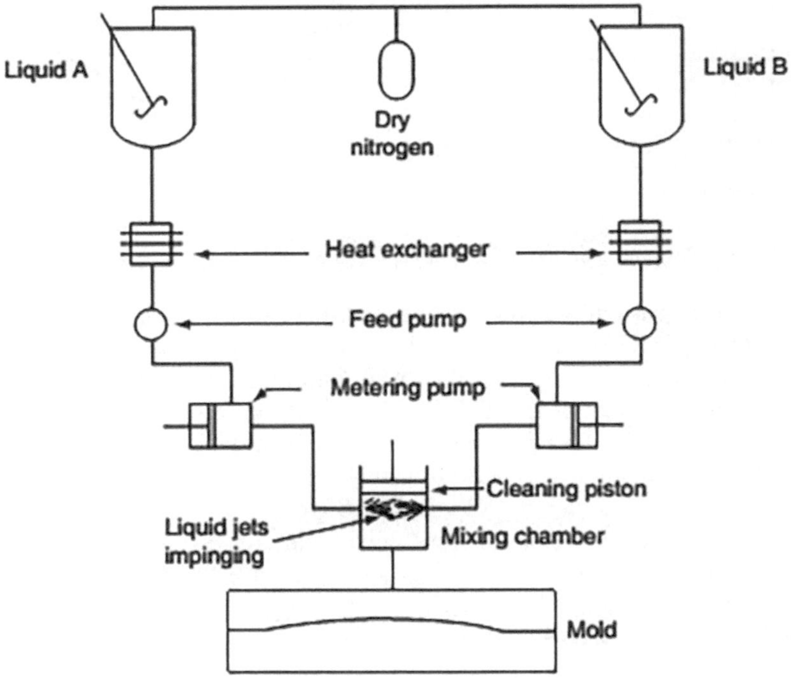

FIGURE 3.10 Structural reaction injection molding process.

The RTM process has the following disadvantages:

a. Matched tooling is expensive and heavy in order to withstand pressures.
b. Generally limited to smaller components.
c. Un-impregnated areas can occur resulting in very expensive scrap parts.

3.1.2 Thermoplastic Matrix Composites

Thermoplastic matrix composites have several advantages and disadvantages over thermoset matrix composites. We first list these and then describe some of the important processes used to form thermoplastic matrix composites. The advantages of thermoplastic matrix composites include the following:

a. Refrigeration is not necessary with a thermoplastic matrix.
b. Parts can be made and joined by heating.
c. Parts can be remolded, and any scrap can be recycled.
d. Thermoplastics have better toughness and impact resistance than thermosets. This can generally also be translated into thermoplastic matrix composites.

The disadvantages include the following:

a. The processing temperatures are generally higher than those with thermosets.
b. Thermoplastics are stiff and boardy; that is, they lack the tackiness of the partially cured epoxies.

The processing techniques for thermoplastic matrix composites have been discussed below.

3.1.2.1 Film Stacking

Laminae of thermoplastic matrix containing fibers with very low resin content (~15 w/o) are used in this process. Low resin content is used because these are very boardy materials. The laminae are stacked alternately with thin films of pure polymer matrix material. This stack of laminae consists of fibers impregnated with insufficient matrix and polymer films of complementary weight to give the desired fiber volume fraction in the end product. These are then consolidated by simultaneous application of heat and pressure. The impregnation of thermoplastic matrix takes place under the simultaneous application of heat and pressure; the magnitude of pressure and temperature must be sufficient to force the polymeric melt to flow into and through the reinforcement preform. The rate of penetration of a fluid into the fibrous preform structure is described by Darcy's law. Darcy's law says that the flow rate is directly proportional to the applied pressure and inversely proportional to the viscosity. Thus, increasing the applied pressure and decreasing the viscosity of the molten polymer (i.e., increasing the temperature) help in the processing. This process is shown in Figure 3.11.

3.1.2.2 Diaphragm Forming

This process involves the sandwiching of freely floating thermoplastic prepreg layers between two diaphragms. The air between the diaphragms is evacuated, and thermoplastic laminate is heated above the melting point of the matrix. Pressure is applied to one side, which deforms the diaphragms and makes them take the shape of the mold. The laminate layers are freely floating and very flexible above the melting point

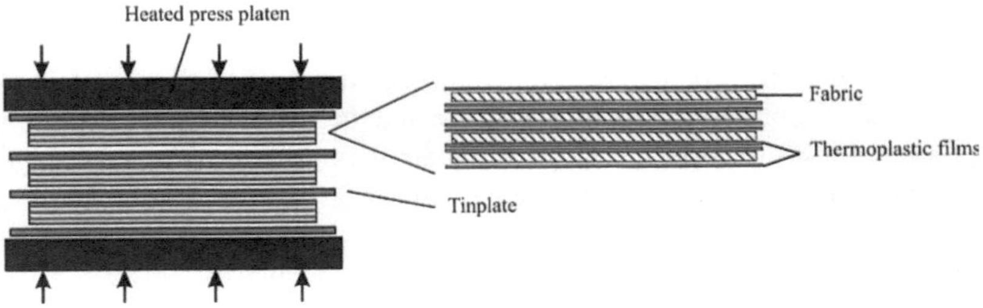

FIGURE 3.11 Film stacking process.

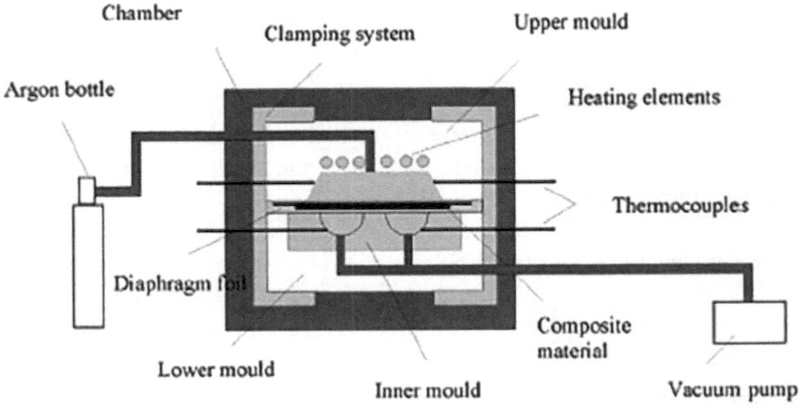

FIGURE 3.12 Diaphragm forming process.

of the matrix; thus, they readily conform to the mold shape. After the completion of the forming process, the mold is cooled, the diaphragms are stripped off, and the composite is obtained. One of the advantages of this technique is that components with double curvatures can be formed. The diaphragms are the key to the forming process, and their stiffness is a very critical parameter. Compliant diaphragms do the job for simple components. For very complex shapes requiring high molding pressures, stiff diaphragms are needed. At high pressures, a significant transverse squeezing flow can result, and this can produce undesirable thickness variations in the final composite. This process is shown in Figure 3.12.

3.1.2.3 Thermoplastic Tape Laying

Thermoplastic tape laying machines are also available, although they are not as common as the thermoset ting tape laying machines. Figure 3.13 shows the schematic of one such machine. A controllable tape head has the tape dispensing and shim dispensing/take-up reels and heating shoes. The hot head dispenses thermoplastic tape from a supply reel. There are three heating and two cooling/compaction shoes. The hot shoes heat the tape to molten state. The cold shoes cool the tape instantly to a solid state.

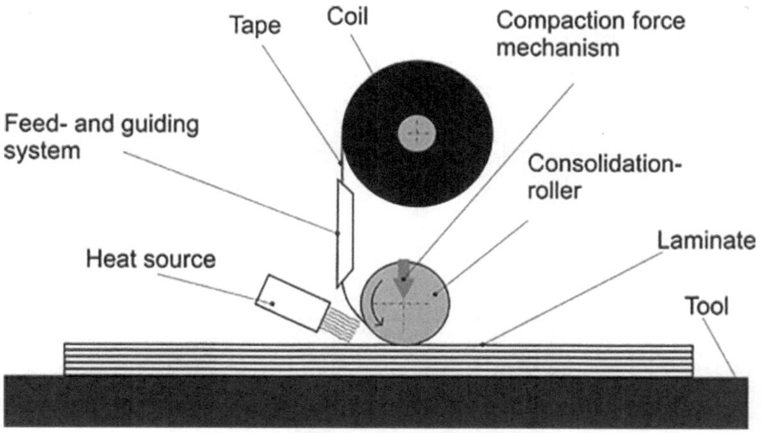

FIGURE 3.13 Tape laying process for thermoplastic matrix composites.

3.1.2.4 Sheet Molding Compound

There are some common PMCs that do not contain long, continuous fibers; hence, we describe them separately in this section. Sheet molding compound (SMC) is the name given to a composite that consists of a polyester resin containing short glass fibers plus some additives called fillers. The additives generally consist of fine calcium carbonate particles and mica flakes. Sometimes calcium carbonate powder is substituted by hollow glass microspheres, which results in a lower density, but makes it more expensive. Figure 3.14 shows a schematic of the SMC processing. Polyester resin can be replaced by vinyl ester to further reduce the weight, but again with a cost penalty. SMC is used in making some auto body parts, such as bumper beams, radiator support panels, and many others. It has been used in the Corvette sports car for many decades. Polypropylene resin can be reinforced with calcium carbonate particles, mica flakes, or glass fibers. Such composites, though structurally not as important as, say, carb on fiber/epoxy composites, do show improved mechanical properties vis-a`-vis unreinforced resin. Characteristics such as strength, stiffness, and service temperature are improved. These materials are used in automotive parts, appliances, electrical components, and so forth.

3.2 METAL MATRIX COMPOSITES

Many processes for fabricating metal matrix composites (MMCs) are available. For the most part, these processes involve processing in the liquid and the solid state. Some processes may involve a variety of deposition techniques or an in situ process of incorporating a reinforcement phase.

3.2.1 Liquid-State Processes

Metals with melting temperatures that are not too high, such as aluminum, can be incorporated easily as a matrix by liquid route. The liquid-state processes have been discussed below.

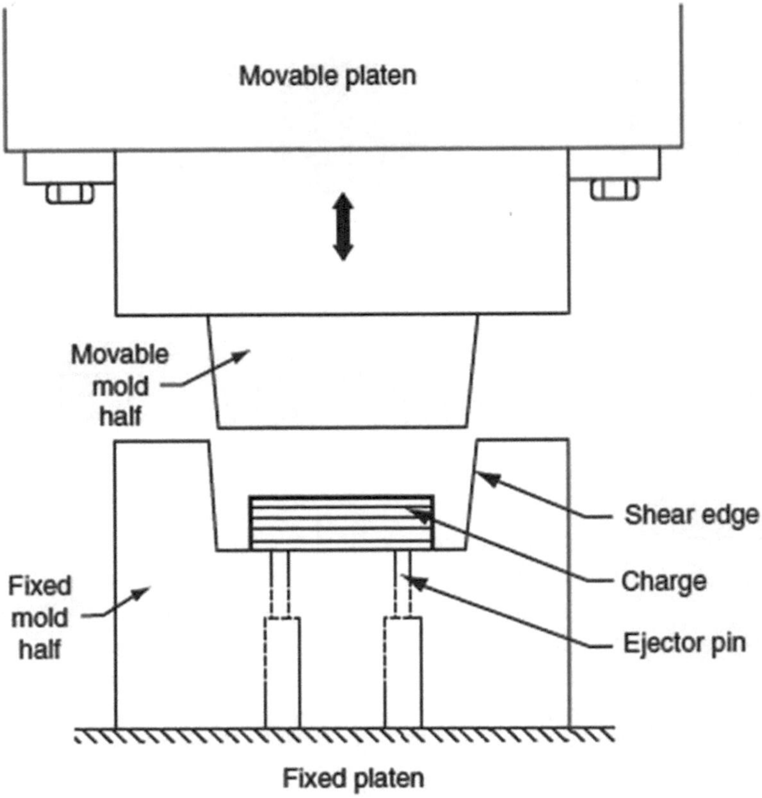

FIGURE 3.14 Sheet molding compound.

3.2.1.1 Casting or Liquid Infiltration

Casting, or liquid infiltration, involves infiltration of a fiber bundle by liquid metal. It is not easy to make MMCs by simple liquid-phase infiltration, mainly because of difficulties with wetting of ceramic reinforcement by the molten metal. When the infiltration of a fiber preform occurs readily, reactions between the fiber and the molten metal can significantly degrade fiber properties. Fiber coatings applied prior to infiltration, which improve wetting and control reactions, have been developed and can result in some improvements. In this case, however, the disadvantage is that the fiber coatings must not be exposed to air prior to infiltration because surface oxidation will alter the positive effects of coating. One commercially successful liquid infiltration process involving particulate reinforcement is the Duralcan process. Figure 3.15 shows a schematic of this process. Ceramic particles and ingot-grade aluminum are mixed and melted. The ceramic particles are given a proprietary treatment. The melt is stirred just above the liquidus temperature – generally between 600°C and 700°C. The melt is then converted into one of the following four forms: extrusion blank, foundry ingot, rolling bloom, or rolling ingot. The Duralcan process of making particulate composites by liquid metal casting involves the use of

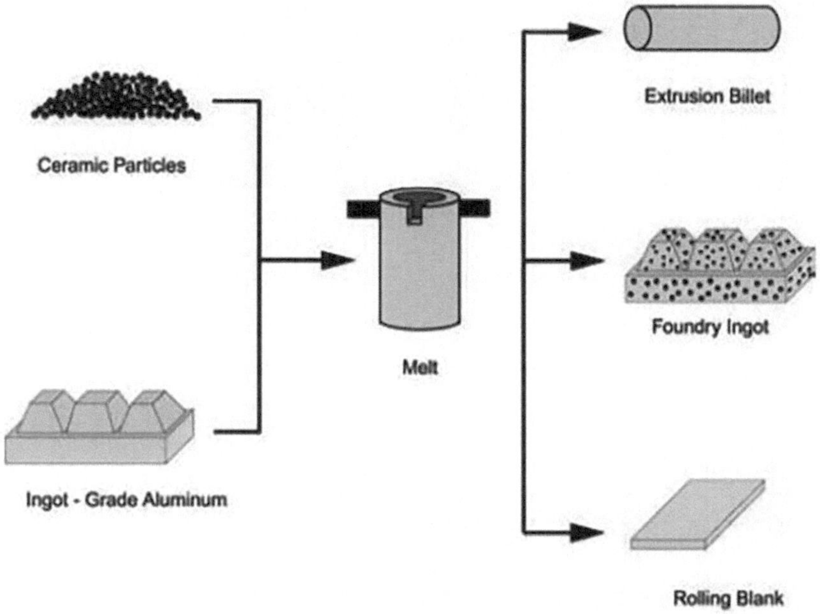

FIGURE 3.15 Duralcan process for metal matrix composites.

8–12 μm particles. Too small particles, for example, 2–3 μm, will result in a very large interface region and thus a very viscous melt. In foundry-grade MMCs, high Si aluminum alloys (e.g., A356) are used, while in wrought MMC, Al–Mg-type alloys (e.g., 6061) are used. Alumina particles are typically used in foundry alloys, while silicon carbide particles are used in the wrought aluminum alloys. For making continuous fiber-reinforced MMCs, tows of fibers are passed through a liquid metal bath, where the individual fibers are wet by the molten metal, excess metal is wiped off, and a composite wire is produced. A bundle of such wires can be consolidated by extrusion to make a composite.

3.2.1.2 Squeeze Casting

Squeeze casting, or pressure infiltration, involves forcing the liquid metal into a fibrous preform. Figure 3.16 shows two processes of making a fibrous preform. In the press-forming process, an aqueous slurry of fibers is well agitated and poured into a mold, pressure is applied to squeeze the water out, and the preform is dried (Figure 3.16a). In other process, suction is applied to a well-agitated mixture of whisker, binder, and water. This is followed by demolding and drying of the fiber preform (Figure 3.16b). A schematic of the squeeze casting process is shown in Figure 3.17. Pressure is applied until the solidification is complete. By forcing the molten metal through small pores of fibrous preform, this method obviates the requirement of good wettability of the reinforcement by the molten metal.

Composites fabricated with this method have a minimal reaction between the reinforcement and molten metal because of short dwell time at high temperature

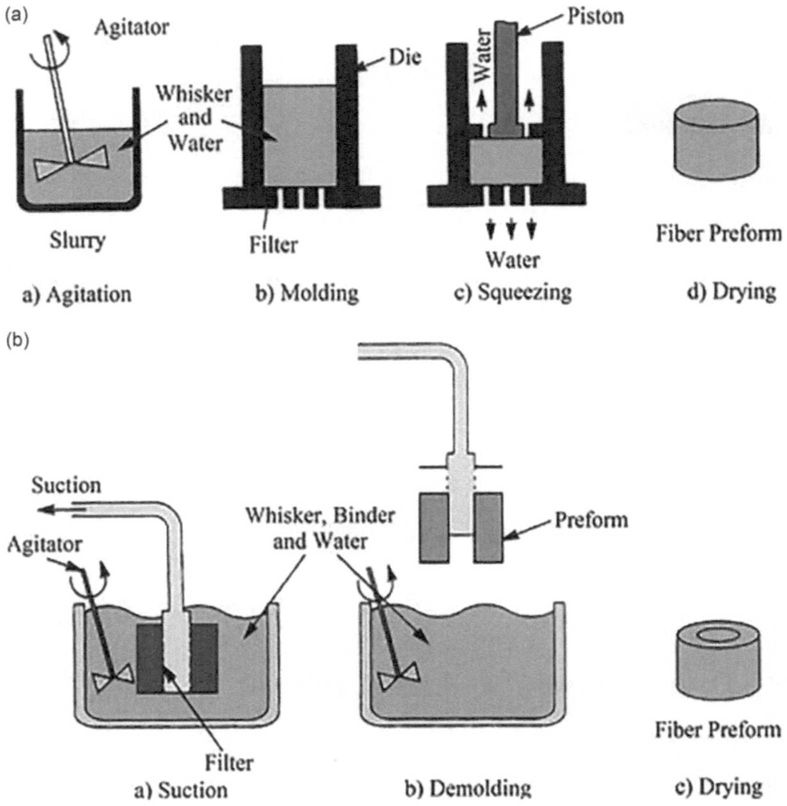

FIGURE 3.16 (a) Press forming and (b) suction forming of a preform.

and are free from common casting defects such as porosity and shrinkage cavities. Squeeze casting is really an old process, also called liquid metal forging in earlier versions. It was developed to obtain pore-free, fine-grained aluminum alloy components with superior properties than conventional permanent mold casting. In particular, the process has been used in the case of aluminum alloys that are difficult to cast by conventional methods, for example, silicon-free alloys used in diesel engine pistons where high-temperature strength is required. Inserts of nickel-containing cast iron, called Ni-resist, in the upper groove area of pistons have also been produced by the squeeze casting technique to provide wear resistance. Use of ceramic fiber-reinforced MMCs at locations of high wear and high thermal stress has resulted in a product much superior to the Ni-resist cast iron inserts.

The squeeze casting technique, as shown in Figure 3.17, has been quite popular in making composites with selective reinforcement. A porous fiber preform (generally of discontinuous Saffil-type Al_2O_3 fibers) is inserted into the die. Molten metal (aluminum) is poured into the preheated die located on the bed of a hydraulic press. The applied pressure (70–100 MPa) makes the molten aluminum penetrate the fiber preform and bond the fibers. Infiltration of a fibrous preform

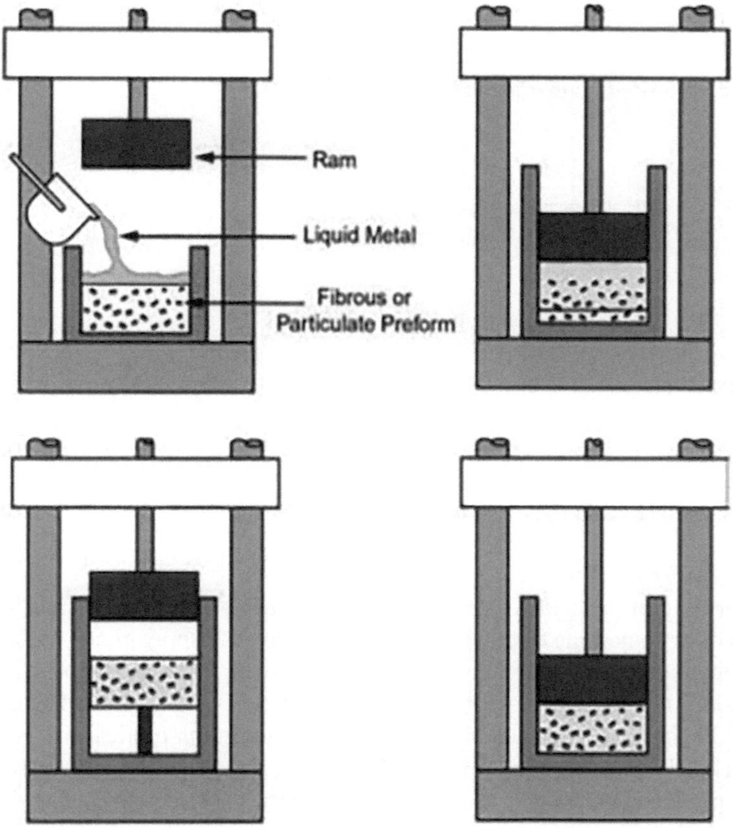

FIGURE 3.17 Squeeze casting process for the fabrication of MMCs.

by means of a pressurized inert gas is another variant of liquid metal infiltration technique. The process is conducted in the controlled environment of a pressure vessel and rather high fiber volume fractions; complex-shaped structures are obtainable. Although commonly, aluminum matrix composites are made by this technique, alumina fiber-reinforced intermetallic matrix composites (e.g., TiAl, Ni_3Al, and Fe_3Al matrix materials) have been prepared by pressure casting. The technique involves melting the matrix alloy in a crucible in vacuum, while the fibrous preform is heated separately. The molten matrix material (at about 100°C above the T_m) is poured onto the fibers, and argon gas is introduced simultaneously. Argon gas pressure forces the melt to infiltrate the preform. The melt generally contains additives to aid in wetting the fibers.

3.2.1.3 Centrifugal Casting

One of the disadvantages of MMCs with ceramic reinforcement is that they are typically more difficult to machine than the unreinforced alloy. In centrifugal casting, optimal placement of the reinforcement can be achieved by inducing a centrifugal

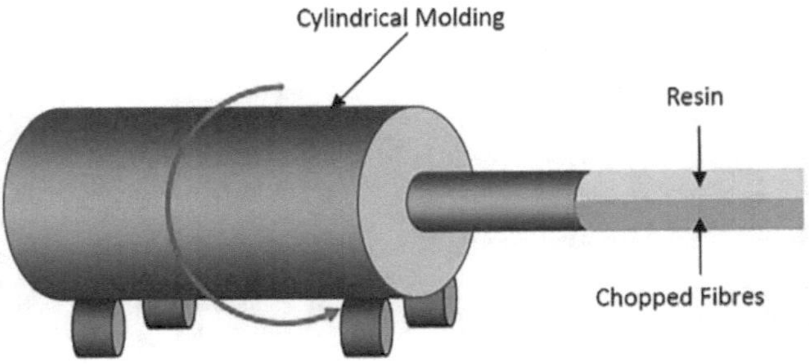

FIGURE 3.18 Centrifugal casting process for MMCs.

force immediately during casting which allows one to intentionally obtain a gradient in reinforcement volume fraction. Figure 3.18 shows the centrifugal casting process. In brake rotors, for example, wear resistance is needed on the rotor face, but not in the hub area. Thus, in areas where reinforcement is not as crucial, such as in the hub area, easier machining may be obtained.

3.2.1.4 Spray Forming

Spray forming of particulate MMCs involves the use of spray techniques that have been used for some time to produce monolithic alloys. A spray gun is used to atomize a molten aluminum alloy matrix. Ceramic particles, such as silicon carbide, are injected into this stream. Usually, the ceramic particles are preheated to dry them. Figure 3.19 shows a schematic of this process. An optimum particle size is required for an efficient transfer. Whiskers, for example, are too fine to be transferred. The preform produced in this way is generally quite porous. The co-sprayed MMC is subjected to scalping, consolidation, and secondary finishing processes, thus making it a wrought material. The process is totally computer-controlled and quite fast. It should also be noted that the process is essentially a liquid metallurgy process. One avoids the formation of deleterious reaction products because the time of flight is extremely short. Silicon carbide particles of an aspect ratio (length/diameter) between 3 and 4 and volume fractions up to 20% have been incorporated into aluminum alloys. A great advantage of the process is the flexibility that it affords in making different types of composites. For example, one can make in situ laminates using two sprayers, or one can have a selective reinforcement. This process, however, is quite expensive, mainly because of the high cost of the capital equipment.

3.2.2 Solid-State Processes

Many solid-state techniques are available. Some of the important ones have been described below.

3.2.2.1 Diffusion Bonding

Diffusion bonding is a common solid-state welding technique used to join similar or dissimilar metals. Interdiffusion of atoms from clean metal surfaces in contact at an elevated temperature leads to welding. There are many variants of the basic diffusion

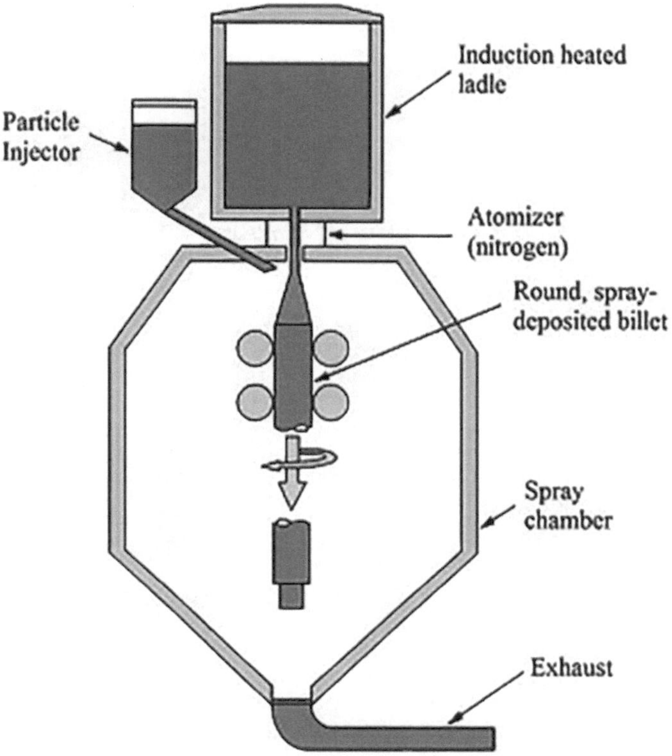

FIGURE 3.19 Spray forming process for MMCs.

bonding process; however, all of them involve a step of simultaneous application of pressure and high temperature. Matrix alloy foil and fiber arrays, composite wire, or monolayer laminae are stacked in a predetermined order. Figure 3.20 shows a schematic of one such diffusion bonding process, also called the foil-fiber-foil process. Vacuum hot pressing is a most important step in the diffusion bonding processes for MMCs. The major advantages of this technique are the ability to process a wide variety of matrix metals and control of fiber orientation and volume fraction. The disadvantages are processing times of several hours, and high processing temperatures and pressures – all of which make the process quite expensive – besides the fact only objects of limited size can be produced. Hot isostatic pressing (HIP), instead of uniaxial pressing, can also be used. In HIP, gas pressure against a can consolidates the composite piece contained inside the can. With HIP, it is relatively easy to apply high pressures at elevated temperatures over variable geometries.

3.2.2.2 Deformation Processing

Deformation processing of MMCs involves mechanical processing (swaging, extrusion, drawing, or rolling) of a ductile two-phase material. The two phases codeform, causing the minor phase to elongate and become fibrous in nature within the matrix.

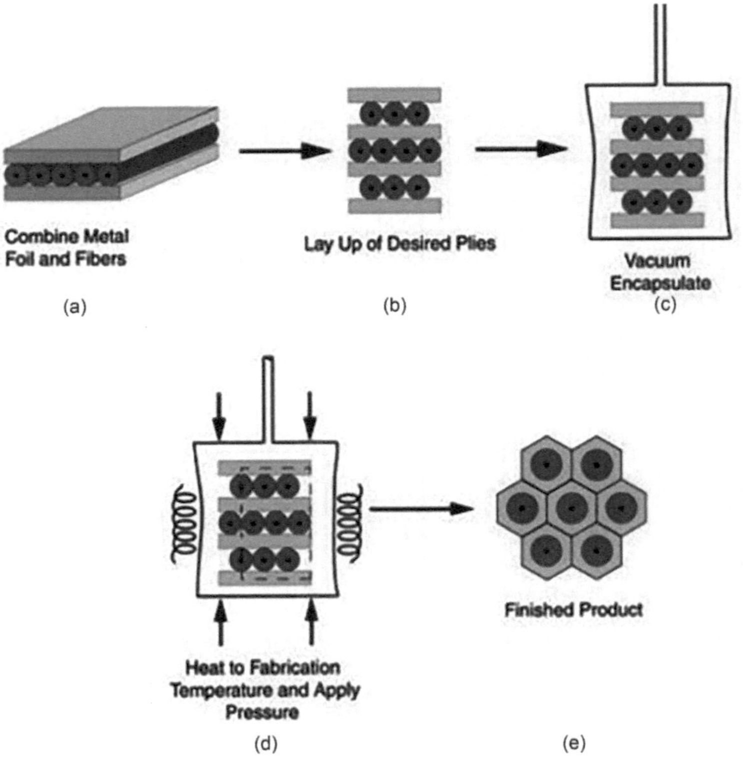

FIGURE 3.20 Diffusion bonding process: (a) Apply metal foil and cut to shape; (b) lay up desired plies; (c) vacuum encapsulate and heat to fabrication temperature; (d) apply pressure and hold for consolidation cycle; and (e) cool, remove, and clean part.

These materials are sometimes referred to as in situ composites. The properties of a deformation-processed composite largely depend on the characteristics of the starting material, which is usually a billet of two-phase alloy that has been prepared by casting or powder metallurgy methods. Roll bonding is a common technique used to produce a laminated composite consisting of different metals in the sheet form. Such composites are called sheet-laminated MMCs. Other examples of deformation-processed MMCs are the niobium-based conventional filamentary superconductors and the high-T_c superconductors. Figure 3.21 shows a roll bonding technique for making a laminated MMC, which produces a metallurgical bond.

3.2.2.3 Powder Processing

Powder processing methods in conjunction with deformation processing are used to fabricate particulate or short-fiber-reinforced composites. This typically involves cold pressing and sintering, or hot pressing, to fabricate primarily particle- or whisker-reinforced MMCs. The matrix and the reinforcement powders are blended to produce a homogeneous distribution. The blending stage is followed by cold pressing to produce what is called a green body, which is about 80% dense and can be

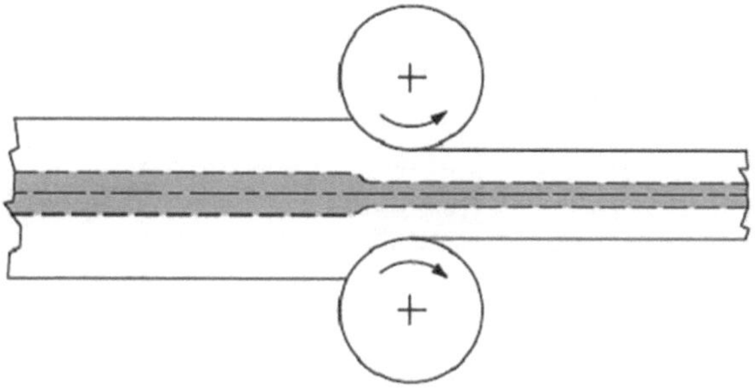

FIGURE 3.21 Roll-bonding process for the fabrication of MMCs.

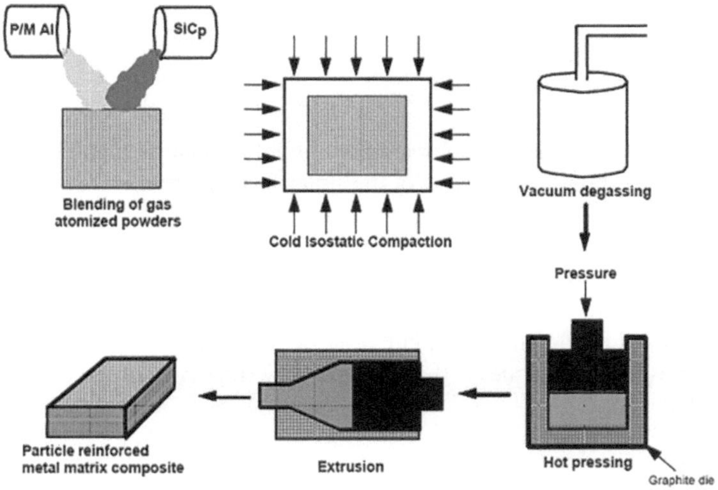

FIGURE 3.22 Powder processing, hot pressing, and extrusion process for fabricating particulate or short-fiber-reinforced MMCs.

easily handled. The cold-pressed green body is canned in a sealed container and degassed to remove any absorbed moisture from the particle surfaces. The material is hot-pressed, uniaxially or isostatically, to produce a fully dense composite and extruded. The rigid particles or fibers cause the matrix to be deformed significantly. The process is shown in Figure 3.22. During hot extrusion, dynamic recrystallization takes place at the particle–matrix interface, yielding randomly oriented grains near the interface, and relatively textured grains far from the interface.

3.2.2.4 Sinter Forging

Sinter forging is a novel and low-cost deformation processing technique. In sinter forging, a powder mixture of reinforcement and matrix is cold-compacted, sintered, and forged to nearly full density. The main advantage of this technique is that forging is conducted to produce a near-net-shaped material, and machining operations and material waste are minimized. The low-cost, sinter-forged composites have tensile

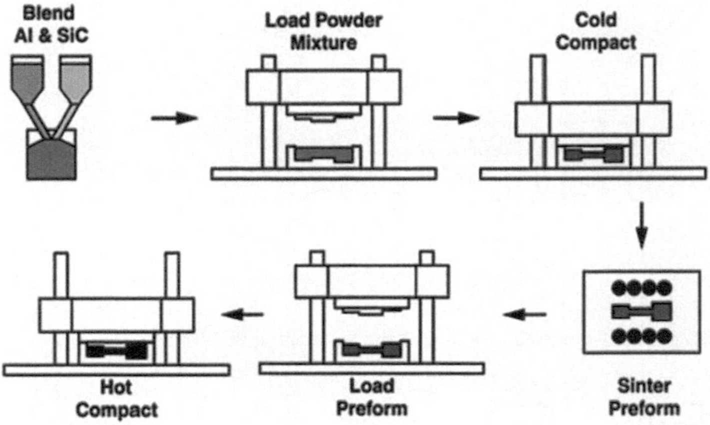

FIGURE 3.23 Sinter-forging process.

and fatigue properties that are comparable to those of materials produced by extrusion. Figure 3.23 shows a schematic of the sinter-forging process.

3.2.2.5 Deposition Techniques

Deposition techniques for MMC fabrication involve coating individual fibers in a tow with the matrix material needed to form the composite followed by diffusion bonding to form a consolidated composite plate or structural shape. The main disadvantage of using deposition techniques is that they are very time-consuming. However, there are several advantages:

a. The degree of interfacial bonding is easily controllable; interfacial diffusion barriers and compliant coatings can be formed on the fiber prior to matrix deposition or graded interfaces can be formed.
b. Thin, monolayer tapes can be produced by filament winding; these are easier to handle and mold into structural shapes than other precursor forms – unidirectional or angle-plied composites can be easily fabricated in this way.

Several deposition techniques are available: immersion plating, electroplating, spray deposition, chemical vapor deposition (CVD), and physical vapor deposition (PVD). Dipping or immersion plating is similar to infiltration casting except that fiber tows are continuously passed through baths of molten metal, slurry, sol, or organometallic precursors. Electroplating produces a coating from a solution containing the ion of the desired material in the presence of an electric current. Fibers are wound on a mandrel, which serves as the cathode, and placed into the plating bath with an anode of the desired matrix material. The advantage of this method is that the temperatures involved are moderate and no damage is done to the fibers. Problems with electroplating involve a void formation between fibers and between fiber layers, adhesion of the deposit to the fibers may be poor, and there are limited numbers of alloy matrices available for this processing.

A spray deposition operation, typically, consists of winding fibers onto a foil-coated drum and spraying molten metal onto them to form a monotape. The source of molten metal may be powder or wire feedstock, which is melted in a flame, arc, or plasma torch. The advantages of spray deposition are easy control of fiber alignment and a rapid solidification of the molten matrix. In a CVD process, a vaporized component decomposes or reacts with another vaporized chemical on the substrate to form a coating on that substrate. The processing is generally carried out at elevated temperatures.

3.2.3 In Situ Processes

In in situ techniques, one forms the reinforcement phase in situ. The composite material is produced in one step from an appropriate starting alloy, thus avoiding the difficulties inherent in combining the separate components as in a typical composite processing. Controlled unidirectional solidification of a eutectic alloy is a classic example of in situ processing. Unidirectional solidification of a eutectic alloy can result in one phase being distributed in the form of fibers or ribbons in the other. One can control the fineness of distribution of the reinforcement phase by simply controlling the solidification rate. The solidification rate in practice, however, is limited to a range of 1–5 cm $hour^{-1}$ because of the need to maintain a stable growth front. The stable growth front requires a high temperature gradient. Figure 3.24 shows the in situ processing by the controlled unidirectional solidification of a eutectic alloy.

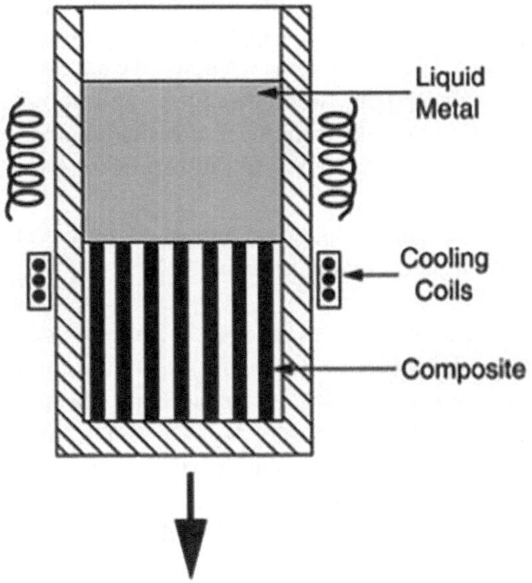

FIGURE 3.24 In situ processing by controlled unidirectional solidification of a eutectic alloy.

3.3 CERAMIC MATRIX COMPOSITES

Ceramic matrix composites (CMCs) can be processed either by conventional powder processing techniques used for making polycrystalline ceramics or by some new techniques developed specifically for making CMCs. Some of the important processing techniques for CMCs have been discussed below.

3.3.1 Cold Pressing and Sintering

Cold pressing of the matrix powder and fiber followed by sintering is a carryover from conventional processing of ceramics. Generally, in the sintering step, the matrix shrinks considerably and the resulting composite has many cracks. In addition to this general problem of shrinkage associated with sintering of any ceramic, certain other problems arise when we put high-aspect ratio (length/diameter) reinforcements in a glass or ceramic matrix material and try to sinter. Fibers and whiskers can form a network that may inhibit the sintering process. Depending on the difference in thermal expansion coefficients of the reinforcement and the matrix, a hydrostatic tensile stress may develop in the matrix on cooling, which will counter the driving force (surface energy minimization) for sintering. Thus, the densification rate of the matrix will, in general, be retarded in the presence of reinforcements. Whiskers or fibers may also give rise to the phenomenon of bridging, which is a function of the orientation and aspect ratio of the reinforcement.

3.3.2 Hot Pressing

Some form of hot pressing is frequently resorted to in the consolidation stage of CMCs. This is because a simultaneous application of pressure and high temperature can accelerate the rate of densification and a pore-free and fine-grained compact can be obtained. A common variant, called the slurry infiltration process, is one of the most important techniques used to produce continuous fiber-reinforced glass and glass–ceramic composites. The slurry infiltration process involves two stages:

i. Incorporation of a reinforcing phase into an unconsolidated matrix
ii. Consolidation of matrix by hot pressing.

Figure 3.25 shows a schematic of this process. In addition to the incorporation of the reinforcing phase, the first stage involves some kind of fiber alignment. A fiber tow or a fiber preform is impregnated with matrix containing slurry by passing it through a slurry tank. The impregnated fiber tow or preform sheets are similar to the prepregs used in PMCs. The slurry consists of the matrix powder, a carrier liquid (water or alcohol), and an organic binder. The organic binder is burned out prior to consolidation. Wetting agents may be added to ease the infiltration of the fiber tow or preform. The impregnated tow or prepreg is wound on a drum and dried. This is followed by cutting and stacking of the prepregs and consolidation in a hot press. The process has the advantage that, as in PMCs, the prepregs can be arranged in a variety of stacking sequences. The slurry infiltration process is well suited for glass or glass–CMCs,

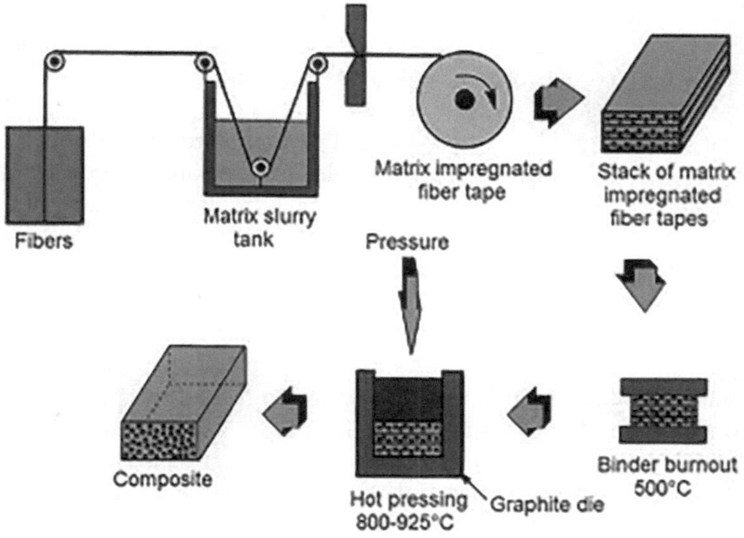

FIGURE 3.25 Slurry impregnation method.

mainly because the processing temperatures for these materials are lower than those used for crystalline matrix materials and glassy phase has good flow properties. Any hot pressing process has certain limitations in producing complex shapes. The fibers should suffer little or no damage during handling. Application of a very high pressure can easily damage fibers.

Refractory particles of a crystalline ceramic can damage fibers by mechanical contact. The reinforcement can also suffer damage from reaction with the matrix at very high processing temperatures. The matrix should have as little porosity as possible in the final product as porosity in a structural ceramic material is highly undesirable. To this end, it is important to completely remove the fugitive binder and use a matrix powder particle smaller than the fiber diameter. The hot pressing operational parameters are also important. Precise control within a narrow working temperature range, minimization of the processing time, and utilization of a pressure low enough to avoid fiber damage are the important factors in this final consolidation part of the process. Fiber damage and any fiber–matrix interfacial reaction, along with its detrimental effect on the bond strength, are unavoidable attributes of the hot pressing operation.

3.3.3 Reaction Bonding

Reaction bonding processes similar to the ones used for monolithic ceramics can be used to make CMCs. Reaction bonding process has the following advantages:

a. Problems with matrix shrinkage during densification can be avoided.
b. Rather large volume fractions of whiskers or fiber can be used.

c. Multidirectional, continuous fiber preforms can be used.
d. The reaction bonding temperatures for most systems are generally lower than the sintering temperatures so that fiber degradation can be avoided.

One great disadvantage of this process is that high porosity is difficult to avoid. A hybrid process involving a combination of hot pressing and reaction bonding technique can also be used. A silicon cloth is prepared by attrition milling of a mixture of silicon powder, a polymer binder, and an organic solvent to obtain a dough of proper consistency. This dough is then rolled to make a silicon cloth of desired thickness. Fiber mats are made by filament winding of silicon carbide with a fugitive binder. The fiber mats and silicon cloth are stacked in an alternate sequence, debinderized (the step of binder removal), and hot-pressed in a molybdenum die in a nitrogen or vacuum environment. The temperature and pressure are adjusted to produce a preform. At this stage, the silicon matrix is converted to silicon nitride by transferring the composite to a nitriding furnace between 1100°C and 1400°C. Typically, the silicon nitride matrix has about 30% porosity, which is not unexpected in the reaction-bonded silicon nitride. Reaction bonding process can also be used to make oxide fiber/oxide matrix composites, which are very attractive for high-temperature applications in air. Specifically, the technique has been applied to alumina and mullite.

3.3.4 Infiltration

Infiltration of a preform made of reinforcement can be done with a matrix material in solid, liquid, or gaseous form.

3.3.4.1 Liquid Infiltration

This technique is very similar to liquid polymer or liquid metal infiltration (Figure 3.26). Proper control of the fluidity of the liquid matrix is, of course, the key to this technique. It yields a high-density matrix, that is, no pores in the matrix. Almost any reinforcement geometry can be used to produce a virtually flaw-free composite. The temperatures involved, however, are much higher than those encountered in polymer or metal processing. Processing at such high temperatures can lead to deleterious chemical reactions between the reinforcement and the matrix. Thermal expansion mismatch between the reinforcement and the matrix, the rather large temperature interval between the processing temperature and the room temperature, and the low strain to failure of ceramics can add up to a formidable set of problems in producing a crack-free CMC. Viscosities of ceramic melts are generally quite high, which makes the infiltration of preforms rather difficult. Wettability of the reinforcement by the molten ceramic is another item to be considered. A preform made of reinforcement in any form (e.g., fiber, whisker, or particle) having a network of pores can be infiltrated by a ceramic melt by using capillary pressure. Application of pressure or processing in vacuum can aid in the infiltration process. Penetration will also be easier if the contact angle is low (i.e., better wettability), and the surface energy and the pore radius are large. If the radius of the channel is made too large, the capillarity effect will be lost. The advantages are as follows:

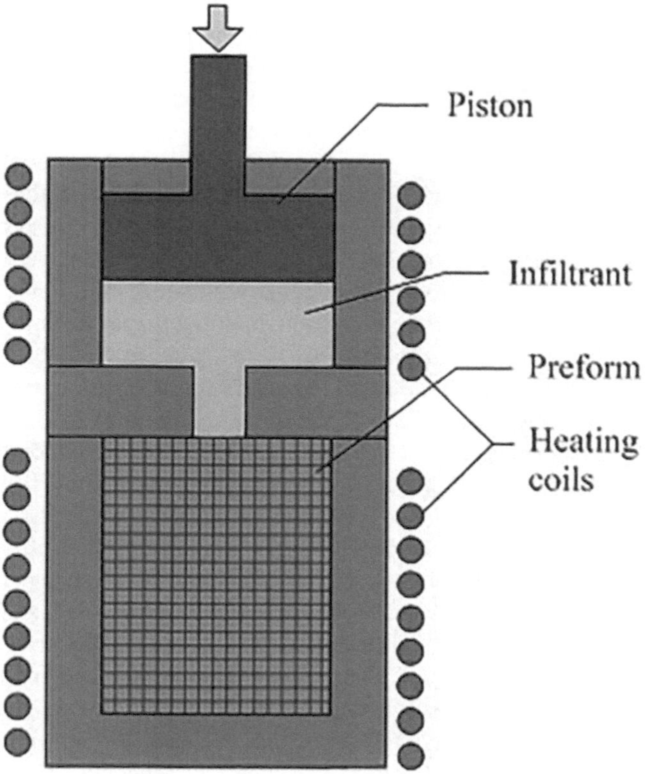

FIGURE 3.26 Melt infiltration process.

a. The matrix is formed in a single processing step.
b. A homogeneous matrix can be obtained.

The disadvantages of infiltration techniques are as follows:

a. High melting points of ceramics mean a greater likelihood of reaction between the melt and the reinforcement.
b. Ceramics have higher melt viscosities than metals; therefore, infiltration of preforms is relatively difficult.
c. The matrix is likely to crack because of the differential shrinkage between the matrix and the reinforcement on solidification. This can be minimized by choosing components with nearly equal coefficients of thermal expansion.

3.3.4.2 Gaseous Infiltration

A reactive gas mixture deposits a ceramic material within a preform of the reinforcement, commonly known as chemical vapor deposition (CVD) or chemical vapor infiltration (CVI), which is typically performed at high temperature – all of which have particular advantages or disadvantages for different types of matrix material. For those matrix materials that can be produced by reaction between gases, such as carbon and silicon carbide, a gas-phase route can be substituted for the liquid impregnation route. This tends to be a slow process, because if deposition is allowed to occur too quickly, it mostly occurs at the external surface, blocking penetration. Developments in the technique have concentrated on ensuring that deposition occurs internally to the component, by forcing reactive gas flow through the walls of the component (CVI), using temperature gradients, and on speeding the process up, for example, by using microwave heating, which also has the advantage of improving deposition internally to the preform. However, as with liquid infiltration, full densification of the matrix never occurs as access to remaining pores becomes blocked, but the method can be used for complex shapes.

In gas-phase routes, that is, the so-called chemical vapor infiltration (CVI) process, the reinforcement (usually as a multidirectional preform) is densified by the matrix deposited from a gaseous precursor, for example, a hydrocarbon for carbon or a mixture of methyltrichlorosilane and hydrogen for silicon carbide. There are several versions of the CVI process. It is now well established that a fiber coating, referred to as the interphase, has to be deposited on the fiber prior to the infiltration of the matrix in order to control the fiber–matrix bonding and the mechanical behavior of the composite. CVI is a method of infiltrating fiber architectures with matrix particles via the vapor phase. Although this is a similar process to CVD in terms of gas reaction, decomposition conditions in CVI are chosen for in-depth decomposition rather than coating the surface of the substrate. Several techniques have been developed to introduce the reaction gases into the fiber architecture, such as temperature gradient, pressure gradient, and pulse CVI. Figure 3.27 shows the schematic of CVI technique.

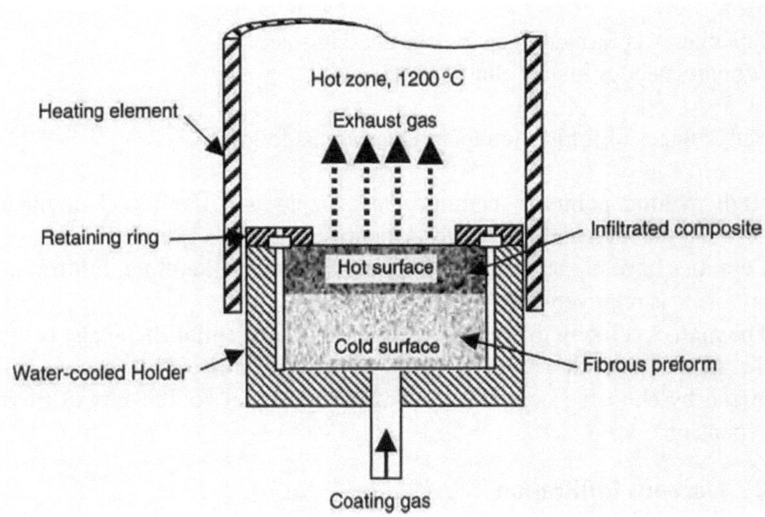

FIGURE 3.27 Chemical vapor impregnation technique.

3.3.5 Polymer Infiltration and Pyrolysis

Just as polymeric precursors can be used to make ceramic fibers, we can use polymeric precursors to form a ceramic matrix in a composite. This technique is called polymer infiltration and pyrolysis (PIP). It is an attractive processing route because of its relatively low cost, small amounts of residual porosity, and minimal degradation of the fibers. Moreover, this approach allows near-net-shaped molding and fabrication technology that is able to produce nearly fully dense composites. In PIP, the fibers are infiltrated with an organic polymer, which is heated to fairly high temperatures and pyrolyzed to form a ceramic matrix. Due to the relatively low yield of polymer to ceramic, multiple infiltrations are used to densify the composite. Polymeric precursors for ceramic matrices allow one to use conventional polymer composite fabrication technology that is readily available, and take advantage of processes used to make PMCs. These include complex shape forming and fabrication. Furthermore, by processing and pyrolyzing at lower temperatures (compared to sintering and hot pressing, for example), one can avoid fiber degradation and the formation of unwanted reaction products at the fiber–matrix interface. The desirable characteristics of a preceramic polymer are as follows:

a. High ceramic yield from polymer precursor
b. Precursor that yields a ceramic with low free carbon content (which will oxidize at high temperatures)
c. Controllable molecular weight, which allows for solvent solubility and control over viscosity for fabrication purposes
d. Low-temperature cross-linking of the polymer, which allows resin to harden and maintain its dimensions during the pyrolysis process
e. Low cost and toxicity.

Most preceramic polymer precursors are formed from chloroorganosilicon compounds to form poly(silanes), poly(carbosilanes), poly(silazanes), poly(borosilanes), poly(silsesquioxanes), and poly(carbosiloxanes). The synthesis reaction involves the dechlorination of the chlorinated silane monomers. Since a lot of the chlorosilane monomers are formed as byproducts in the silicone industry, they are inexpensive and readily available. The monomers can be further controlled by an appropriate amount of branching, which controls important properties such as the viscosity of the precursor as well as the amount of ceramic yield. All silicon-based polymer precursors lead to an amorphous ceramic matrix, where silicon atoms are tetrahedrally arranged with nonsilicon atoms. This arrangement is similar to that found in amorphous silica. High-temperature treatments typically lead to crystallization and slight densification of the matrix, which results in shrinkage. At high temperatures, the amorphous ceramic begins to form small domains of crystalline phase, which are more thermodynamically stable. Typically, the range of the molecular weight of the polymer is tailored, followed by shaping of the product. The polymer is then cross-linked and finally pyrolyzed in an inert or reactive atmosphere (e.g., NH_3)

at temperatures between 1000°C and 1400°C. The pyrolysis step can be further subdivided into three steps:

a. In the first step, between 550°C and 880°C, an amorphous hydrogenated compound of the type $Si(C_aO_bN_cB_d)$ is formed.
b. The second step involves a nucleation of crystalline precipitates such as SiC, Si_3N_4, and SiO_2 at temperatures between 1,200 and 1,600°C. Grain coarsening may also result from the consumption of any residual amorphous phase and reduction of the amount of oxygen due to vaporization of SiO and CO. Porosity is typically on the order of 5–20 vol.% with pore sizes of the order of 1–50 nm. It should be noted that the average pore size and volume fraction of pores decrease with increasing pyrolysis temperature, since the amount of densification (and shrinkage) becomes irreversible at temperatures above the maximum pyrolysis temperature. Figure 3.28 shows a schematic of the PIP process.

Some of the disadvantages of PIP are as follows:

a. Low yield accompanies the polymer-to-ceramic transformation.
b. Large shrinkage, which typically causes cracking in the matrix during fabrication. Due to shrinkage and weight loss during pyrolysis, residual porosity after a single impregnation is on the order of 20%–30%.
c. To reduce the amount of residual porosity, multiple impregnations are needed. Reimpregnation is typically conducted with a very low-viscosity prepolymer so that the slurry may wet and infiltrate the small micropores that exist in the preform. Usually, reimpregnation is done by immersing the part in the liquid polymer in a vacuum bag, while higher-viscosity polymers require pressure impregnation. Typically, the amount of porosity will reduce from 35% to less than 10% after about five impregnations. Significant gas evolution also occurs during pyrolysis. Thus, it is advisable

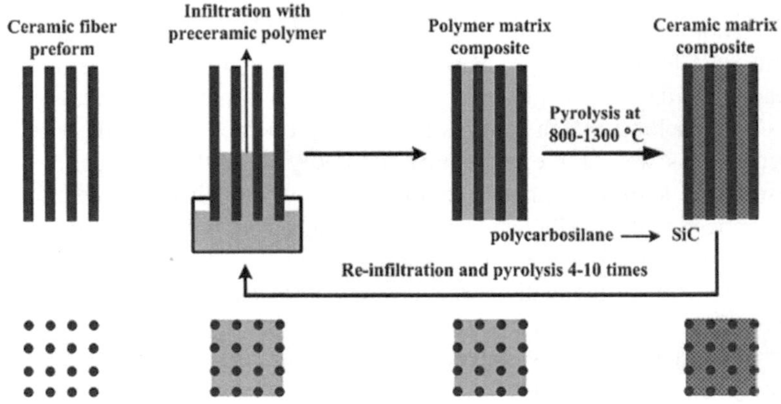

FIGURE 3.28 Polymer infiltration and pyrolysis technique.

to allow these volatile gases to slowly diffuse out of the matrix, especially for thicker parts. Typically, pyrolysis cycles ramp to somewhere between 800°C and 1400°C over periods of 1–2 days, to avoid delamination. Recall that pyrolysis must be done at a temperature below the crystallization temperature of the matrix (or large volume changes will occur) and below the degradation temperature of the reinforcing fibers. The pyrolysis atmosphere is most commonly argon or nitrogen, although in ammonia, we obtain a pure amorphous silicon nitride with low amounts of free carbon.

3.4 MISCELLANEOUS TECHNIQUES

3.4.1 Resin Film Infusion

In the resin film infusion (RFI) process, a precatalyzed resin film placed under the dry fiber preform provides the liquid resin that flows through the preform and on curing, becomes the matrix. The process starts by covering the mold surface with the resin film and then placing the dry fiber preform on top of the resin film (Figure 3.29). The thickness of the resin film depends on the quantity of resin needed to completely infiltrate the preform. RFI can be carried out using the bag-molding technique described earlier. In that case, the assembly of resin film and dry fiber preform is covered with a vacuum bag and placed inside an autoclave. The full vacuum is applied at the beginning to remove the trapped air from the preform. As the temperature is increased in the autoclave, the resin viscosity decreases and the resin starts to flow through the dry fiber preform. Pressure is applied to force the liquid resin to infiltrate the preform and wet out the fibers. With the temperature now raised to the prescribed curing temperature, the curing reaction begins and the liquid resin starts to gel. If an epoxy film is used, the curing cycle may take several minutes to several hours depending on the resin type and the curing conditions used.

3.4.2 Elastic Reservoir Molding

In elastic reservoir molding (ERM), a sandwich of liquid resin-impregnated open-celled foam and face layers of dry continuous strand mat, woven roving, or cloth placed in a heated mold (Figure 3.30) are pressed with a molding pressure

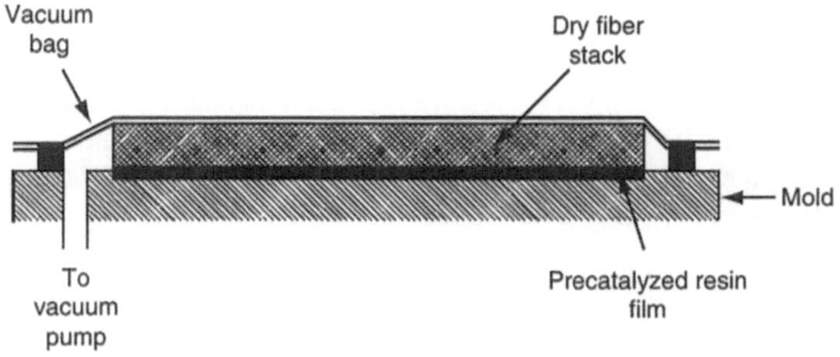

FIGURE 3.29 Resin film infusion process.

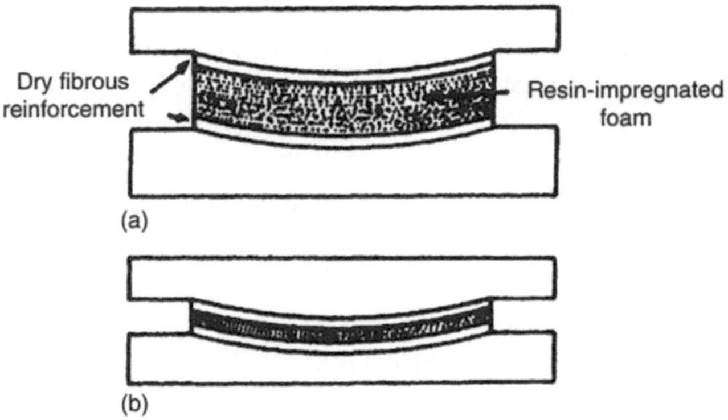

FIGURE 3.30 (a and b) Elastic reservoir molding process.

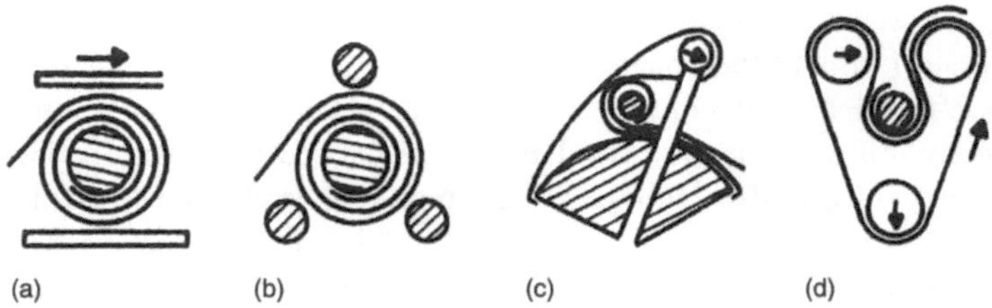

FIGURE 3.31 (a–d) Various tube-rolling methods.

of 520–1030 kPa (75–150 psi). The foam in the center of the sandwich is usually a flexible polyurethane that acts as an elastic reservoir for the catalyzed liquid resin. As the foam is compressed, the resin flows out vertically and wets the face layers. On curing, a sandwich of low-density core and fiber-reinforced skins is formed. The advantages of an ERM process are low-cost tooling, better control of properties (since there is no horizontal flow), and a better stiffness–weight ratio (due to the sandwich construction). It is generally restricted to molding thin panels of simple geometry. Examples of ERM applications are bus roof panels, radar-reflecting surfaces, automotive body panels, and luggage carriers.

3.4.3 Tube Rolling

Circular tubes for space truss or bicycle frame, for example, are often fabricated from prepregs using the tube-rolling technique. In this process, precut lengths of a prepreg are rolled onto a removable mandrel, as illustrated in Figure 3.31. The uncured tube is wrapped with a heat-shrinkable film or sleeve and cured at elevated temperatures in an air-circulating oven. As the outer wrap shrinks tightly on the rolled prepreg, air entrapped between the layers is squeezed out through the ends. For a better surface finish, the curing operation can be performed in a close-fitting steel tube or a split steel mold. The outer steel mold also prevents the mandrel from sagging at the high temperatures used for curing. After curing, the mandrel is removed and a hollow tube is formed. The advantages of the tube-rolling process over the filament winding process are low tooling cost, simple operation, better control of resin content and resin distribution, and faster production rate. However, this process is generally more suitable for simple layups containing 0° and 90° layers.

3.4.4 Compocasting

When a liquid metal is vigorously stirred during solidification by slow cooling, it forms a slurry of fine spheroidal solids floating in the liquid. Stirring at high speeds creates a high shear rate, which tends to reduce the viscosity of the slurry even at solid fractions that are as high as 50%–60% by volume. The process of casting such a slurry is called rheocasting. The slurry can also be mixed with particulates, whiskers, or short fibers before casting. This modified form of rheocasting to produce near-net-shaped MMC parts is called compocasting. The melt-reinforcement slurry can be cast by gravity casting, die-casting, centrifugal casting, or squeeze casting. The reinforcements have a tendency to either float to the top or segregate near the bottom of the melt due to the differences in their density from that of the melt. Therefore, a careful choice of the casting technique as well as the mold configuration is of great importance in obtaining a uniform distribution of reinforcements in a compocast MMC. Compocasting allows a uniform distribution of reinforcement in the matrix as well as a good wet-out between the reinforcement and the matrix. Continuous stirring of the slurry creates an intimate contact between them. Good bonding is achieved by reducing the slurry viscosity as well as increasing the mixing time. The slurry viscosity is reduced by increasing the shear rate as well as increasing the slurry temperature. Increasing mixing time provides longer interaction between the reinforcement and the matrix.

3.4.5 Spark Plasma Sintering

Recently, spark plasma sintering (SPS) has attracted attention as an efficient solid-state fabrication technique for highly thermally conductive materials. SPS is a pressure-assisted sintering technique which is based on electrical spark discharge, where a pulsed DC current passes through a conductive powder compact (the application to nonconductive materials is also possible, but limited). Figure 3.32 shows a schematic diagram of this process. In the case of conventional hot pressing, the heat is provided by external heating. In contrast, in SPS, the heat is generated internally between the powder particles. The sintering times for SPS are very short, approximately 5–20 minutes. When carbon fibers are included in the metal powder, heat will be generated between the fibers and the metal particles by both spark discharge and Joule heating, because carbon fiber is a conductive material. This means that only the surfaces of the metal particles will be melted, resulting in good bonding with the carbon fibers, and a decrease in thermal resistance between the carbon fibers and metal particles. Recently, scale-like graphite particles which have high thermal conductivity in two dimensions have been developed. Scale-like graphite particle/aluminum composites were fabricated by SPS and showed very high thermal conductivity [1]. A key aspect of fabricating these composites is how to distribute the

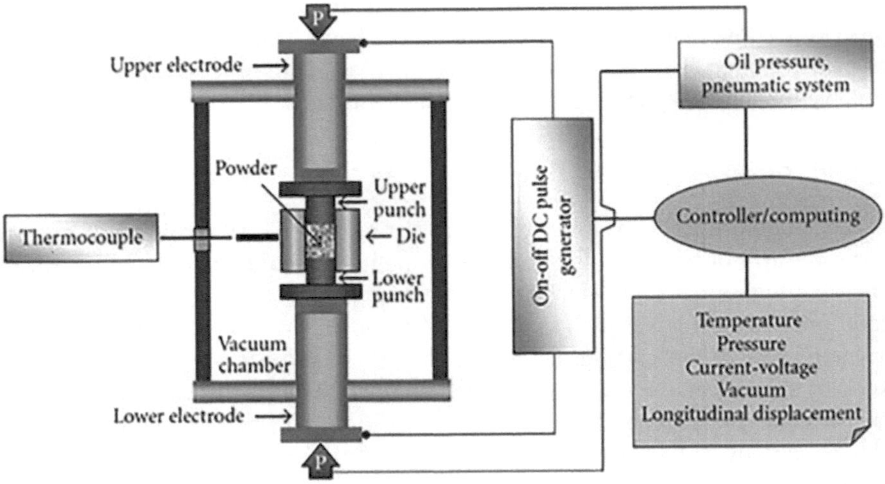

FIGURE 3.32 Spark plasma sintering process.

scale-like graphite particles horizontally on a plane. This scale-like graphite particle/aluminum composite is likely to perform better than the carbon fiber/aluminum composites, because the carbon fiber/aluminum composites only have high thermal conductivity in one dimension.

3.4.6 Vortex Addition Technique

As shown in Figure 3.33, a vortex forms when molten metal is strongly agitated. Reinforcements are then added into the vortex, which is maintained until the reinforcements wet well. The mixture is then poured into a mold. This process is called the "vortex addition technique," because the reinforcements are incorporated using a vortex. In conjunction with this technique, chemical techniques such as the addition of some elements into the matrix or coating the reinforcement surface are used to improve the wetting of ceramic particles by molten aluminum. The effective additive elements are Ca, Mg, or Li at levels of about 1 wt.% [2]. As an example of chemical modification, ceramic particles chemically plated with nickel or copper are wetted easily and incorporated into molten aluminum alloys by agitating. In particular, graphite particles coated with nickel can be easily dispersed in aluminum alloys or copper alloys. The key to the effectiveness of the nickel chemical plating is a reaction product, nickel tetracarbonyl ($Ni(CO)_4$), between the carbon and the nickel coating. $Ni(CO)_4$ mediates wetting between carbon particles and aluminum. However, if long agitation times are used with aluminum alloys, aluminum carbide (Al_4C_3) will be formed on the surface of graphite particles, which means that care must be taken when these particles are mixed into molten metal. These graphite particle dispersed alloys show good oil-less wear resistance [3,4]. A crucial aspect of the vortex addition technique is the atmosphere. If the ceramic particles or fibers are added into molten aluminum in air, the surface oxide film of the molten aluminum will be mixed

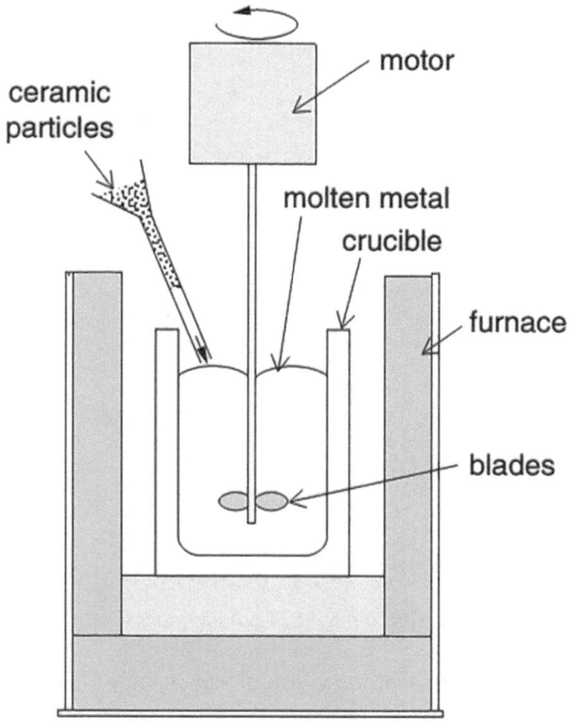

FIGURE 3.33 Vortex addition technique.

into the composite along with the fibers, and the properties of the composite will be degraded. Nitrogen or argon atmospheres are preferred.

Ceramic particles are wet easily by magnesium alloys, so it is possible to disperse these particles into magnesium alloys to produce composites without needing a vortex. However, an inert gas atmosphere is still required. By this process, when molten metal is agitated, the metal and ceramic particles move together. It is difficult to fix only the particles and allow only the metal to move. Therefore, most of the mechanical energy added by agitation is exhausted in making the metal form a vortex, and a small fraction of the energy is used for wetting of the particle–metal interface. This process is not energy efficient. If we make a narrow space between the crucible wall and the agitator to allow shear stress to do work in the molten metal, wetting will be improved. The addition of some elements to the metal is also effective at increasing viscosity and shear stress. Duralcan is an example of a composite fabricated by the vortex addition technique.

3.4.7 Pressureless Infiltration Process

If molten metal infiltrates spontaneously into a preform, a porous material made of ceramic fibers or particles, a composite will be easily produced. Such a pressureless infiltration process has been developed in the USA and is named the

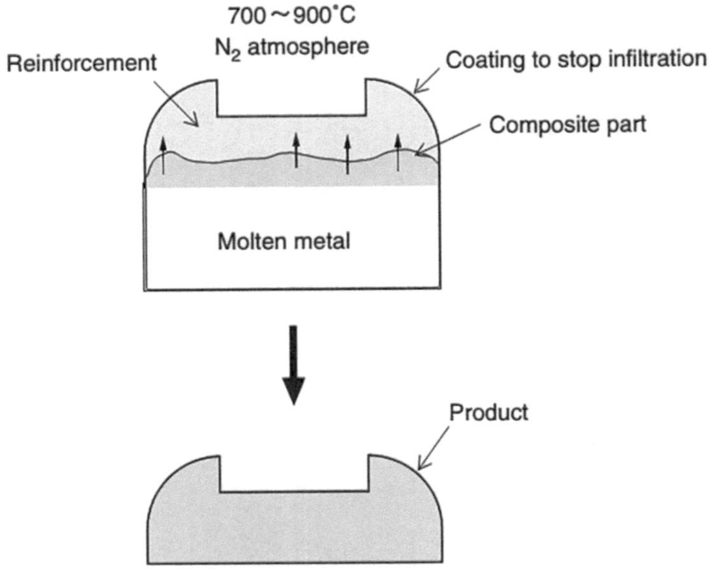

FIGURE 3.34 Pressureless infiltration by the Lanxide process.

"Lanxide process" [5]. In this process, a preform made of SiC particles which do not wet well with molten aluminum is treated with magnesium. The preform is set on top of molten aluminum under a nitrogen atmosphere. Then, magnesium nitride is formed on the surface of the SiC particles and molten aluminum spontaneously infiltrates into the preform without pressurization. The infiltration continues until the preform fills with molten aluminum. This process is schematically shown in Figure 3.34. The preform is made in the same shape as the desired product, and its outer surface (except for the base which will be in contact with the molten aluminum) is coated so that infiltration stops once the molten aluminum reaches the outer surface. Therefore, after infiltration, the composite has the near-net shape of the product. The application of this process is not wide because appropriate combinations of reinforcements and molten matrix metal are limited. In addition, it is not easy to achieve the same extent of filling with molten metal in this process as can be obtained with pressurized infiltration processes.

3.4.8 Ultrasonic Infiltration

The term "ultrasonic wave" brings to mind the ultrasonic cleaner which is very useful to remove stains. This cleaning mechanism is an interesting phenomenon because ultrasonic vibration may be related to the wetting between water and stains. Nakanishi et al. [6] investigated the effect of ultrasonic vibration on the contact angle between liquid and solid [6]. They showed that the apparent contact angle of a water droplet on a paraffin-coated substrate is decreased greatly by ultrasonic vibration and reaches a stable value. Further, they applied the ultrasonic vibration to the infiltration of molten aluminum into an alumina fiber preform and showed the feasibility of ultrasonic infiltration for the fabrication of composites [7]. The ultrasonic vibration reduced the threshold pressure for the infiltration of molten aluminum into the alumina fiber preform. The reduction of the threshold pressure was about 140 kPa for an alumina fiber preform with $V_f = 0.17$ at a resonant frequency of 20.5 kHz [7]. When an ultrasonic wave propagates through a material, the energy travels through the atomic structure by a series of compression (dense zone) and expansion movements (rough zone). As the dense zone is under high pressure, when the dense zone passes through the reinforcement/molten metal interface, the equilibrium of forces may be broken and the molten metal may advance as shown in Figure 3.35a.

In another application of ultrasonic waves, Deming et al. [8] developed "preform wires" (which are aluminum composite wires reinforced with continuous SiC fibers), by ultrasonic infiltration. They showed that it is possible to infiltrate molten aluminum into the fiber bundle using ultrasonic vibration without fiber coating or fiber pretreatment. Preform wires of carbon fiber/aluminum composites by ultrasonic infiltration without fiber surface treatments were developed by Cheng et al. [9] although the wettability of molten aluminum with carbon fiber is very poor. Matsunaga et al. [10–13] determined the optimum conditions (such as preheating temperature of fibers, fabrication speed, ultrasonic power, among others) for the fabrication of carbon fiber/Al–Mg alloy preform wires by ultrasonic infiltration. The effect of high temperature holding on the carbon fiber/Al–Mg alloy preform wire composites fabricated by supersonic infiltration was reported by Mizoguchi et al. [14] and Yamaguchi et al. [15]. Research on the interface reaction of carbon fiber/Al alloy preform wire composites was performed by Mikuni et al. [16]. The effect of Mg content on the tensile strength of the carbon fiber/Al–Mg alloy preform wire composites was clarified by Matsunaga et al. [17].

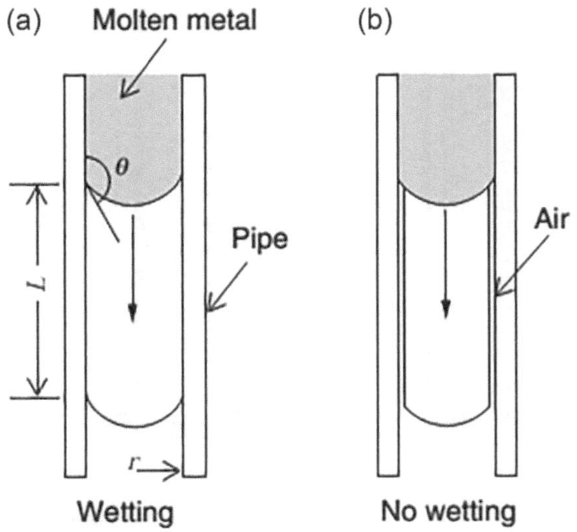

FIGURE 3.35 Schematic of molten metal flow in a ceramic pipe, (a) wetting by the advance of molten metal, (b) no wetting by the advance of molten metal.

3.4.9 Chemical Vapor Deposition

During CVD, flowing gas-phase materials react on the surface of a hot solid material (substrate) and solid reaction products deposit on the surface of the substrate. Three major processes have been used: (1) reduction of metal halides by hydrogen with a catalyst, (2) thermal decomposition of gas phases, and (3) the reaction of the substrate with gas phases. The deposition rate or crystallization rate mainly depends on the temperature of the substrate. Therefore, the microstructure of the deposited layer also depends on the substrate temperature. Generally, at high temperature, several kinds of single crystals grow, and at lower temperature, polycrystalline or amorphous phases grow. Fine polycrystalline deposits having dense and homogeneous mechanical properties are desirable for coating purposes. An example of the apparatus for this process, which has a heating system around the furnace inner wall, is shown in Figure 3.36. This equipment is suitable for coating of many small substrates at the same time and is called a "hot wall-type" apparatus. Alternatively, when high-frequency induction heating is used, the furnace wall is not hot; this is the "cold wall type" of apparatus.

When the difference in thermal expansion coefficient between the coating material and the substrate is large, cracks may form at the substrate–coating-layer interface after CVD. To prevent such cracking, the following methods are employed:

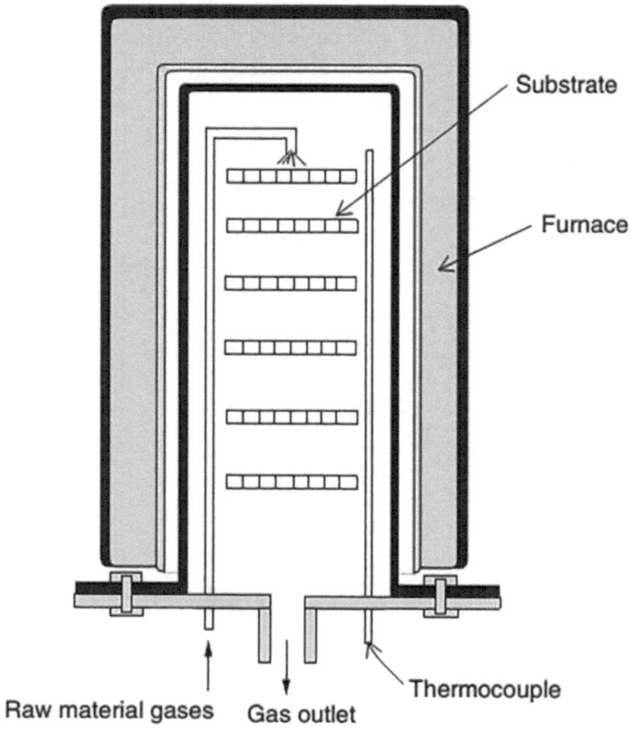

FIGURE 3.36 Hot-wall CVD system.

1. Using a thinner coating and sacrificing some of the properties of the coating material.
2. Using an intermediate phase or diffusion layer containing both materials between the coating material and the substrate.

3.4.10 Physical Vapor Deposition

The simple image of PVD is that a material is evaporated by electrical resistive heating or electron beam heating in a vacuum and is deposited onto the surface of a substrate. Many different PVD processes exist; sputtering is a typical example. In the sputtering process, atoms or clusters are ejected from a material (target) by ion bombardment and deposited onto the surfaces of substrates. Another variation on this process, which includes additional materials (such as C_2H_2 or NH_3) in the atmosphere, has been developed to produce carbides or nitrides [18]. Here, we introduce two types of sputtering.

3.4.10.1 Conventional Sputtering

In an argon, oxygen, or nitrogen plasma at a pressure of 1–10 Pa, atoms are ejected from the negatively charged target (the material to be used for the coating) by the bombardment of positive ions from the plasma and then deposited onto the surface of a substrate. The potential difference used in this technique is several thousand volts. During this process, the substrate temperature rises to several hundred °C, because the substrate is exposed to high-speed electrons. The deposition rate is about 0.1 μm $minute^{-1}$.

3.4.10.2 Ion Beam Sputtering

An ion beam is injected from a separately equipped ion source chamber at a pressure of 0.1–1 Pa and irradiates the target. The ejected atoms from the target are deposited onto a substrate. With this equipment, because the substrate is not exposed to an electron beam, the substrate temperature can be lower, allowing better control of the deposition conditions.

3.5 BASICS OF CURING

Transformation of uncured or partially cured fiber-reinforced thermoset polymers into composite parts or structures involves curing the material at elevated temperatures and pressures for a predetermined length of time. High cure temperatures are required to initiate and sustain the chemical reaction that transforms the uncured or partially cured material into a fully cured solid. High pressures are used to provide the force needed for the flow of the highly viscous resin or fiber–resin mixture in the mold, as well as for the consolidation of individual unbonded plies into a bonded laminate. The magnitude of these two important process parameters, as well as their duration, significantly affects the quality and performance of the molded product. The length of time required to properly cure a part is called the cure cycle. Since the cure cycle determines the production rate for a part, it is desirable to achieve the proper cure in the shortest amount of time. It should be noted that the cure cycle depends on a number of factors, including resin chemistry, catalyst reactivity, cure temperature, and the presence of inhibitors or accelerators.

3.5.1 Degree of Curing

A number of investigators have experimentally measured the heat evolved in a curing reaction and related it to the degree of cure achieved at any time during the curing process. Experiments are performed in a differential scanning calorimeter (DSC) in which a small sample, weighing a few milligrams, is heated either isothermally (i.e., at constant temperature) or dynamically (i.e., with uniformly increasing temperature). The instrumentation in DSC monitors the rate of heat generation as a function of time and records it. Figure 3.37 schematically illustrates the rate of heat generation curves for isothermal and dynamic heating.

The total heat generation to complete a curing reaction (i.e., 100% degree of cure) is equal to the area under the rate of heat generation–time curve obtained in a dynamic heating experiment. It is expressed as:

$$H_R = \int_0^{t_f} \left(\frac{dQ}{dt} \right)_d dt, \tag{3.1}$$

where

H_R = heat of reaction

$\left(\frac{dQ}{dt} \right)_d$ = rate of heat generation in a dynamic experiment

t_f = time required to complete the reaction.

The amount of heat released in time t at a constant curing temperature T is determined from isothermal experiments. The area under the rate of heat generation–time curve obtained in an isothermal experiment is expressed as:

$$H = \int_0^{t} \left(\frac{dQ}{dt} \right)_i dt, \tag{3.2}$$

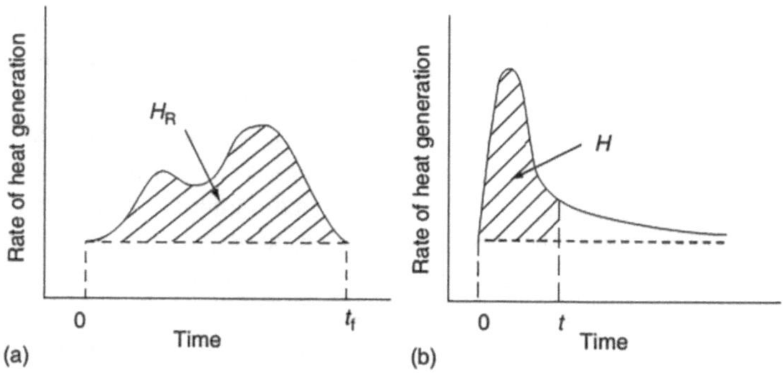

FIGURE 3.37 Schematic representation of the rate of heat generation in (a) dynamic and (b) isothermal heating of a thermoset polymer in a differential scanning calorimeter (DSC).

where H is the amount of heat released in time t and $(dQ/dt)_i$ is the rate of heat generation in an isothermal experiment conducted at a constant temperature T. The degree of cure α_c at any time t is defined as:

$$\alpha_c = \frac{H}{H_R}. \tag{3.3}$$

Figure 3.38 shows a number of curves relating the degree of cure α_c to cure time for a vinyl ester resin at various cure temperatures. From this figure, it can be seen that α_c increases with both time and temperature; however, the rate of cure, $d\alpha_c/dt$, is decreased as the degree of cure attains asymptotically a maximum value. If the cure temperature is too low, the degree of cure may not reach a 100% level for any reasonable length of time. The rate of cure $d\alpha_c/dt$, obtained from the slope of α_c vs. t curve and plotted in Figure 3.39, exhibits a maximum value at 10%–40% of the total cure achieved. Higher cure temperatures increase the rate of cure and produce the maximum degree of cure in shorter periods of time. On the other hand, the addition of a low-profile agent, such as a thermoplastic polymer, to a polyester or a vinyl ester resin decreases the cure rate.

3.5.2 Curing Cycle

In general, the higher the temperature during curing, the shorter the cure time (short of burning the material, of course). Heat is required because (1) some catalysts and/or hardeners do not react below a critical temperature; (2) molecular mobility is

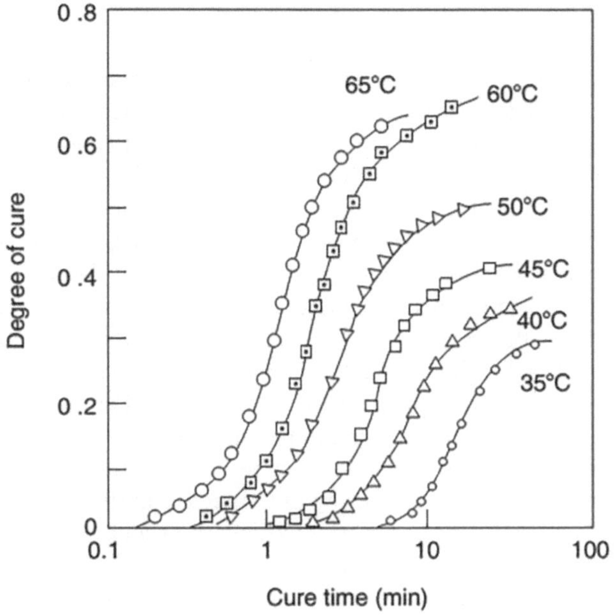

FIGURE 3.38 Degree of cure for a vinyl ester resin at various cure temperatures.

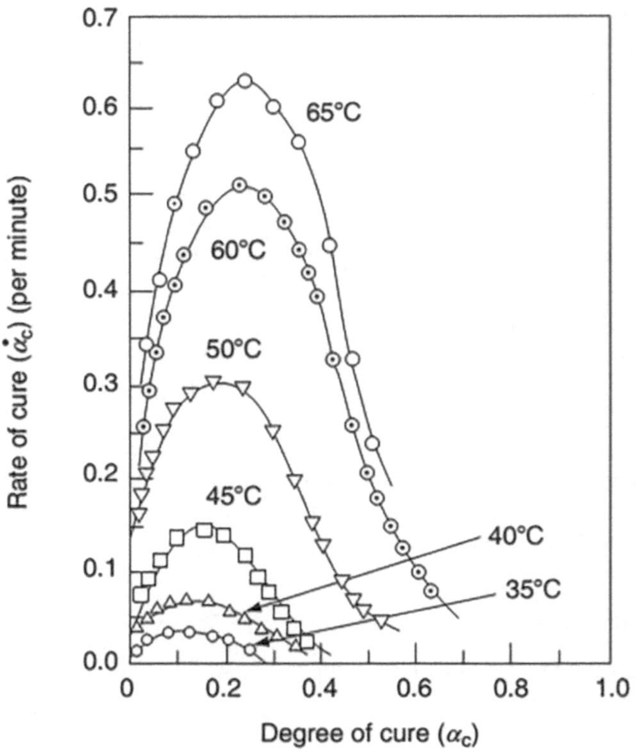

FIGURE 3.39 Rate of cure for a vinyl ester resin at various cure temperatures.

necessary for contact of reactive chemical groups; (3) heat drives off volatiles from solvents and water (otherwise, voids occur; note that volatiles will not outgas if pressure is also being applied); and (4) resin flows more easily to obtain a uniform distribution. Pressure is required to consolidate (debulk) the fiber and matrix system and to squeeze out excess resin. A typical curing cycle of temperature vs. time of epoxy resin is shown in Figure 3.40. The timescale is several hours, and the temperature scale is hundreds of °F (also hundreds of °C). The curing cycle starts with a gradual temperature increase under vacuum conditions so that volatiles and water (vapor) can be driven off. Then, the temperature is gradually increased to the maximum curing temperature which is held for a couple of hours to develop a high degree of cross-linking along with pressure application to consolidate the laminae.

Before curing, the initial form of the laminate is laminae laid adjacently in a B-staged condition (partially cured to reduce resin flow during lamination or molding). The resin is a semi-solid with negligible strength and stiffness. As the temperature is gradually increased, resin cross-linking begins and is significant when the gel temperature is reached (the temperature at which the viscosity is so high that no further dimensional change occurs). The progressive cross-linking causes solidification, but the elevated temperature causes softening and hence lowers stiffness. At the highest temperature reached (if the proper prescribed cure cycle is followed, which

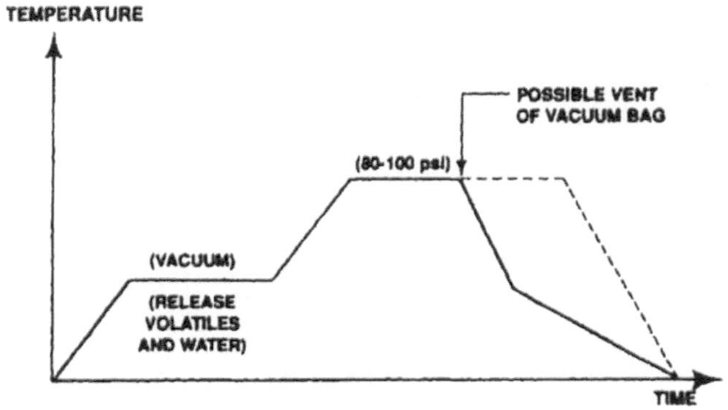

FIGURE 3.40 Cure cycle of epoxy resin.

also means that the temperature must be held for a specified time), cross-linking is nearly complete. The resin is now solidified, but is of low stiffness because of the high temperature. Then, the temperature is gradually decreased to room temperature over a period of about an hour to avoid thermal shock. The pressure can be released quickly. If post-curing is performed, no further cross-linking occurs unless the previous maximum temperature is exceeded and held for at least an hour (presuming the previous maximum temperature was held for an hour or so). Curing can be performed in several devices: heated mold, hot press (heated plates that are forced together), and an autoclave, which is essentially a very large version of an ordinary kitchen pressure cooker.

The curing process for thermoplastic matrix materials does not involve cross-linking but only melting and cooling. That is, a thermoplastic is already a solid that, like metals, can be heated to soften and cooled to stiffen. For some thermoplastic materials, a small degree of cross-linking occurs, so such thermoplastics cannot be cycled more than a few times through a heating–cooling cycle. Also, the time at elevated temperature [usually nearly 1000°F (500°C)] need not be but a few moments. The term "co-curing" means that two parts that must be fastened together are cured simultaneously and in contact to achieve permanent bonding between them. The process applies equally to thermoset matrix composite materials and to thermoplastic matrix composite materials (except the co-curing of two thermoplastic matrix parts is not, of course, permanent).

3.5.3 Viscosity

Viscosity of a fluid is a measure of its resistance to flow under shear stresses. Low molecular weight fluids, such as water and motor oil, have low viscosities and flow readily. High molecular weight fluids, such as polymer melts, have high viscosities and flow only under high stresses. The two most important factors determining the viscosity of a fluid are the temperature and shear rate. For all fluids, the viscosity decreases with increasing temperature. Shear rate does not have any influence on the viscosity of low molecular weight fluids, whereas it tends to either increase

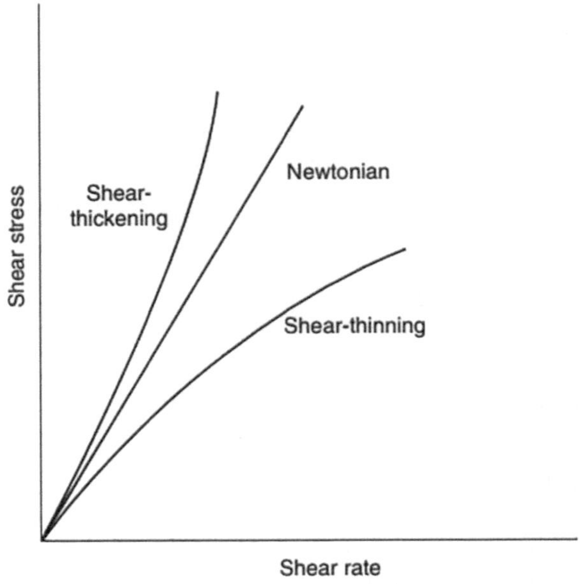

FIGURE 3.41 Shear stress vs. shear rate curves for various types of liquids. Note that the viscosity is defined as the slope of the shear stress–shear rate curve.

(shear thickening) or decrease (shear thinning) the viscosity of a high molecular weight fluids (Figure 3.41). Polymer melts, in general, are shear-thinning fluids since their viscosity decreases with increasing intensity of shearing.

The starting material for a thermoset resin is a low-viscosity fluid. However, its viscosity increases with curing and approaches a very large value as it transforms into a solid mass. Variation of viscosity during isothermal curing of an epoxy resin is shown in Figure 3.42. Similar viscosity–time curves are also observed for polyester and vinyl ester resins. In all cases, the viscosity increases with increasing cure time and temperature. The rate of viscosity increase is low at the early stage of curing. After a threshold degree of cure is achieved, the resin viscosity increases at a very rapid rate. The time at which this occurs is called the gel time. The gel time is an important molding parameter since the flow of resin in the mold becomes increasingly difficult at the end of this time period.

A number of important observations can be made from the viscosity data reported in the literature:

i. A B-staged or a thickened resin has a much higher viscosity than the neat resin at all stages of curing.
ii. The addition of fillers, such as $CaCO_3$, to the neat resin increases its viscosity, and the rate of viscosity increases during curing. On the other hand, the addition of thermoplastic additives (such as those added in low-profile polyester and vinyl ester resins) tends to reduce the rate of viscosity increase during curing.

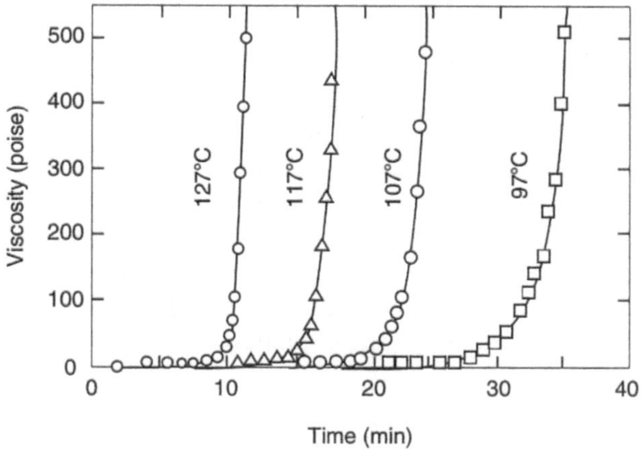

FIGURE 3.42 Variation of viscosity during isothermal curing of an epoxy resin.

iii. The increase in viscosity with cure time is less if the shear rate is increased. This phenomenon, known as shear thinning, is more pronounced in B-staged or thickened resins than in neat resins. Fillers and thermoplastic additives also tend to increase the shear-thinning phenomenon.

iv. The viscosity η of a thermoset resin during the curing process is a function of cure temperature T, shear rate $\dot{\gamma}$, and the degree of cure α_c:

$$\eta = \eta(T, \dot{\gamma}, \alpha_c). \tag{3.4}$$

The viscosity function for thermosets is significantly different from that for thermoplastics. Since no in situ chemical reaction occurs during the processing of a thermoplastic polymer, its viscosity depends on temperature and shear rate.

v. At a constant shear rate and for the same degree of cure, the η vs. $1/T$ plot is linear (Figure 3.43). This suggests that the viscous flow of a thermoset polymer is an energy-activated process. Thus, its viscosity as a function of temperature can be written as:

$$\eta = \eta_0 \exp\left(\frac{E}{RT}\right), \tag{3.5}$$

where
η = viscosity (Pa s or poise)
E = flow activation energy (cal/g mol)
R = universal gas constant
T = cure temperature (°K)
η_0 = constant.

The activation energy for viscous flow increases with the degree of cure and approaches a very high value near the gel point.

3.5.4 Resin Flow

Proper flow of resin through a dry fiber network (in liquid composite molding [LCM]) or a prepreg layup (in bag molding) is critical in producing void-free parts and good fiber wet-out. In thermoset resins, curing may take place simultaneously with resin flow, and if the resin viscosity rises too rapidly due to curing, its flow may be inhibited, causing voids and poor interlaminar adhesion.

Resin flow through fiber network has been modeled using Darcy's equation, which was derived for flow of Newtonian fluids through a porous medium. This equation relates the volumetric resin flow rate q per unit area to the pressure gradient that causes the flow to occur. For one-dimensional flow in the x direction:

$$q = \frac{P_0}{\eta}\left(\frac{dP}{dx}\right), \tag{3.6}$$

where

q = volumetric flow rate per unit area (m s^{-1}) in the x direction
P_0 = permeability (m^2)
η = viscosity (N s/m^2)
dp/dx = pressure gradient (N m^{-3}), which is negative in the direction of flow (positive x direction).

The permeability is determined by the following equation known as the Kozeny–Carman equation:

$$P_0 = \frac{d_f^2}{16K}\frac{(1-v_f)^3}{v_f^2}, \tag{3.7}$$

where

d_f = fiber diameter
v_f = fiber volume fraction
K = Kozeny constant.

Equations (3.6) and (3.7), although simplistic, have been used by many investigators in modeling resin flow from prepregs in bag-molding process and mold filling in RTM. Equation 3.7 assumes that the porous medium is isotropic, and the pore size and distribution are uniform. However, fiber networks are nonisotropic, and therefore, the Kozeny constant, K, is not the same in all directions. For example, for a fiber network with unidirectional fiber orientation, the Kozeny constant in the transverse direction (K_{22}) is an order of magnitude higher than the Kozeny constant in the longitudinal direction (K_{11}). This means that the resin flow in the transverse direction is much lower than that in the longitudinal direction. Furthermore, the fiber packing in a fiber network is not uniform, which also affects the Kozeny constant, and therefore the resin flow.

Equation (3.7) works well for predicting resin flow in the fiber direction. However, Eq. (3.7) is not valid for resin flow in the transverse direction, since according to this equation resin flow between the fibers does not stop even when the fiber volume fraction reaches the maximum value at which the fibers touch each other and there are no gaps between them. Gebart [19] derived the following permeability equations in the fiber direction and normal to the fiber direction for unidirectional continuous fiber network with regularly arranged, parallel fibers.

In the fiber direction:

$$P_{11} = \frac{2d_f^2}{C_1} \frac{\left(1 - v_f^3\right)}{v_f^2} \tag{3.8}$$

Normal to the fiber direction:

$$P_{22} = C_2 \left(\sqrt{\frac{v_{f,\max}}{v_f}} - 1 \right)^{5/2} \frac{d_f^2}{4}, \tag{3.9}$$

where

C_1 = hydraulic radius between the fibers

C_2 = a constant

$v_{f,\max}$ = maximum fiber volume fraction (i.e., at maximum fiber packing).

The parameters C_1, C_2, and $v_{f,\max}$ depend on the fiber arrangement in the network. For a square arrangement of fibers, $C_1 = 57$, $C_2 = 0.4$, and $v_{f,\max} = 0.785$. For a hexagonal arrangement of fibers, $C_1 = 53$, $C_2 = 0.231$, and $v_{f,\max} = 0.906$. Note that Eq. (3.8) for resin flow parallel to the fiber direction has the same form as the Kozeny–Carman Eq. (3.7). According to Eq. (3.9), which is applicable for resin flow transverse to the flow direction, $P_{22} = 0$ at $v_f = v_{f,\max}$, and therefore, the transverse resin flow stops at the maximum fiber volume fraction.

The permeability equations assume that the fiber distribution is uniform, the gaps between the fibers are the same throughout the network, the fibers are perfectly aligned, and all fibers in the network have the same diameter. These assumptions are not valid in practice, and therefore, the permeability predictions using Eq. (3.7) or (3.8, 3.9) can only be considered approximate.

3.5.5 Consolidation

Consolidation of layers in a fiber network or a prepreg layup requires good resin flow and compaction; otherwise, the resulting composite laminate may contain a variety of defects, including voids, interply cracks, resin-rich areas, or resin-poor areas. Good resin flow by itself is not sufficient to produce good consolidation [20]. Both resin flow and compaction require the application of pressure during processing in a direction normal to the dry fiber network or prepreg layup. The pressure is applied to squeeze out the trapped air or volatiles, as the liquid resin flows through the fiber network or prepreg layup, suppresses voids, and attains a uniform fiber volume fraction. Gutowski et al. [21] developed a model for consolidation in which it is assumed that the applied pressure is shared by the fiber network and the resin so that:

$$p = \sigma + \overline{p_r}, \tag{3.10}$$

where

p = applied pressure
σ = average effective stress on the fiber network
$\overline{p_r}$ = average pressure on the resin.

The average effective pressure on the fiber network increases with increasing fiber volume fraction and is given by:

$$\sigma = A \frac{1 - \sqrt{\frac{v_f}{v_0}}}{\left(\sqrt{\frac{v_a}{v_f}} - 1 \right)^4}, \tag{3.11}$$

where

A is a constant.
v_0 is the initial fiber volume fraction in the fiber network (before compaction).
v_f is the fiber volume fraction at any instant during compaction.
v_a is the maximum possible fiber volume fraction.

The constant A in Eq. (3.11) depends on the fiber stiffness and the fiber waviness, and is a measure of the deformability of the fiber network. Since the fiber volume fraction, v_f, increases with increasing compaction, Eq. (3.11) predicts that σ also increases with increasing compaction; that is, the fiber network begins to take up an increasing amount of the applied pressure. On the other hand, the average pressure on the resin decreases with increasing compaction, which can lead to void formation.

3.5.6 Gel-Time Test

The curing characteristics of a resin–catalyst combination are frequently determined by the gel-time test. In this test, a measured amount (10 g) of a thoroughly mixed resin–catalyst combination is poured into a standard test tube. The temperature rise in the material is monitored as a function of time by means of a thermocouple while the test tube is suspended in a 82°C (180°F) water bath. A typical temperature–time curve (also known as exotherm curve) obtained in a gel-time test is illustrated in Figure 3.44. On this curve, point A indicates the time required for the resin–catalyst mixture to attain the bath temperature. The beginning of temperature rise indicates the initiation of the curing reaction. As the curing reaction begins, the liquid mix begins to transform into a gel-like mass. Heat generated by the exothermic curing reaction increases the mix temperature, which in turn causes the catalyst to decompose at a faster rate and the reaction to proceed at a progressively increasing speed. Since the rate of heat generation is higher than the rate of heat loss to the surrounding medium, the temperature rises rapidly to high values. As the curing reaction nears completion, the rate of heat generation is reduced and a decrease in temperature follows. The exothermic peak temperature observed in a gel-time test is a function of the resin chemistry (level of unsaturation) and the resin–catalyst ratio. The slope of

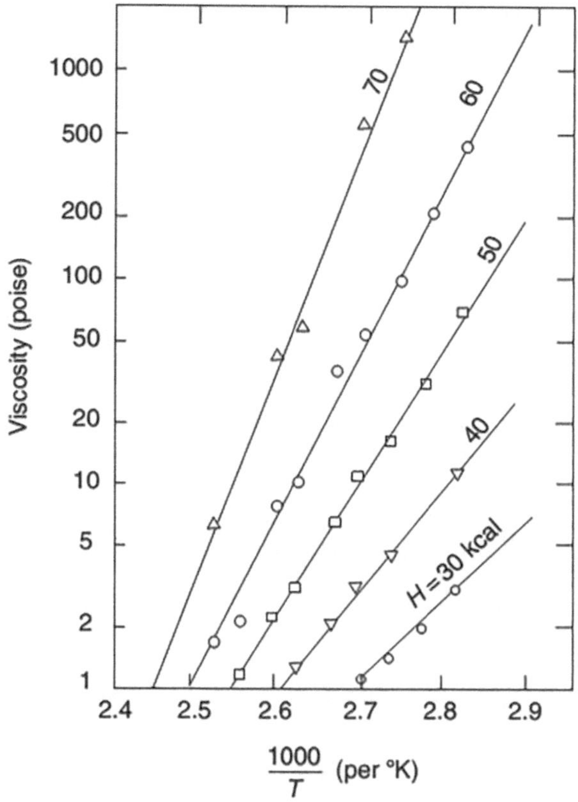

FIGURE 3.43 Viscosity–temperature relationships for an epoxy resin at different levels of cure.

the exotherm curve is a measure of cure rate, which primarily depends on the catalyst reactivity.

Shortly after the curing reaction begins at point A, the resin viscosity increases very rapidly owing to the increasing number of cross-links formed by the curing reaction. The time at which a rapid increase in viscosity ensues is called the gel time and is indicated by point B in Figure 3.44. According to one standard, the time at which the exotherm temperature increases by 5.5°C (10°F) above the bath temperature is considered the gel time. It is sometimes measured by probing the surface of the reacting mass with a clean wooden applicator stick every 15 seconds until the reacting material no longer adheres to the end of a clean stick.

3.5.7 Shrinkage

Shrinkage is the reduction in volume or linear dimensions caused by curing as well as thermal contraction. Curing shrinkage occurs because of the rearrangement of polymer molecules into a more compact mass as the curing reaction proceeds. The thermal shrinkage occurs during the cooling period that follows the curing reaction and

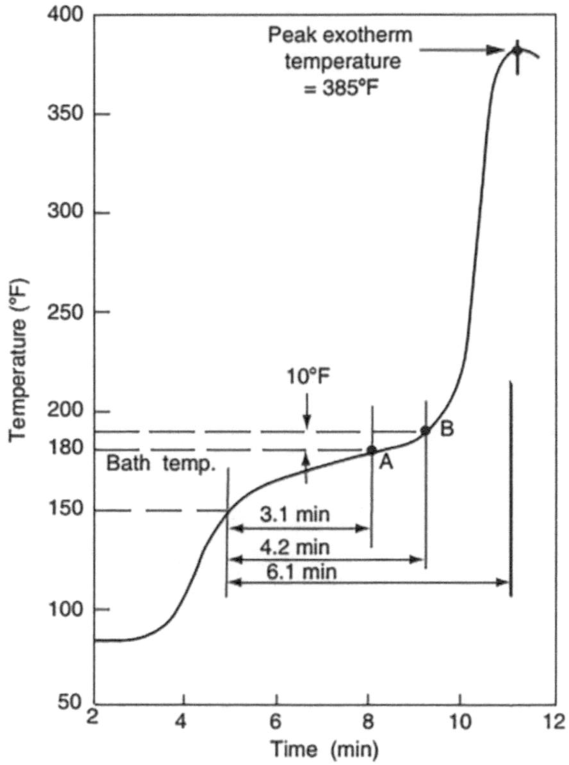

FIGURE 3.44 Typical temperature–time curve obtained in a gel-time test.

may take place both inside and outside the mold. The volumetric shrinkage for cast-epoxy resins is of the order of 1%–5%, and that for polyester and vinyl ester resins may range from 5% to 12%. The addition of fibers or fillers reduces the volumetric shrinkage of a resin. However, in the case of unidirectional fibers, the reduction in shrinkage in the longitudinal direction is higher than in the transverse direction.

High shrinkage in polyester or vinyl ester resins can be reduced significantly by the addition of low shrink additives (also called low-profile agents), which are thermoplastic polymers, such as polyethylene, polymethyl acrylate, polyvinyl acetate, and polycaprolactone. These thermoplastic additives are usually mixed in styrene monomer during blending with the liquid resin. On curing, the thermoplastic polymer becomes incompatible with the cross-linked resin and forms a dispersed second phase in the cured resin. High resin shrinkage is desirable for easy release of the part from the mold surface; however, at the same time, high resin shrinkage can contribute to many molding defects, such as warpage and sink marks.

3.5.8 Voids

Among the various defects produced during the molding of a composite laminate, the presence of voids is considered the most critical defect in influencing its mechanical properties. The most common cause for void formation is the inability of the resin to displace air from the fiber surface during the time fibers are coated with the liquid resin. The rate at which the fibers are pulled through the liquid resin, the resin viscosity, the relative values of fiber and resin surface energies, and the mechanical manipulation of fibers in the liquid resin affect air entrapment at the fiber–resin interface. Voids may also be caused by air bubbles and volatiles entrapped in the liquid resin. Solvents used for resin viscosity control, moisture, and chemical contaminants in the resin, as well as styrene monomer, may remain dissolved in the resin mix and volatilize during elevated temperature curing. In addition, air is also entrapped between various layers during the lamination process.

Much of the air or volatiles entrapped at the premolding stages can be removed by (1) degassing the liquid resin, (2) applying vacuum during the molding process, and (3) allowing the resin mix to flow freely in the mold, which helps in carrying the air and volatiles out through the vents in the mold. The various process parameters controlling the resin flow are described in later sections. The presence of large volume fractions of voids in a composite laminate can significantly reduce its tensile, compressive, and flexural strengths. Large reductions in interlaminar shear strength are observed even if the void content is only 2%–3% by volume (Figure 3.45). The presence of voids generally increases the rate and amount of moisture absorption in a humid environment, which in turn increases the physical dimensions of the part and reduces its matrix-dominated properties.

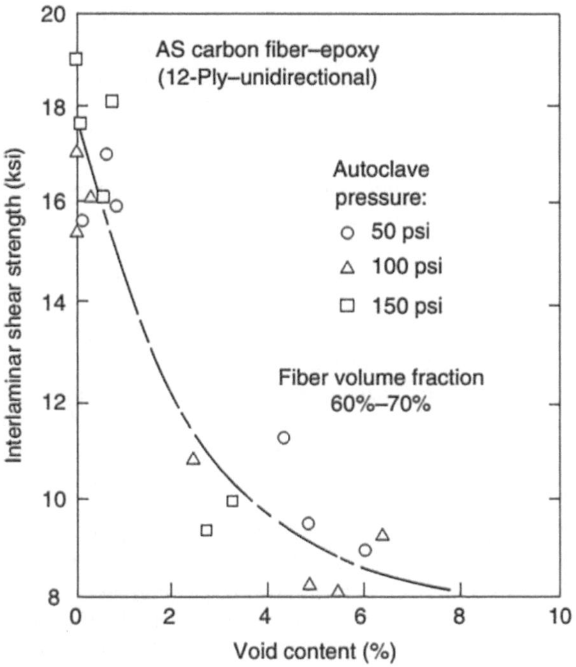

FIGURE 3.45 Effect of void volume fraction on the interlaminar shear strength of a composite laminate.

REFERENCES

1. Ueno T. and Yoshioka H. Japanese Patent JP 4441768.
2. Hikosaka T., Miki K., Nishida Y. (1989), "Mechanical properties of aluminum-alumina particle composites fabricated by vortex method", *Imono-Journal of Japan Foundry Engineering Society*, Vol. 61, pp. 780–786.
3. Badia F.A. and Rohatgi P.K. (1969), "Dispersion of graphite particles in aluminium castings through injection of melt", *AFS Transactions*, Vol. 77, pp. 402–406.
4. Suwa M., Komuro K., Soeno K. (1976), "Mechanical properties and wear resistance of graphite dispersed Al–Si alloys", *Journal of the Japan Institute of Metals and Materials*, Vol. 40, pp. 1074–1081.
5. Gebart B.R. (1992), "Permeability of unidirectional reinforcements in RTM", *Journal of Composite Materials*, Vol. 26, pp. 1100–1133.
6. Nagelberg A.S., Antolin S., Urquhart A.W. (1992), "Formation of Al_2O_3/metal composites by the directed oxidation of molten aluminum–magnesium–silicon alloys: part II, growth kinetics", *Journal of the American Ceramic Society*, Vol. 75, pp. 455–462.
7. Nakanishi H., Tsunekawa Y., Mohri N., Okumiya M., Niimi I. (1993), "Ultrasonic infiltration in alumina particle/molten aluminum system", *Journal of Japan Institute of Light Metals*, Vol. 43, pp. 14–19.
8. Nakanishi H., Tsunekawa Y., Okumiya M., Mohri N. (1993), "Ultrasonic infiltration in alumina fiber/molten aluminum system", *Materials transactions JIM*, Vol. 34, pp. 62–68.
9. Deming Y., Xinfang Y., Jin P. (1993), "Continuous yarn fibre-reinforced aluminium composites prepared by the ultrasonic liquid infiltration method", *Journal of Materials Science Letters*, Vol. 12, pp. 252–253.
10. Cheng H.M., Lin Z.H., Zhou B.L., Zhen Z.G., Kobayashi K., Uchiyama Y. (1993), "Preparation of carbon fibre reinforced aluminum via ultrasonic liquid infiltration technique", *Materials Science and Technology*, Vol. 9, pp. 609–614.
11. Matsunaga T., Matsuda K., Hatayama T., Shinozaki K., Amanuma S., Jin P., Yoshida M. (2006), "Development in manufacturing of carbon fiber reinforced aluminum preform wires using ultrasonic infiltration method", *Journal of Japan Institute of Light Metals*, Vol. 56, pp. 28–33.
12. Matsunaga T., Ogata K., Hatayama T., Shinozaki K., Yoshida M. (2006), "Infiltration mechanism of molten aluminum alloys into bundle of carbon fibers using ultrasonic infiltration method", *Journal of Japan Institute of Light Metals*, Vol. 56, pp. 226–232.
13. Matsunaga T., Ogata K., Hatayama T., Shinozaki K., Yoshida M. (2007), "Effect of acoustic cavitation on ease of infiltration of molten aluminum alloys into carbon fiber bundles using ultrasonic infiltration method", *Composites Part A*, Vol. 38, pp. 771–778.
14. Matsunaga T., Matsuda K., Hatayama T., Shinozaki K., Yoshida M. (2007), "Fabrication of continuous carbon fiber-reinforced aluminum–magnesium alloy composite wires using ultrasonic infiltration", *Composites Part A*, Vol. 38, pp. 1902–1911.
15. Mizoguchi I., Yamaguchi S., Yachi S., Yoshida M. (2010), "Influence of high temperature holding on tensile strength of pitch-based carbon fiber reinforced Al–Mg alloy composites fabricated by ultrasonic infiltration method", *Journal of Japan Institute of Light Metals*, Vol. 60, pp. 396–402.
16. Yamaguchi S., Mikuni J., Mizoguchi I., Matsunaga T., Shinozaki K., Yoshida M. (2009), "Influence of high temperature holding on tensile strength of PAN-based carbon fiber reinforced aluminum–magnesium alloy composites fabricated by ultrasonic infiltration method", *Journal of Japan Institute of Light Metals*, Vol. 59, pp. 241–247.
17. Mikuni J., Nonokawa K., Matsunaga T., Shinozaki K., Yoshida M. (2008), "Influence of interfacial chemical reaction for tensile strength of carbon fiber reinforced aluminum–magnesium alloy composites", *Journal of Japan Institute of Light Metals*, Vol. 58, pp. 27–32.

18. Matsunaga T., Matsuda K., Hatayama T., Shinozaki K., Amanuma S., Yoshida M. (2006), "Effect of magnesium content on tensile strength of carbon-fiber-reinforced aluminum–magnesium alloy composite wires fabricated by ultrasonic infiltration method", *Journal of Japan Institute of Light Metals*, Vol. 56, pp. 105–111.
19. Solzbacher F. (2005), "Physical vapor deposition," in *Semiconductor Manufacturing Handbook* (H. Geng, ed.), McGraw-Hill, New York (Chapter 13).
20. Hubert P. and Poursartip A. (1998), "A review of flow and compaction modelling relevant to thermoset matrix laminate processing", *Journal of Reinforced Plastics and Composites*, Vol. 17, pp. 286–318.
21. Gutowski T.G., Morgaki T., Cai Z. (1987), "The consolidation of laminate composites", *Journal of Composite Materials*, Vol. 21, pp. 172–188.

4 Mechanics of Composites

Composite materials have many mechanical behavior characteristics that are different from those of more conventional engineering materials. Some characteristics are merely modifications of conventional behavior; others are totally new and require new analytical and experimental procedures. Most common engineering materials are both homogeneous and isotropic.

i. A homogeneous body has uniform properties throughout; that is, the properties are independent of position in the body.
ii. An isotropic body has material properties that are the same in every direction at a point in the body; that is, the properties are independent of orientation at a point in the body.

Bodies with temperature-dependent isotropic material properties are not homogeneous when subjected to a temperature gradient, but still are isotropic. In contrast, composite materials are often both inhomogeneous (or nonhomogeneous or heterogeneous – the three terms can be used interchangeably) and non-isotropic (orthotropic or, more generally, anisotropic, but the words are not interchangeable).

i. An inhomogeneous body has nonuniform properties over the body; that is, the properties depend on position in the body.
ii. An orthotropic body has material properties that are different in three mutually perpendicular directions at a point in the body and, further, has three mutually perpendicular planes of material property symmetry. Thus, the properties depend on orientation at a point in the body.
iii. An anisotropic body has material properties that are different in all directions at a point in the body. No planes of material property symmetry exist. Again, the properties depend on orientation at a point in the body.

Some composite materials have very simple forms of inhomogeneity. For example, laminated safety glass has three layers, each of which is homogeneous and isotropic; thus, the inhomogeneity of the composite material is a step function in the direction perpendicular to the plane of the glass. Also, some particulate composite materials are inhomogeneous, yet isotropic, although some are orthotropic and others are anisotropic. Other composite materials are typically more complex, especially those with fibers placed at many angles in space. Because of the inherently heterogeneous nature of composite materials, they are conveniently studied from two points of view: micromechanics and macromechanics.

Micromechanics is the study of composite material behavior wherein the interaction of the constituent materials is examined on a microscopic scale to determine their effect on the properties of the composite material.

Macromechanics is the study of composite material behavior wherein the material is presumed homogeneous and the effects of the constituent materials are detected only as averaged apparent macroscopic properties of the composite material.

Use of the two concepts of macromechanics and micromechanics allows the tailoring of a composite material to meet a particular structural requirement with little waste of material capability. The ability to tailor a composite material to its job is one of the most significant advantages of a composite material over an ordinary material. Perfect tailoring of a composite material yields only the stiffness and strength required in each direction, no more. In contrast, an isotropic material is, by definition, constrained to have excess strength and stiffness in any direction other than that of the largest required strength or stiffness. The inherent anisotropy (most often only orthotropy) of composite materials leads to mechanical behavior characteristics that are quite different from those of conventional isotropic materials. The behavior of isotropic, orthotropic, and anisotropic materials under loadings of normal stress and shear stress is shown in Figure 4.1 and discussed in the following paragraphs.

For isotropic materials, application of normal stress causes extension in the direction of the stress and contraction in the perpendicular directions, but no shearing deformation. Also, application of shear stress causes only shearing deformation, but no extension or contraction in any direction. Only two material properties, Young's modulus (the extensional modulus or slope of the material's stress–strain curve) and Poisson's ratio (the negative ratio of lateral contraction strain to axial extensional strain caused by axial extensional stress), are needed to quantify the deformations. The shear modulus (the ratio of shear stress to shear strain at a point) could be used as an alternative to either Young's modulus or Poisson's ratio. For orthotropic materials, like isotropic materials, application of normal stress in a principal material direction (along one of the intersections of three orthogonal planes of material symmetry) results in extension in the direction of the stress and contraction perpendicular

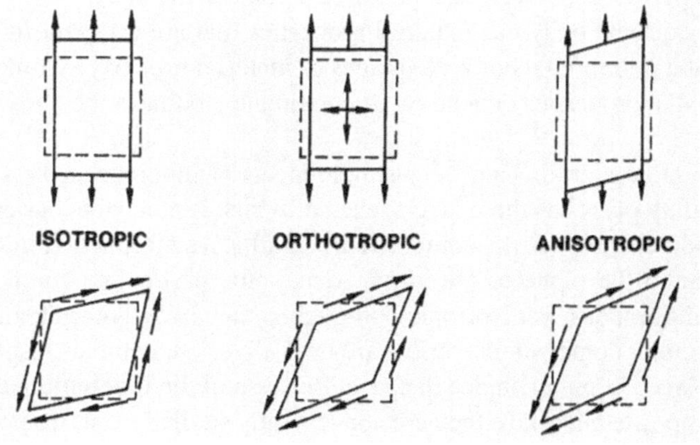

FIGURE 4.1 Mechanical behavior of various materials.

to the stress. The magnitude of the extension in one principal material direction under normal stress in that direction is different from that of the extension in another principal material direction under the same normal stress in that other direction. Thus, different Young's moduli exist in the various principal material directions. In addition, because of different properties in the two principal material directions, the contraction can be either more or less than the contraction of a similarly loaded isotropic material with the same elastic modulus in the direction of the load. Thus, different Poisson's ratios are associated with different pairs of principal material directions (and with the order of the coordinate direction numbers designating the pairs). Application of shear stress causes shearing deformation, but the magnitude of the shearing deformation is totally independent of the various Young's moduli and Poisson's ratios. That is, the shear modulus of an orthotropic material is, unlike isotropic materials, not dependent on other material properties. Thus, at least five material properties are necessary to describe the mechanical behavior of orthotropic materials.

For anisotropic materials, application of a normal stress leads not only to extension in the direction of the stress and contraction perpendicular to it, but also to shearing deformation. Conversely, application of shearing stress causes extension and contraction in addition to the distortion of shearing deformation. This coupling between both loading modes and both deformation modes, that is, shear–extension coupling, is also characteristic of orthotropic materials subjected to normal stress in a non-principal material direction. For example, cloth is an orthotropic material composed of two sets of interwoven fibers at right angles to each other. If cloth is subjected to a normal stress at 45° to a fiber direction, both stretching and distortion occur, as can easily be demonstrated by the reader. Even more material properties than for orthotropic materials are necessary to describe the mechanical behavior of anisotropic materials because of the additional response characteristics.

Coupling between deformation modes and types of loading creates problems that are not easily overcome and, at the very least, cause a reorientation of thinking. For example, the conventional American Society for Testing and Materials (ASTM) dog-bone tensile specimen shown in Figure 4.2 obviously cannot be used to determine the tensile moduli of orthotropic materials loaded in non-principal material directions (nor of anisotropic materials). For an isotropic material, loading on a dog-bone specimen is actually a prescribed lengthening that is only coincidentally a prescribed stress because of the symmetry of an isotropic material.

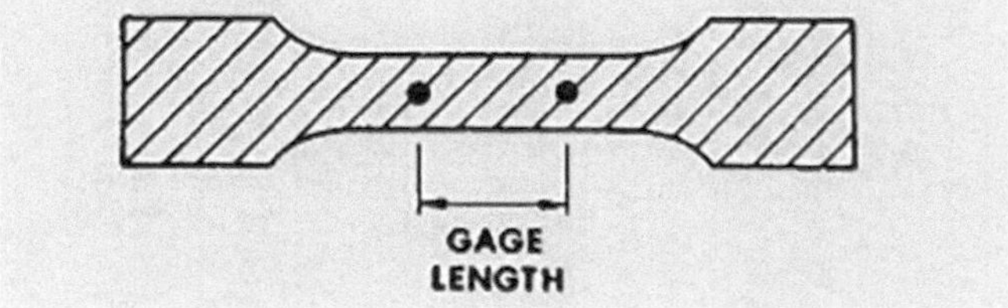

FIGURE 4.2 ASTM dog-bone specimen for tensile testing.

However, for an off-axis-loaded orthotropic material or an anisotropic material, only the prescribed lengthening occurs because of the lack of symmetry of the material about the loading axis and the clamped ends of the specimen. Accordingly, shearing stresses result in addition to normal stresses in order to counteract the natural tendency of the specimen to shear. Furthermore, the specimen has a tendency to bend. Thus, the strain measured in the specimen gage length in Figure 4.2 cannot be used with the axial load to determine the axial stiffness or modulus. Accordingly, techniques more sophisticated than the ASTM dog-bone test must typically be used to determine the mechanical properties of a composite material.

4.1 LAMINAE

The basic building block of a laminate is a lamina which is a flat (sometimes curved as in a shell) arrangement of unidirectional fibers or woven fibers in a matrix. Two typical flat laminae along with their principal material axes that are parallel and perpendicular to the fiber direction are shown in Figure 4.3. The fibers are the principal reinforcing or load-carrying agent and are typically strong and stiff. The matrix can be organic, metallic, ceramic, or carbon. The function of the matrix is to support and protect the fibers and to provide a means of distributing load among, and transmitting load between, the fibers. The latter function is especially important if a fiber breaks as in Figure 4.4. There, load from one portion of a broken fiber is transferred to the matrix and, subsequently, to the other portion of the broken fiber as well as to adjacent fibers. The mechanism for load transfer is the shearing stress developed in the matrix; the shearing stress resists the pulling out of the broken fiber. This load-transfer mechanism is the means by which whisker-reinforced composite materials carry any load at all above the inherent matrix strength.

The properties of the lamina constituents, the fibers and the matrix, have been only briefly discussed so far. Their stress–strain behavior is typified as one of the four classes depicted in Figure 4.5. Fibers generally exhibit a linear elastic behavior, although reinforcing steel bars in concrete are more nearly elastic–perfectly plastic. Aluminum, as well as many polymers, and some composite materials exhibit elastic–plastic behavior that is really nonlinear elastic behavior if there is no unloading. Commonly, resinous matrix materials are viscoelastic if not viscoplastic; that is, they have strain rate dependence and linear or nonlinear stress–strain behavior. The various stress–strain relations are sometimes referred to as constitutive relations because

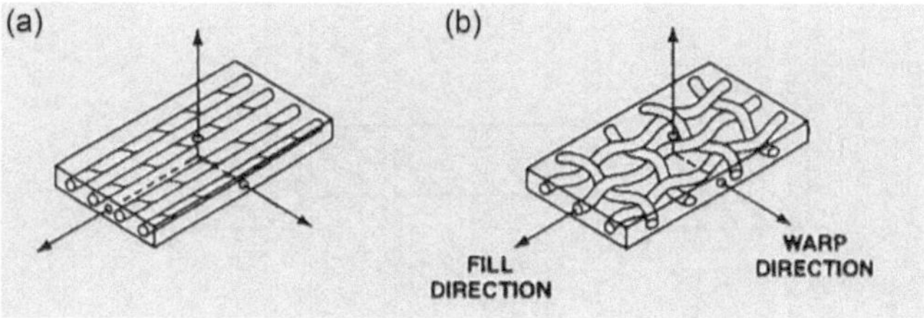

FIGURE 4.3 Two principal types of laminae: (a) Lamina with unidirectional fibers and (b) lamina with woven fibers.

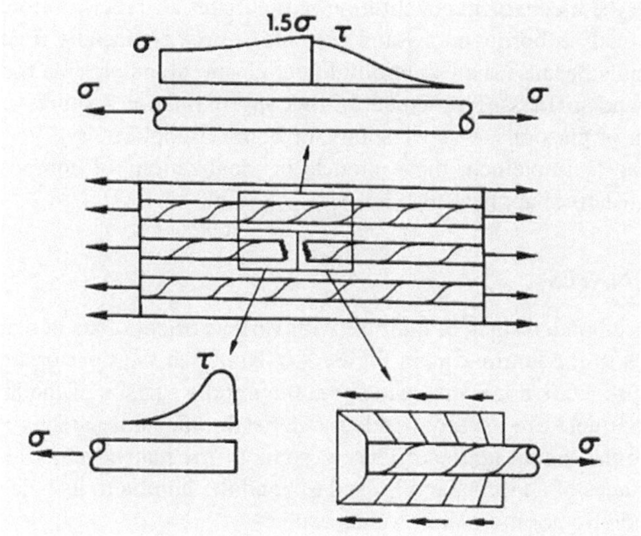

FIGURE 4.4 Effect of broken fiber on matrix and fiber stresses.

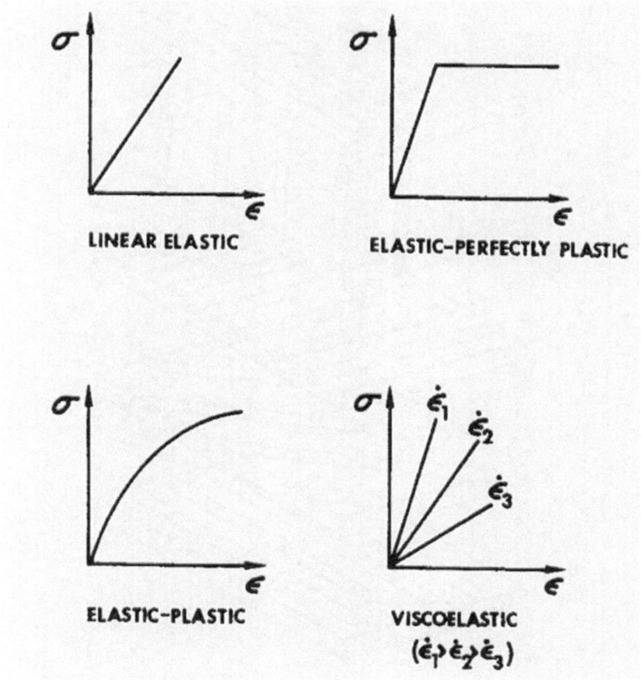

FIGURE 4.5 Various stress–strain behaviors.

they describe the mechanical constitution of the material. Fiber-reinforced composite materials such as boron–epoxy and graphite–epoxy are usually treated as linear elastic materials because the essentially linear elastic fibers provide the majority of the strength and stiffness. Refinement of that approximation requires consideration of some form of plasticity, viscoelasticity, or both (viscoplasticity). Very little work has been done to implement those models or idealizations of composite material behavior in structural applications.

4.2 LAMINATES

A laminate is a bonded stack of laminae with various orientations of principal material directions in the laminae as in Figure 4.6. Note that the fiber orientation of the layers in Figure 4.6 is not symmetric about the middle surface of the laminate. The layers of a laminate are usually bonded together by the same matrix material that is used in the individual laminae. That is, some of the matrix material in a lamina coats the surfaces of a lamina and is used to bond the lamina to its adjacent laminae without the addition of more matrix material.

Laminates can be composed of plates of different materials or, in the present context, layers of fiber-reinforced laminae. A laminated circular cylindrical shell can

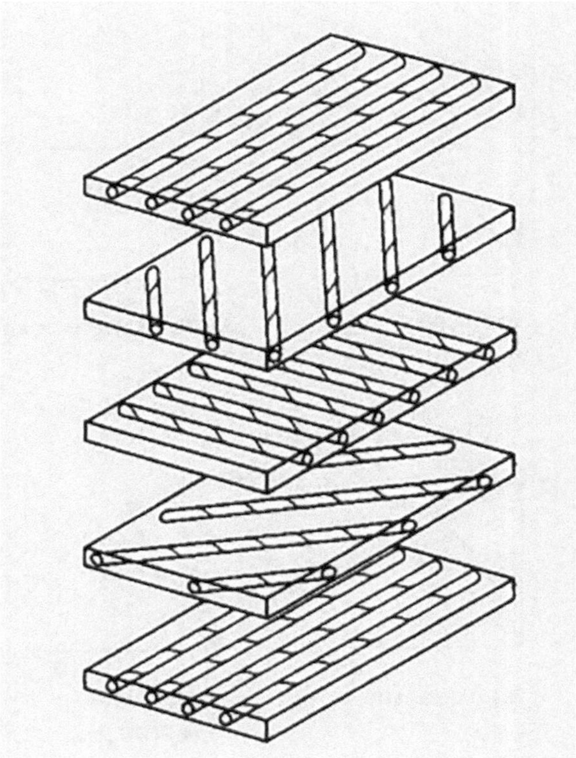

FIGURE 4.6 Unbonded view of laminate construction.

be constructed by winding resin-coated fibers on a removable core structure called a mandrel first with one orientation to the shell axis, then another, and so on until the desired thickness is achieved.

A major purpose of lamination is to tailor the directional dependence of strength and stiffness of a composite material to match the loading environment of the structural element. Laminates are uniquely suited to this objective because the principal material directions of each layer can be oriented according to need. For example, six layers of a ten-layer laminate could be oriented in one direction and the other four at 90° to that direction; the resulting laminate then has a strength and extensional stiffness roughly 50% higher in one direction than in the other. The ratio of the extensional stiffnesses in the two directions is approximately 6:4, but the ratio of bending stiffnesses is unclear because the order of lamination is not specified in the example. Moreover, if the laminae are not arranged symmetrically about the middle surface of the laminate, the result is stiffnesses that represent coupling between bending and extension.

4.3 TENSORS

Tensors are mathematical representations of physical quantities. They have components that change from one coordinate system to another (e.g., unprimed coordinates to primed coordinates in Figure 4.7) according to the so-called transformation equations. We will be concerned only with rectangular Cartesian coordinates and hence Cartesian tensors. Tensors usually are written using index notation, with the order of a tensor indicated by the number of live (non-repeated) subscripts. If a subscript is repeated, summation over the range of that subscript is implied, unless otherwise indicated. For rectangular Cartesian coordinates, the subscripts have a range of 1, 2, 3 in three dimensions and a range of 1, 2 in two dimensions. Examples of the types of tensor quantities used in this book are presented in Table 4.1.

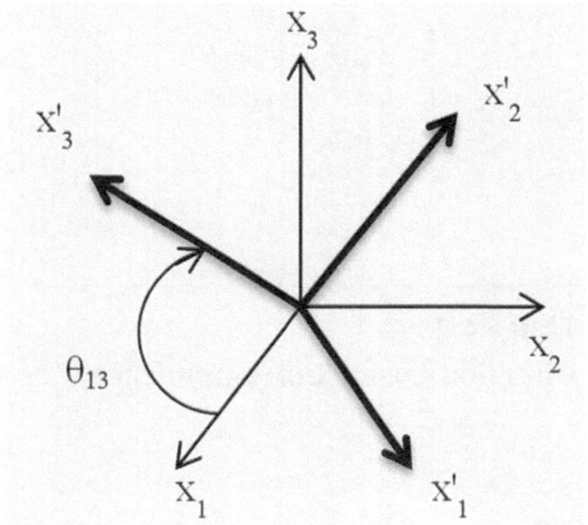

FIGURE 4.7 Rectangular Cartesian coordinates.

TABLE 4.1
Examples of Tensor Quantities

Example	Quantity	Live Subscripts
F	Scalar, zeroth-order tensor	0
σ_{ii}	Scalar: $(\sigma_{11} + \sigma_{22} + \sigma_{33})$	0
u_i	Vector: first-order tensor	1
σ_{ij}	Second-order tensor	2
C_{ijkl}	Fourth-order tensor	4

The equations for transforming a tensor quantity from one coordinate system to another are written in terms of the direction cosines a_{ij} $(i, j = 1, 2, 3)$ of the angles measured from the unprimed axes, x_i to the primed axes, x'_i (e.g., $a_{13} = \cos\theta_{13}$ in Figure 4.7). In this book, we shall use the convention that the first subscript (i) of a_{ij} corresponds to the initial, unprimed axes and the second subscript (j) corresponds to the final, primed axes. (The reader is forewarned that some authors use the opposite convention for a_{ij}; that is, the first subscript corresponds to the final, primed axes, and the second subscript corresponds to the initial, unprimed axes.) For the fully three-dimensional case, the direction cosines for coordinate transformation can be depicted conveniently in Table 4.2.

The a_{ij} for transformation in a plane (coordinate rotation through an angle θ about an axis normal to the plane) are depicted in Figure 4.8. From this figure, it is apparent that $\angle(x_1 \text{ to } x'_2) \neq \angle(x'_1 \text{ to } x_2)$, and hence, it is important that the convention for a_{ij} be known and understood. The individual a_{ij} for this planar rotation of coordinate systems about the x_3 axis are listed in Figure 4.8. All components of a_{ij} are also given in terms of the sine or cosine of the angle, $\theta = \theta_{11}$, for this special case of rotation about an axis.

In matrix form, the transformation coefficients a_{ij} for rotation about the three axes are:

$$a_{ij} = \begin{bmatrix} \cos\theta & -\sin\theta & 0 \\ \sin\theta & \cos\theta & 0 \\ 0 & 0 & 1 \end{bmatrix}. \tag{4.1}$$

TABLE 4.2
Direction Cosine Correspondence

		x'_2	x'_3
x_1	a_{11}	a_{12}	a_{13}
x_2	a_{21}	a_{22}	a_{23}
x_3	a_{31}	a_{32}	a_{33}

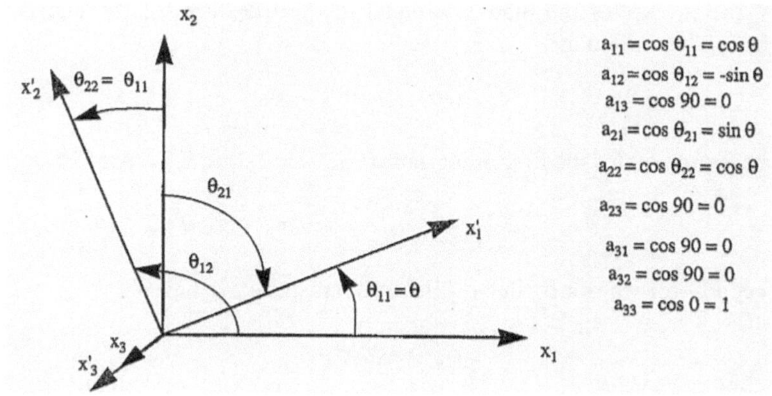

FIGURE 4.8 Transformation about an axis.

It is common practice to define $m = \cos\theta$ and $n = \sin\theta$, and thus, (4.1) can be written as:

$$a_{ij} = \begin{bmatrix} m & -n & 0 \\ n & m & 0 \\ 0 & 0 & 1 \end{bmatrix}. \tag{4.2}$$

Using the definitions of a_{ij} in the previous paragraphs, the transformation equation for a vector V_i' (in a rotated coordinate system) in terms of the components V_i in the original, unrotated coordinate system is:

$$V_i' = a_{ji}V_j. \tag{4.3}$$

Expanding Eq. (4.3) for the repeated subscript j, over the range 1,2,3 gives:

$$V_i' = a_{1i}V_1 + a_{2i}V_2 + a_{3i}V_3\,. \tag{4.4}$$

And writing the transformation equations explicitly in terms of direction cosines, we have:

$$V_i' = \cos\theta_{1i}V_1 + \cos\theta_{2i}V_2 + \cos\theta_{3i}V_3\,. \tag{4.5}$$

Equation (4.3) can be inverted and written as:

$$V_j = a_{ji}^{-1}V_i'. \tag{4.6}$$

Also, from Eq. (4.1), we have:

$$a_{ji}^{-1} = a_{ij} = \left(a_{ji}\right)^T. \tag{4.7}$$

That is, the inverse of the matrix is equal to the transform of the matrix. Thus, Eq. (4.6) can be written as:

$$V_j = a_{ij}V_i'. \tag{4.8}$$

The transformation for the first-order tensor u_i $(i, j = 1,2)$ in 2-D space is:

$$u_i' = a_{ji}u_j = a_{1i}u_1 + a_{2i}u_2 = \cos\theta_{1i}u_1 + \cos\theta_{2i}u_2 . \tag{4.9}$$

For a second-order tensor σ_{ij} in two-dimensional space, we have:

$$\sigma_{ij}' = a_{ki}a_{lj}\sigma_{kl} \tag{4.10}$$

$$\sigma_{ij}' = a_{1i}a_{1j}\sigma_{11} + a_{1i}a_{2j}\sigma_{12} + a_{2i}a_{1j}\sigma_{21} + a_{2i}a_{2j}\sigma_{22}$$

$$\sigma_{ij}' = \cos\theta_{1i}\cos\theta_{1j}\sigma_{11} + \cos\theta_{1i}\cos\theta_{2j}\sigma_{12} + \cos\theta_{2i}\cos\theta_{1j}\sigma_{21} + \cos\theta_{2i}\cos\theta_{2j}\sigma_{22}.$$

The following gives the expanded transformation expression for a second-order tensor σ_{ij} in three-dimensional space (i, j-1, 2, 3) for a rotational transformation about an axis:

$$\begin{aligned}\sigma_{ij}' &= a_{ki}a_{lj}\sigma_{kl}\\ &= a_{1i}a_{1j}\sigma_{11} + a_{1i}a_{2j}\sigma_{12} + a_{1i}a_{3j}\sigma_{13} + a_{2i}a_{1j}\sigma_{21} + a_{2i}a_{2j}\sigma_{22} + a_{2i}a_{3j}\sigma_{23}\\ &\quad + a_{3i}a_{1j}\sigma_{31} + a_{3i}a_{2j}\sigma_{32} + a_{3i}a_{3j}\sigma_{33}.\end{aligned} \tag{4.11}$$

It follows from the definition of Eq. (4.1) that

$$a_{ik}a_{jk} = \delta_{ij} = \begin{cases} 1 & (i = j) \\ 0 & (i \neq j) \end{cases}. \tag{4.12}$$

The term δ_{ij}, known as the Kronecker delta, has the value = 1 on the diagonal $(i = j)$ and = 0 on the off-diagonal $(i \neq j)$.

The transformation equations for a fourth-order tensor C_{ijkl} can be written as:

$$C_{ijkl}' = a_{mi}a_{nj}a_{rk}a_{sl}C_{mnrs}. \tag{4.13}$$

4.4 DEFORMATION

Under the action of forces, a body S may translate and rotate as a rigid body as well as deform to occupy a new region S' as indicated in Figure 4.9. We define the displacements u_i of any point P in the body to its new location P' in terms of the three components of the vector u_i (in a rectangular Cartesian coordinate system) as $u_i = (u_1, u_2, u_3)$. An equivalent notation for displacements that will be used is $u_i = (u, v, w)$. Displacement is a vector or first-order tensor quantity.

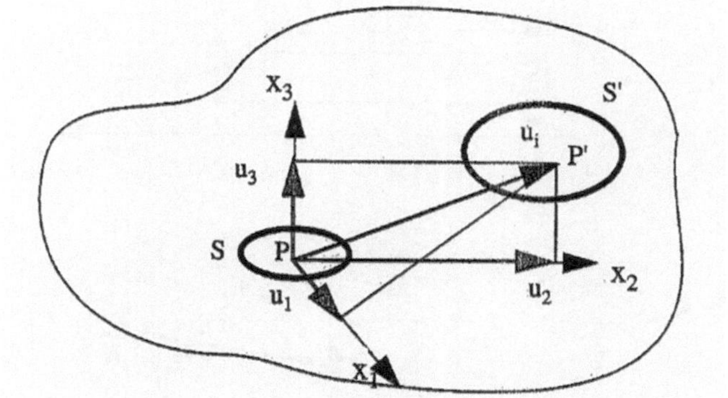

FIGURE 4.9 Displacement components.

4.5 STRAIN

If the gradients of the displacements are so small that products of partial derivatives of u_i are negligible compared with linear (first-order) derivative terms, then the (infinitesimal) strain tensor ε_{ij} is given by

$$\varepsilon_{ij} = \frac{1}{2}\left(u_{i,j} + u_{j,i}\right), \tag{4.14}$$

where the notation $u_{i,j}$ is defined as

$$u_{i,j} = \frac{\partial u_i}{\partial x_j}. \tag{4.15}$$

From Eq. (4.14), strain is a second-order symmetric tensor; that is, $\varepsilon_{ij} = \varepsilon_{ji}$. In expanded form, the strains are:

$$\begin{aligned}
&\varepsilon_{11} = \frac{\partial u_1}{\partial x_1},\ \varepsilon_{22} = \frac{\partial u_2}{\partial x_2},\ \varepsilon_{33} = \frac{\partial u_3}{\partial x_3} \\
&\varepsilon_{12} = \varepsilon_{21} = \frac{1}{2}\left(\frac{\partial u_1}{\partial x_2} + \frac{\partial u_2}{\partial x_1}\right) \\
&\varepsilon_{13} = \varepsilon_{31} = \frac{1}{2}\left(\frac{\partial u_1}{\partial x_3} + \frac{\partial u_3}{\partial x_1}\right) \\
&\varepsilon_{23} = \varepsilon_{32} = \frac{1}{2}\left(\frac{\partial u_2}{\partial x_3} + \frac{\partial u_3}{\partial x_2}\right).
\end{aligned} \tag{4.16}$$

The normal components of strain (the change in length per unit length, Figure 4.10a) correspond to the terms $i = j$, and the shear components (one-half the change in an original right angle, Figure 4.10b) correspond to the terms $i \neq j$. We also note that it

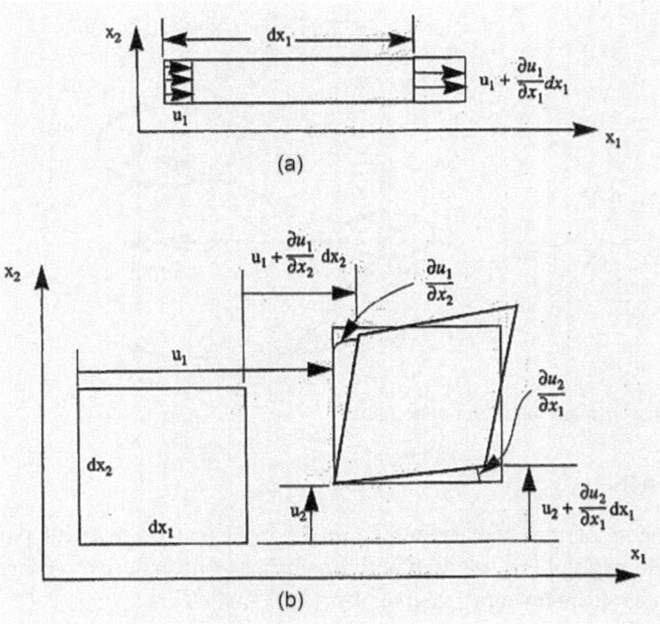

FIGURE 4.10 Components of strain: (a) Normal strain: $\varepsilon_{11} = \dfrac{\partial u_1}{\partial x_1}$, and (b) engineering shear strain $\gamma_{12} = \dfrac{\partial u_2}{\partial x_1} + \dfrac{\partial u_1}{\partial x_2}$.

is common in the study of mechanics of materials to make use of the engineering shear strain $\gamma_{ij} = 2\varepsilon_{ij}$, for $i \neq j$. The engineering shear strain is introduced because of its relationship with the shear modulus.

The engineering shear strains are written as:

$$\begin{aligned}
\gamma_{12} = \gamma_{21} &= \left(\frac{\partial u_1}{\partial x_2} + \frac{\partial u_2}{\partial x_1} \right) \\
\gamma_{13} = \gamma_{31} &= \left(\frac{\partial u_1}{\partial x_3} + \frac{\partial u_3}{\partial x_1} \right) \\
\gamma_{23} = \gamma_{32} &= \left(\frac{\partial u_2}{\partial x_3} + \frac{\partial u_3}{\partial x_2} \right).
\end{aligned} \tag{4.17}$$

In matrix notation, the second-order tensor is written as:

$$[\varepsilon] = \begin{bmatrix} \varepsilon_{11} & \varepsilon_{12} & \varepsilon_{13} \\ \varepsilon_{12} & \varepsilon_{22} & \varepsilon_{23} \\ \varepsilon_{13} & \varepsilon_{23} & \varepsilon_{33} \end{bmatrix}. \tag{4.18}$$

Or equivalently, using the engineering shear strains

$$[\varepsilon]=\begin{bmatrix} \varepsilon_{11} & \gamma_{12}/2 & \gamma_{13}/2 \\ \gamma_{12}/2 & \varepsilon_{22} & \gamma_{23}/2 \\ \gamma_{13}/2 & \gamma_{23}/2 & \varepsilon_{33} \end{bmatrix}. \tag{4.19}$$

4.6 STRESS

The components of stress at a point are the forces per unit area (in the limit) acting on planes passing through the point. The second-order stress tensor can be expressed in terms of the components acting on three mutually perpendicular planes aligned with the orthogonal coordinate directions as indicated in Figure 4.11. The tensor notation for stress is σ_{ij} (i, j= 1, 2, 3), where the first subscript corresponds to the direction of the normal to the plane of interest and the second subscript corresponds to the direction of the stress. Tensile normal stresses ($i = j$) are positive, and shear stresses ($i \neq j$) are defined to be positive when the normal to the plane and the stress component directions are either positive or negative. All components of stress depicted in Figure 4.11 have a positive sense.

In matrix notation, the second-order stress tensor is written as:

$$[\sigma]=\begin{bmatrix} \sigma_{11} & \sigma_{12} & \sigma_{13} \\ \sigma_{12} & \sigma_{22} & \sigma_{23} \\ \sigma_{13} & \sigma_{23} & \sigma_{33} \end{bmatrix}, \tag{4.20}$$

where we have used the fact that force and moment equilibrium of the element in Figure 4.11 requires that the stress tensor be symmetric (i.e., $\sigma_{ij} = \sigma_{ji}$). The proof is left as an exercise. It is common practice to express quantities in terms of a global

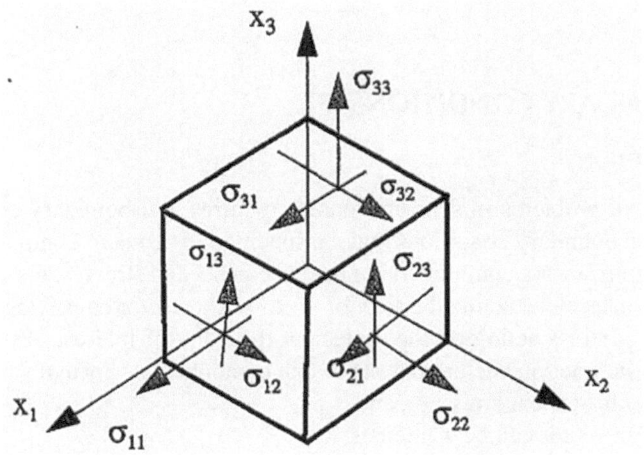

FIGURE 4.11 Components of stress.

x-y-z coordinate system, and it is also common to use the notation for shear stresses. Thus, the stress tensor will also be written as:

$$[\sigma]=\begin{bmatrix} \sigma_{xx} & \tau_{xy} & \tau_{xz} \\ \tau_{xy} & \sigma_{yy} & \tau_{yz} \\ \tau_{xz} & \tau_{yz} & \sigma_{zz} \end{bmatrix}. \tag{4.21}$$

4.7 EQUILIBRIUM

For a body in static equilibrium, the equations of equilibrium at a point are written in tensor notation as:

$$\sigma_{ij,j}+F_i=0, \tag{4.22}$$

where F_i is the body force per unit volume. Body forces will be considered to be negligible for the applications in this book, and hence, the equilibrium equations take the simple form:

$$\sigma_{ij,j}=0. \tag{4.23}$$

The expanded form of the equilibrium equations, when written in terms of a global x-y-z coordinate system, is:

$$\begin{aligned} \frac{\partial\sigma_{xx}}{\partial x}+\frac{\partial\sigma_{xy}}{\partial y}+\frac{\partial\sigma_{xz}}{\partial z}&=0 \\ \frac{\partial\sigma_{xy}}{\partial x}+\frac{\partial\sigma_{yy}}{\partial y}+\frac{\partial\sigma_{yz}}{\partial z}&=0 \\ \frac{\partial\sigma_{xz}}{\partial x}+\frac{\partial\sigma_{yz}}{\partial y}+\frac{\partial\sigma_{zz}}{\partial z}&=0. \end{aligned} \tag{4.24}$$

4.8 BOUNDARY CONDITIONS

4.8.1 Tractions

The solution of problems in solid mechanics requires that boundary conditions be specified. The boundary conditions may be specified in terms of components of displacement or stress, or a combination of displacements and stresses. For any point on an arbitrary surface, we define the traction T_i, to be the vector consisting of the three components of stress acting on the surface at the point of interest. As indicated in Figure 4.12, the traction vector consists of one component of normal stress, σ_{nn}, and two components of shear stress, σ_{nt} and σ_{ns}.

The traction vector can be written as:

$$T_i=\sigma_{ji}n_j, \tag{4.25}$$

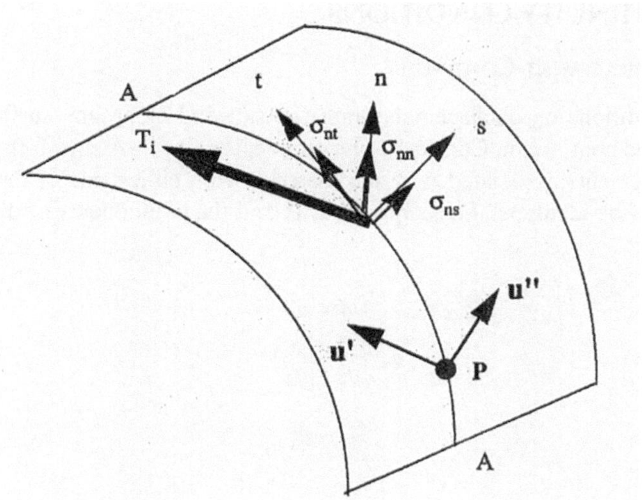

FIGURE 4.12 Traction vector and displacement continuity.

where n_j is the unit normal to the surface at the point under consideration. For a plane perpendicular to the x_1 axis (Figure 4.11), $n_i = (1, 0, 0)$ and the components of the traction are:

$$T_1 = \sigma_{11}$$

$$T_2 = \sigma_{12} \tag{4.26}$$

$$T_3 = \sigma_{13}.$$

4.8.2 Free Surface Boundary Conditions

The condition that a surface be free of stress is equivalent to all components of traction (Figure 4.12) being zero, that is,

$$T_n = \sigma_{nn} = 0 \tag{4.27}$$

$$T_t = \sigma_{nt} = 0$$

$$T_n = \sigma_{ns} = 0.$$

It is possible, of course, that only selected components of the traction be zero, while others are nonzero. For example, pure pressure loading would correspond to nonzero normal stress and zero shear stress.

4.9 CONTINUITY CONDITIONS

4.9.1 Displacement Continuity

Certain conditions on displacements must be satisfied along any surface in a perfectly bonded continuum. Consider, for example, the line A-A shown in Figure 4.12. The displacements associated with the material from either side of the line at any point **P** must be identical. Thus, $u_i' = u_i''$ at **P**, and the components of displacements must satisfy:

$$\begin{aligned} u_1' &= u_1'' \\ u_2' &= u_2'' \\ u_3' &= u_3''. \end{aligned} \tag{4.28}$$

The continuity condition (4.28) must be satisfied at every point in a perfectly bonded continuum. However, continuity is not required in the presence of debonding or sliding between regions or phases of a material.

4.9.2 Traction Continuity

Equilibrium (action and reaction) requires that the tractions T_i must be continuous across any surface. Mathematically, this is stated:

$$T_i' = T_i''. \tag{4.29}$$

Or using Eq. (4.25),

$$\sigma_{ji}' n_j' = \sigma_{ji}'' n_j''. \tag{4.30}$$

In terms of individual stress components, this is written as:

$$\begin{aligned} \sigma_{nn}' &= \sigma_{nn}'' \\ \sigma_{nt}' &= \sigma_{nt}'' \\ \sigma_{ns}' &= \sigma_{ns}''. \end{aligned} \tag{4.31}$$

Thus, the normal and shear components of stress acting on a surface (Figure 4.13) must be continuous across that surface. We emphasize that there are no continuity requirements on the other three components of stress. In fact, they can be discontinuous. Thus, it is permissible that:

$$\sigma_{tt}' = \sigma_{tt}'' \tag{4.32}$$

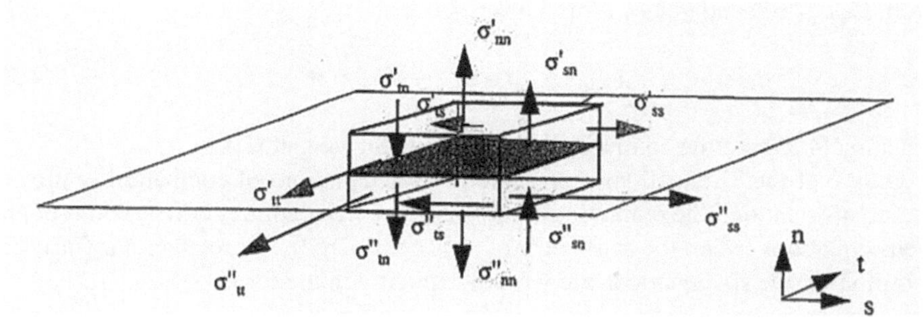

FIGURE 4.13 Stress continuity.

$$\sigma'_{ss} = \sigma''_{ss}$$

$$\sigma'_{ts} = \sigma''_{ts}.$$

For heterogeneous materials (including laminated composites), the lack of continuity of the two normal and one shear components of stress indicated in Eq. (4.32) is very common. This is true because the material properties are discontinuous across phase (or layer) boundaries.

4.10 COMPATIBILITY

The strain displacement Eqs. (4.14) provide six equations for only three unknown displacements u_i. Thus, integration of the equations to determine the unknown displacements will not have a singled-valued solution unless the strains ε_{ij} satisfy certain conditions. Arbitrary specification of the ε_{ij} could result in a situation in which there were discontinuities (gaps) in the material or overlapping regions of material.

The conditions that are necessary to provide single-valued displacements are called the compatibility conditions. They are conditions to be satisfied by the strains ε_{ij} and are obtained by eliminating u_i from Eq. (4.14). Differentiation of Eq. (4.14) gives:

$$\varepsilon_{ij,kl} = \frac{1}{2}\left(u_{i,jkl} + u_{j,ikl}\right). \tag{4.33}$$

This expression can be rewritten by interchanging subscripts with the results:

$$\varepsilon_{kl,ij} = \frac{1}{2}\left(u_{k,lij} + u_{l,kij}\right) \tag{4.34}$$

$$\varepsilon_{jl,ik} = \frac{1}{2}\left(u_{j,lik} + u_{l,jik}\right)$$

$$\varepsilon_{ik,jl} = \frac{1}{2}\left(u_{i,kjl} + u_{k,ijl}\right).$$

From Eqs. (4.33) and (4.44), it can be verified that:

$$\varepsilon_{ij,kl} + \varepsilon_{kl,ij} - \varepsilon_{ik,jl} - \varepsilon_{jl,ik} = 0. \tag{4.35}$$

Equation (4.35) are the compatibility equations (St. Venant [1]).

Only 6 of the 81 equations represented by the equations of compatibility provide useful information. The remaining equations are either identities or repetitions of previous equations (when the symmetry of ε_{ij} is considered). For rectangular Cartesian coordinates, the six equations are written explicitly in the form:

$$\varepsilon_{xx,zy} = -\varepsilon_{yz,xx} + \varepsilon_{zx,yx} + \varepsilon_{xy,zx} \tag{4.36}$$

$$\varepsilon_{yy,xz} = -\varepsilon_{zx,yy} + \varepsilon_{xy,zy} + \varepsilon_{yz,xy}$$

$$\varepsilon_{zz,yx} = -\varepsilon_{xy,zz} + \varepsilon_{yz,xz} + \varepsilon_{zx,yz}$$

$$2\varepsilon_{xy,yx} = \varepsilon_{xx,yy} + \varepsilon_{yy,xx}$$

$$2\varepsilon_{zx,xz} = \varepsilon_{zz,xx} + \varepsilon_{xx,zz}.$$

4.11 CONSTITUTIVE EQUATIONS

Constitutive equations provide the relationship between stress and strain. We shall be primarily concerned with linear elastic material response in this book. Thus, the relationships between stress and strain are linear, with the most general form being:

$$\sigma_{ij} = C_{ijkl}\varepsilon_{kl}, \tag{4.37}$$

where C_{ijkl} are the elastic constants called elastic moduli or stiffnesses. This relation is called the generalized Hooke's law (Hooke [2], Love [3]). We will be concerned with a variety of material types, distinguished by their degree of anisotropy. The number of independent constants ranges from 2 to 21 depending upon the degree of anisotropy. This topic is discussed in detail in the following paragraphs.

Inverting (4.37) gives the strains in terms of the stresses in the form:

$$\varepsilon_{ij} = C_{ijkl}^{-1}\sigma_{kl} = S_{ijkl}\sigma_{kl}, \tag{4.38}$$

where we have defined the compliance S_{ijkl} as the inverse of the stiffness.

As will be shown later in this chapter, the constitutive equations for an orthotropic material (a material with three mutually perpendicular planes of material symmetry) can be written in terms of nine independent stiffness coefficients in the form:

$$\varepsilon_{11} = S_{1111}\sigma_{11} + S_{1122}\sigma_{22} + S_{1133}\sigma_{33}$$

$$\varepsilon_{22} = S_{2211}\sigma_{11} + S_{2222}\sigma_{22} + S_{2233}\sigma_{33}$$

$$\varepsilon_{33} = S_{3311}\sigma_{11} + S_{3322}\sigma_{22} + S_{3333}\sigma_{33} \tag{4.39}$$

$$\varepsilon_{23} = S_{2323}\sigma_{23}$$

$$\varepsilon_{31} = S_{3131}\sigma_{31}$$

$$\varepsilon_{12} = S_{1212}\sigma_{12}.$$

4.12 PLANE STRESS

Plane stress corresponds to a condition in which all three out-of-plane components of stress are identically zero throughout the region. For the plane stress case shown in Figure 4.14, the normal stress σ_{zz} and both out-of-plane shear components σ_{zx} and σ_{zy} are zero for plane stress in the *x-y* plane. It is noted that the *z* components of strain are not necessarily zero for plane stress in the *x-y* plane. The magnitudes of the *z* components of strain depend on the constitutive equation as expressed in (4.38). The out-of-plane normal strain ε_{zz} is nonzero for materials having a nonzero Poisson's ratio.

For the case of plane stress in the *x-y* plane, the equilibrium equations (4.24) reduce to the simpler form:

$$\frac{\partial \sigma_{xx}}{\partial x} + \frac{\partial \sigma_{xy}}{\partial y} = 0$$

$$\frac{\partial \sigma_{xy}}{\partial x} + \frac{\partial \sigma_{yy}}{\partial y} = 0 \tag{4.40}$$

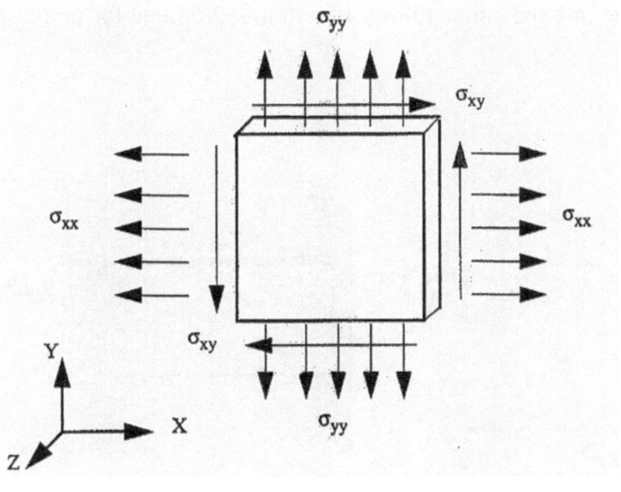

FIGURE 4.14 Plane stress in the *x-y* plane.

and the compatibility equations (4.36) reduce to a single equation:

$$2\varepsilon_{xy,yx} = \varepsilon_{xx,yy} + \varepsilon_{yy,xx.} \tag{4.41}$$

4.13 PLANE STRAIN

Plane strain corresponds to a condition in which all three out-of-plane components of strain (ε_{zz}, ε_{zx}, ε_{zy}) are zero (see Figure 4.15) and the stresses are, at most, functions of x and y. As for the plane stress case, the out-of-plane components of stress depend upon the form of the constitutive equations, and the normal stress σ_{zz} is generally nonzero.

Equilibrium Eq. (4.24) reduces to:

$$\frac{\partial \sigma_{xx}}{\partial x} + \frac{\partial \sigma_{xy}}{\partial y} = 0$$

$$\frac{\partial \sigma_{xy}}{\partial x} + \frac{\partial \sigma_{yy}}{\partial y} = 0 \tag{4.42}$$

$$\sigma_{zz} = f(x, y).$$

With the out-of-plane strain components (ε_{zz}, ε_{zx}, ε_{zy}) zero and strains independent of z through Eq. (4.38) (since the stresses are independent of z), compatibility Eq. (4.36) reduces to the single equation:

$$2\varepsilon_{xy,yx} = \varepsilon_{xx,yy} + \varepsilon_{yy,xx}. \tag{4.43}$$

Thus, we see that the compatibility equation is the same for plane stress and plane strain.

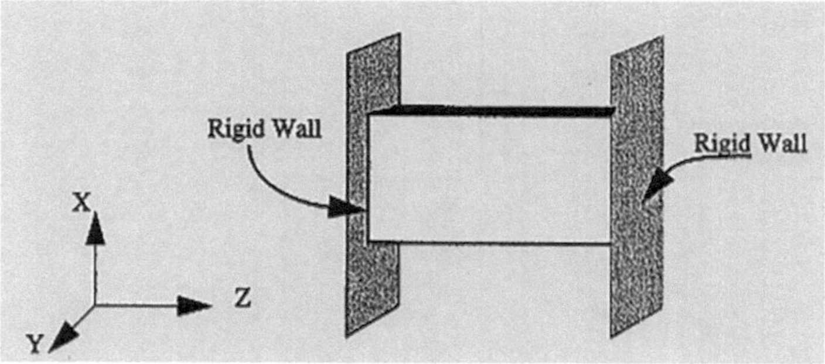

FIGURE 4.15 Plane strain in x-y plane.

4.14 GENERALIZED PLANE PROBLEMS

Generalized plane problems are those in which all six components of stress and strain are (may be) nonzero, but stresses and strains do not vary along a prescribed direction (say the z direction). The state of stress and strain can then be represented by that in a generic plane (Lekhnitskii, 1963). The compatibility equations then reduce to the single equation (4.43), and equilibrium (4.24) reduces to:

$$\frac{\partial \sigma_{xx}}{\partial x} + \frac{\partial \sigma_{xy}}{\partial y} = 0$$

$$\frac{\partial \sigma_{xy}}{\partial x} + \frac{\partial \sigma_{yy}}{\partial y} = 0 \tag{4.44}$$

$$\frac{\partial \sigma_{xz}}{\partial x} + \frac{\partial \sigma_{yz}}{\partial y} = 0.$$

4.15 STRAIN ENERGY DENSITY

The strain energy density is the stored internal energy per unit volume in a body upon which work has been done. If the internal energy is zero in the unstressed state, the strain energy density in the stressed state is:

$$W = \frac{1}{2}\sigma_{ij}\varepsilon_{ij}. \tag{4.45}$$

Expansion of (4.45) in a global *x-y-z* coordinate system, invoking the symmetry of stress σ_{ij} and the strain ε_{ij}, gives the expression:

$$W = \frac{1}{2}\left(\sigma_{xx}\varepsilon_{xx} + 2\sigma_{xy}\varepsilon_{xy} + 2\sigma_{yz}\varepsilon_{yz} + 2\sigma_{xz}\varepsilon_{xz} + \sigma_{yy}\varepsilon_{yy} + \sigma_{zz}\varepsilon_{zz}\right). \tag{4.46}$$

It is noted that the strain energy density is a scalar function. Integration of the strain energy density over a given volume provides the strain energy of the body.

4.16 MINIMUM PRINCIPLES

Minimum principles, which have their foundation in the calculus of variations, have proven to be very effective for obtaining approximate solutions to problems in solid mechanics. The principles are based upon the concept that a specific function, chosen from a class of admissible functions that satisfy some, but not all, of the requirements of the complete solution, minimizes a certain functional expression, defined for this class of functions, when the specific admissible function satisfies the remaining requirements for the complete solution.

For the general statement of these principles, it is convenient to consider the boundary of the body under consideration to be composed of a displacement portion,

S_D, and a remaining traction portion, S_T. A portion of the boundary is S_D if, in each of three independent directions, either the component of displacement is given or the corresponding component of traction vanishes. A portion of the boundary is S_T if, in each of three independent directions, either the component of traction is given or the corresponding component of displacement vanishes (Hodge [4]). The minimum principles are expressed in terms of statically admissible stress states σ_{ij}^0 which satisfy the equations of equilibrium and the traction boundary condition, or kinematically admissible displacements $\overset{*}{u}_i$, which are single-valued continuous displacement fields that satisfy the displacement boundary conditions. Designating the strains associated (through the constitutive equations) with the statically admissible stresses as ε_{ij}^0, the stresses associated with the kinematically admissible strain as $\overset{*}{\sigma}_{ij}$, and the material elastic compliance as S_{ijkl}, two minimum principles in the theory of elasticity are the principle of minimum potential energy and the principle of minimum complementary energy. They are defined in the following sections.

4.16.1 Minimum Potential Energy

The principle of minimum potential energy states that among all kinematically admissible displacements fields $\overset{*}{u}_i$, the actual solution to the problem minimizes the functional $\overset{*}{\Pi}$ defined as:

$$\overset{*}{\Pi} = \frac{1}{2}\int_V S_{ijkl}\ \overset{*}{\sigma}_{ij}\ \overset{*}{\sigma}_{kl}\, dV - \int_{S_T} T_i \overset{*}{u}_i\, dS. \tag{4.47}$$

The proof proceeds along the following lines. For a given boundary value problem with tractions T_i specified on S_T, let:

- u_i be the actual displacement field which is the exact solution to the problem.
- σ_{ij} be the actual stresses.
- $\overset{*}{u}_i$ be any kinematically admissible displacement field.
- $\overset{*}{\sigma}_{ij}$ be the stresses associated with the kinematically admissible displacements.
- Π be the actual potential energy.
- $\overset{*}{\Pi}$ be the potential energy associated with the admissible displacement field.

Now define $\Delta\Pi$ as the difference between the admissible and actual potential energies.

$$\Delta\Pi = \overset{*}{\Pi} - \Pi. \tag{4.48}$$

In order to prove the theorem, we must show that:

$$\Delta\Pi \geq 0 \tag{4.49}$$

for all admissible displacement fields with the equality holding if and only if $\overset{*}{u}_i = u_i$.

Substituting the actual and admissible quantities in Eqs. (4.47) and (4.48), we can write:

$$\Delta\Pi = \frac{1}{2}\int_V S_{ijkl}\left(\sigma_{ij}^* \ \sigma_{kl}^* - \sigma_{ij} \ \sigma_{kl}\right)dV - \int_{S_T} T_i\left(u_i^* - u_i\right)dS. \tag{4.50}$$

Now we can write:

$$\int_{S_T} T_i\left(u_i^* - u_i\right)dS = \int_S T_i\left(u_i^* - u_i\right)dS. \tag{4.51}$$

Since $S = S_T + S_D$ and $u_i^* = u_i$ on S_D, giving:

$$\int_{S_D} T_i\left(u_i^* - u_i\right)dS = 0. \tag{4.52}$$

Using the divergence theorem and the definition of Eq. (4.25) for tractions in terms of stresses, we can write the surface integral as a volume integral:

$$\int_S T_i\left(u_i^* - u_i\right)dS = \int_V \frac{\partial\left[\sigma_{ij}\left(u_i^* - u_i\right)\right]}{\partial x_j}dV. \tag{4.53}$$

Now, performing the partial differentiation indicated in Eq. (4.53), and using equilibrium Eq. (4.23), strain displacement Eq. (4.14) and the constitutive Eq. (4.38), we can write:

$$\int_S T_i\left(u_i^* - u_i\right)dS = \int_V S_{ijkl}\sigma_{ij}\left(\sigma_{kl}^* - \sigma_{kl}\right)dV. \tag{4.54}$$

Combining Eqs. (4.50) and (4.54) and using the symmetry of S_{ijkl}, we write:

$$\Delta\Pi = \frac{1}{2}\int_V S_{ijkl}\left(\sigma_{ij}^* - \sigma_{ij}\right)\left(\sigma_{kl}^* - \sigma_{kl}\right)dV. \tag{4.55}$$

The integrand in Eq. (4.55) is positive definite for all nonzero $(\sigma_{ij}^* - \sigma_{ij})$ since S_{ijkl} is symmetric and the strain energy density is assumed to be positive definite (see Kreyszig [5]). Thus,

$$\Delta\Pi \geq 0\,, \tag{4.56}$$

with $\Delta\Pi = 0$ if and only if $\sigma_{ij}^* = \sigma_{ij}$.

This implies that the displacements satisfy $u_i^* = u_i$ (to within a rigid body motion). Hence, the theorem is proven.

4.16.2 Minimum Complementary Energy

The principle of minimum complementary energy states that among all statically admissible stress fields σ_{ij}^{0}, the actual solution to the problem minimizes the admissible complementary energy (Π_c^0), defined as:

$$\Pi_c^0 = \frac{1}{2}\int_V S_{ijkl}\sigma_{ij}^0\sigma_{kl}^0\, dV - \int_{S_D}\left(T_i^0 u_i\right) dS. \tag{4.57}$$

In order to prove the theorem, it is necessary to show that:

$$\Pi_c^0 - \Pi_c \geq 0, \tag{4.58}$$

where Π_c is the complimentary energy of the exact solution.

The proof is similar to that for potential energy.

4.16.3 Bounds and Uniqueness

From the definitions of Eqs. (4.47) and (4.57), the sum of the potential and complementary energies for the actual solution must satisfy the principle of virtual work, that is,

$$\Pi + \Pi_c = 0. \tag{4.59}$$

For a proof of the principle of virtual work, see Fung [6]. It can be combined with Eqs. (4.48), (4.49), and (4.58) with the result:

$$-\Pi_c^0 \leq -\Pi_c = \Pi \leq \overset{*}{\Pi}. \tag{4.60}$$

Equation (4.60) shows that the complementary energy and potential energy of admissible states can be used to provide bounds on the exact solution.

Uniqueness of the solution is proved by considering two distinct complete solutions whose potential energies are $\Pi^{(1)}$ and $\Pi^{(2)}$. By definition, complete solutions must also be kinematically admissible. Therefore, considering $\Pi^{(1)}$ as the complete solution and $\Pi^{(2)}$ as the kinematically admissible solution, we can write:

$$\Pi^{(2)} \geq \Pi^{(1)}. \tag{4.61}$$

Reversing the procedure and taking $\Pi^{(2)}$ as the complete solution and $\Pi^{(1)}$ as the kinematically admissible solution, we write:

$$\Pi^{(1)} \geq \Pi^{(2)}. \tag{4.62}$$

The only way for both equations to be satisfied is for:

$$\Pi^{(1)} = \Pi^{(2)}. \tag{4.63}$$

Hence, the solution is unique. This is a most important principle of elasticity theory. If a solution that satisfies all the requirements of an elasticity solution is determined, it is the only solution.

4.17 EFFECTIVE PROPERTY CONCEPT

In this section, we consider ways to write the linear elastic constitutive equations for the effective (or average) response of fibrous composite materials. Our goal is to develop equations for predicting the elastic constants required for the average stress–strain relationship. This relationship is usually referred to as Hooke's law. We will consider three-dimensional constitutive equations in this section. We want to write constitutive equations for the effective or smeared properties of the composite because an analysis of the exact configuration of the individual fibers and matrix in a composite structure would be impractical for structural analysis. To do so, we model the properties of the composite as the properties of a repeating representative volume element (RVE), which is small enough to be representative of local material response yet large enough to represent "average" material response. In a fibrous composite, the length scale over which the properties can be averaged in a meaningful manner must be several times that of the mean fiber spacing. If this dimension is small compared with the characteristic dimension of the structure, the material can be modeled as effectively homogeneous. Figure 4.16 shows an example RVE as well as volume elements that are too small to be representative of the composite.

4.18 GENERALIZED HOOKE'S LAW

As indicated in previous sections, the most general linear relationship between stresses and strains such that the stresses vanish when the strains are zero is:

$$\sigma_{ij} = C_{ijkl}\varepsilon_{kl}, \tag{4.64}$$

where C_{ijkl} is a fourth-order tensor with 81 (3^4) elastic "constants." This linear elastic stress–strain constitutive relationship is called the generalized Hooke's law. Invoking symmetry of the stress tensor gives $C_{ijkl} = C_{jikl}$. Similarly, symmetry of the strain tensor gives $C_{ijkl} = C_{jilk}$. Thus, the number of independent constants is reduced from 81

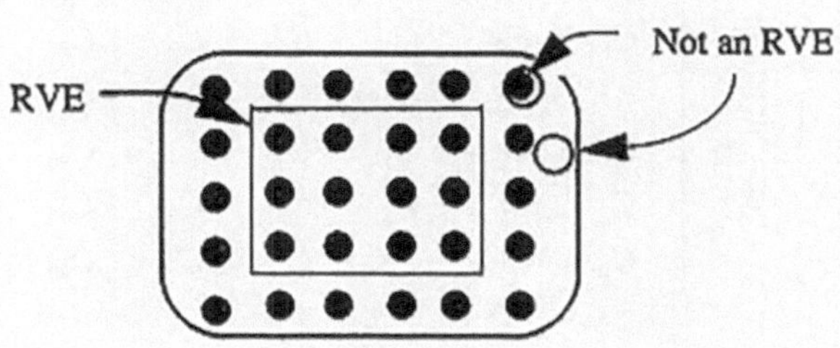

FIGURE 4.16 Representative volume element.

to 36 when the stresses and strains are symmetric. In view of this reduced number of constants, Hooke's law can be written in contracted notation as:

$$\sigma_i = C_{ij}\varepsilon_j \quad (i, j = 1, 2, 3\ldots, 6), \tag{4.65}$$

where the contracted notation for stress and strain is defined by the following equivalences:

$$\sigma_1 \equiv \sigma_{11}, \sigma_2 \equiv \sigma_{22}, \sigma_3 \equiv \sigma_{33}, \sigma_4 \equiv \sigma_{23}, \sigma_5 \equiv \sigma_{31}, \sigma_6 \equiv \sigma_{12}$$

$$\varepsilon_1 \equiv \varepsilon_{11}, \varepsilon_2 \equiv \varepsilon_{22}, \varepsilon_3 \equiv \varepsilon_{33}, \varepsilon_4 \equiv 2\varepsilon_{23}, \varepsilon_5 \equiv 2\varepsilon_{31}, \varepsilon_6 \equiv 2\varepsilon_{12}. \tag{4.66}$$

We note here that, in the contracted notation, the shear strain terms are the engineering shear strains, for example, $\varepsilon_4 = 2\varepsilon_{23} = \gamma_{23}$.

The C_{ij} are referred to by a variety of names, including elastic constants, moduli, and stiffness coefficients. In order for Eq. (4.65) to be solvable for the strains in terms of the stresses, the determinant of the stiffness matrix must be nonzero, that is, $|C_{ij}| \neq 0$. The number of independent elastic constants can be reduced further if there exists a strain energy density function W such that

$$W = \frac{1}{2} C_{ij}\varepsilon_i\varepsilon_j. \tag{4.67}$$

With the property,

$$\sigma_i = \frac{\partial W}{\partial \varepsilon_i}. \tag{4.68}$$

The quadratic form Eq. (4.67) exists such that Eq. (4.68) is true, then C_{ij} is symmetric and the number of constants is reduced accordingly, that is,

$$C_{ij} = C_{ji}. \tag{4.69}$$

The symmetric stiffness, C_{ij}, can be written in matrix notation as:

$$[C_{ij}] = \begin{bmatrix} C_{11} & C_{12} & C_{13} & C_{14} & C_{15} & C_{16} \\ C_{12} & C_{22} & C_{23} & C_{24} & C_{25} & C_{26} \\ C_{13} & C_{23} & C_{33} & C_{34} & C_{35} & C_{36} \\ C_{14} & C_{24} & C_{34} & C_{44} & C_{45} & C_{46} \\ C_{15} & C_{25} & C_{35} & C_{45} & C_{55} & C_{56} \\ C_{16} & C_{26} & C_{36} & C_{46} & C_{56} & C_{66} \end{bmatrix}. \tag{4.70}$$

This symmetric matrix has 21 independent constants; that is, counting the independent C_{ij} by row gives $6+5+4+3+2+1 = 21$.

The existence of the function W is based upon the first and second laws of thermodynamics. It was first proposed by Green [7]. An in-depth discussion of the existence and positive definiteness of the strain energy function is given in Fung [6]. Experimental results are consistent with this theory for the elastic response of materials.

The inverted form of Hooke's law (4.65) can be written as:

$$\varepsilon_i = S_{ij}\sigma_j. \tag{4.71}$$

The coefficients S_{ij} in Eq. (4.71) are called the compliance coefficients. They will be discussed in more detail in the following sections. From Eq. (3.2) and Eq. (3.8), it is evident that the compliance matrix is the inverse of the stiffness matrix; thus:

$$S_{ij} = C_{ij}^{-1}. \tag{4.72}$$

We note here that since the stiffness matrix is symmetric, the compliance matrix is also symmetric.

A material with 21 independent constants is called anisotropic. The maximum number of independent constants for an anisotropic material is limited by the symmetry of the stress and strain tensors and the assumption of the existence of the strain energy density function. Materials that exhibit material symmetry about planes passing through the material have fewer independent constants. Different degrees of material symmetry are discussed in the following section.

4.19 MATERIAL SYMMETRY

4.19.1 Monoclinic Material

Consider a material which is symmetric about the x_1–x_2 plane (Figure 4.17); that is, the elastic constants in the primed and the unprimed coordinate systems are identical

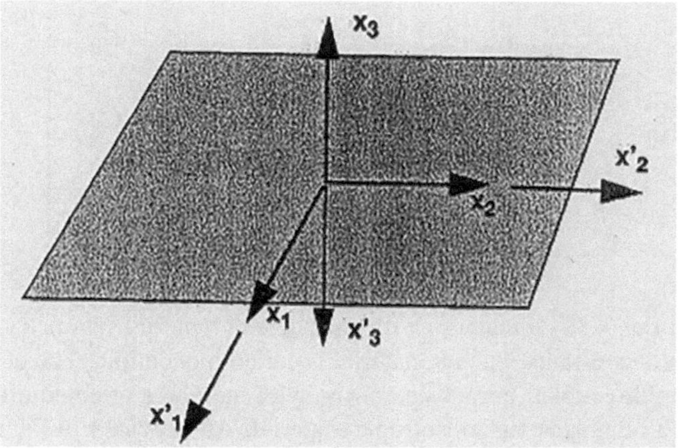

FIGURE 4.17 Symmetry about $x_1 - x_2$ plane.

$C_{ij} = C_{ij}'$. The implications of this symmetry can be deduced by considering Hooke's law in each coordinate system. In the unprimed coordinates,

$$\sigma_i = C_{ij}\varepsilon_j. \tag{4.73}$$

In primed coordinates,

$$\sigma_i' = C_{ij}'\varepsilon_j'. \tag{4.74}$$

From the definition of coordinate systems, it is apparent that:

$$\sigma_1' = \sigma_1, \sigma_2' = \sigma_2, \sigma_3' = \sigma_3, \sigma_4' = -\sigma_4, \sigma_5' = -\sigma_5, \sigma_6' = \sigma_6 \tag{4.75}$$

$$\varepsilon_1' = \varepsilon_1, \varepsilon_2' = \varepsilon_2, \varepsilon_3' = \varepsilon_3, \varepsilon_4' = -\varepsilon_4, \varepsilon_5' = -\varepsilon_5, \varepsilon_6' = \varepsilon_6. \tag{4.76}$$

Expressing the first stress equality $\sigma_1' = \sigma_1$ in Eq. (4.75) in terms of strains and stiffnesses using the constitutive equations (4.73) and (4.74) gives:

$$C_{11}'\varepsilon_1' + C_{12}'\varepsilon_2' + \ldots + C_{16}'\varepsilon_6' = C_{11}\varepsilon_1 + C_{12}\varepsilon_2 + \cdots + C_{16}\varepsilon_6. \tag{4.77}$$

Now substituting for the ε_i' in terms of the ε_i from Eq. (4.76) and comparing like terms in Eq. (4.77) and noting that $C_{ij} = C_{ij}'$, we obtain:

$$C_{14} = C_{15} = 0. \tag{4.78}$$

In a similar fashion, the remaining equations in (3.12) lead to the results:

$$C_{24} = C_{25} = C_{34} = C_{35} = C_{46} = C_{56} = 0. \tag{4.79}$$

For material symmetry about the $(x_1 - x_2)$ plane, the stiffness matrix reduces to:

$$\left[C_{ij}\right] = \begin{bmatrix} C_{11} & C_{12} & C_{13} & 0 & 0 & C_{16} \\ C_{12} & C_{22} & C_{23} & 0 & 0 & C_{26} \\ C_{13} & C_{23} & C_{33} & 0 & 0 & C_{36} \\ 0 & 0 & 0 & C_{44} & C_{45} & 0 \\ 0 & 0 & 0 & C_{45} & C_{55} & 0 \\ C_{16} & C_{26} & C_{36} & 0 & 0 & C_{66} \end{bmatrix}. \tag{4.80}$$

As shown in Eq. (4.80), a material with one plane of material symmetry has 13 independent elastic constants. Such a material is called monoclinic. The effective properties of a unidirectional fibrous composite with the fibers oriented off-axis can be modeled as a homogeneous anisotropic material. As indicated in Figure 4.18, two Cartesian coordinate systems are identified: the 1–2–3 system and the *x-y-z* system. The 1-direction is the fiber direction, with the 2- and 3-directions perpendicular

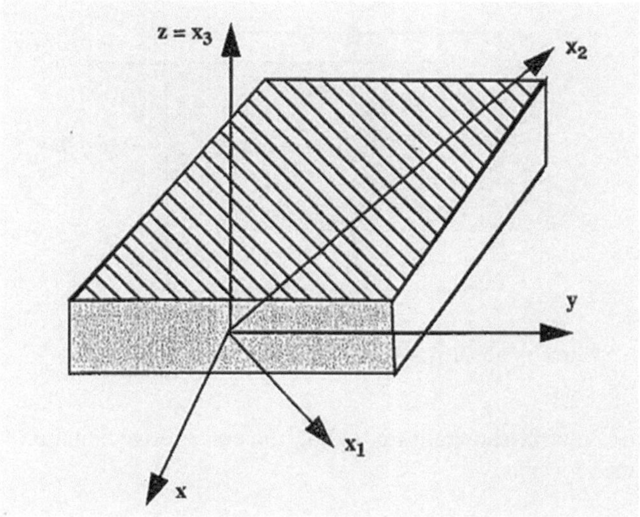

FIGURE 4.18 Off-axis unidirectional lamina.

to the fiber direction. The *x-y-z* coordinates are obtained by a rotation about the 3 axes. The material exhibits symmetry with respect to the *x-y* plane, but not with respect to the *x-z* or *y-z* plane. Such a material with one plane of symmetry is called a monoclinic material. It has a stiffness matrix of the form (4.80) when referred to the global *x-y-z* coordinate system. The symmetry that such a material possesses when viewed in the 1–2–3 coordinate system which is parallel and perpendicular to the fiber direction will be discussed in the following section.

4.19.2 Orthotropic Material

If the material under consideration has a second plane of material symmetry, say the $x_2 - x_3$ plane (Figure 4.19), the σ_1, σ_2, σ_3, and σ_5 stress components must be equal to the corresponding stresses in the primed coordinate system. Following the same procedure as in the previous section gives:

$$C_{16} = C_{26} = C_{36} = C_{45} = 0. \tag{4.81}$$

The stiffness matrix then has the form of an orthotropic material with nine independent constants, that is,

$$\left[C_{ij}\right] = \begin{bmatrix} C_{11} & C_{12} & C_{13} & 0 & 0 & 0 \\ C_{12} & C_{22} & C_{23} & 0 & 0 & 0 \\ C_{13} & C_{23} & C_{33} & 0 & 0 & 0 \\ 0 & 0 & 0 & C_{44} & 0 & 0 \\ 0 & 0 & 0 & 0 & C_{55} & 0 \\ 0 & 0 & 0 & 0 & 0 & C_{66} \end{bmatrix}. \tag{4.82}$$

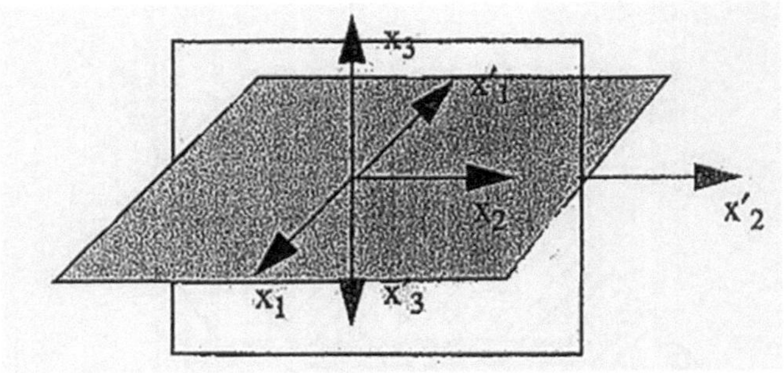

FIGURE 4.19 Symmetry about $x_1 - x_2$ and $x_2 - x_3$ planes.

For the stiffness matrix shown in Eq. (4.80), the constitutive equation (4.65) can be written in matrix form as:

$$\begin{Bmatrix} \sigma_1 \\ \sigma_2 \\ \sigma_3 \\ \tau_{23} \\ \tau_{31} \\ \tau_{12} \end{Bmatrix} = \begin{bmatrix} C_{11} & C_{12} & C_{13} & 0 & 0 & 0 \\ C_{12} & C_{22} & C_{23} & 0 & 0 & 0 \\ C_{13} & C_{23} & C_{33} & 0 & 0 & 0 \\ 0 & 0 & 0 & C_{44} & 0 & 0 \\ 0 & 0 & 0 & 0 & C_{55} & 0 \\ 0 & 0 & 0 & 0 & 0 & C_{66} \end{bmatrix} \begin{Bmatrix} \varepsilon_1 \\ \varepsilon_2 \\ \varepsilon_3 \\ \gamma_{23} \\ \gamma_{31} \\ \gamma_{12} \end{Bmatrix}. \quad (4.83)$$

It is left as an exercise to show that a material with two perpendicular planes of material symmetry also exhibits symmetry about the third, mutually perpendicular plane. Thus, an orthotropic material has three mutually perpendicular planes of symmetry. An orthotropic fibrous material is depicted in Figure 4.20, where the cross section of the fibers is oval.

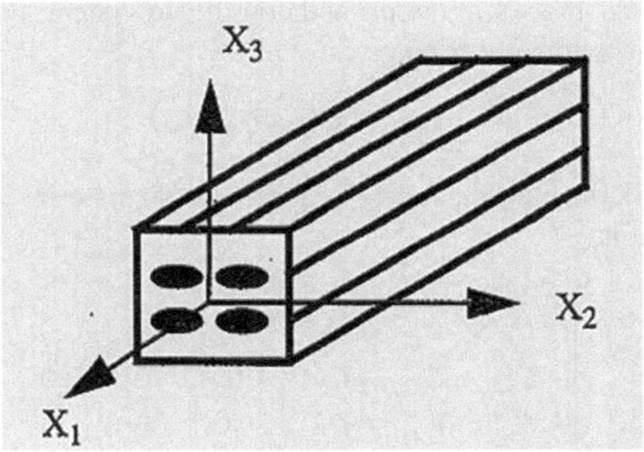

FIGURE 4.20 Orthotropic composite.

4.19.3 Transversely Isotropic Material

A transversely isotropic material is defined to be a material whose effective properties are isotropic (independent of direction) in one of its planes. A unidirectional fibrous composite consisting of a random array of fibers exhibits isotropic properties in the plane transverse to the fibers, the 2–3 plane in Figure 4.21. Transverse isotropy results in the following relations for stiffness coefficients:

$$C_{22} = C_{33} \quad C_{12} = C_{13} \quad C_{55} = C_{66} \quad C_{44} = (C_{22} - C_{23})/2. \tag{4.84}$$

The relations Eq. (4.84) can be deduced directly from the equality of properties in the x_2 and x_3 directions and the E, G, v relationship ($G = E/2(1+v)$) in the plane of isotropy (see, e.g., Popov [8]). The stiffness matrix for a transversely isotropic material then has the form:

$$[C] = \begin{bmatrix} C_{11} & C_{12} & C_{13} & 0 & 0 & 0 \\ C_{12} & C_{22} & C_{23} & 0 & 0 & 0 \\ C_{12} & C_{23} & C_{22} & 0 & 0 & 0 \\ 0 & 0 & 0 & \dfrac{C_{22} - C_{23}}{2} & 0 & 0 \\ 0 & 0 & 0 & 0 & C_{66} & 0 \\ 0 & 0 & 0 & 0 & 0 & C_{66} \end{bmatrix}. \tag{4.85}$$

It is evident from Eq. (4.85) that a transversely isotropic material has only five independent elastic constants.

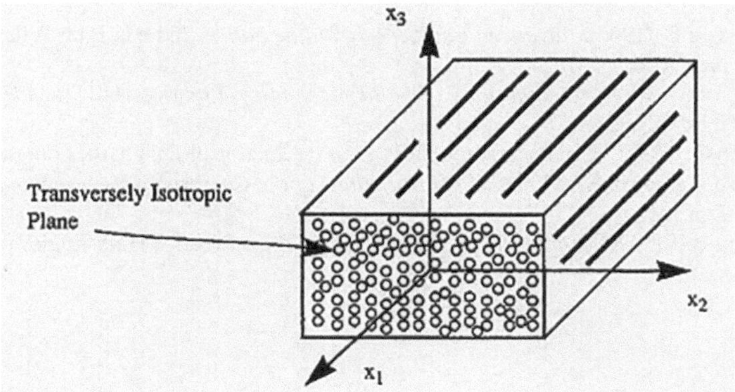

FIGURE 4.21 Transversely isotropic material.

4.19.4 Isotropic Material

An isotropic material is one whose properties are independent of direction and whose planes are all planes of symmetry. In this case, we have the following additional relations between stiffness coefficients:

$$C_{11} = C_{22} \quad C_{12} = C_{23} \quad C_{66} = (C_{22} - C_{23})/2 = (C_{11} - C_{12})/2. \tag{4.86}$$

Thus, an isotropic material has only two independent elastic constants, and the stiffness matrix can be written as:

$$[C] = \begin{bmatrix} C_{11} & C_{12} & C_{12} & 0 & 0 & 0 \\ C_{12} & C_{11} & C_{12} & 0 & 0 & 0 \\ C_{12} & C_{12} & C_{11} & 0 & 0 & 0 \\ 0 & 0 & 0 & \dfrac{C_{11} - C_{12}}{2} & 0 & 0 \\ 0 & 0 & 0 & 0 & \dfrac{C_{11} - C_{12}}{2} & 0 \\ 0 & 0 & 0 & 0 & 0 & \dfrac{C_{11} - C_{12}}{2} \end{bmatrix}. \tag{4.87}$$

REFERENCES

1. B. St. de Venant (1864), in his edition of Navier's *Resume des sur l'application de la Mecanique*, app. 3, Carilian-Goeury, Paris.
2. Hooke R. (1678), *De Potentia Restitutiva*, London.
3. Love A.E.H. (1892), *A Treatise on the Mathematical Theory of Elasticity*, Cambridge University Press, London (also 4th ed., 1944, Dover Publications, Inc., New York).
4. Hodge P.G. Jr. (1958), "The mathematical theory of plasticity," in *Elasticity and Plasticity* (J. N. Goodier and P. G. Hodge, Jr., eds.), John Wiley & Sons, Inc., New York, p. 81.
5. Kreyszig E. (1967), *Advanced Engineering Mathematics*, 2nd ed., John Wiley & Sons, Inc., New York, p. 416.
6. Fung Y.C. (1965), *Foundations of Solid Mechanics*, Prentice-Hall, Inc., Englewood Cliffs, NJ, p. 285.
7. Green G. (1839), "On the laws of reflexion and refraction of light at the common surface of two non-crystallized media", *Transactions of the Cambridge Philosophical Society*, Vol. 7, p. 121.
8. Popov E.P. (1990), *Engineering Mechanics of Solids*, Prentice Hall, Englewood Cliffs, NJ.

5 Linear Elastic Stress–Strain Characteristics of Fiber-Reinforced Composites

The study of the mechanics of fiber-reinforced composites could begin at several points. Because the fiber plays a key role in the performance of the material, one logical starting point would be to study the interaction of the fiber with the matrix. We might address a host of problems: stresses in the fiber, stresses in the matrix around the fiber, adhesive stresses at the interface, breaking of the fiber, cracking of the matrix, the interaction of two or more fibers, the effects on stresses in the matrix of moving two fibers farther apart or closer together, the effect on neighboring fibers of a broken fiber, or local yielding of either the matrix or the fiber. This localized look at the interaction of the fibers with the matrix is called micromechanics. It is indeed a logical starting point for a study of the mechanics of fiber-reinforced composites. On the other end of the spectrum, interest centers on the behavior of structures made using fiber-reinforced material. Deflections, maximum allowable loads, vibration frequencies, vibration damping, energy absorption, buckling loads, the effects of geometric discontinuities such as holes and notches, and many other more global responses are all of interest. In between these two extremes, the response of an individual layer or the response of a group of layers can be of interest.

How does an individual layer respond when subject to stresses? How much does it deform? How much load can it sustain? Similarly, how does a laminate respond when subject to stresses? How much does it deform? How much load can it sustain? What is the influence of neighboring or adjacent layers on any particular layer? What is the effect on the laminate of changing the fiber orientation of any particular layer? What is the effect of changing the material properties of any one particular layer? The list of questions is almost endless. With the continuous introduction of new fibers having greater strength and stiffness, old questions have to be re-answered and new ones are asked. Moreover, new polymer matrix materials are introduced frequently, and issues need to be re-examined in the context of these new materials.

In this study of the mechanics of composite materials and structures, we will present methodologies that will enable engineers, scientists, and designers to answer some of these questions. The starting point will be an examination of the deformations of an element of material taken from a single layer. The element, though small, is assumed to contain many fibers. Rather than starting with the examination of fiber and matrix interaction, this starting point is used because it allows an orderly and smooth transition to the analysis of composite structures, the final products for any fiber-reinforced material. Also, in the analysis of a complete structure, it is impossible to include the response of every fiber to the surrounding matrix material.

Computers are not big enough to allow this. On the other hand, what is responsible for the failure of composite structures is the breaking of the fibers, the breaking of the fiber–matrix interface bond, and the breaking of the matrix. Therefore, micromechanics cannot be overlooked. A glimpse of micromechanics will be provided in Chapter 6 when the issue of estimating the elastic properties of a composite is addressed.

5.1 STRESSES AND DEFORMATION

In discussing the mechanics of fiber-reinforced materials, it is convenient to use an orthogonal coordinate system that has one axis aligned with the fiber direction. We will do so here and identify the system as the 1–2–3 coordinate system or the principal material coordinate system. Figure 5.1 illustrates an isolated layer and the orientation of the principal material coordinate system. The 1 axis is aligned with the fiber direction, the 2 axis is in the plane of the layer and perpendicular to the fibers, and the 3 axis is perpendicular to the plane of the layer and thus also perpendicular to the fibers. The 1 direction is the fiber direction, while the 2 and 3 directions are the matrix directions. The direction perpendicular to the fibers is also called the transverse direction. The terms "matrix direction" and "transverse direction" are somewhat ambiguous because there are two directions that fit either of these descriptions. We shall use matrix direction, and the specific direction (i.e., 2 or 3) will be made clear from the context of the problem.

Stresses, strains, and strengths will ultimately be referred to the principal material coordinate system. The study of the stress–strain response of a single layer is equivalent to determining the relations between the stresses applied to the bounding surfaces of the layer and the deformations of the layer as a whole. The strain of an individual fiber or element of matrix is of no consequence at this level of analysis [1]. The effect of the fiber reinforcement is smeared over the volume of material, and we assume that the two-material fiber–matrix system is replaced by a single

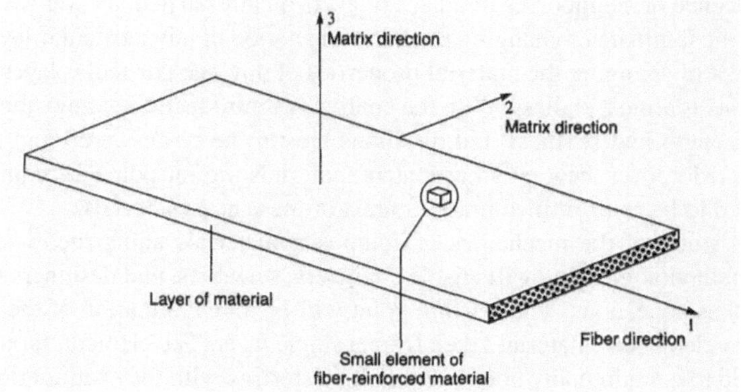

FIGURE 5.1 Principal material coordinate system.

homogenous material. This is an important concept because it makes the analysis of a fiber-reinforced composite easier. Equally important is the fact that this single material does not have the same properties in all directions. It is obviously stronger and stiffer in the 1 direction than in the 2 or 3 direction. In addition, just because the 2 and 3 directions are both perpendicular to the fiber direction, the properties in the 2 and 3 directions are not necessarily equal to each other. A material with different properties in three mutually perpendicular directions is called an orthotropic material. As a result, a layer is said to be orthotropic. The 1–2, 1–3, and 2–3 are three planes, and the material properties are symmetric with respect to each of these planes.

A material with the same properties in all directions is said to be isotropic. Figure 5.2 illustrates a small element of smeared fiber-reinforced material subject to stresses on its six bounding surfaces. As Figure 5.1 shows, this small volume of material has been considered removed from a layer. The normal stress acting on the element face with its outward normal in the 1 direction is denoted as σ_1. The shear stress acting in the 2 direction on that face is denoted as τ_{12}, and the shear stress acting in the 3 direction on that face is denoted as τ_{13}. The normal and shear stresses acting on the other faces are similarly labeled. The extensional strain responses of the element as referenced in the 1–2–3 coordinate system are denoted as ε_1, ε_2, and ε_3, while the engineering shearing strain responses are denoted as γ_{12}, γ_{23}, and γ_{13}, respectively. With this notation, ε_1 is the stretching of the element in the fiber direction, γ_{12} is the change of right angle in the 1–2 plane, and so on [2]. The stress–strain relation for

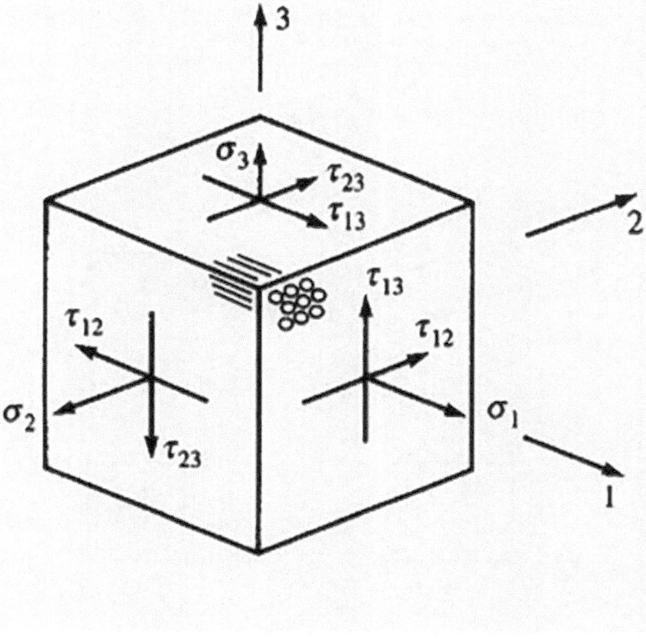

FIGURE 5.2 Stresses acting on a small element of fiber-reinforced material.

the small element of material will be constructed by considering the response of the element to each of the six stress components. As only linear elastic response is to be considered, superposition of the responses will be used to determine the response of the element to a complex or combined stress state.

Figure 5.3a illustrates the element subjected to only a tensile normal stress in the 1 direction, σ_1. Figure 5.3b–d illustrates three views of the element indicating how it would be deformed by this tensile stress. The tensile normal stress σ_1 causes extension of the element in the 1 direction and, due to Poisson effects, contraction in the 2 and 3 directions. There is no a priori reason to believe that the contractions in the 2 and 3 directions are the same. In addition, nothing has been said so far to indicate that the element of material actually contracts. It could expand. In reality, this is not the case for a single layer. However, laminates can be made that will expand rather than contract. Laminate Poisson's ratios will be discussed later. The extensional strain in the 1 direction is related to the tensile normal stress in the 1 direction by the tensile, or extensional, modulus of the equivalent smeared material in the fiber direction E_1. The relation between these quantities is:

$$\varepsilon_1 = \frac{\sigma_1}{E_1}. \tag{5.1}$$

If Poisson's ratio relating contraction in the 2 direction to extension in the 1 direction is defined to be:

$$\nu_{12} = -\frac{\varepsilon_2}{\varepsilon_1}, \tag{5.2}$$

then:

$$\varepsilon_2 = -\nu_{12}\ \varepsilon_1 = -\nu_{12}\frac{\sigma_1}{E_1}. \tag{5.3}$$

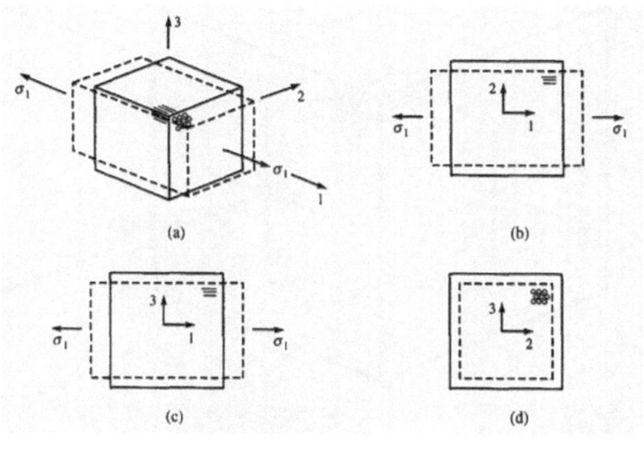

FIGURE 5.3 Deformation of an element due to σ_1. (a) Overall view, (b) as viewed in 1–2 plane, (c) as viewed in 1–3 plane, and (d) as viewed in 2–3 plane.

The subscripts, and their order, on Poisson's ratio are important. The first subscript refers to the direction of the applied tensile stress and the resulting extensional strain. The second subscript refers to the direction of the contraction. According to this convention, the contraction in the 3 direction is related to the extension in the 1 direction by ν_{13}, specifically:

$$\nu_{13} = -\frac{\varepsilon_3}{\varepsilon_1}. \tag{5.4}$$

We rewrite this equation as:

$$\varepsilon_3 = -\nu_{13}\ \varepsilon_1 = -\nu_{13}\frac{\sigma_1}{E_1}. \tag{5.5}$$

If instead of applying a tensile normal stress in the 1 direction, a tensile normal stress is applied in the 2 direction, then the element of smeared material will deform as in Figure 5.4. Because of the softness of the material perpendicular to the fibers, the element will extend more easily in the 2 direction than in the 1 direction. Because the stiff fibers tend to counter any Poisson effect, the contraction in the 1 direction will be minimal. In contrast, the contraction in the 3 direction will be large; there is only matrix and the soft diametral direction of the fiber-resisting deformation.

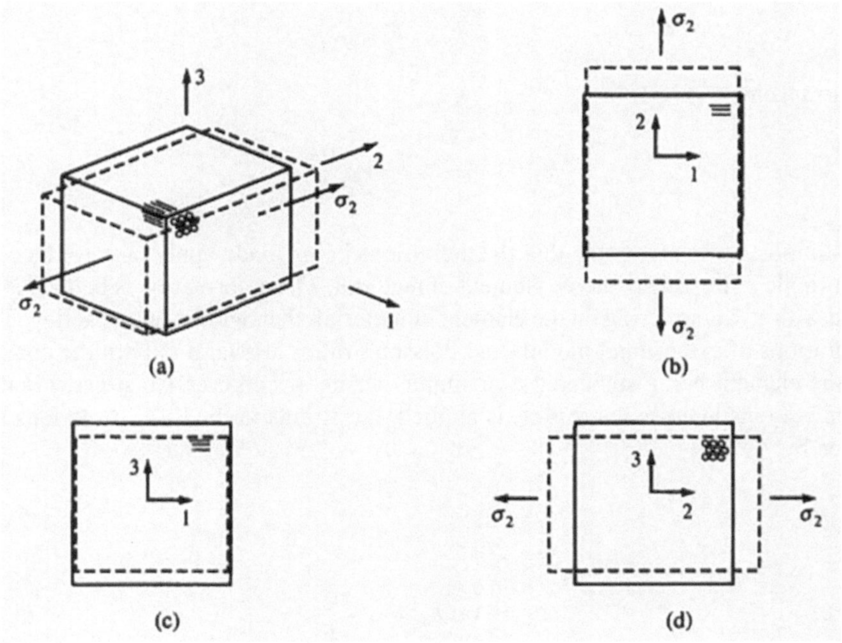

FIGURE 5.4 Deformation of an element due to σ_2. (a) Overall view, (b) as viewed in 1–2 plane, (c) as viewed in 1–3 plane, and (d) as viewed in 2–3 plane.

For the loading of Figure 5.4, the tensile normal stress in the 2 direction, σ_2, and the extensional strain in the 2 direction are related by another extensional modulus through the relation:

$$\varepsilon_2 = \frac{\sigma_2}{E_2}. \tag{5.6}$$

As might be expected, because the stress is acting perpendicularly to the fibers, E_2 is much smaller than E_1. Using the subscript convention established for Poisson's ratios, the contraction in the 1 direction is related to the extension in the 2 direction by yet another Poisson's ratio, namely:

$$\nu_{21} = -\frac{\varepsilon_1}{\varepsilon_2}. \tag{5.7}$$

As a result, due to σ_2:

$$\varepsilon_1 = -\nu_{21}\ \varepsilon_2 = -\nu_{21}\frac{\sigma_2}{E_2}. \tag{5.8}$$

Similarly, the contraction in the 3 direction is related to the extension in the 2 direction by ν_{23}, which is defined as:

$$\nu_{23} = -\frac{\varepsilon_3}{\varepsilon_2}. \tag{5.9}$$

Rearrangement gives:

$$\varepsilon_3 = -\nu_{23}\ \varepsilon_2 = -\nu_{23}\frac{\sigma_2}{E_2}. \tag{5.10}$$

It is important to recognize that the definitions being made apply only to the case of a single stress acting on the element of material. The deformation is being examined with σ_1 alone acting on the element of material, then with σ_2 alone acting. The definitions of extensional moduli and Poisson's ratios are valid only in the context of the element being subjected to a simple tensile or compressive stress. Finally, if only a tensile normal stress σ_3 is applied, the strains in the three directions are given by:

$$\begin{aligned} \varepsilon_3 &= \frac{\sigma_3}{E_3} \\ \varepsilon_2 &= -\nu_{32}\ \varepsilon_3 = -\nu_{32}\frac{\sigma_3}{E_3} \\ \varepsilon_1 &= -\nu_{31}\ \varepsilon_3 = -\nu_{31}\frac{\sigma_3}{E_3} \end{aligned} \tag{5.11}$$

In the above, E_3 is the extensional modulus in the 3 direction, ν_{32} relates contraction in the 2 direction and extension in the 3 direction, and ν_{31} relates contraction in the 1 direction and extension in the 3 direction. This is for the case of only stress σ_3 being applied. Again, due to the relative stiffness in the fiber and matrix directions, σ_3 stress will not cause much contraction in the 1 direction. Contraction in the 2 direction will be much larger, as Figure 5.5 shows.

If all three tensile stresses are applied simultaneously, the strain in any one direction is a result of the combined effects, namely:

$$\begin{Bmatrix} \varepsilon_1 \\ \varepsilon_2 \\ \varepsilon_3 \end{Bmatrix} = \begin{bmatrix} \dfrac{1}{E_1} & \dfrac{-\nu_{21}}{E_2} & \dfrac{-\nu_{31}}{E_3} \\ \dfrac{-\nu_{12}}{E_1} & \dfrac{1}{E_2} & \dfrac{-\nu_{32}}{E_3} \\ \dfrac{-\nu_{13}}{E_1} & \dfrac{-\nu_{23}}{E_2} & \dfrac{1}{E_3} \end{bmatrix} \begin{Bmatrix} \sigma_1 \\ \sigma_2 \\ \sigma_3 \end{Bmatrix}. \quad (5.12)$$

Note that a given component of extensional strain, say, ε_2, is a result of the combined effects of the three components of normal stress. The normal stresses and the extensional strains are completely coupled; that is, the matrix is full. Poisson's ratio ν_{12} is generally referred to as the major Poisson's ratio. The effects of the shearing stress

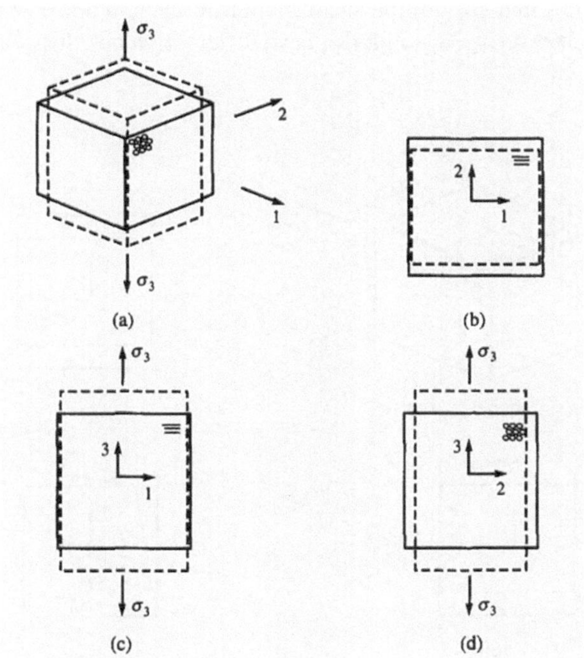

FIGURE 5.5 Deformation of an element due to σ_3. (a) Overall view, (b) as viewed in 1–2 plane, (c) as viewed in 1–3 plane, and (d) as viewed in 2–3 plane.

are less complicated. For an orthotropic material, there is no coupling among the three shear deformations. Figure 5.6 illustrates the deformation of the small element of composite subjected to a shear stress τ_{23}, with the shearing stress causing right angles in the 2–3 plane to change. All other angles of the element remain orthogonal. It is important to note the sense of the change in right angles in the various corners of the element due to a positive shear stress, with some right angles decreasing and some increasing. As the engineering shear stress y denotes the change in right angle, the relation between the applied shear stress in the 2–3 plane and the change in right angle in the 2–3 plane is given by:

$$\gamma_{23} = \frac{\tau_{23}}{G_{23}}. \tag{5.13}$$

The quantity G_{23} is called the shear modulus in the 2–3 plane and γ_{23} is the engineering shear strain in the 2–3 plane. The convention established for shearing in the 2–3 plane can be easily extended to the 1–3 and 1–2 planes. As Figure 5.7 illustrates, shear stress τ_{13} causes right angles in the 1–3 plane to change but all other angles in the cube remain orthogonal. Similarly, as in Figure 5.8, a shear stress τ_{12} causes only the angles in the 1–2 plane to deform. As a result,

$$\gamma_{13} = \frac{\tau_{13}}{G_{13}} = \frac{\tau_{12}}{G_{12}}. \tag{5.14}$$

The quantities G_{13} and G_{12} are the shear moduli in the 1–3 and 1–2 planes, respectively, and in general G_{12}, G_{13}, and G_{23} have different values. It is easy to envision,

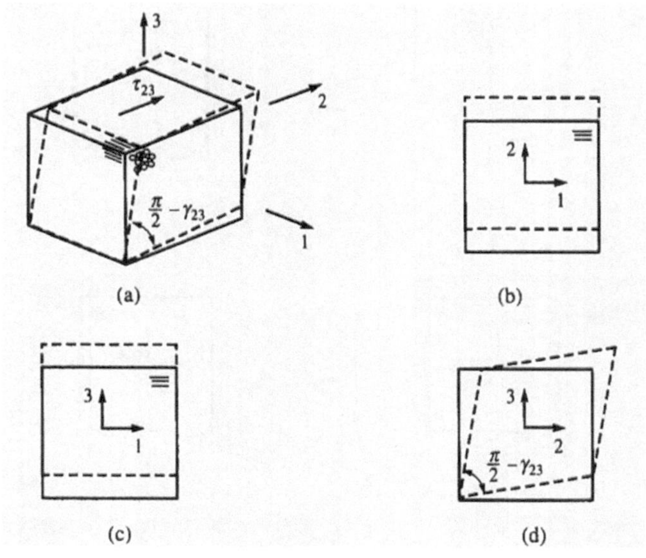

FIGURE 5.6 Deformation of an element due to τ_{23}. (a) Overall view, (b) as viewed in 1–2 plane, (c) as viewed in 1–3 plane, and (d) as viewed in 2–3 plane.

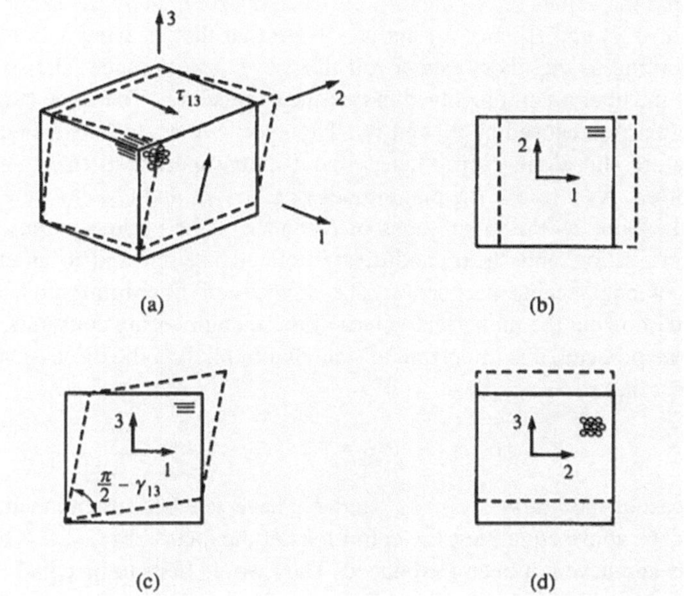

FIGURE 5.7 Deformation of an element due to τ_{13}. (a) Overall view, (b) as viewed in 1–2 plane, (c) as viewed in 1–3 plane, and (d) as viewed in 2–3 plane.

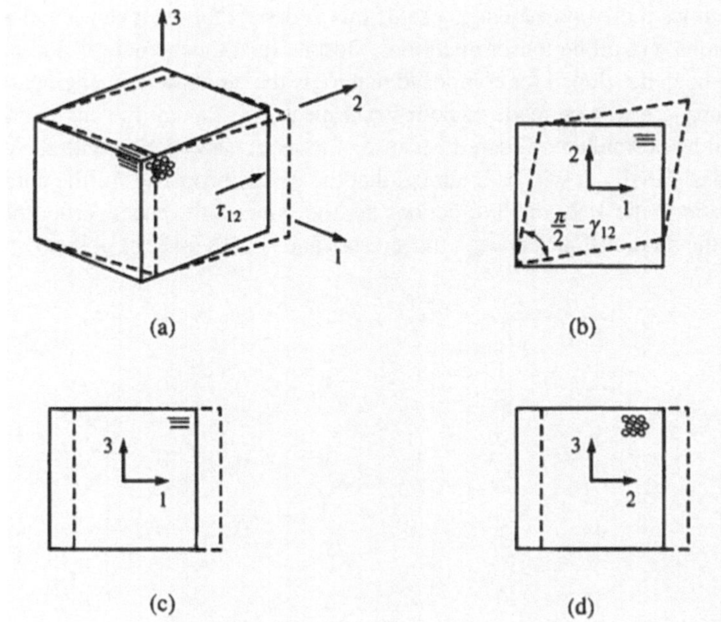

FIGURE 5.8 Deformation of an element due to τ_{12}. (a) Overall view, (b) as viewed in 1–2 plane, (c) as viewed in 1–3 plane, and (d) as viewed in 2–3 plane.

however, that the values of G_{12} and G_{13} could be approximately the same. The same can be said of E_2 and E_3, and ν_{12} and ν_{13}. Note that the shearing action caused by τ_{23} does nothing to the fibers except roll them over one another. The stiffness and strength of the fiber are not involved in this shearing action. The same can be said of the shearing action caused by τ_{13} and τ_{12}. Though the fibers are not rolled over each other, they are slid along each other, without their superior stiffness and strength being involved. As a result, the magnitudes of G_{12}, G_{13}, and G_{23} can be expected to be about the same as the magnitudes of E_2 and E_3. The various extensional moduli, Poisson's ratios, and shear moduli are collectively referred to as engineering constants, or engineering properties. The analysis of fiber-reinforced composites depends on knowing the numerical values of these engineering constants.

Before we proceed, it is important to state that implicit in the discussion so far has been the fact that:

$$\tau_{21} = \tau_{12} \quad \tau_{31} = \tau_{13} \quad \tau_{32} = \tau_{23}. \tag{5.15}$$

For this reason, the stresses τ_{21}, τ_{31}, and τ_{32} have not been mentioned explicitly. Because of the above equalities, the definitions of the quantities G_{21}, G_{31}, and G_{32} are superfluous and have not been introduced. They would have to be equal to G_{12}, G_{13}, and G_{23}, respectively. Also, in the application of the theory of elasticity [3,4] to the analysis of composite materials, the definitions of the tensor shear strains, namely:

$$\varepsilon_{12} = \frac{\gamma_{12}}{2}, \quad \varepsilon_{13} = \frac{\gamma_{13}}{2}, \quad \varepsilon_{23} = \frac{\gamma_{23}}{2}, \tag{5.16}$$

are sometimes more convenient to use. If this is done, then the extensional and shear deformations will all be tensor quantities. Despite this, much analysis and nomenclature have been developed for composite materials that are based on engineering shear strain. Here, use will be made of both strain measures, though the engineering shear strain will be normally considered because of its direct association with the change in right angles. Finally, it will be assumed that the elastic properties of the composite in compression in the 1, 2, and 3 directions are the same as the elastic properties in tension. All the relationships between the stresses and strains take the collective form:

$$\begin{Bmatrix} \varepsilon_1 \\ \varepsilon_2 \\ \varepsilon_3 \\ \gamma_{23} \\ \gamma_{13} \\ \gamma_{12} \end{Bmatrix} = \begin{bmatrix} \frac{1}{E_1} & \frac{-\nu_{21}}{E_2} & \frac{-\nu_{31}}{E_3} & 0 & 0 & 0 \\ \frac{-\nu_{12}}{E_1} & \frac{1}{E_2} & \frac{-\nu_{32}}{E_3} & 0 & 0 & 0 \\ \frac{-\nu_{13}}{E_1} & \frac{-\nu_{23}}{E_2} & \frac{1}{E_3} & 0 & 0 & 0 \\ 0 & 0 & 0 & \frac{1}{G_{23}} & 0 & 0 \\ 0 & 0 & 0 & 0 & \frac{1}{G_{13}} & 0 \\ 0 & 0 & 0 & 0 & 0 & \frac{1}{G_{12}} \end{bmatrix} \begin{Bmatrix} \sigma_1 \\ \sigma_2 \\ \sigma_3 \\ \tau_{23} \\ \tau_{13} \\ \tau_{12} \end{Bmatrix}. \tag{5.17}$$

The square six-by-six matrix of material properties is called the compliance matrix, commonly denoted by S. In terms of S, the stress–strain relations are written as given by Eq. (5.18):

$$\begin{Bmatrix} \varepsilon_1 \\ \varepsilon_2 \\ \varepsilon_3 \\ \gamma_{23} \\ \gamma_{13} \\ \gamma_{12} \end{Bmatrix} = \begin{bmatrix} S_{11} & S_{12} & S_{13} & 0 & 0 & 0 \\ S_{21} & S_{22} & S_{23} & 0 & 0 & 0 \\ S_{31} & S_{32} & S_{33} & 0 & 0 & 0 \\ 0 & 0 & 0 & S_{44} & 0 & 0 \\ 0 & 0 & 0 & 0 & S_{55} & 0 \\ 0 & 0 & 0 & 0 & 0 & S_{66} \end{bmatrix} \begin{Bmatrix} \sigma_1 \\ \sigma_2 \\ \sigma_3 \\ \tau_{23} \\ \tau_{13} \\ \tau_{12} \end{Bmatrix}. \tag{5.18}$$

With this notation:

$$\begin{aligned} &S_{11} = \frac{1}{E_1} \quad S_{12} = \frac{-\nu_{21}}{E_2} \quad S_{13} = \frac{-\nu_{31}}{E_3} \\ &S_{21} = \frac{-\nu_{12}}{E_1} \quad S_{22} = \frac{1}{E_2} \quad S_{23} = \frac{-\nu_{32}}{E_3} \\ &S_{31} = \frac{-\nu_{13}}{E_1} \quad S_{32} = \frac{-\nu_{23}}{E_2} \quad S_{33} = \frac{1}{E_3} \\ &S_{44} = \frac{1}{G_{23}} \quad S_{55} = \frac{1}{G_{13}} \quad S_{66} = \frac{1}{G_{12}} \end{aligned} . \tag{5.19}$$

The inverse of the compliance matrix is called the stiffness matrix, sometimes called the modulus matrix or the elasticity matrix, and is commonly denoted by C. With the inverse defined, the stress–strain relations become:

$$\begin{Bmatrix} \sigma_1 \\ \sigma_2 \\ \sigma_3 \\ \tau_{23} \\ \tau_{31} \\ \tau_{12} \end{Bmatrix} = \begin{bmatrix} C_{11} & C_{12} & C_{13} & 0 & 0 & 0 \\ C_{21} & C_{22} & C_{23} & 0 & 0 & 0 \\ C_{31} & C_{32} & C_{33} & 0 & 0 & 0 \\ 0 & 0 & 0 & C_{44} & 0 & 0 \\ 0 & 0 & 0 & 0 & C_{55} & 0 \\ 0 & 0 & 0 & 0 & 0 & C_{66} \end{bmatrix} \begin{Bmatrix} \varepsilon_1 \\ \varepsilon_2 \\ \varepsilon_3 \\ \gamma_{23} \\ \gamma_{31} \\ \gamma_{12} \end{Bmatrix}. \tag{5.20}$$

Clearly, the C_{ij} can be written in terms of the S_{ij}, and ultimately in terms of the engineering constants. For shorthand notation, the relations between stress and strain will be abbreviated by:

$$\{\varepsilon\}_1 = [S]\{\sigma\}_1 \;\; \{\sigma\}_1 = [C]\{\varepsilon\}_1, \tag{5.21}$$

where

$$\{\varepsilon\}_1 = \begin{Bmatrix} \varepsilon_1 \\ \varepsilon_2 \\ \varepsilon_3 \\ \gamma_{23} \\ \gamma_{31} \\ \gamma_{12} \end{Bmatrix} \text{ and } \{\sigma\}_1 = \begin{Bmatrix} \sigma_1 \\ \sigma_2 \\ \sigma_3 \\ \tau_{23} \\ \tau_{31} \\ \tau_{12} \end{Bmatrix}. \tag{5.22}$$

The subscript 1 outside the brackets means that the stresses and strains are referred to the 1–2–3 coordinate system.

As seen by Eq. (5.19), the compliance matrix involves 12 engineering properties: three extensional moduli (E_1, E_2, and E_3), six Poisson's ratios (ν_{12}, ν_{21}, ν_{13}, ν_{31}, ν_{32}, and ν_{23}), and three shear moduli (G_{23}, G_{13}, and G_{12}). As a result, the stiffness matrix also depends on 12 engineering constants. However, the 12 engineering properties are not all independent. This is a very important point. There are actually only nine independent material properties. The so-called reciprocity relationships can be established among the extensional moduli and the Poisson's ratios. As a result of these reciprocity relationships, the compliance and stiffness matrices are symmetric. To establish these relations, it is convenient to use the Maxwell–Betti reciprocal theorem [5,6]. In the next two sections, the theorem is briefly reviewed and used to establish the reciprocity relations.

5.2 MAXWELL–BETTI RECIPROCAL THEOREM

Consider an elastic body acted upon by two sets of loads, $\vec{P}_1, \vec{P}_2, \ldots, \vec{P}_M$ and $\vec{p}_1, \vec{p}_2, \ldots, \vec{p}_N$, where the overbar arrow denotes a vector quantity. These two sets of loads act at different locations on the body. For this discussion, the two sets of loads are to be applied to the body in two specific sequences. Figure 5.9a illustrates a body with the two sets of loads, $\vec{P}_m$, $i = 1, M$ and $\vec{p}_n$, $n = 1, N$. Assume, as in Figure 5.9b, load set $\vec{P}_m$ is first applied to the body. The body deforms. These deformations are of no consequence here. Subsequent application of load set $\vec{p}_n$, as in Figure 5.9c, causes the body to deform further. In particular, application of the load set $\vec{p}_n$ causes displacements at points of application of load set $\vec{P}_m$. Denoting these displacements by $\vec{d}_m$, the work done by load set $\vec{P}_m$ due to these displacements is given by the dot product:

$$W_{P/p} = \sum_{m=1}^{M} \vec{P}_m \cdot \vec{d}_m. \tag{5.23}$$

The subscript on W denotes the fact that the work is due to loads $\vec{P}_m$ moving through the displacements caused by the application of loads $\vec{p}_n$. Obviously, load set $\vec{P}_m$ does work as it is initially applied to the body and the body deforms. Like the initial deformations caused by $\vec{P}_m$, this work is not involved in this discussion. Conversely, as in Figure 5.9d, assume that load set $\vec{p}_n$ is first applied to the body. The body deforms. Neither these

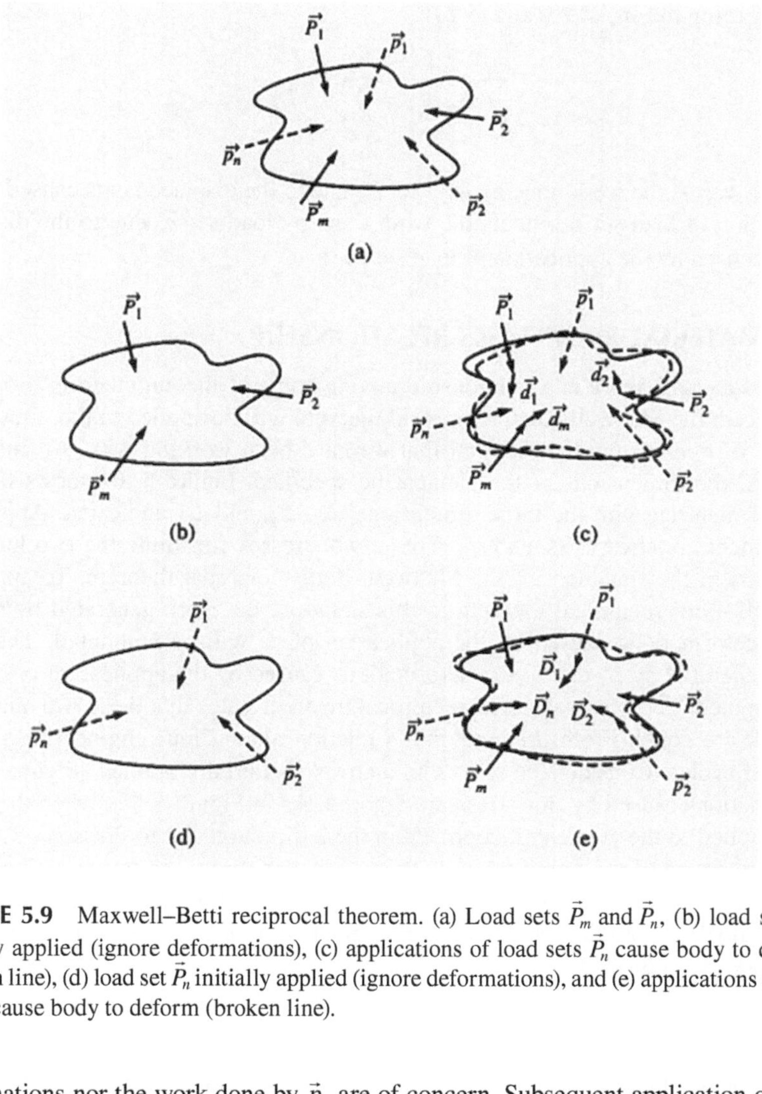

FIGURE 5.9 Maxwell–Betti reciprocal theorem. (a) Load sets $\vec{P}_m$ and $\vec{P}_n$, (b) load sets $\vec{P}_m$ initially applied (ignore deformations), (c) applications of load sets $\vec{P}_n$ cause body to deform (broken line), (d) load set $\vec{P}_n$ initially applied (ignore deformations), and (e) applications of load set $\vec{P}_m$ cause body to deform (broken line).

deformations nor the work done by $\vec{p}_n$ are of concern. Subsequent application of load set $\vec{P}_m$, as shown in Figure 5.9e, causes further deformation. In particular, the body deforms at points of application of the load set $\vec{p}_n$. These displacements are denoted by $\vec{D}_n$; the work done by load set $\vec{p}_n$ due to these displacements is given by:

$$W_{p/P} = \sum_{n=1}^{N} \vec{p}_n \cdot \vec{D}_n. \tag{5.24}$$

The Maxwell–Betti reciprocal theorem states that these two quantities of work are equal; that is:

$$W_{P/p} = W_{p/P}. \tag{5.25}$$

Or, according to Eqs. (5.23) and (5.24):

$$\sum_{m=1}^{M} \vec{P}_m \cdot \vec{d}_m = \sum_{n=1}^{N} \vec{p}_n \cdot \vec{D}_n. \tag{5.26}$$

In other words, the work done by load set $\vec{P}_m$ due to the displacements caused by the application of load set $\vec{p}_n$ equals the work done by load set $\vec{p}_n$ due to the displacements caused by the application of load set $\vec{P}_m$.

5.3 MATERIAL PROPERTIES RELATIONSHIP

To illustrate that there is a relationship among some of the engineering properties introduced, the Maxwell–Betti reciprocal theorem will be applied to the same small volume of fiber-reinforced material that we have been working with. At this time, however, the dimensions of the element be specified. Figure 5.10a shows the element of material with the three dimensions, Δ_1, Δ_2, and Δ_3, indicated. Applied to the element are stresses σ_1 and σ_2. These two stresses constitute the two load sets discussed in the statement of the Maxwell–Betti reciprocal theorem. To apply the Maxwell–Betti reciprocal theorem to this situation, the work generated by σ_2 due to the deformations caused by the application of σ_1 will be computed. Then, the work generated by σ_1 due to the deformations caused by the application of σ_2 will be computed. The Maxwell–Betti reciprocal theorem states that these two quantities of work are equal. From this equality, a relation among four engineering properties will evolve. Consider the case with a stress σ_2 initially applied (Figure 5.10b). Deformations caused by this stress are ignored. As in Figure 5.10c, when the stress σ_1 is applied to the element, it contracts in the 2 direction due to Poisson's ratio ν_{12}. The application of σ_1 causes σ_2 to do work. Since σ_2 moves in the direction opposite to which it is acting, this work is actually negative. (For simplicity, Figure 5.10c shows all the contraction occurring at the upper component of σ_2.) Let us compute this work. By definition:

$$\varepsilon_2 = -\nu_{12}\varepsilon_1. \tag{5.27}$$

The actual displacement of σ_2, denoted by $\delta\Delta_2$, is the strain times the element dimension in the 2 direction. That is:

$$\delta\Delta_2 = \varepsilon_2\Delta_2. \tag{5.28}$$

If we substitute from Eq. (5.27):

$$\delta\Delta_2 = -\nu_{12}\varepsilon_1\Delta_2. \tag{5.29}$$

The strain ε_1 is caused by σ_1 and is given by Eq. (5.1), namely:

$$\varepsilon_1 = \frac{\sigma_1}{E_1}. \tag{5.30}$$

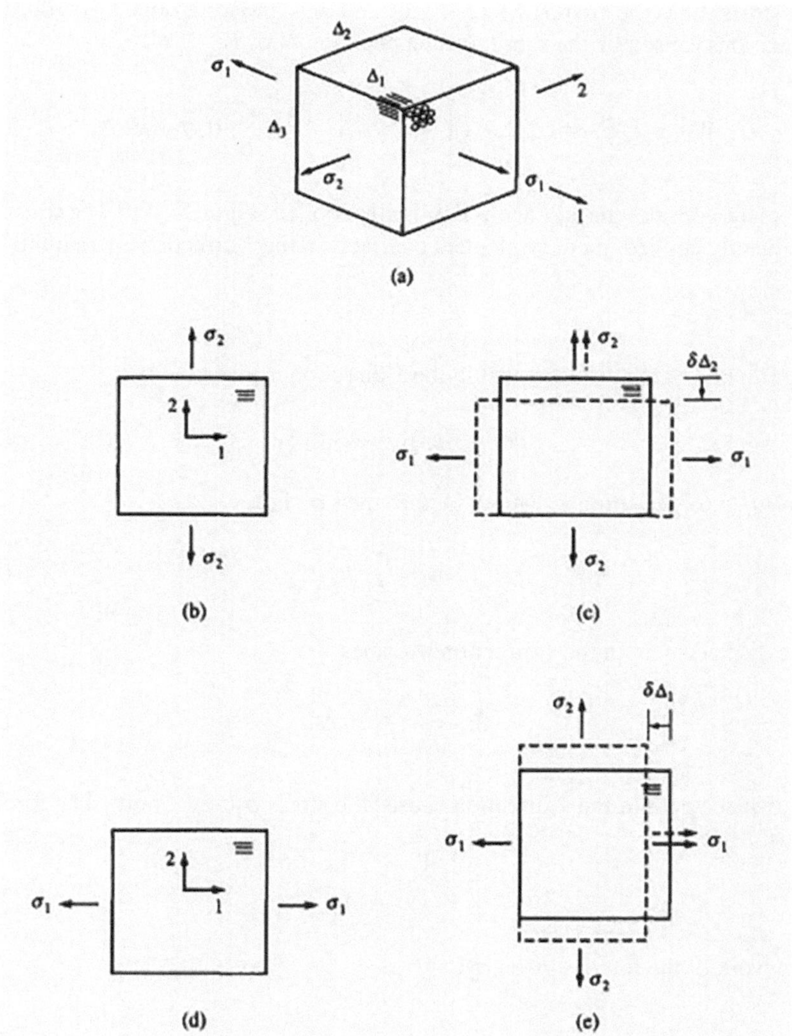

FIGURE 5.10 Application of the Maxwell–Betti reciprocal theorem to an element of fiber-reinforced material. (a) Element with σ_1 and σ_2 applied, (b) σ_2 initially applied, (c) deformations caused by the application of σ_1, (d) σ_1 initially applied, and (e) deformations caused by the application of σ_2.

Therefore, the displacement of σ_2 due to the application of σ_1 can be expressed as:

$$\delta\Delta_2 = -\frac{v_{12}}{E_1}\sigma_1\Delta_2. \tag{5.31}$$

The force due to σ_2 is:

$$F_2 = \sigma_2\Delta_1\Delta_3 \tag{5.32}$$

or the stress times the area on which it acts, and thus, the work done by σ_2 due to the deformations caused by the subsequent application of σ_1 is:

$$W_{2/1} = F_2\delta\Delta_2 = (\sigma_2\Delta_1\Delta_3)\left(-\frac{\nu_{12}}{E_1}\sigma_1\Delta_2\right) = -\frac{\nu_{12}}{E_1}\sigma_1\sigma_2\Delta_1\Delta_2\Delta_3. \tag{5.33}$$

Conversely, consider the case with σ_1, initially applied (Figure 5.10d). If a stress σ_2 is subsequently applied, then the element contracts in the 1 direction an amount:

$$\varepsilon_1 = -\nu_{21}\varepsilon_2. \tag{5.34}$$

Using Eq. (5.34), the displacement in the 1 direction is given by:

$$\delta\Delta_1 = -\varepsilon_1\Delta_1 = -\nu_{21}\varepsilon_2\Delta_1. \tag{5.35}$$

From Eq. (5.6), the strain ε_2 caused by the stress σ_2 is:

$$\varepsilon_2 = \frac{\sigma_2}{E_2}. \tag{5.36}$$

So the displacement in the 1 direction becomes:

$$\delta\Delta_1 = -\frac{\nu_{21}}{E_2}\sigma_2\Delta_1. \tag{5.37}$$

This displacement in the 1 direction causes the stress σ_1 to do work. The force due to σ_1 is:

$$F_1 = \sigma_1\Delta_2\Delta_3 \tag{5.38}$$

So the work of the force is given by:

$$W_{1/2} = F_1\delta\Delta_1 = (\sigma_1\Delta_2\Delta_3)\left(-\frac{\nu_{21}}{E_2}\sigma_2\Delta_1\right) = -\frac{\nu_{21}}{E_2}\sigma_1\sigma_2\Delta_1\Delta_2\Delta_3. \tag{5.39}$$

By the Maxwell–Betti reciprocal theorem:

$$W_{2/1} = W_{1/2}. \tag{5.40}$$

Or, using the results of Eqs. (5.33) and (5.39):

$$-\frac{\nu_{12}}{E_1}\sigma_1\sigma_2\Delta_1\Delta_2\Delta_3 = -\frac{\nu_{21}}{E_2}\sigma_1\sigma_2\Delta_1\Delta_2\Delta_3. \tag{5.41}$$

After we simplify, a relation among the material properties involved is immediately obvious, namely:

$$\frac{\nu_{12}}{E_1} = \frac{\nu_{21}}{E_2}. \tag{5.42}$$

From this equation, it is clear that the two extensional moduli E_1 and E_2, and the two Poisson's ratios ν_{12} and ν_{21} for the material are not completely arbitrary. If one knows any three of the properties, the fourth one can be determined. Similar considerations for work by pairs of stresses σ_1 and σ_3, and σ_2 and σ_3 lead to reciprocity relations for the other extensional moduli and Poisson's ratios, namely:

$$\frac{\nu_{13}}{E_1} = \frac{\nu_{31}}{E_3} \quad \frac{\nu_{23}}{E_2} = \frac{\nu_{32}}{E_3}. \tag{5.43}$$

Because of the three reciprocity relations, only nine independent constants are needed to describe the linear elastic behavior of a fiber-reinforced material. Also, because of the reciprocity relations, from Eq. (5.19):

$$\begin{aligned} S_{21} &= \frac{-\nu_{12}}{E_1} = \frac{-\nu_{21}}{E_2} = S_{12} \\ S_{31} &= \frac{-\nu_{13}}{E_1} = \frac{-\nu_{31}}{E_3} = S_{13}\,. \\ S_{32} &= \frac{-\nu_{23}}{E_2} = \frac{-\nu_{32}}{E_3} = S_{23} \end{aligned} \tag{5.44}$$

As a result, the compliance matrix and therefore the stiffness matrix are symmetric. The symmetry of these two matrices is an important property. After we incorporate the reciprocity relations, the stress–strain relations in terms of the compliances are:

$$\begin{Bmatrix} \varepsilon_1 \\ \varepsilon_2 \\ \varepsilon_3 \\ \gamma_{23} \\ \gamma_{13} \\ \gamma_{12} \end{Bmatrix} = \begin{bmatrix} S_{11} & S_{12} & S_{13} & 0 & 0 & 0 \\ S_{12} & S_{22} & S_{23} & 0 & 0 & 0 \\ S_{13} & S_{23} & S_{33} & 0 & 0 & 0 \\ 0 & 0 & 0 & S_{44} & 0 & 0 \\ 0 & 0 & 0 & 0 & S_{55} & 0 \\ 0 & 0 & 0 & 0 & 0 & S_{66} \end{bmatrix} \begin{Bmatrix} \sigma_1 \\ \sigma_2 \\ \sigma_3 \\ \tau_{23} \\ \tau_{13} \\ \tau_{12} \end{Bmatrix}. \tag{5.45}$$

In the above equation:

$$\begin{aligned} S_{11} &= \frac{1}{E_1} \quad S_{12} = \frac{-\nu_{12}}{E_1} \quad S_{13} = \frac{-\nu_{13}}{E_1} \\ S_{22} &= \frac{1}{E_2} \quad S_{23} = \frac{-\nu_{23}}{E_2} \quad S_{33} = \frac{1}{E_3}\;. \\ S_{44} &= \frac{1}{G_{23}} \quad S_{55} = \frac{1}{G_{13}} \quad S_{66} = \frac{1}{G_{12}} \end{aligned} \tag{5.46}$$

The inverse relations are:

$$\begin{Bmatrix} \sigma_1 \\ \sigma_2 \\ \sigma_3 \\ \tau_{23} \\ \tau_{31} \\ \tau_{12} \end{Bmatrix} = \begin{bmatrix} C_{11} & C_{12} & C_{13} & 0 & 0 & 0 \\ C_{12} & C_{22} & C_{23} & 0 & 0 & 0 \\ C_{13} & C_{23} & C_{33} & 0 & 0 & 0 \\ 0 & 0 & 0 & C_{44} & 0 & 0 \\ 0 & 0 & 0 & 0 & C_{55} & 0 \\ 0 & 0 & 0 & 0 & 0 & C_{66} \end{bmatrix} \begin{Bmatrix} \varepsilon_1 \\ \varepsilon_2 \\ \varepsilon_3 \\ \gamma_{23} \\ \gamma_{31} \\ \gamma_{12} \end{Bmatrix}. \tag{5.47}$$

In terms of the compliances, the components of the stiffness matrix are given by:

$$\begin{aligned} C_{11} &= \frac{S_{22}S_{33} - S_{23}S_{23}}{S} & C_{12} &= \frac{S_{13}S_{23} - S_{12}S_{33}}{S} \\ C_{22} &= \frac{S_{33}S_{11} - S_{13}S_{13}}{S} & C_{13} &= \frac{S_{12}S_{23} - S_{13}S_{22}}{S} \\ C_{33} &= \frac{S_{11}S_{22} - S_{12}S_{12}}{S} & C_{23} &= \frac{S_{12}S_{13} - S_{23}S_{11}}{S} \\ C_{44} &= \frac{1}{S_{44}} \quad C_{55} = \frac{1}{S_{55}} & C_{66} &= \frac{1}{S_{66}}, \end{aligned} \tag{5.48}$$

where

$$S = S_{11}S_{22}S_{33} - S_{11}S_{23}S_{23} - S_{22}S_{13}S_{13} - S_{33}S_{12}S_{12} + 2S_{12}S_{23}S_{13}. \tag{5.49}$$

If needed, the compliances in Eq. (5.48) can be written in terms of the engineering properties, and therefore, the stiffnesses can be expressed directly in terms of the engineering properties. Despite the reciprocity relations among some of the material properties, an element of fiber-reinforced material still has different material properties in each of the three principal material directions. As we mentioned previously, such a material is called orthotropic. The zero entries in the upper-right and lower-left portions of both the compliance and stiffness matrices characterize orthotropic behavior. If a material is orthotropic and the stress–strain relations are written in the principal coordinate material system, then the compliance and stiffness matrices will always have these zero entries. We shall see at a later time that if the material is orthotropic, but the stress–strain relations are written in a coordinate system other than the principal one, then some of the zero entries become nonzero. It is possible to find materials that have nonzero entries in the upper-right and lower-left portions of their compliance and stiffness matrices for every coordinate system. Such a material is said to be anisotropic. We choose here to concentrate on an orthotropic material because it is the building block of most composite materials.

For isotropic materials:

$$\begin{aligned} &E_1 = E_2 = E_3 = E \\ &\nu_{23} = \nu_{13} = \nu_{12} = \nu \\ &G_{23} = G_{13} = G_{12} = G = \frac{E}{2(1+\nu)} \end{aligned} \quad . \tag{5.50}$$

As a result, the compliance matrix is:

$$\begin{bmatrix} S_{11} & S_{12} & S_{12} & 0 & 0 & 0 \\ S_{12} & S_{11} & S_{12} & 0 & 0 & 0 \\ S_{12} & S_{12} & S_{11} & 0 & 0 & 0 \\ 0 & 0 & 0 & S_{44} & 0 & 0 \\ 0 & 0 & 0 & 0 & S_{44} & 0 \\ 0 & 0 & 0 & 0 & 0 & S_{44} \end{bmatrix}, \tag{5.51a}$$

where

$$S_{11} = \frac{1}{E} \quad S_{12} = \frac{-\nu}{E} \quad S_{44} = \frac{1}{G} = \frac{2(1+\nu)}{E}. \tag{5.51b}$$

The lack of any directional dependence is reflected in the fact that the off-diagonal terms are identical, and the on-diagonal terms for the three components of shear are identical, as are the other three on-diagonal terms. Likewise, the stiffness matrix becomes:

$$\begin{bmatrix} C_{11} & C_{12} & C_{12} & 0 & 0 & 0 \\ C_{12} & C_{11} & C_{12} & 0 & 0 & 0 \\ C_{12} & C_{12} & C_{11} & 0 & 0 & 0 \\ 0 & 0 & 0 & C_{44} & 0 & 0 \\ 0 & 0 & 0 & 0 & C_{44} & 0 \\ 0 & 0 & 0 & 0 & 0 & C_{44} \end{bmatrix}, \tag{5.52a}$$

where

$$C_{11} = \frac{E}{(1+\nu)(1+2\nu)} \quad C_{12} = \frac{(1-\nu)E}{(1+\nu)(1+2\nu)} \quad C_{44} = G = \frac{E}{2(1+\nu)}. \tag{5.52b}$$

Between orthotropic material behavior and isotropic material behavior lies a third type of material behavior, namely, transversely isotropic behavior. For an element of fiber-reinforced material, it is often assumed that the material behavior in the 2 direction is identical to the material behavior in the 3 direction. As these directions are both perpendicular to the fiber direction, assuming identical properties in these directions is understandable. For this situation:

$$E_2 = E_3 \quad \nu_{12} = \nu_{13} \quad G_{12} = G_{13}, \tag{5.53a}$$

and more importantly:

$$G_{23} = \frac{E_2}{2(1+\nu_{23})}. \tag{5.53b}$$

If Eqs. (5.53a) and (5.53b) are true, then the material is said to be isotropic in the 2–3 plane, or transversely isotropic in the 2–3 plane. Any particular property is independent of direction within that plane. With this characteristic, the compliance matrix takes the shape as given by Eq. (5.54a):

$$\begin{bmatrix} S_{11} & S_{12} & S_{12} & 0 & 0 & 0 \\ S_{12} & S_{22} & S_{23} & 0 & 0 & 0 \\ S_{12} & S_{23} & S_{22} & 0 & 0 & 0 \\ 0 & 0 & 0 & S_{44} & 0 & 0 \\ 0 & 0 & 0 & 0 & S_{55} & 0 \\ 0 & 0 & 0 & 0 & 0 & S_{55} \end{bmatrix}, \tag{5.54a}$$

where

$$\begin{aligned} S_{11} &= \frac{1}{E_1} \quad S_{12} = \frac{-\nu_{12}}{E_1} \\ S_{22} &= \frac{1}{E_2} \quad S_{23} = \frac{-\nu_{23}}{E_2} \\ S_{44} &= \frac{1}{G_{23}} = \frac{2(1+\nu_{23})}{E_2} \quad S_{55} = \frac{1}{G_{12}} \end{aligned} \tag{5.54b}$$

And the stiffness matrix becomes:

$$\begin{bmatrix} C_{11} & C_{12} & C_{12} & 0 & 0 & 0 \\ C_{12} & C_{22} & C_{23} & 0 & 0 & 0 \\ C_{12} & C_{23} & C_{22} & 0 & 0 & 0 \\ 0 & 0 & 0 & C_{44} & 0 & 0 \\ 0 & 0 & 0 & 0 & C_{55} & 0 \\ 0 & 0 & 0 & 0 & 0 & C_{55} \end{bmatrix}. \tag{5.55}$$

For a transversely isotropic material, there are five independent material properties: E_1, E_2, ν_{12}, ν_{23}, and G_{12}.

5.4 TYPICAL PROPERTIES OF MATERIALS

Representative numerical values of the engineering properties of two common fiber-reinforced composite materials are given in Table 5.1, namely, an intermediate-modulus graphite-reinforced polymeric material and an S-glass-reinforced polymeric material [7,8]. For consistency, the numerical values in the table will be used throughout this book. The numerical values of aluminum are provided for comparison and future reference. Table 5.1 includes the values of coefficients of thermal expansion (CTE) and coefficients of moisture expansion for the materials. These will be discussed shortly. As can be seen from the table, for the graphite-reinforced polymer, the extensional modulus in the fiber direction is about ten times greater than the extensional modulus perpendicular to the fibers. For glass-reinforced materials, the difference is not as great. For both fiber-reinforced materials, the three shear moduli are similar.

As an example of the magnitude of the components of the compliance and stiffness matrices, for the graphite-reinforced material in Table 5.1, according to Eq. (5.46):

$$\begin{aligned} &S_{11} = 6.45(\text{TPa})^{-1} \quad S_{12} = -1.6 \quad S_{13} = -1.6 \\ &S_{22} = 82.6 \quad S_{23} = -37.9 \quad S_{33} = 82.6 \\ &S_{44} = 312 \quad S_{55} = 227 \quad S_{66} = 227. \end{aligned} \tag{5.56}$$

TABLE 5.1
Engineering Properties of Several Materials (Layer Thickness = 0.150 mm)

	Graphite–Polymer Composite	Glass–Polymer Composite	Aluminum
E_1	155 GPa	50 GPa	72.4 GPa
E_2	12.10 GPa	15.20 GPa	72.4 GPa
E_3	12.10 GPa	15.20 GPa	72.4 GPa
ν_{23}	0.458	0.428	0.300
ν_{13}	0.248	0.254	0.300
ν_{12}	0.248	0.254	0.300
G_{23}	3.20 GPa	3.28 GPa	$G = \frac{E}{2(1+\nu)}$
G_{13}	4.40 GPa	4.70 GPa	$G = \frac{E}{2(1+\nu)}$
G_{12}	4.40 GPa	4.70 GPa	$G = \frac{E}{2(1+\nu)}$
α_1	$-0.01800 \times 10^{-6}/°C$	$6.34 \times 10^{-6}/°C$	$22.5 \times 10^{-6}/°C$
α_2	$24.30 \times 10^{-6}/°C$	$23.30 \times 10^{-6}/°C$	$24.30 \times 10^{-6}/°C$
α_3	$24.30 \times 10^{-6}/°C$	$23.30 \times 10^{-6}/°C$	$24.30 \times 10^{-6}/°C$
β_1	$146.0 \times 10^{-6}/\%M$	$434 \times 10^{-6}/\%M$	0
β_2	$4770 \times 10^{-6}/\%M$	$6320 \times 10^{-6}/\%M$	0
β_3	$4770 \times 10^{-6}/\%M$	$6320 \times 10^{-6}/\%M$	0

Using Eq. (5.48), the stiffnesses for the graphite-reinforced material are:

$$\begin{aligned} &C_{11} = 158\text{ GPa} \quad C_{12} = 5.64 \\ &C_{22} = 15.51 \quad C_{13} = 5.64 \quad C_{33} = 15.51 \\ &C_{23} = 7.21 \\ &C_{44} = 3.20 \quad C_{55} = 4.40 \quad C_{66} = 4.40 \end{aligned} \tag{5.57}$$

For aluminum which is generally isotropic:

$$\begin{aligned} &S_{11} = S_{22} = S_{33} = 13.81(\text{TPa}) \quad S_{12} = S_{13} = S_{23} = -4.14 \\ &S_{44} = S_{55} = S_{66} = 36. \end{aligned} \tag{5.58}$$

Also,

$$\begin{aligned} &C_{11} = C_{22} = C_{33} = 97.5\text{ GPa} \quad C_{12} = C_{13} = C_{23} = 41.8 \\ &C_{44} = C_{55} = C_{66} = 27.8 \end{aligned} \tag{5.59}$$

In matrix form, for the graphite-reinforced material, the compliance and stiffness matrices are given by Eqs. (5.60) and (5.61) as:

$$[S] = \begin{bmatrix} 6.45 & -1.6 & -1.6 & 0 & 0 & 0 \\ -1.6 & 82.6 & -37.9 & 0 & 0 & 0 \\ -1.6 & -37.9 & 82.6 & 0 & 0 & 0 \\ 0 & 0 & 0 & 312 & 0 & 0 \\ 0 & 0 & 0 & 0 & 227 & 0 \\ 0 & 0 & 0 & 0 & 0 & 227 \end{bmatrix} (\text{TPa})^{-1} \tag{5.60}$$

$$[C] = \begin{bmatrix} 158 & 5.64 & 5.64 & 0 & 0 & 0 \\ 5.64 & 15.51 & 7.21 & 0 & 0 & 0 \\ 5.64 & 7.21 & 15.51 & 0 & 0 & 0 \\ 0 & 0 & 0 & 3.20 & 0 & 0 \\ 0 & 0 & 0 & 0 & 4.40 & 0 \\ 0 & 0 & 0 & 0 & 0 & 4.40 \end{bmatrix} \text{GPa}. \tag{5.61}$$

The degree of directional dependence, or degree of orthotropy, for a material can be evaluated by examining differences among the 11, 22, and 33 terms. For extensional effects in graphite-reinforced material, there is a high degree of directional dependence, or orthotropy. Shear effects exhibit little orthotropy, but Poisson coupling is directionally dependent, for example, S_{13} vs. S_{23}. As will be discussed, thermal expansion characteristics are highly dependent on direction.

5.5 INTERPRETATION OF STRESS–STRAIN RELATIONS

The stress–strain relations just established can be interpreted in several ways. First, the relations can be considered simple algebraic relations between 12 quantities, σ_1, $\sigma_2, \ldots, \tau_{12}$ and $\varepsilon_1, \varepsilon_2, \ldots, \gamma_{12}$. Given any six quantities, the other six can be determined by the rules of algebra. However, there is a second, more important and more physical interpretation. The 12 stresses and strains should be considered in pairs: $\sigma_1 - \varepsilon_1$, $\sigma_2 - \varepsilon_2$, $\sigma_3 - \varepsilon_3$, $\tau_{23} - \gamma_{23}$, $\tau_{13} - \gamma_{13}$, $\tau_{12} - \gamma_{12}$. Though it is possible to prescribe any six and solve for the other six, physically one can only prescribe either a stress or a strain from each pair, but not both. Though they are two of the six prescribed quantities, it is not physically correct to prescribe both σ_3 and ε_3, for example. Only one or the other can be prescribed. This is obvious when the shear response is considered. Because:

$$\tau_{12} = G_{12}\gamma_{12}. \tag{5.62}$$

It is impossible to specify both τ_{12} and γ_{12}. To do so violates the shear stress–shear strain relation. Similarly, to specify both σ_1 and ε_1, or both σ_2 and ε_2, or both σ_3 and ε_3, also violates the stress–strain relation, but not in so obvious a fashion. There is one situation, however, where efforts are made to know both the stress and the strain. For example, if the extensional modulus in the 1 direction is to be determined from an experiment, then both σ_1 and ε_1 have to be known. If all the stresses are zero except σ_1, and the strain ε_1 is measured, then the extensional modulus can be computed directly from Eq. (5.1). Similarly for E_2, E_3, and the other engineering properties. However, even in this case of determining material properties, σ_1 is specified from the level of applied load and cross-sectional area, but ε_1 is not really specified also. The strain ε_1 is measured and E_1 is inferred. As an example of the above comments regarding pairing of stress and strain components, and to illustrate the use and some of the implications of the stress–strain relations for fiber-reinforced materials, consider the following.

A 50-mm cube of graphite-reinforced material, shown in Figure 5.11a, is subjected to a 125-kN compressive force perpendicular to the fiber direction, specifically in the 2 direction. In one situation, shown in Figure 5.11b, the cube is free to expand or contract; in the second situation, shown in Figure 5.11c, the cube is constrained against expansion in the 3 direction; and in the third case, shown in Figure 5.11d, the cube is constrained against expansion in the 1 direction. Of interest are the changes in the 50-mm dimensions in each of these three cases, and the stresses, and hence forces, required to provide the constraints in the latter two cases. To begin, we shall assume that the compressive force is uniformly distributed over the 2-direction faces. Further, we shall assume that stress resulting from the force is distributed uniformly throughout the volume of the cube. Then,

$$\sigma_2 = \frac{P}{\Delta_1\Delta_3} = \frac{-12{,}5000 \text{ N}}{(0.050 \text{ m})(0.050 \text{ m})} = -50{,}000{,}000 \text{ Pa} = -50 \text{ MPa}. \tag{5.63}$$

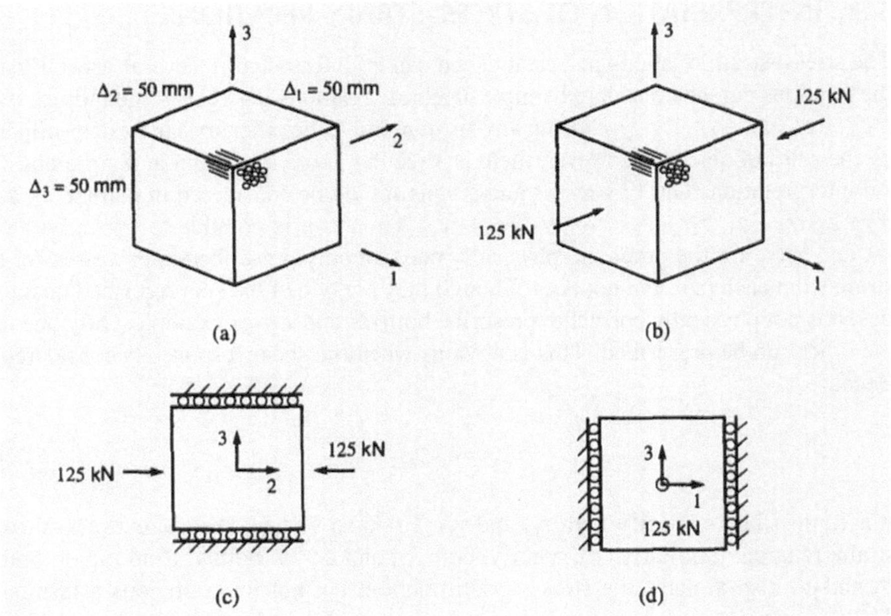

FIGURE 5.11 Cube of material subjected to compressive stress in 2 direction. (a) A 50-mm cube of graphite–epoxy, (b) cube subjected to compression in 2 directions no constraints, (c) constrained in 3 directions, and (d) constrained in 1 direction.

For the first case, shown in Figure 5.11b, the cube is free from any stress on the other faces. Thus, throughout the volume of the cube:

$$\sigma_2 = -50 \quad \sigma_1 = \sigma_3 = \tau_{23} = \tau_{13} = \tau_{12} = 0. \tag{5.64}$$

Because of the previous discussions of pairing of stresses and strains, nothing can be said regarding $\varepsilon_1, \varepsilon_2, \varepsilon_3, \gamma_{23}, \gamma_{13}$, or γ_{12}. These are all unknown and will be determined. It is most convenient to use the compliance form of the stress–strain relation, shown in Eq. (5.45). Specifically, if we substitute Eq. (5.64) in Eq. (5.45):

$$\begin{Bmatrix} \varepsilon_1 \\ \varepsilon_2 \\ \varepsilon_3 \\ \gamma_{23} \\ \gamma_{13} \\ \gamma_{12} \end{Bmatrix} = \begin{bmatrix} S_{11} & S_{12} & S_{13} & 0 & 0 & 0 \\ S_{12} & S_{22} & S_{23} & 0 & 0 & 0 \\ S_{13} & S_{23} & S_{33} & 0 & 0 & 0 \\ 0 & 0 & 0 & S_{44} & 0 & 0 \\ 0 & 0 & 0 & 0 & S_{55} & 0 \\ 0 & 0 & 0 & 0 & 0 & S_{66} \end{bmatrix} \begin{Bmatrix} 0 \\ \sigma_2 \\ 0 \\ 0 \\ 0 \\ 0 \end{Bmatrix}. \tag{5.65}$$

Expanding,

$$\varepsilon_1 = S_{12}\sigma_2 \quad \varepsilon_2 = S_{22}\sigma_2 \quad \varepsilon_3 = S_{23}\sigma_2 \quad \gamma_{23} = \gamma_{13} = \gamma_{12} = 0. \tag{5.66}$$

As the stresses are uniform throughout the cube, the strains are also uniform throughout the cube. By definition:

$$\varepsilon_1 = \frac{\delta\Delta_1}{\Delta_1} \quad \varepsilon_2 = \frac{\delta\Delta_2}{\Delta_2} \quad \varepsilon_3 = \frac{\delta\Delta_3}{\Delta_3}, \tag{5.67}$$

where $\delta\Delta_1$, $\delta\Delta_2$, and $\delta\Delta_3$ denote the change in the 50 mm length in the 1, 2, and 3 directions, respectively. Rearranging Eq. (5.67), and using Eq. (5.66) and values from Eq. (5.56):

$$\delta\Delta_1 = \Delta_1\varepsilon_1 = \Delta_1 S_{12}\sigma_2 = (0.050)(-1.6\times10^{-12})(-50\times10^6) = 0.00400\text{ mm}$$

$$\delta\Delta_2 = \Delta_2\varepsilon_2 = \Delta_2 S_{22}\sigma_2 = (0.050)(82.6\times10^{-12})(-50\times10^6) = -0.207\text{ mm} \tag{5.68}$$

$$\delta\Delta_3 = \Delta_3\varepsilon_3 = \Delta_3 S_{23}\sigma_2 = (0.050)(-37.9\times10^{-12})(-50\times10^6) = 0.0946\text{ mm}.$$

In this situation, the fibers provide considerable constraint and prevent the expansion in the 1 direction from being anywhere near as large as the expansion in the 3 direction. Of course, there is compression in the 2 direction, and obviously:

$$\gamma_{23} = \gamma_{13} = \gamma_{12} = 0. \tag{5.69}$$

Consider the second case, shown in Figure 5.11c. For this situation, we do not know the normal stress in the 3 direction that is providing the constraint – because the displacement in the 3 direction is restrained to be zero along two sides of the cube, $\delta\Delta_3 = 0$, and because the strains are assumed to be uniform throughout the cube, $\varepsilon_3 = 0$. Therefore, instead of Eq. (5.66), we have:

$$\sigma_2 = -50\text{ MPa} \quad \sigma_1 = \varepsilon_3 = \tau_{23} = \tau_{13} = \tau_{12} = 0. \tag{5.70}$$

And we must solve for $\varepsilon_1, \varepsilon_2, \sigma_3, \gamma_{23}, \gamma_{13}$, or γ_{12}, the other half of each of the pairs involved in Eq. (5.70). Again referring to the compliance form of the stress–strain relations:

$$\begin{Bmatrix} \varepsilon_1 \\ \varepsilon_2 \\ 0 \\ \gamma_{23} \\ \gamma_{13} \\ \gamma_{12} \end{Bmatrix} = \begin{bmatrix} S_{11} & S_{12} & S_{13} & 0 & 0 & 0 \\ S_{12} & S_{22} & S_{23} & 0 & 0 & 0 \\ S_{13} & S_{23} & S_{33} & 0 & 0 & 0 \\ 0 & 0 & 0 & S_{44} & 0 & 0 \\ 0 & 0 & 0 & 0 & S_{55} & 0 \\ 0 & 0 & 0 & 0 & 0 & S_{66} \end{bmatrix} \begin{Bmatrix} 0 \\ \sigma_2 \\ \sigma_3 \\ 0 \\ 0 \\ 0 \end{Bmatrix} \tag{5.71}$$

$$\text{Or; } \varepsilon_1 = S_{12}\sigma_2 + S_{13}\sigma_3 \quad \varepsilon_2 = S_{22}\sigma_2 + S_{23}\sigma_3 \quad 0 = S_{23}\sigma_2 + S_{33}\sigma_3. \tag{5.72}$$

For this case, we must use the third equation to solve for a relation between σ_2 and σ_3, namely:

$$\sigma_3 = -\frac{S_{23}}{S_{33}}\sigma_2. \tag{5.73}$$

Solving the first and second equations, and substituting for σ_3 result in:

$$\begin{aligned}\varepsilon_1 &= S_{12}\sigma_2 + S_{13}\sigma_3 = \left(S_{12} - \frac{S_{13}S_{23}}{S_{33}}\right)\sigma_2\\ \varepsilon_2 &= S_{22}\sigma_2 + S_{23}\sigma_3 = \left(S_{22} - \frac{S_{23}S_{23}}{S_{33}}\right)\sigma_2.\end{aligned} \tag{5.74}$$

For this case also:

$$\gamma_{23} = \gamma_{13} = \gamma_{12} = 0. \tag{5.75}$$

We often refer to the combination of compliances in Eq. (5.74) as apparent or reduced compliances. In the second of Eq. (5.74), the term "$\left(\frac{S_{23}S_{23}}{S_{33}}\right)$" subtracts from S_{22} and indicates the material behaves in a less compliant manner than determined by S_{22} alone. Certainly, it makes physical sense that a constraint of any kind, here a constraint against expansion in the 3 direction, makes the system less compliant. Equation (5.67) relates the change of length to the strain and so, using Eq. (5.74) and values from Eq. (5.56):

$$\begin{aligned}\delta\Delta_1 &= \Delta_1\varepsilon_1 = \Delta_1\left(S_{12} - \frac{S_{13}S_{23}}{S_{33}}\right)\sigma_2 = (0.050)\left(-2.33\times10^{-12}\right)\left(-50\times10^{6}\right)\\ &= 0.00400583\ \text{mm}\\ \delta\Delta_2 &= \Delta_2\varepsilon_2 = \Delta_2\left(S_{22} - \frac{S_{23}S_{23}}{S_{33}}\right)\sigma_2 = (0.050)\left(65.3\times10^{-12}\right)\left(-50\times10^{6}\right) = -0.1633\ \text{mm}\\ \delta\Delta_3 &= 0\end{aligned} \tag{5.76}$$

The stress to constrain σ_3 to be zero is given by Eq. (5.73), namely:

$$\sigma_3 = -\frac{S_{23}}{S_{33}}\quad\sigma_2 = -22.9\ \text{MPa}. \tag{5.77}$$

If we compare the deformations for the case of Figure 5.11c, given by Eq. (5.76), with the deformations for the case of Figure 5.11b, given by Eq. (5.68), we can see that with the constraint in the 3 direction, the cube expands about 50% more in the fiber direction, and compresses about 25% less in the 2 direction. The constraint in the

3 direction makes the cube stiffer in the 2 direction and forces the inevitable volume change to be reflected with increased expansion in the fiber direction.

Finally, consider the third case, shown in Figure 5.11d. For this situation, the normal stress in the 3 direction is again zero and the extensional strain in that direction is unknown. The extensional strain in the fiber direction is known to be zero, but the corresponding stress, σ_1, is not known. With this:

$$\sigma_2 = -50\text{ MPa} \quad \varepsilon_1 = \sigma_3 = \tau_{23} = \tau_{13} = \tau_{12} = 0, \tag{5.78}$$

and we must solve for: $\sigma_1, \varepsilon_2, \sigma_3, \gamma_{23}, \gamma_{13}$ and γ_{12}. Equation (5.45) becomes:

$$\begin{Bmatrix} 0 \\ \varepsilon_2 \\ \varepsilon_3 \\ \gamma_{23} \\ \gamma_{13} \\ \gamma_{12} \end{Bmatrix} = \begin{bmatrix} S_{11} & S_{12} & S_{13} & 0 & 0 & 0 \\ S_{12} & S_{22} & S_{23} & 0 & 0 & 0 \\ S_{13} & S_{23} & S_{33} & 0 & 0 & 0 \\ 0 & 0 & 0 & S_{44} & 0 & 0 \\ 0 & 0 & 0 & 0 & S_{55} & 0 \\ 0 & 0 & 0 & 0 & 0 & S_{66} \end{bmatrix} \begin{Bmatrix} \sigma_1 \\ \sigma_2 \\ 0 \\ 0 \\ 0 \\ 0 \end{Bmatrix}. \tag{5.79}$$

$$\text{Or, } 0 = S_{11}\sigma_1 + S_{12}\sigma_2 \quad \varepsilon_2 = S_{12}\sigma_1 + S_{22}\sigma_2 \quad \varepsilon_3 = S_{13}\sigma_1 + S_{23}\sigma_2. \tag{5.80}$$

If we use the first equation:

$$\sigma_1 = -\frac{S_{12}}{S_{11}}\sigma_2. \tag{5.81}$$

The second and third equations become:

$$\begin{aligned} \varepsilon_2 &= S_{12}\sigma_1 + S_{22}\sigma_2 = \left(S_{22} - \frac{S_{12}S_{12}}{S_{11}}\right)\sigma_2 \\ \varepsilon_3 &= S_{13}\sigma_1 + S_{23}\sigma_2 = \left(S_{23} - \frac{S_{12}S_{13}}{S_{11}}\right)\sigma_2 \end{aligned} \tag{5.82}$$

$$\gamma_{23} = \gamma_{13} = \gamma_{12} = 0. \tag{5.83}$$

If we use Eq. (5.67) and numerical values,

$$\begin{aligned} &\delta\Delta_1 = 0 \\ &\delta\Delta_2 = \Delta_2\varepsilon_2 = \Delta_2\left(S_{22} - \frac{S_{12}S_{12}}{S_{11}}\right)\sigma_2 = (0.050)(82.2\times10^{-12})(-50\times10^6) = -0.206\text{ mm} \\ &\delta\Delta_3 = \Delta_3\varepsilon_3 = \Delta_3\left(S_{23} - \frac{S_{12}S_{13}}{S_{11}}\right)\sigma_2 = (0.050)(-38.2\times10^{-12})(-50\times10^6) = 0.0956\text{ mm} \end{aligned} \tag{5.84}$$

The stress in the fiber direction is given by Eq. (5.81) as:

$$\sigma_1 = -12.4 \text{ MPa}. \quad (5.85)$$

It is important to note that the deformations in the 2 and 3 directions for this case are not very different from the case of no restraint, shown in Eq. (5.68). This is because the fibers constrain deformation in the 1 direction to a considerable degree, and adding a constraint in the 1 direction has less influence than adding a constraint in the 3 direction. Likewise, the stress necessary to constrain the deformation in the fiber direction is, by Eq. (5.85), −12.4 MPa, whereas the stress to constrain deformation perpendicular to the fibers is, by Eq. (5.77), −22.9 MPa, about twice as much.

It is instructive to repeat the same exercise, but using aluminum in place of graphite-reinforced composite. This will provide a feel for the effects of the lack of orthotropy, and a comparison of the magnitude of the deformations of aluminum and of a common fiber-reinforced composite material. Before studying the case of aluminum, however, it is important to comment on the physical aspects of the example just discussed. First, the 50.0 MPa compressive stress in the 2 direction is about 25% of the compressive failure stress of graphite-reinforced material in the 2 direction and is the failure strength in the 2 direction in tension. Thus, the 50.0 MPa in the example is a realistic level of stress. Second, it is important to note that the magnitudes of the deformations $\delta\Delta_1, \delta\Delta_2$, and $\delta\Delta_3$ are quite small – submillimeter in size. This level of deformation cannot be detected with the eye. Special instrumentation is needed to measure these small deformations and in practice that is what is used. Finally, it is important to become familiar with realistic strain levels associated with composite materials. In the unconstrained case, shown in Figure 5.11b, the strains ε_1, ε_2, and ε_3 are given in Eq. (5.66) as:

$$\varepsilon_1 = S_{12}\sigma_2 \quad \varepsilon_2 = S_{22}\sigma_2 \quad \varepsilon_3 = S_{23}\sigma_2. \quad (5.86)$$

If we use numerical values:

$$\varepsilon_1 = \left(-1.6\times10^{-12}\right)\left(-50\times10^{6}\right) = 80\times10^{-6}\text{ m m}^{-1} = 80\times10^{-6}\text{ mm mm}^{-1}$$
$$= 80\ \mu\text{mm mm}^{-1}$$
$$\varepsilon_2 = \left(82.6\times10^{-12}\right)\left(-50\times10^{6}\right) = -4130\times10^{-6}\text{ m m}^{-1} = -4130\times10^{-6}\text{ mm mm}^{-1}$$
$$= -4130\ \mu\text{mm mm}^{-1}$$
$$\varepsilon_3 = \left(-37.8\times10^{-12}\right)\left(-50\times10^{6}\right) = -1893\times10^{-6}\text{ m m}^{-1} = -1893\times10^{-6}\text{ mm mm}^{-1}$$
$$= -1893\ \mu\text{mm mm}^{-1} \quad (5.87)$$

In the above section, we noted that the computations lead to extensional strains in m m^{-1}, but as that is a dimensionless quantity, mm/mm is just as valid. There is no standard notation for reporting strain. Reporting ε_1 as 0.000080 can be done, or it can be reported as 0.008%. Because the displacements that result from the strains are usually small, if units are to be assigned, then mm/mm is more appropriate, although light-years/light-years is valid, but ridiculous. In the above section, the notation for micro has also been introduced; that is:

$$\mu = 10^{-6}. \tag{5.88}$$

Herein, the 10^{-6} or the $\mu\text{m mm}^{-1}$ forms will be used. For the unconstrained element of material, then, according to Eq. (5.87), the 50 MPa stress in the 2 direction produces over 4000 $\mu\text{m mm}^{-1}$ compressive strain in the 2 direction, and through Poisson effects, about 100 $\mu\text{m mm}^{-1}$ elongation strain in the fiber direction and about 2000 $\mu\text{m mm}^{-1}$ elongation strain in the 3 direction. These are realistic strain levels and as more examples are discussed, familiarity with strain levels associated with other applied stress levels will be established. If instead of subjecting an element of graphite-reinforced material to the 50 MPa compressive stress, an element of aluminum is compressed, the deformations of the unconstrained element are given by the analog of Eq. (5.68), namely:

$$\begin{aligned}
\delta\Delta_1 &= \Delta_1\varepsilon_1 = \Delta_1 S_{12}\sigma_2 = (0.050)\left(-4.14\times10^{-12}\right)\left(-50\times10^{6}\right) = 0.01036\ \text{m} \\
\delta\Delta_2 &= \Delta_2\varepsilon_2 = \Delta_2 S_{22}\sigma_2 = (0.050)\left(13.8\times10^{-12}\right)\left(-50\times10^{6}\right) = -0.0345\ \text{mm}\ . \\
\delta\Delta_3 &= \Delta_3\varepsilon_3 = \Delta_3 S_{23}\sigma_2 = (0.050)\left(-4.14\times10^{-12}\right)\left(-50\times10^{6}\right) = 0.01036\ \text{mm}
\end{aligned} \tag{5.89}$$

The compliances used in the above calculation are given in Eq. (5.58), and the use of a 1–2–3 coordinate system is only to provide a reference system. With aluminum, there is no difference in material properties in the three coordinate directions. As a result, due to the compressive stress in the 2 direction, the increase in length in the 1 direction of the aluminum cube is identical to the increase in length in the 3 direction. This was certainly not the case for the fiber-reinforced material. By the results of Eq. (5.58), we see that for the graphite-reinforced material, expansion in the 3 direction is about 25 times the expansion in the 1 direction!

If the aluminum is constrained in the 3 direction, as in the second case, the deformations are given by the analog to Eq. (5.76), namely:

$$\begin{aligned}
\delta\Delta_1 &= \Delta_1\varepsilon_1 = \Delta_1\left(S_{12} - \frac{S_{13}S_{23}}{S_{33}}\right)\sigma_2 = (0.050)\left(-5.39\times10^{-12}\right)\left(-50\times10^{6}\right) \\
&= 0.01347\ \text{mm} \\
\delta\Delta_2 &= \Delta_2\varepsilon_2 = \Delta_2\left(S_{22} - \frac{S_{23}S_{23}}{S_{33}}\right)\sigma_2 = (0.050)\left(12.57\times10^{-12}\right)\left(-50\times10^{6}\right) \\
&= -0.0314\ \text{mm} \\
\delta\Delta_3 &= 0
\end{aligned} \tag{5.90}$$

$$\sigma_3 = -\frac{S_{23}}{S_{33}} \quad \sigma_2 = -15 \text{ MPa} \tag{5.91}$$

Finally, for a constraint in the 1 direction, the analog to Eq. (5.84) is used. Because the material is isotropic, we can immediately write, referring to Eqs. (5.90) and (5.91):

$$\begin{aligned} \delta\Delta_1 &= 0 \\ \delta\Delta_2 &= -0.0314 \text{ mm} \\ \delta\Delta_3 &= 0.01347 \text{ mm} \\ \sigma_1 &= -15 \text{ MPa} \end{aligned} \tag{5.92}$$

If we use Eqs. (2.83) and (2.84) as a check:

$$\begin{aligned} \sigma_1 &= -\frac{S_{12}}{S_{11}} \quad \sigma_2 = -15 \text{ MPa} \\ \delta\Delta_1 &= 0 \\ \delta\Delta_2 &= \Delta_2\varepsilon_2 = \Delta_2\left(S_{22} - \frac{S_{12}S_{12}}{S_{11}}\right)\sigma_2 \\ &= (0.050)\left(12.57\times10^{-12}\right)\left(-50\times10^{6}\right) = -0.0134 \text{ mm} \\ \delta\Delta_3 &= \Delta_3\varepsilon_3 = \Delta_3\left(S_{23} - \frac{S_{12}S_{13}}{S_{11}}\right)\sigma_2 \\ &= (0.050)\left(-5.39\times10^{-12}\right)\left(-50\times10^{6}\right) = 0.01347 \text{ mm} \end{aligned} \tag{5.93}$$

These examples provide an indication of the strains and deformations that result from applying a force to a fiber-reinforced material. More importantly, the results indicate how important the fibers are in controlling the response of composite materials, even though the fibers may not be loaded directly.

5.6 FREE THERMAL STRAINS

To this point, the only deformations we have studied have been those that are a result of an applied load. However, when a fiber-reinforced composite material is heated or cooled, just as with an isotropic material, the material expands or contracts. This is a deformation that takes place independently of any applied load. If a load is applied, then we must contend with the deformations due to applied load and thermal expansion effects. Unlike an isotropic material, the thermal expansion of a fiber-reinforced material is different in each of the three principal material directions. Graphite fibers contract along their length when heated (see Table 2.6). Polymers, aluminum, boron, ceramics, and most other matrix materials expand when heated. Therefore, when heated, a composite will expand, contract, or possibly exhibit no change in length in the fiber direction, depending on the relative effects of the fiber and matrix materials. In the other two directions, due to the dominance of the properties of the matrix, the composite will expand. Figure 5.12a shows our small element of material with its temperature at some reference temperature. The element is not part of any structure; rather, it is isolated in space and free of stresses on its six faces. The temperature is uniform within the element, and at this reference temperature, the element has dimensions Δ_1, Δ_2, and Δ_3. As the temperature of the element is changed, the element changes dimensions slightly in the fiber direction and moderately in the other two directions. The change in length of the element per unit original length is defined to be the thermal strain. Because the element has no tractions acting on any of its six faces, the strain is termed "free thermal strain." The free thermal strains in the three coordinate directions will be denoted as:

$$\varepsilon_1^T(T,T_{\text{ref}}) \quad \varepsilon_2^T(T,T_{\text{ref}}) \quad \varepsilon_3^T(T,T_{\text{ref}}). \tag{5.94}$$

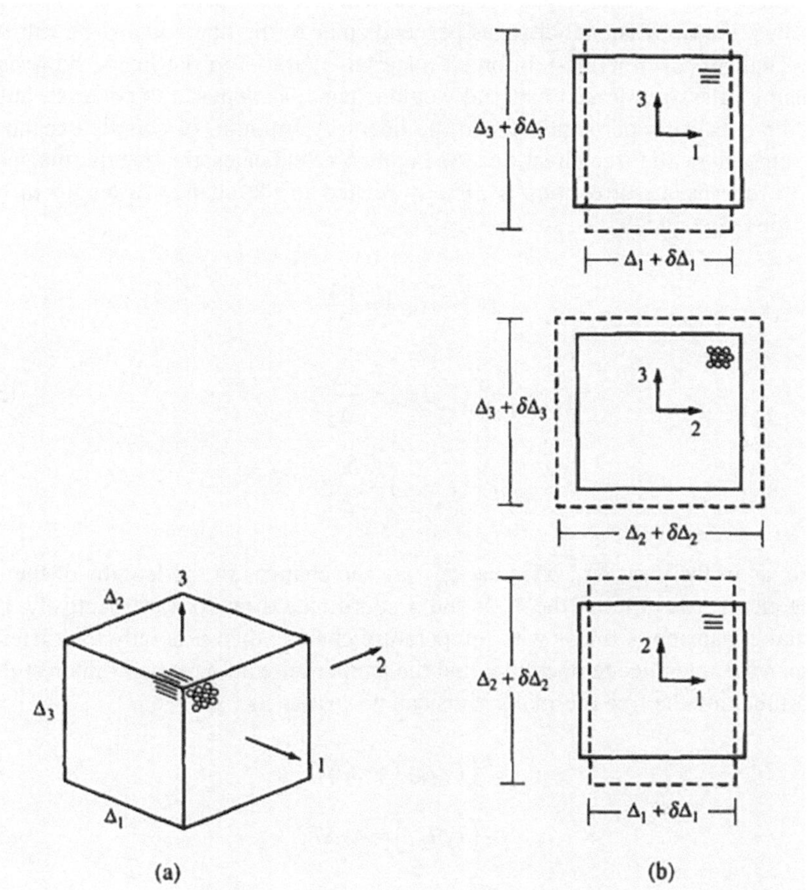

FIGURE 5.12 Thermal expansion of an element of fiber-reinforced material. (a) Element at reference temperature and (b) deformations due to ΔT.

The superscript T indicates that the strain is a free thermal strain. T_{ref} is the temperature at the reference state, and T is the temperature at the state of interest, a temperature above or below the reference temperature. The two arguments T_{ref} and T emphasize the fact that the free thermal strain involves both of these temperatures. The reference state is, of course, arbitrary and depends on the specific problem being studied. For polymer-based fiber-reinforced composite materials, the reference state is often taken to be the stress-free processing condition. The reference temperature would then be the temperature corresponding to that condition. The physical interpretation of these free thermal strains is indicated in Figure 5.12b, which shows the three views of the dimensional changes experienced by an element of graphite-reinforced material as the temperature is increased. Since, as mentioned before, graphite contracts when heated, the length of the element in the fiber direction is assumed to decrease. The dimensions in the directions perpendicular to the fibers are assumed to increase. In addition, the increases in the dimensions perpendicular to the fibers are much larger than the decrease in the fiber direction, reflecting the overwhelming influence of polymer expansion perpendicular to the fibers. For glass-reinforced materials, the dimensional changes perpendicular to the fibers would be similar to those changes for graphite-reinforced materials. Parallel to the fibers, however, an element of glass-reinforced material would expand, as opposed to contract, but less than the expansion perpendicular to the fibers. Aluminum, of course, expands the same amount in all three directions. As Figure 5.12 indicates, the free thermal expansion in the various directions is directly related to the change in length in those directions. Specifically:

$$\varepsilon_1^T\left(T, T_{ref}\right) = \frac{\delta\Delta_1}{\Delta_1}$$
$$\varepsilon_2^T\left(T, T_{ref}\right) = \frac{\delta\Delta_2}{\Delta_2}, \tag{5.95}$$
$$\varepsilon_3^T\left(T, T_{ref}\right) = \frac{\delta\Delta_3}{\Delta_3}$$

where, as in the past, $\delta\Delta_1$, $\delta\Delta_2$, and $\delta\Delta_3$ are the changes in the lengths of the sides of the element parallel to the 1, 2, and 3 coordinate directions, respectively. If the thermal expansion is linear with temperature change, then it is only the difference between the reference temperature and the temperature at the state of interest that is important, and the free thermal strains can be written as:

$$\varepsilon_1^T\left(T, T_{ref}\right) = \alpha_1 \Delta T$$
$$\varepsilon_2^T\left(T, T_{ref}\right) = \alpha_2 \Delta T, \tag{5.96}$$
$$\varepsilon_3^T\left(T, T_{ref}\right) = \alpha_3 \Delta T$$

$$\text{where } \Delta T = T - T_{ref}. \tag{5.97}$$

The quantities α_1, α_2, and α_3 are referred to as the coefficients of thermal expansion, or CTE for short. They have units of 1/°C, or to emphasize they are related to strains, mm/mm/°C. Obviously, ΔT is the temperature difference between the reference temperature pertinent to the problem and the temperature of interest. It is implicit that the thermal strains are defined to be zero at the reference temperature. A positive ΔT is associated with a temperature increase.

Several important points should be made relative to the above formulation. First, strictly speaking, α_1, α_2, and α_3 are referred to as the linear coefficients of thermal expansion. Expansional effects are linearly proportional to the temperature change, and as will be seen, the thermally induced stresses in a laminate will double if the temperature change doubles. Second, if expansional effects are not linearly proportional to the temperature change, then the linear CTE is meaningless. To study thermally induced deformation effects in these cases, it is necessary to have explicit forms of ε_1^T, ε_2^T, and ε_3^T as a function of temperature. These functional forms are usually derived from experiments by the least-squares fitting of simple polynomial functions of temperature to the data. As implied by the notation of Eq. (5.94), in these cases, the reference temperature and the temperature of interest must be part of the functional form for a complete description of the free thermal expansion effects.

A third important point is rather obvious but it should be mentioned explicitly. In the principal material coordinate system, the free thermal strains do not involve any shearing deformations. The strains are strictly dilatational, not distortional. Dilatation effects are those identified with a change in volume of an element, whereas distortional effects are those identified with a change in shape of an element. Finally, for an element of material that is isolated in space, the strains that accompany any temperature change do not result in stresses on any of the six surfaces. This is counter to the stress–strain relations already presented. In the previous formulations, the extensional strains ε_1, ε_2, and ε_3 are accompanied by normal stresses σ_1, σ_2, and σ_3. Clearly, the stress–strain relations must be modified to account for free thermal strains that do not, in an unrestrained and completely free element, produce stresses. Typical values for the CTE for graphite- and glass-reinforced composites, and aluminum, are presented in Table 5.1. For graphite-reinforced materials, the coefficient in the fiber direction can be slightly negative, while the coefficient perpendicular to the fibers is similar to the expansion of aluminum. Parallel to the fibers, the stiffness of the fibers dominates, and the free thermal strains are governed by the contraction properties of the fibers. Perpendicular to the fiber direction, the free thermal strains are governed by the large expansion properties of the matrix and the lack of stiffness of the fibers. In linear thermal expansion, the changes in length of the three dimensions of the element of material in Figure 5.12 are given by:

$$\delta\Delta_1 = \Delta_1\alpha_1\Delta T \quad \delta\Delta_2 = \Delta_2\alpha_2\Delta T \quad \delta\Delta_3 = \Delta_3\alpha_3\Delta T, \tag{5.98}$$

where Eqs. (5.95) and (5.96) have been employed. For a 50 mm × 50 mm × 50mm element of graphite-reinforced composite material, if the element is heated by 50°C, the changes in dimensions are given by, from Table 5.1 and Eq. (5.98):

$$\delta\Delta_1 = (50)\left(-0.018\times10^{-6}\right)(50) = -0.0000450 \text{ mm} \tag{5.99}$$

$$\delta\Delta_2 = (50)\left(24.3\times10^{-6}\right)(50) = 0.0608 \text{ mm}$$

$$\delta\Delta_3 = (50)\left(24.3\times10^{-6}\right)(50) = 0.0608 \text{ mm}.$$

Thus, the dimensions of the heated element are: in the 1 direction

$$\Delta_1 + \delta\Delta_1 = 49.999955 \text{ mm} \tag{5.100a}$$

In the 2 direction:

$$\Delta_2 + \delta\Delta_2 = 50.0608 \text{ mm} \tag{5.100b}$$

and in the 3 direction

$$\Delta_3 + \delta\Delta_3 = 50.0608 \text{ mm}. \tag{5.100c}$$

5.7 EFFECT OF FREE THERMAL STRAINS ON STRESS–STRAIN RELATIONS

Free thermal strain refers to the fact that the smeared element is free of any stresses if the temperature is changed. When one considers an unsmeared material and deals with the individual fibers and the surrounding matrix, a temperature change can create significant stresses in the fiber and matrix. Clearly, if graphite contracts in the fiber direction and polymers expand, and the materials are combined, such stresses will certainly exist. However, when the stresses are integrated (i.e., smeared) over the volume of the element, the net result is zero. To accommodate the fact that in a smeared element of material with no constraints on its bounding surfaces free thermal strains do not cause stresses, the stress–strain relation as it has been presented to this point has to be reinterpreted. The simplest interpretation, and the one that is consistent with the definitions of stress, strain, and free thermal strain, is that the strains in the stress–strain relations are the mechanical strains. In the context of free thermal strains, this is interpreted to mean that mechanical strains are the total strains minus the free thermal strains. The total strains are a measure of the change in dimensions of an element of material, specifically the change in length per unit length. In the context of our element of material, these changes in length per unit length are given by: $\frac{\delta\Delta_1}{\Delta_1}$, $\frac{\delta\Delta_2}{\Delta_2}$, and $\frac{\delta\Delta_3}{\Delta_3}$.

Thus:

$$\varepsilon_1 \equiv \frac{\delta\Delta_1}{\Delta_1} \quad \varepsilon_2 \equiv \frac{\delta\Delta_2}{\Delta_2} \quad \varepsilon_3 \equiv \frac{\delta\Delta_3}{\Delta_3}, \tag{5.101}$$

where it is implied the strains are the total strains. In the previous section, referring to Figure 5.12, the total strains were equal to the thermal strains (because there were no stresses) and hence the notation of Eq. (5.95). However, in the presence of stresses on the faces of an element of material, Eq. (5.95) is not valid. The concept of mechanical strain is something of an artifact. Unlike the dimensional changes associated with free expansion and those associated with total strains, no specific dimensional changes can be associated with mechanical strains. Nonetheless, the concept is useful for gaining insight into a problem if it is realized that these so-called mechanical strains are the key elements in the stress–strain relations. If we use the concept of mechanical strains to accommodate the free thermal strains, the stress–strain relations can be rewritten as:

$$\begin{Bmatrix} \varepsilon_1 - \varepsilon_1^T(T,T_{\text{ref}}) \\ \varepsilon_2 - \varepsilon_2^T(T,T_{\text{ref}}) \\ \varepsilon_3 - \varepsilon_3^T(T,T_{\text{ref}}) \\ \gamma_{23} \\ \gamma_{13} \\ \gamma_{12} \end{Bmatrix} = \begin{bmatrix} S_{11} & S_{12} & S_{13} & 0 & 0 & 0 \\ S_{12} & S_{22} & S_{23} & 0 & 0 & 0 \\ S_{13} & S_{23} & S_{33} & 0 & 0 & 0 \\ 0 & 0 & 0 & S_{44} & 0 & 0 \\ 0 & 0 & 0 & 0 & S_{55} & 0 \\ 0 & 0 & 0 & 0 & 0 & S_{66} \end{bmatrix} \begin{Bmatrix} \sigma_1 \\ \sigma_2 \\ \sigma_3 \\ \tau_{23} \\ \tau_{13} \\ \tau_{12} \end{Bmatrix}, \tag{5.102}$$

where the total strains are:

$$\begin{Bmatrix} \varepsilon_1 \\ \varepsilon_2 \\ \varepsilon_3 \\ \gamma_{23} \\ \gamma_{13} \\ \gamma_{12} \end{Bmatrix} \tag{5.103}$$

And the mechanical strains are:

$$\begin{Bmatrix} \varepsilon_1^{\text{mech}} \\ \varepsilon_2^{\text{mech}} \\ \varepsilon_3^{\text{mech}} \\ \gamma_{23}^{\text{mech}} \\ \gamma_{13}^{\text{mech}} \\ \gamma_{12}^{\text{mech}} \end{Bmatrix} = \begin{Bmatrix} \varepsilon_1 - \varepsilon_1^T(T,T_{\text{ref}}) \\ \varepsilon_2 - \varepsilon_2^T(T,T_{\text{ref}}) \\ \varepsilon_3 - \varepsilon_3^T(T,T_{\text{ref}}) \\ \gamma_{23} \\ \gamma_{13} \\ \gamma_{12} \end{Bmatrix}. \tag{5.104}$$

In the principal material system, the mechanical shear strains and the total shear strains are identical. If the free thermal strain is linearly dependent on temperature, the stress–strain equations take the form:

$$\begin{Bmatrix} \varepsilon_1 - \alpha_1 \Delta T \\ \varepsilon_2 - \alpha_2 \Delta T \\ \varepsilon_3 - \alpha_3 \Delta T \\ \gamma_{23} \\ \gamma_{13} \\ \gamma_{12} \end{Bmatrix} = \begin{bmatrix} S_{11} & S_{12} & S_{13} & 0 & 0 & 0 \\ S_{12} & S_{22} & S_{23} & 0 & 0 & 0 \\ S_{13} & S_{23} & S_{33} & 0 & 0 & 0 \\ 0 & 0 & 0 & S_{44} & 0 & 0 \\ 0 & 0 & 0 & 0 & S_{55} & 0 \\ 0 & 0 & 0 & 0 & 0 & S_{66} \end{bmatrix} \begin{Bmatrix} \sigma_1 \\ \sigma_2 \\ \sigma_3 \\ \tau_{23} \\ \tau_{13} \\ \tau_{12} \end{Bmatrix}. \tag{5.105}$$

The inverse relations for this linear thermal expansion case are:

$$\begin{Bmatrix} \sigma_1 \\ \sigma_2 \\ \sigma_3 \\ \tau_{23} \\ \tau_{13} \\ \tau_{12} \end{Bmatrix} = \begin{bmatrix} C_{11} & C_{12} & C_{13} & 0 & 0 & 0 \\ C_{12} & C_{22} & C_{23} & 0 & 0 & 0 \\ C_{13} & C_{23} & C_{33} & 0 & 0 & 0 \\ 0 & 0 & 0 & C_{44} & 0 & 0 \\ 0 & 0 & 0 & 0 & C_{55} & 0 \\ 0 & 0 & 0 & 0 & 0 & C_{66} \end{bmatrix} \begin{Bmatrix} \varepsilon_1 - \alpha_1 \Delta T \\ \varepsilon_2 - \alpha_2 \Delta T \\ \varepsilon_3 - \alpha_3 \Delta T \\ \gamma_{23} \\ \gamma_{13} \\ \gamma_{12} \end{Bmatrix}, \tag{5.106}$$

where the mechanical strains are given by:

$$\begin{Bmatrix} \varepsilon_1^{\text{mech}} \\ \varepsilon_2^{\text{mech}} \\ \varepsilon_3^{\text{mech}} \\ \gamma_{23}^{\text{mech}} \\ \gamma_{13}^{\text{mech}} \\ \gamma_{12}^{\text{mech}} \end{Bmatrix} = \begin{Bmatrix} \varepsilon_1 - \alpha_1 \Delta T \\ \varepsilon_2 - \alpha_2 \Delta T \\ \varepsilon_3 - \alpha_3 \Delta T \\ \gamma_{23} \\ \gamma_{13} \\ \gamma_{12} \end{Bmatrix}. \tag{5.107}$$

These definitions of strains are the most general and include the cases of stresses with no thermal effects, thermal effects but no stresses (i.e., free thermal strains), and thermal effects with stresses. For stresses with no temperature change, $\Delta T = 0$ and Eqs. (5.105) and (5.106) reduce directly to Eqs. (5.45) and (5.47). For a temperature change but no stresses, the free thermal strain case,

$$\sigma_1 = \sigma_2 = \sigma_3 = \tau_{23} = \tau_{13} = \tau_{12} = 0. \tag{5.108}$$

And by using either Eqs. (5.105) or (5.106), the total strains are:

$$\varepsilon_1 = \alpha_1 \Delta T \quad \varepsilon_2 = \alpha_2 \Delta T \quad \varepsilon_3 = \alpha_3 \Delta T \quad \gamma_{23} = \gamma_{13} = \gamma_{12} = 0. \tag{5.109}$$

For this case, if we use Eqs. (5.107) and (5.109), the mechanical strains are given by:

$$\begin{aligned}
\varepsilon_1^{\text{mech}} &= \varepsilon_1 - \alpha_1 \Delta T = \alpha_1 \Delta T - \alpha_1 \Delta T = 0 \\
\varepsilon_2^{\text{mech}} &= \varepsilon_2 - \alpha_2 \Delta T = \alpha_2 \Delta T - \alpha_2 \Delta T = 0 \\
\varepsilon_3^{\text{mech}} &= \varepsilon_3 - \alpha_3 \Delta T = \alpha_3 \Delta T - \alpha_3 \Delta T = 0 \\
\gamma_{23}^{\text{mech}} &= \gamma_{23} = 0 \\
\gamma_{13}^{\text{mech}} &= \gamma_{13} = 0 \\
\gamma_{12}^{\text{mech}} &= \gamma_{12} = 0.
\end{aligned} \tag{5.110}$$

That is, the mechanical strains are all zero. On the other hand, if an element of material is fully restrained against deformation:

$$\delta\Delta_1 = \delta\Delta_2 = \delta\Delta_3 = 0. \tag{5.111}$$

By Eq. (5.101):

$$\varepsilon_1 = \varepsilon_2 = \varepsilon_3 = 0. \tag{5.112}$$

Also:

$$\gamma_{23} = \gamma_{13} = \gamma_{12} = 0. \tag{5.113}$$

If the temperature is changed an amount ΔT, the stresses caused by the temperature change are:

$$\begin{Bmatrix} \sigma_1 \\ \sigma_2 \\ \sigma_3 \\ \tau_{23} \\ \tau_{13} \\ \tau_{12} \end{Bmatrix} = \begin{bmatrix} C_{11} & C_{12} & C_{13} & 0 & 0 & 0 \\ C_{12} & C_{22} & C_{23} & 0 & 0 & 0 \\ C_{13} & C_{23} & C_{33} & 0 & 0 & 0 \\ 0 & 0 & 0 & C_{44} & 0 & 0 \\ 0 & 0 & 0 & 0 & C_{55} & 0 \\ 0 & 0 & 0 & 0 & 0 & C_{66} \end{bmatrix} \begin{Bmatrix} -\alpha_1 \Delta T \\ -\alpha_2 \Delta T \\ -\alpha_3 \Delta T \\ 0 \\ 0 \\ 0 \end{Bmatrix}. \tag{5.114}$$

In this case, if we use Eqs. (5.107), (5.112), and (5.113), the mechanical strains are given by:

$$\begin{aligned}
&\varepsilon_1^{\text{mech}} = -\alpha_1 \Delta T \quad \varepsilon_2^{\text{mech}} = -\alpha_2 \Delta T \quad \varepsilon_3^{\text{mech}} = -\alpha_3 \Delta T \\
&\gamma_{23}^{\text{mech}} = \gamma_{13}^{\text{mech}} = \gamma_{12}^{\text{mech}} = 0.
\end{aligned} \tag{5.115}$$

The concept of mechanical strain is an important one. As an example, consider a fully restrained element of graphite-reinforced material. Substituting numerical values for the stiffnesses from Eq. (5.57), and numerical values for the CTE from Table 5.1, we find that if the temperature is raised 50°C from some reference temperature, the changes in the stresses in a fully restrained material are, from Eq. (5.114):

$$\begin{aligned} \sigma_1 &= -(\alpha_1 C_{11} + \alpha_2 C_{12} + \alpha_3 C_{13})\Delta T = -13.55 \text{ MPa} \\ \sigma_2 &= -(\alpha_1 C_{12} + \alpha_2 C_{22} + \alpha_3 C_{33})\Delta T = -27.6 \text{ MPa} \\ \sigma_3 &= -(\alpha_1 C_{13} + \alpha_2 C_{23} + \alpha_3 C_{33})\Delta T = -27.6 \text{ MPa} \\ \tau_{23} &= \tau_{13} = \tau_{12} = 0 \end{aligned} \quad . \tag{5.116}$$

Though these normal stresses do not seem significant, a lowering of the temperature would cause the sign of the stresses to become positive, and the roughly 30 MPa stress perpendicular to the fiber direction (i.e., σ_2 or σ_3) is a substantial percentage of the stress to cause failure of the material in tension in these two directions. In addition, due to the interaction of the three components of normal stress with the three components of extensional strain, the stress in the fiber direction, σ_1, is compressive. This is the case despite the fact that in the context of free thermal strains, the material contracts in the fiber direction when heated. Contraction in a direction when heated would lead one to believe the thermally induced stress would be tensile in that direction. The compressive stress in the fiber direction is a result of Poisson effects coupling the fiber direction stress with the other two stresses. The interaction of thermal and Poisson effects is subtle but quite important. The mechanical strains for this fully constrained case are, from Eq. (5.115):

$$\begin{aligned} \varepsilon_1^{\text{mech}} &= 0.900 \times 10^{-6} \text{ mm mm}^{-1} = 0.900 \,\mu\text{mm mm}^{-1} \\ \varepsilon_2^{\text{mech}} &= -1215 \times 10^{-6} \text{ mm mm}^{-1} = -1215 \,\mu\text{mm mm}^{-1} \\ \varepsilon_1^{\text{mech}} &= -1215 \times 10^{-6} \text{ mm mm}^{-1} = -1215 \,\mu\text{mm mm}^{-1} \\ \gamma_{23}^{\text{mech}} &= \gamma_{13}^{\text{mech}} = \gamma_{12}^{\text{mech}} = 0. \end{aligned} \tag{5.117}$$

A significant point should be made at this time. In reality, the elastic properties of a fiber-reinforced material are dependent on temperature. It is the absolute temperature that is important, not the temperature relative to some reference temperature. Then, in the stress–strain relations, the stiffnesses and compliances should be considered as functions of temperature; that is:

$$C_{ij} = C_{ij}(T) \quad i,j = 1,6 \tag{5.118}$$

$$S_{ij} = S_{ij}(T) \quad i,j = 1,6 \tag{5.119}$$

The interpretation in this case is that at a particular temperature T, the stresses and strains are related in accordance with Eq. (5.102), where it has been assumed that the free thermal expansion is a function of T and of the reference temperature, T_{ref}.

How the material is heated to the particular temperature or what the material properties are at other temperatures are not important. The relationship is path independent, and the material properties depend only on the current temperature, and for the free thermal strain, also on the reference temperature. For moderate increases of temperature relative to room temperature, the elastic properties can be assumed to be independent of temperature. As the temperature increase approaches the processing temperature, however, the elastic properties are a function of temperature. At temperatures lower than room temperature, the elastic properties for any particular material should also be checked for temperature dependence.

5.8 EFFECT OF FREE MOISTURE STRAINS ON STRESS–STRAIN RELATIONS

When exposed to a liquid, polymers absorb a certain amount of that liquid and, in general, expand. The amount of liquid absorbed is not significant, however. Weight gains in excess of 3% or 4% are unusual. As a result, it is not the weight gain that is important; rather, it is the expansion, which is similar to the expansion that accompanies a temperature increase, which is the issue. The level of the expansion is close to being linear with the amount of liquid absorbed. Consequently, in analogy to the CTE, defining a coefficient of moisture expansion represents a viable model for characterizing the free moisture expansion of a polymer. For polymer matrix composites, the use of a linear expansion model is also applicable. As might be expected, however, the moisture expansion in the fiber direction is small, at least for graphite fibers; the fibers usually do not absorb moisture, and the stiffness of the fibers overcomes any tendency for the polymer to expand in that direction. On the other hand, expansion perpendicular to the fibers is significant. Using the analogy to linear thermal expansion, Eq. (5.96), the free moisture strains are given by:

$$\varepsilon_1^M\left(M, M_{\text{ref}}\right) = \beta_1 \Delta M \quad \varepsilon_2^M\left(M, M_{\text{ref}}\right) = \beta_2 \Delta M$$
$$\varepsilon_3^M\left(M, M_{\text{ref}}\right) = \beta_3 \Delta M. \tag{5.120}$$

The superscript M identifies that the strain is moisture-induced and the dependence on both M and M_{ref} signifies that the free moisture strain is measured relative to some reference moisture state. However, here it is assumed that it is only the change in moisture relative to that state that determines the strain relative to that state. The free moisture strains in the reference state are defined to be zero. The β_i are referred to as the coefficients of moisture expansion. Generally, β_1 can be taken to be zero, and β_2 and β_3 are taken to be equal. For the most fiber-reinforced materials, β_2 ranges from 0.003% to 0.005/% moisture. With a 3% moisture change, say, from the dry state, this can result in expansional strains upward of 0.015 perpendicular to the fibers. This is a large amount of strain. Table 5.1 shows the values of the moisture expansion coefficients for graphite- and glass-reinforced composite materials.

Using the values of β_1, β_2, and β_3 from Table 5.1, we can determine the dimensional changes of a 50 mm × 50 mm × 50mm element of graphite-reinforced material which has absorbed 0.5% moisture by using the moisture analogy to Eq. (5.95), namely:

$$\delta\Delta_1 = \Delta_1\varepsilon_1^M\left(M, M_{\text{ref}}\right) = \Delta_1\beta_1\Delta M = 50\times\left(146\times10^{-6}\right)\times0.5 = 0.00365\text{ mm} \tag{5.121a}$$

$$\delta\Delta_2 = \Delta_2\varepsilon_2^M\left(M, M_{\text{ref}}\right) = \Delta_2\beta_2\Delta M = 50\times\left(4770\times10^{-6}\right)\times0.5 = 0.1193\text{mm} \tag{5.121b}$$

$$\delta\Delta_3 = \Delta_3\varepsilon_3^M\left(M, M_{\text{ref}}\right) = \Delta_3\beta_3\Delta M = 50\times\left(4770\times10^{-6}\right)\times0.5 = 0.1193\text{ mm}. \tag{5.121c}$$

The dimensional changes in the 2 and 3 directions are significant, and restraining them leads to large moisture-induced stresses. The manner and the rate at which polymers absorb moisture is an important topic in the study of the mechanics of fiber-reinforced material. The study parallels the characteristics of heat transfer within a solid. For the moment, however, it is only necessary to assume that the composite has absorbed moisture and that the moisture has produced a strain. Incorporating these free moisture strains into the stress–strain relations, in the same manner as the free thermal strains were accounted for, leads to, in analogy to Eqs. (5.105) and (5.106):

$$\begin{Bmatrix} \varepsilon_1 - \alpha_1\Delta T - \beta_1\Delta M \\ \varepsilon_2 - \alpha_2\Delta T - \beta_2\Delta M \\ \varepsilon_3 - \alpha_3\Delta T - \beta_3\Delta M \\ \gamma_{23} \\ \gamma_{13} \\ \gamma_{12} \end{Bmatrix} = \begin{bmatrix} S_{11} & S_{12} & S_{13} & 0 & 0 & 0 \\ S_{12} & S_{22} & S_{23} & 0 & 0 & 0 \\ S_{13} & S_{23} & S_{33} & 0 & 0 & 0 \\ 0 & 0 & 0 & S_{44} & 0 & 0 \\ 0 & 0 & 0 & 0 & S_{55} & 0 \\ 0 & 0 & 0 & 0 & 0 & S_{66} \end{bmatrix} \begin{Bmatrix} \sigma_1 \\ \sigma_2 \\ \sigma_3 \\ \tau_{23} \\ \tau_{13} \\ \tau_{12} \end{Bmatrix} \tag{5.122}$$

$$\begin{Bmatrix} \sigma_1 \\ \sigma_2 \\ \sigma_3 \\ \tau_{23} \\ \tau_{13} \\ \tau_{12} \end{Bmatrix} = \begin{bmatrix} C_{11} & C_{12} & C_{13} & 0 & 0 & 0 \\ C_{12} & C_{22} & C_{23} & 0 & 0 & 0 \\ C_{13} & C_{23} & C_{33} & 0 & 0 & 0 \\ 0 & 0 & 0 & C_{44} & 0 & 0 \\ 0 & 0 & 0 & 0 & C_{55} & 0 \\ 0 & 0 & 0 & 0 & 0 & C_{66} \end{bmatrix} \begin{Bmatrix} \varepsilon_1 - \alpha_1\Delta T - \beta_1\Delta M \\ \varepsilon_2 - \alpha_2\Delta T - \beta_2\Delta M \\ \varepsilon_3 - \alpha_3\Delta T - \beta_3\Delta M \\ \gamma_{23} \\ \gamma_{13} \\ \gamma_{12} \end{Bmatrix}. \tag{5.123}$$

The mechanical strains, the key strains in the stress–strain relations, are given by:

$$\begin{aligned} \varepsilon_1^{\text{mech}} &= \varepsilon_1 - \alpha_1\Delta T - \beta_1\Delta M \\ \varepsilon_2^{\text{mech}} &= \varepsilon_2 - \alpha_2\Delta T - \beta_2\Delta M \\ \varepsilon_3^{\text{mech}} &= \varepsilon_3 - \alpha_3\Delta T - \beta_3\Delta M, \end{aligned} \tag{5.124}$$

where, as has consistently been the case, the dimensional changes are reflected in:

$$\delta\Delta_1 = \varepsilon_1\Delta_1$$
$$\delta\Delta_2 = \varepsilon_2\Delta_2 \tag{5.125}$$
$$\delta\Delta_3 = \varepsilon_3\Delta_3.$$

With both free thermal strains and free moisture strains present in various combinations, some unusual effects and stress states are possible. The combination of moisture and heat is, in general, detrimental to polymer matrix composite materials, and it represents an important class of problems for these materials. As a numerical example, using the values of β_1, β_2, and β_3 from Table 5.1, we find that a fully restrained element of graphite-reinforced composite that has absorbed 0.5% moisture but has experienced no temperature change requires the following stresses to prevent it from expanding:

$$\sigma_1 = -\left(C_{11}\beta_1 + C_{12}\beta_2 + C_{13}\beta_3\right)\Delta M = -38.4\,\text{MPa}$$
$$\sigma_2 = -\left(C_{12}\beta_1 + C_{22}\beta_2 + C_{23}\beta_3\right)\Delta \text{M} = -54.6\ \text{MPa}. \tag{5.126}$$
$$\sigma_3 = -\left(C_{13}\beta_1 + C_{23}\beta_2 + C_{33}\beta_3\right)\Delta M = -54.6\,\text{MPa}$$

The stresses generated by moisture absorption are substantial. Again, since β_1 is small, the stress generated in the fiber direction is due to a coupling of the 1, 2, and 3 directions through Poisson effects. Clearly, one needs to be very aware of Poisson effects in materials, particularly in fiber-reinforced materials. The mechanical strains for this fully restrained isothermal case are:

$$\varepsilon_1^{\text{mech}} = -\beta_1\Delta M = -73\times10^{-6}\,\text{mm}\,\text{mm}^{-1} = -73\,\mu\text{mm}\,\text{mm}^{-1}$$
$$\varepsilon_2^{\text{mech}} = -\beta_2\Delta M = -2380\times10^{-6}\ \text{mm}\,\text{mm}^{-1} = -2380\,\mu\,\text{mm}\,\text{mm}^{-1}$$
$$\varepsilon_3^{\text{mech}} = -\beta_3\Delta M = -2380\times10^{-6}\,\text{mm}\,\text{mm}^{-1} = -2380\,\mu\text{mm}\,\text{mm}^{-1} \quad . \tag{5.127}$$
$$\gamma_{23}^{\text{mech}} = \gamma_{13}^{\text{mech}} = \gamma_{12}^{\text{mech}} = 0$$

Like temperature, moisture influences the elastic properties of a fiber-reinforced material. For graphite-reinforced materials, moisture has little influence on properties in the fiber direction. However, moisture does influence other properties, and it can influence properties in the fiber direction for some polymeric fibers. To account for moisture-induced property changes, the stiffnesses and the compliances can be made dependent on the absolute moisture content. As with temperature-dependent material properties, the interpretation in this case is that at particular moisture content M, the stresses and strains are related in accordance with Eq. (5.122) or (5.123). How the moisture content reaches a particular level or what the material properties are at other moisture levels are not important. The relationship is path independent, and the material properties depend only on the current moisture content, and the free moisture strains depend only on the reference moisture state and the current moisture state.

REFERENCES

1. de St. Venant B. (1864), in his edition of Navier's *Resume des sur l'application de la Mecanique*, app. 3, Carilian-Goeury, Paris.
2. Hooke R. (1678), De Potentia Restitutiva, Printed for John Martyn Printer to the Royal Society, London.
3. Love A.E.H. (1892), *A Treatise on the Mathematical Theory of Elasticity*, Cambridge University Press, London (also 4th ed., 1944, Dover Publications, Inc., New York).
4. Hodge P.G. Jr. (1958), "The mathematical theory of plasticity," in *Elasticity and Plasticity* (J. N. Goodier and P. G. Hodge, Jr., eds.), John Wiley & Sons, Inc., New York, p. 81.
5. Kreyszig E. (1967), *Advanced Engineering Mathematics*, 2nd ed., John Wiley & Sons, Inc., New York, p. 416.
6. Fung Y.C. (1965), *Foundations of Solid Mechanics*, Prentice-Hall, Inc., Englewood Cliffs, NJ, p. 285.
7. Green G. (1839), "On the laws of reflexion and refraction of light at the common surface of two non-crystallized media", *Transactions of the Cambridge Philosophical Society*, Vol. 7, p. 121.
8. Popov E.P. (1990), *Engineering Mechanics of Solids*, Prentice Hall, Englewood Cliffs, NJ.

6 Micromechanics

In Chapter 5, the stress–strain relationships and engineering constants were developed using elastic moduli, strength parameters, two coefficients of thermal expansion (CTE), and two coefficients of moisture expansion (CME) for a unidirectional lamina. These parameters can be found experimentally by conducting several tension, compression, shear, and hygrothermal tests on unidirectional lamina (laminates). However, unlike in isotropic materials, experimental evaluation of these parameters is quite costly and time-consuming because they are functions of several variables: the individual constituents of the composite material, fiber volume fraction, packing geometry, processing etc. Thus, the need and motivation for developing analytical models to find these parameters are very important. In this chapter, we will develop simple relationships for these parameters in terms of the stiffnesses, strengths, coefficients of thermal and moisture expansion of the individual constituents of a composite, fiber volume fraction, packing geometry etc. An understanding of this relationship, called micromechanics of lamina, helps the designer to select the constituents of a composite material for use in a laminated structure.

6.1 VOLUME AND MASS FRACTIONS

The concept of volume and mass fractions is important because theoretical formulas for finding the stiffness, strength, and hygrothermal properties of a unidirectional lamina are a function of fiber volume fraction. Measurements of the constituents are generally based on their mass, so fiber mass fractions must also be defined. Moreover, defining the density of a composite also becomes necessary because its value is used in the experimental determination of fiber volume and void fractions of a composite.

6.1.1 Volume Fractions

Consider a composite consisting of fiber and matrix. Take the following symbol notations:

$$v_{c,f,m} = \text{Volume of composite, fiber, and matrix, respectively}$$

$$\rho_{c,f,m} = \text{Density of composite, fiber, and matrix, respectively}$$

The fiber and matrix volume fraction are defined as:

$$\text{Fiber volume fraction}: V_f = \frac{v_f}{v_c} = \frac{\text{Volume of fiber}}{\text{Volume of composite}} \tag{6.1a}$$

$$\text{Matrix volume fraction}: V_m = \frac{v_m}{v_c} = \frac{\text{Volume of matrix}}{\text{Volume of composite}}. \tag{6.1b}$$

Note that the sum of volume fractions is unity, that is,

$$V_f + V_m = 1.$$

Using Eqs. (6.1a) and (6.1b):

$$v_f + v_m = v_c.$$

6.1.2 Mass Fractions

Consider a composite consisting of fiber and matrix, and take the following notation:

$$w_{c,f,m} = \text{Mass of composite, fiber, and matrix, respectively.}$$

The mass fractions (weight fraction) of the fibers (W_f) and the matrix (W_m) are defined as:

$$W_f = \frac{w_f}{w_c} \tag{6.2a}$$

$$W_m = \frac{w_m}{w_c}. \tag{6.2b}$$

The sum of mass fractions is unity, that is,

$$W_f + W_m = 1.$$

Using Eqs. (6.2a) and (6.2b):

$$w_f + w_m = w_c.$$

From the definition of the density of a single material:

$$w_c = \rho_c v_c \tag{6.3a}$$

$$w_f = \rho_f v_f \tag{6.3b}$$

$$w_m = \rho_m v_m. \tag{6.3c}$$

Substituting Eq. (6.3) in Eq. (6.2), the mass fractions and volume fractions are related as:

$$W_f = \frac{\rho_f}{\rho_c} V_f \tag{6.4a}$$

$$W_m = \frac{\rho_m}{\rho_c} V_m. \tag{6.4b}$$

In terms of individual constituent properties, the mass fractions and volume fractions are related by:

$$W_f = \frac{\frac{\rho_f}{\rho_m}}{\frac{\rho_f}{\rho_m} V_f + V_m} V_f \tag{6.5a}$$

$$W_m = \frac{1}{\frac{\rho_f}{\rho_m}(1 - V_m) + V_m} V_m. \tag{6.5b}$$

6.2 DENSITY

The mass of composite w_c is the sum of the mass of the fibers w_f and the mass of the matrix w_m as:

$$w_f + w_m = w_c. \tag{6.6}$$

Substituting Eq. (6.3) into Eq. (6.6) yields:

$$\rho_c v_c = \rho_f v_f + \rho_m v_m.$$

Therefore,

$$\rho_c = \rho_f \frac{v_f}{v_c} + \rho_m \frac{v_m}{v_c}. \tag{6.7}$$

Using Eq. (6.1):

$$\rho_c = \rho_f V_f + \rho_m V_m. \tag{6.8}$$

The volume of a composite v_c is the sum of the volumes of the fiber v_f and the matrix (v_m):

$$v_f + v_m = v_c. \tag{6.9}$$

The density of the composite in terms of mass fractions can be found as:

$$\frac{1}{\rho_c} = \frac{W_m}{\rho_f} + \frac{W_m}{\rho_m}. \tag{6.10}$$

6.3 VOID CONTENT

During the manufacture of a composite, voids are introduced in the composite. This causes the theoretical density of the composite to be higher than the actual density. Also, the void content of a composite is detrimental to its mechanical properties. These detriments include lower:

i. Shear stiffness and strength
ii. Compressive strengths
iii. Transverse tensile strengths
iv. Fatigue resistance
v. Moisture resistance.

A decrease of 2%–10% in the preceding matrix-dominated properties generally takes place with every 1% increase in the void content. For composites with a certain volume of voids v_v, the volume fraction of voids V_v is defined as:

$$V_v = \frac{v_v}{v_c}. \tag{6.11}$$

The total volume of a composite (v_c) with voids is given by:

$$v_f + v_m + v_v = v_c. \tag{6.12}$$

Using the experimental density ρ_{ce} of a composite, the actual volume of the composite is:

$$v_c = \frac{w_c}{\rho_{ce}}. \tag{6.13}$$

Using the definition of the theoretical density ρ_{ct} of the composite, the theoretical volume of the composite is:

$$v_f + v_m = \frac{w_c}{\rho_{ct}}. \tag{6.14}$$

Substituting Eqs. (6.13) and (6.14) into Eq. (6.12):

$$\frac{w_c}{\rho_{ce}} = \frac{w_c}{\rho_{ct}} + v_v.$$

The volume of the void is given by:

$$v_v = \frac{w_c}{\rho_{ce}}\left(\frac{\rho_{ct} - \rho_{ce}}{\rho_{ct}}\right). \tag{6.15}$$

Substituting Eqs. (6.13) and (6.15) into Eq. (6.11), the volume fraction of the voids is:

$$V_v = \frac{v_v}{v_c} = \left(\frac{\rho_{ct} - \rho_{ce}}{\rho_{ct}} \right). \tag{6.16}$$

6.4 EVALUATION OF ELASTIC MODULI

An important result that can be obtained from the concept of the unit cell and the use of finite elements, or the use of elasticity solutions [1,2], is an estimate of the overall elastic and thermal expansion properties of the composite. With these models, it is possible to evaluate how the overall properties are influenced by fiber volume fraction, fiber properties, matrix properties, and the assumptions of how the fibers are packed (i.e., square- or hexagonal-packed arrays). While the use of finite element representations of unit cells provides a detailed information about the stresses in the fibers, in the matrix, and at the interface between the fiber and the matrix, often it is only the elastic or thermal expansion property estimates that are of interest. This was the case in early micromechanics studies, whereby strength-of-materials approaches were used to provide insight into the elastic properties. These approaches can be considered to be at the opposite end of the spectrum from the finite element or elasticity approaches.

6.4.1 Strength-of-Materials Approach

The strength-of-materials approaches do not concern themselves with the details of the stresses at the fiber–matrix interface, the packing arrangements, or the many other characteristics that can be considered with unit cell finite element models. However, reasonable estimates of some of the elastic and thermal expansion properties of composite materials can be obtained with these approaches. They shall be studied next to provide a contrast to the unit cell finite element methods. In addition, the strength-of-materials approaches result in rather simple algebraic expressions for the elastic and thermal expansion properties of the composite as a function of fiber and matrix properties. These algebraic expressions can be conveniently used for parametric studies or for embedding within other analysis.

6.4.1.1 Model for E_1 and ν_{12}

The strength-of-materials models, sometimes called *rule-of-mixtures models*, also rely on what could be termed a "unit cell." The unit cell used in the strength-of-materials models is quite different from the unit cells of the previous sections, and the particular unit cell considered depends on the composite property being studied. To study E_1 and ν_{12} for the composite, consider a section cut from a single layer of fiber-reinforced material. The section consists of side-by-side alternating regions of fiber and matrix; the fibers are arranged in parallel arrays, as in Figure 6.1; and the widths of each of the regions of fiber and matrix are denoted by W_f and W_m, respectively.

Figure 6.1 shows the 1 and 2 principal material directions. The thickness of the layer is not important at the moment, and can be taken as unity. In fact, in these

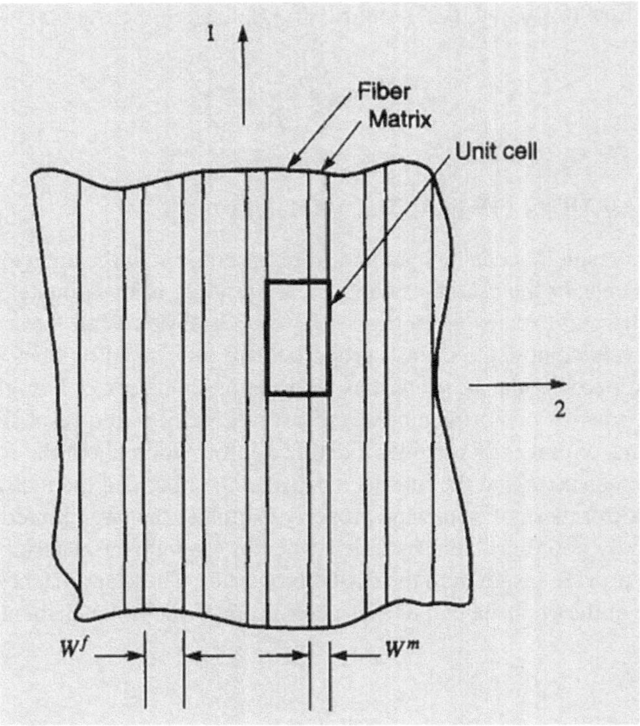

FIGURE 6.1 Unit cell representation.

rule-of-mixtures models, the cross-sectional shape of the fibers is not important. They can be considered circular, square, elliptical, or any other shape. For simplicity, assume they are square. As it will turn out, only the cross-sectional areas of the fiber and the matrix will be important. Figure 6.2a shows the details of a "unit cell" cut from a single layer, the length of the cell denoted by L, and the cross-sectional areas of the fiber and the matrix denoted as A^f and A^m, respectively. Assume as in Figure 6.2b that the unit cell is subjected to a stress σ_1 such that it stretches in the 1 direction and, due to Poisson effects, contracts in the 2 direction. Because the fiber and the matrix are bonded together, they both stretch the same amount in the 1 direction, namely, ΔL; the strain in the 1 direction in both the fiber and the matrix is given by:

$$\varepsilon_1^f = \varepsilon_1^m = \frac{\Delta L}{L}. \tag{6.17}$$

Because the fiber and the matrix have different Poisson's ratios, they will not contract the same amount in the widthwise, or 2, direction. The combined contraction of the fiber and matrix results in the overall contraction of the composite in the 2 direction. Treating the two constituents as if they were each in a one-dimensional state of stress, we find the stresses in the fiber and matrix are:

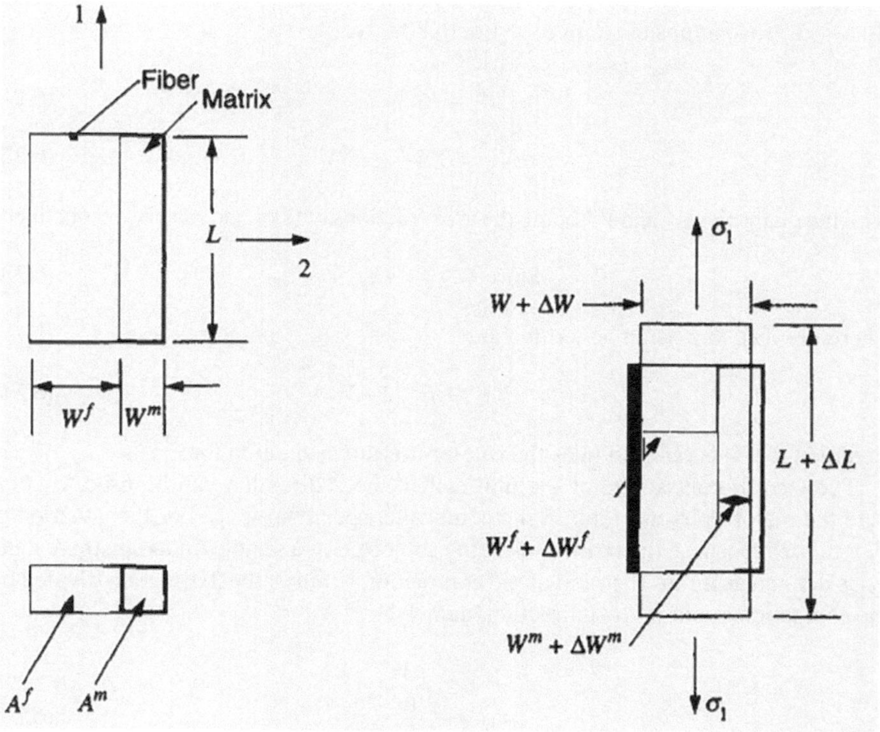

FIGURE 6.2 Model for predicting E_1 and ν_{12}: (a) Geometry and nomenclature and (b) unit cell subjected to σ_1.

$$\sigma_1^f = E_1^f \varepsilon_1^f = E_1^f \frac{\Delta L}{L}$$
$$\sigma_1^m = E^m \varepsilon_1^m = E^m \frac{\Delta L}{L}, \tag{6.18}$$

and accordingly, the forces in the 1 direction in the fiber and matrix are given by:

$$F_1^f = \sigma_1^f A^f = E_1^f \frac{\Delta L}{L} A^f$$
$$F_1^m = \sigma_1^m A^m = E^m \frac{\Delta L}{L} A^m. \tag{6.19}$$

The total force in the 1 direction divided by the total cross-sectional area of the unit cell, A, where $A = A^f + A^m$, is defined as the composite stress σ_1, namely:

$$\sigma_1 = \frac{F_1^f + F_1^m}{A} = \left(E_1^f \frac{A^f}{A} + E^m \frac{A^m}{A} \right) \frac{\Delta L}{L}. \tag{6.20}$$

But $\frac{\Delta L}{L}$ is the composite strain ε_1 so Eq. (6.20) gives:

$$\sigma_1 = E_1\varepsilon_1, \tag{6.21}$$

$$\text{where } E_1 = E_1^f V^f + E^m V^m \tag{6.22}$$

with the quantities V^f and V^m being the area fractions of fiber and matrix, respectively.

$$\text{Since } V^m = 1 - V^f. \tag{6.23}$$

Therefore, Eq. (6.22) can be written as:

$$E_1 = E_1^f V^f + E^m\left(1 - V^f\right). \tag{6.24}$$

Equation (6.24) is referred to as the rule-of-mixtures equation for E_1.

The overall contraction of the unit cell in the 2 direction can be used to compute the major Poisson's ratio of the composite, specifically ν_{12}. For the situation in Figure 6.2b, because the unit cell is being subjected to a simple uniaxial stress state, ν_{12} is defined as minus the ratio of the contraction strain in the 2 direction divided by the elongation strain in the 1 direction, namely:

$$\nu_{12} = -\frac{\frac{\Delta W}{W}}{\frac{\Delta L}{L}}. \tag{6.25}$$

From Figure 6.2b,

$$W^f + W^m = W \tag{6.26}$$

$$\text{Also, } \Delta W = \Delta W^f + \Delta W^m. \tag{6.27}$$

Using the definition of Poisson's ratio for each constituent and the fact that each constituent is assumed to be in a state of uniaxial stress, we find the contraction of each constituent is:

$$\begin{aligned} \frac{\Delta W^f}{W^f} &= -\nu_{12}^f \frac{\Delta L}{L} \\ \frac{\Delta W^m}{W^m} &= -\nu^m \frac{\Delta L}{L}. \end{aligned} \tag{6.28}$$

or

$$\begin{aligned} \Delta W^f &= -\nu_{12}^f W^f \frac{\Delta L}{L} \\ \Delta W^m &= -\nu^m W^m \frac{\Delta L}{L}. \end{aligned} \tag{6.29}$$

Substituting these relations into Eq. (6.27) and dividing both sides by W result in:

$$\frac{\Delta W}{W} = -\left(\nu_{12}^{f}\frac{W^{f}}{W} + \nu^{m}\frac{W^{m}}{W}\right)\frac{\Delta L}{L}. \tag{6.30}$$

The fiber and the matrix volume fractions can be identified, and as a result, Eq. (6.30) becomes:

$$\frac{\Delta W}{W} = -\left(\nu_{12}^{f}V^{f} + \nu^{m}V^{m}\right)\frac{\Delta L}{L}. \tag{6.31}$$

Using the definition of Poisson's ratio in Eq. (6.25):

$$\nu_{12} = \nu_{12}^{f}V^{f} + \nu^{m}V^{m}. \tag{6.32}$$

Also, using Eq. (6.23):

$$\nu_{12} = \nu_{12}^{f}V^{f} + \nu^{m}\left(1 - V^{f}\right). \tag{6.33}$$

This is the rule-of-mixtures expression for the major Poisson's ratio ν_{12}. It is very similar to the rule-of-mixtures expression for the modulus E_1 in that it is linear in all of the variables.

6.4.1.2 Model for E_2

We can approach one of the most basic considerations for the determination of E_2 by studying the unit cell of Figure 6.1 when it is subjected to a transverse stress, σ_2, as shown in Figure 6.3. Isolating the fiber and matrix elements, we can argue by

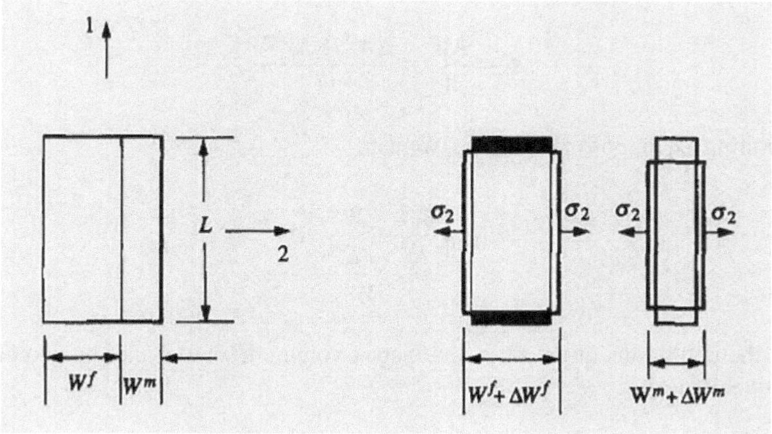

FIGURE 6.3 Rule-of-mixtures model for E_2: (a) Geometry and nomenclature and (b) unit cell subjected to σ_2.

equilibrium that each element is subjected to the same transverse stress, σ_2. If this is the case, then the transverse strains in the fiber and matrix are, respectively:

$$\varepsilon_2^f = \frac{\Delta W^f}{W^f}$$
$$\varepsilon_2^m = \frac{\Delta W^m}{W^m}. \tag{6.34}$$

Considering again a one-dimensional stress state, we find the stress and strain in the fiber and matrix are related by:

$$\sigma_2^f = \sigma_2 = E_2^f \varepsilon_2^f = E_2^f \frac{\Delta W^f}{W^f}$$
$$\sigma_2^m = \sigma_2 = E^m \varepsilon_2^m = E^m \frac{\Delta W^m}{W^m} \tag{6.35}$$

These equations can be rearranged and written as:

$$\Delta W^f = \frac{W^f}{E_2^f}\sigma_2$$
$$\Delta W^m = \frac{W^m}{E^m}\sigma_2. \tag{6.36}$$

The overall change in the transverse dimension of the unit cell is:

$$\Delta W = \Delta W^f + \Delta W^m. \tag{6.37}$$

The definition of the overall transverse strain is:

$$\varepsilon_2 = \frac{\Delta W}{W} = \frac{\Delta W^f + \Delta W^m}{W}. \tag{6.38}$$

Substituting Eq. (6.36) in Eq. (6.38), we get:

$$\varepsilon_2 = \frac{\left(\frac{W^f}{E_2^f} + \frac{W^m}{E^m}\right)\sigma_2}{W}. \tag{6.39}$$

Using the definitions of the fiber and matrix volume fractions and the geometry of the unit cell yields:

$$\varepsilon_2 = \left(\frac{V^f}{E_2^f} + \frac{V^m}{E^m}\right)\sigma_2. \tag{6.40}$$

Comparing Eq. (6.40) with the equation of the form:

$$\varepsilon_2 = \frac{\sigma_2}{E_2}. \tag{6.41}$$

We can thus write:

$$\frac{1}{E_2} = \frac{V^f}{E_2^f} + \frac{V^m}{E^m} \tag{6.42}$$

$$\text{Or, } \frac{1}{E_2} = \frac{V^f}{E_2^f} + \frac{\left(1 - V^f\right)}{E^m}. \tag{6.43}$$

This equation is the rule-of-mixtures relation for the transverse modulus as a function of the transverse moduli of the fiber and matrix, and the fiber volume fraction. Like rule-of-mixtures relations for E_1 and ν_{12}, this relation is a simple linear relation, in this case among the inverse moduli and volume fractions of the two constituents.

The one-dimensional state of stress may not be accurate because, like the situation depicted in Figure 6.2, the fiber and matrix elements are bonded together and hence change length together in the 1 direction. Considering the one-dimensional stress state, we find the diagram of Figure 6.3 indicates that due to the different Poisson's ratios, the element of fiber is allowed to contract in the 1 direction differently than the element of matrix. Thus, a modification of the model would be to have the length change in the 1 direction of the fiber element and the matrix element be the same, an approach to be taken shortly. However, another difficulty with the simplification of Figure 6.3 is the diagram assuming that both the fiber and the matrix are subjected to transverse stress σ_2. The transverse stress σ_2 in the fiber and matrix is *partitioned* differently than is implied by Figure 6.3. To correct the rule-of-mixtures model for E_2, a so-called stress-partitioning factor is often introduced into Eq. (6.43) to account for the error in the assumption that both the fiber and the matrix are subjected to the full value σ_2. The stress-partitioning factor accounts for a more proper division of the stress in each of the two constituents. To incorporate a partitioning factor, consider Eq. (6.39) rewritten in slightly different form, specifically:

$$\varepsilon_2 = \frac{\left(\dfrac{W^f}{E_2^f} + \dfrac{W^m}{E^m}\right)\sigma_2}{W^f + W^m}. \tag{6.44}$$

Dividing the numerator and denominator by W and again using the geometry of the unit cell as it relates to the fiber and matrix volume fractions, we find Eq. (6.44) becomes:

$$\varepsilon_2 = \frac{\left(\dfrac{V^f}{E_2^f} + \dfrac{V^m}{E^m}\right)\sigma_2}{V^f + V^m}. \tag{6.45}$$

Now consider that instead of V^m being the volume fraction of matrix that is subjected to stress level σ_2, assume the volume fraction is less than that, namely, ηV^m, where η will be referred to as a partitioning factor and $0 < \eta < 1$. The volume fraction of fiber that is subjected to stress level σ_2 is still V^f. As a result of this new nomenclature, the total effective volume of fiber and matrix is now:

$$V^f + \eta V^m. \tag{6.46}$$

Equation (6.45) takes the form:

$$\varepsilon_2 = \frac{\left(\dfrac{V^f}{E_2^f} + \dfrac{\eta V^m}{E^m}\right)\sigma_2}{V^f + \eta V^m}. \tag{6.47}$$

By analogy with Eq. (6.41), the composite modulus E_2 is given as:

$$\frac{1}{E_2} = \frac{\left(\dfrac{V^f}{E_2^f} + \dfrac{\eta V^m}{E^m}\right)}{V^f + \eta V^m} \tag{6.48}$$

$$\frac{1}{E_2} = \frac{\left(\dfrac{V^f}{E_2^f} + \dfrac{\eta\left(1 - V^f\right)}{E^m}\right)}{V^f + \eta\left(1 - V^f\right)}. \tag{6.49}$$

This expression for E_2 is referred to as the modified rule-of-mixtures model for E_2. The stress-partitioning factor η generally must be determined empirically. If it can be determined for a specific material by measuring E_2 at a particular volume fraction, and if the elastic properties of the fiber and matrix are known, then the value of η can be determined and used for parameter studies involving other fiber volume fractions. Note that when $\eta = 1$ in Eq. (6.49), the original rule-of-mixtures relation, Eq. (6.43), is recovered.

The rule-of-mixtures model for E_2 violates intuition regarding the response of the fiber and matrix elements in that the fiber element is allowed to change length independently of the change in length of the matrix element [3,4]. Figure 6.3 can be modified to account for the elements changing length by the same amount when subjected to a transverse stress σ_2. This will imply, of course, that each element is subjected to a stress in the 1 direction, σ_1. The free-body diagrams of the elements with the two stress components are shown in Figure 6.4, and the deformed transverse widths of the fiber and matrix elements are, respectively:

$$W^f + \Delta W^f \quad \text{and} \quad W^m + \Delta W^m. \tag{6.50}$$

The deformed length of *both* elements is:

$$L + \Delta L \tag{6.51}$$

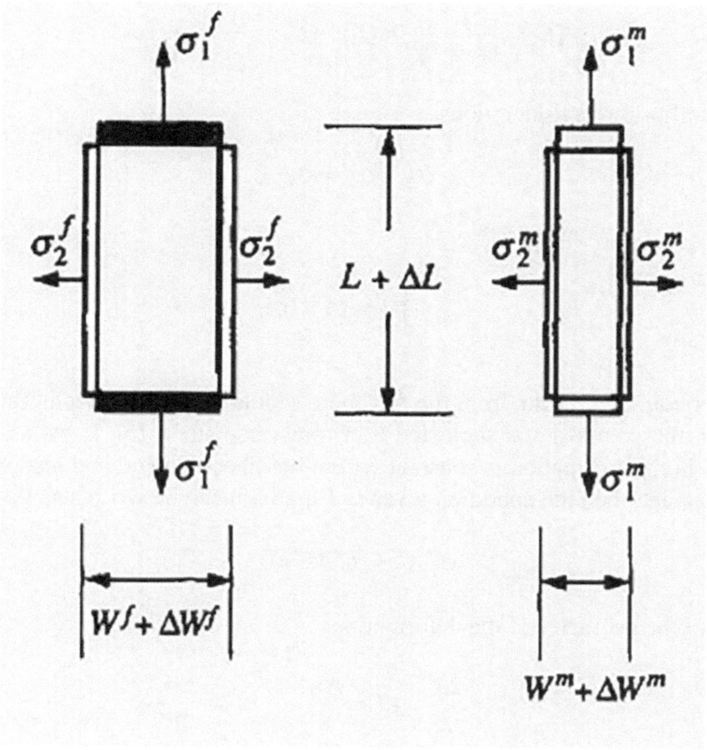

FIGURE 6.4 Alternate rule-of-mixtures model for E_2.

Considering each element to be in a state of stress such that:

$$\sigma_3 = \tau_{32} = \tau_{31} = \tau_{12} = 0. \tag{6.52}$$

The stress–strain relations for the fiber element are:

$$\begin{aligned} \sigma_1^f &= Q_{11}^f \varepsilon_1^f + Q_{12}^f \varepsilon_2^f \\ \sigma_2^f &= Q_{12}^f \varepsilon_1^f + Q_{22}^f \varepsilon_2^f . \end{aligned} \tag{6.53a}$$

For the matrix element:

$$\begin{aligned} \sigma_1^m &= Q_{11}^m \varepsilon_1^m + Q_{12}^m \varepsilon_2^m \\ \sigma_2^m &= Q_{12}^m \varepsilon_1^m + Q_{22}^m \varepsilon_2^m . \end{aligned} \tag{6.53b}$$

In Eqs. (6.53a) and (6.53b):

$$Q_{11}^f = \frac{E_1^f}{1 - \nu_{12}^f \nu_{21}^f} \quad Q_{22}^f = \frac{E_2^f}{1 - \nu_{12}^f \nu_{21}^f} \quad Q_{12}^f = \frac{\nu_{12}^f E_2^f}{1 - \nu_{12}^f \nu_{21}^f} = \frac{\nu_{21}^f E_1^f}{1 - \nu_{12}^f \nu_{21}^f} \tag{6.54a}$$

$$Q_{11}^m = \frac{E^m}{1-(\nu^m)^2} = Q_{22}^m \quad Q_{12}^m = \frac{\nu^m E^m}{1-(\nu^m)^2}. \tag{6.54b}$$

Due to equilibrium considerations:

$$\sigma_2^f = \sigma_2^m = \sigma_2. \tag{6.55}$$

Also,

$$\int_{A^m+A^f} \sigma_1 dA = 0. \tag{6.56}$$

This latter equation results from the fact there should be no net force in the 1 direction when the composite is subjected to a transverse stress [5]. If we assume that the stress in the 1 direction is constant within the fiber element, and also within the matrix element, then the condition given in Eq. (6.56) can be written as:

$$\sigma_1^f A^f + \sigma_1^m A^m = 0. \tag{6.57}$$

Because of the geometry of the deformation:

$$\varepsilon_1^f = \varepsilon_1^m = \frac{\Delta L}{L} \quad \varepsilon_2^f = \frac{\Delta W^f}{W^f} \quad \varepsilon_2^m = \frac{\Delta W^m}{W^m}. \tag{6.58}$$

Using Eqs. (6.55) and (6.58) in the stress–strain relations Eqs. (6.53a) and (6.53b), as well as using Eq. (6.57), results in equations that can be used to find ΔW^f and ΔW^m, and ultimately E_2. With some rearrangement, these equations are:

$$\begin{gathered}
Q_{22}^f \frac{\Delta W^f}{W^f} + Q_{12}^f \frac{\Delta L}{L} = \sigma_2 \\
Q_{22}^m \frac{\Delta W^m}{W^m} + Q_{12}^m \frac{\Delta L}{L} = \sigma_2 \\
\left(Q_{11}^f \frac{\Delta L}{L} + Q_{12}^f \frac{\Delta W^f}{W^f} \right) A^f + \left(Q_{11}^m \frac{\Delta L}{L} + Q_{12}^m \frac{\Delta W^m}{W^m} \right) A^m = 0.
\end{gathered} \tag{6.59}$$

Solving for ΔW^f and ΔW^m and using the basic definition of ε_2, namely, Eq. (6.38):

$$\varepsilon_2 = \left(\frac{\eta^f V^f}{E_2^f} + \frac{\eta^m V^m}{E^m} \right) \sigma_2. \tag{6.60}$$

Dividing by σ_2 yields the expression for E_2, namely:

$$\frac{1}{E_2} = \left(\frac{\eta^f V^f}{E_2^f} + \frac{\eta^m V^m}{E^m} \right) \tag{6.61}$$

$$\frac{1}{E_2} = \frac{\eta^f V^f}{E_2^f} + \frac{\eta^m \left(1 - V^f\right)}{E^m}. \tag{6.62}$$

To arrive at Eq. (6.62), we've made use of the fact that for the geometry of Figure 6.4, A^f and A^m are directly related to the volume fractions V^f and V^m, respectively. Also, the fiber and matrix partitioning factors are given by:

$$\eta^f = \frac{E_1^f V^f + \left[\left(1 - \nu_{12}^f \nu_{21}^f\right) E^m + \nu^m \nu_{21}^f E_1^f\right]\left(1 - V^f\right)}{E_1^f V^f + E^m \left(1 - V^f\right)}$$

$$\eta^m = \frac{E^m V^m + \left[\left(1 - (\nu^m)^2\right) E_1^f - \left(1 - \nu^m \nu_{12}^f\right) E^m\right] V^f}{E_1^f V^f + E^m \left(1 - V^f\right)}. \tag{6.63}$$

Equation (6.62) is an alternative version of the rule of mixtures. Assuming the empirically derived stress-partitioning factor η is known, we find that the modified rule-of-mixtures model of Eq. (6.49) appears to be most accurate for the incorporation of a simple formula into a parameter study. However, if an empirically derived stress-partitioning factor is not available, the alternative rule-of mixtures model can be used for improved accuracy relative to the rule-of-mixtures model, and there is a physical basis for the model [6].

6.4.1.3 Model for G_{12}

The rule-of-mixtures model for the axial, or fiber direction, shear modulus G_{12} is similar to the rule-of-mixtures model for E_2. The fiber and matrix elements are each considered to be subjected to shear stress τ_{12}, as shown in Figure 6.5. By equilibrium considerations [7], the shear stress on the fiber element has to be the same as the shear stress on the matrix element, and thus, the shear strains in the elements of fiber and matrix as shown in Figure 6.5 are given by:

$$\gamma_{12}^f = \frac{\tau_{12}}{G_{12}^f} \tag{6.64a}$$

$$\gamma_{12}^m = \frac{\tau_{12}}{G^m}, \tag{6.64b}$$

where the shear modulus in the fiber and that in the matrix are given by, respectively, G_{12}^f and G^m. By the geometry of the deformation and the definition of shear strain:

$$\Delta^f = \gamma_{12}^f W^f \tag{6.65a}$$

$$\Delta^m = \gamma_{12}^m W^m. \tag{6.65b}$$

By considering the fiber and matrix elements joined together, as shown in Figure 6.5c, the total deformation of the unit cell is:

$$\Delta = \Delta^f + \Delta^m. \tag{6.66}$$

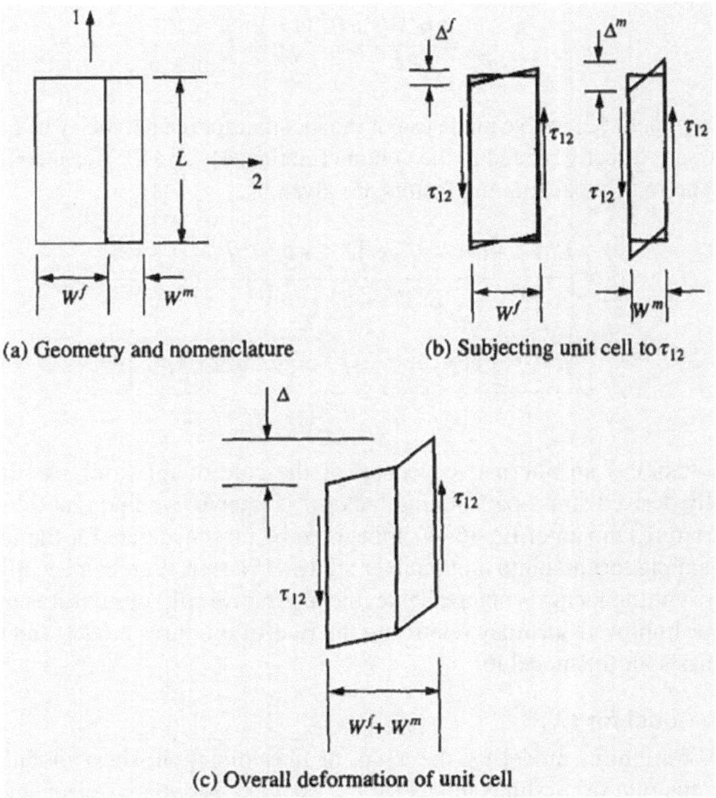

FIGURE 6.5 Rule-of-mixtures model for shear modulus G_{12}.

The average shear strain for the unit cell is:

$$\gamma_{12} = \frac{\Delta}{W^f + W^m}. \tag{6.67}$$

Substituting Eq. (6.66) into Eq. (6.67), we find:

$$\gamma_{12} = \frac{\Delta^f + \Delta^m}{W^f + W^m}. \tag{6.68}$$

Using Eq. (6.65):

$$\gamma_{12} = \frac{\gamma_{12}^f W^f + \gamma_{12}^m W^m}{W}. \tag{6.69}$$

Substituting the stress–strain relations, Eq. (6.64), and recognizing the definition of the volume fractions, we find Eq. (6.69) becomes:

$$\gamma_{12} = \left(\frac{V^f}{G_{12}^f} + \frac{V^m}{G^m} \right) \tau_{12}. \tag{6.70}$$

By analogy:

$$\gamma_{12} = \frac{\tau_{12}}{G_{12}}, \tag{6.71}$$

with G_{12} being the axial shear modulus of the composite, and from Eq. (6.70):

$$\frac{1}{G_{12}} = \left(\frac{V^f}{G_{12}^f} + \frac{V^m}{G^m}\right). \tag{6.72}$$

or

$$\frac{1}{G_{12}} = \left(\frac{V^f}{G_{12}^f} + \frac{1 - V^f}{G^m}\right). \tag{6.73}$$

This is the rule-of-mixtures expression for G_{12}.

As with the transverse modulus, E_2, we can modify the partitioning assumption for the shear stress in the fiber and the matrix, and develop a modified rule-of-mixtures model for G_{12}, resulting in:

$$\frac{1}{G_{12}} = \frac{\left(\frac{V^f}{G_{12}^f} + \frac{\eta'\left(1 - V^f\right)}{G^m}\right)}{V^f + \eta'\left(1 - V^f\right)}, \tag{6.74}$$

where η' is the partitioning factor for the shear stresses.

6.4.2 Semi-Empirical Models

The values obtained for transverse Young's modulus and in-plane shear modulus using the strength-of-materials approach do not agree well with the experimental results. This establishes a need for better modeling techniques. These techniques include numerical methods, such as finite element and finite difference, and boundary element methods, elasticity solution, and variational principal models. Unfortunately, these models are available only as complicated equations or in graphical form. Due to these difficulties, semi-empirical models have been developed for design purposes. The most useful of these models include those of Halphin and Tsai because they can be used over a wide range of elastic properties and fiber volume fractions. Halphin and Tsai developed their models as simple equations by curve fitting to results that are based on elasticity. The equations are semi-empirical in nature because the involved parameters in the curve fitting carry physical meaning.

6.4.2.1 Longitudinal Young's Modulus

The Halphin–Tsai equation for the longitudinal Young's modulus, E_1, is the same as that obtained through the strength-of-materials approach – that is:

$$E_1 = E^f V^f + E^m V^m \tag{6.75}$$

6.4.2.2 Transverse Young's Modulus

The transverse Young's modulus, E_2, is given by:

$$\frac{E_2}{E^m} = \frac{1+\xi\eta V^f}{1-\eta V^f}, \tag{6.76}$$

where

$$\eta = \frac{\dfrac{E^f}{E^m} - 1}{\dfrac{E^f}{E^m} + \xi}. \tag{6.77}$$

The term ξ is called the reinforcing factor, and it depends on the following:

i. Fiber geometry
ii. Packing geometry
iii. Loading conditions.

Halphin and Tsai obtained the value of the reinforcing factor ξ by comparing Eqs. (6.76) and (6.77) to the solutions obtained from the elasticity solutions. For example, for a fiber geometry of circular fibers in a packing geometry of a square array, $\xi = 2$. For a rectangular fiber cross section of length a and width b in a hexagonal array, $\xi = 2(a/b)$, where b is in the direction of loading.

6.4.2.3 In-plane Shear Modulus

The Halphin–Tsai equation for the in-plane shear modulus G_{12} is:

$$\frac{G_{12}}{G^m} = \frac{1+\xi\eta V^f}{1-\eta V^f}, \tag{6.78}$$

where

$$\eta = \frac{\dfrac{G^f}{G^m} - 1}{\dfrac{G^f}{G^m} + \xi}. \tag{6.79}$$

The value of the reinforcing factor, ξ, depends on fiber geometry, packing geometry, and loading conditions. For example, for circular fibers in a square array, $\xi = 1$. For a rectangular fiber cross-sectional area of length a and width b in a hexagonal array, $\xi = \sqrt{3}\log_e\left(\frac{a}{b}\right)$, where a is the direction of loading. The value of $\xi = 1$ for circular fibers in a square array gives reasonable results only for fiber volume fractions of up to 0.5.

6.4.3 Elasticity Approach

When fiber-reinforced materials were first used, numerical methods were not as readily available as they have become. Therefore, some of the early approaches to studying the response of composite materials at the micromechanics level were based on classical elasticity solutions. Solutions to elasticity problems can be quite difficult to determine, and without simplifying assumptions, obtaining solutions is sometimes impossible. As many of the early elasticity solutions were derived for the purpose of determining composite properties from the properties of the constituents, as opposed to studying the details of the stresses at, say, the fiber–matrix interface, some of the simplifying assumptions were not too limiting. One of the key simplifying assumptions was that the volume of fibers and matrix in a composite could be filled with an assemblage of cylindrical fibers and surrounding matrix material, with the fibers being of various sizes to the degree that the fiber–matrix combination of cylinders completely filled the volume of the composite. This notion is shown in Figure 6.6 and is called the composite cylinders model or composite cylinders assemblage. For each fiber–matrix combination, the ratio of the diameter of the fiber to the diameter of the surrounding matrix is the same; this ratio represents the volume fraction of fiber in the composite. For a representative fiber–matrix combination with fiber radius and band matrix outer radius c, the fiber volume fraction is given by:

$$V^f = \frac{\pi b^2}{\pi c^2} = \frac{b^2}{c^2}. \tag{6.80}$$

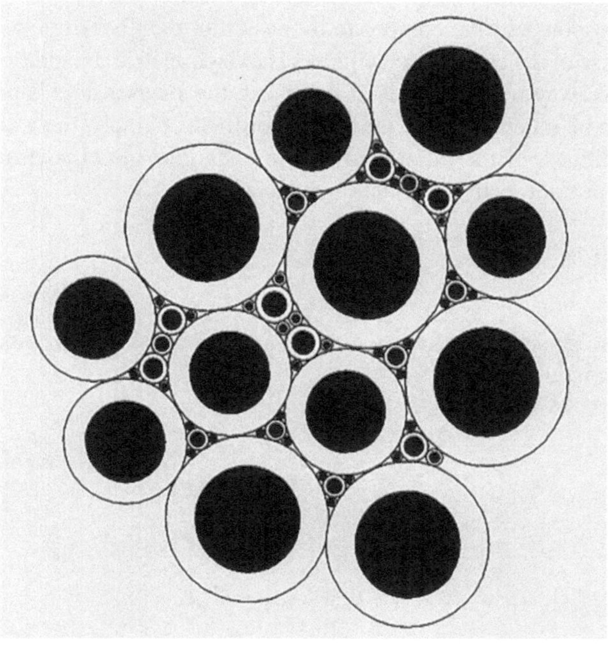

FIGURE 6.6 Concentric cylinders model.

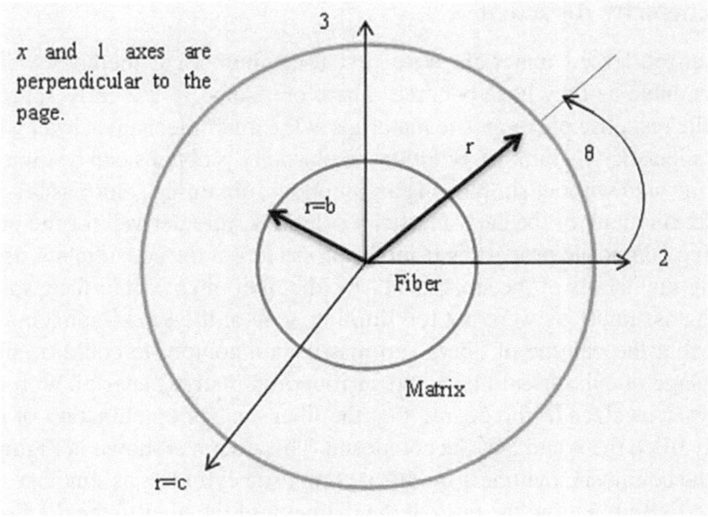

FIGURE 6.7 Concentric cylinders model: Isolated fiber–matrix combination.

The elasticity approaches are concentrated on an isolated fiber–matrix combination from this assemblage. Such a combination is shown in Figure 6.7, with the cylindrical x-θ-r and the composite 1–2–3 principal material coordinate systems indicated.

6.4.3.1 Tension in Fiber Direction

To study the response of the composite to tension in the fiber direction, we assume that the response of the concentric cylinders is axisymmetric. In addition, if attention is concentrated away from the ends of the fibers, the stresses, and hence the strains, are assumed to be independent of the axial coordinate, namely, the x, or 1, direction. For such conditions, of the three equilibrium equations in cylindrical coordinates, only one is important, and that equation reduces to:

$$\frac{d\sigma_r}{dr} + \frac{\sigma_r - \sigma_\theta}{r} = 0. \tag{6.81}$$

As there are no shear stresses for the axisymmetric axial loading case, the stress–strain relations reduce to:

$$\begin{aligned}
\sigma_x &= C_{11}\varepsilon_x + C_{12}\varepsilon_\theta + C_{13}\varepsilon_r \\
\sigma_\theta &= C_{12}\varepsilon_x + C_{22}\varepsilon_\theta + C_{23}\varepsilon_r \\
\sigma_r &= C_{13}\varepsilon_x + C_{23}\varepsilon_\theta + C_{33}\varepsilon_r \\
\sigma_r &= C_{13}\varepsilon_x + C_{23}\varepsilon_\theta + C_{33}\varepsilon_r
\end{aligned} \quad . \tag{6.82}$$

These equations were introduced in Eq. (5.47) in the context of the smeared properties of the composite. However, they are equally valid individually for the fiber and for the matrix. Here, we will assume that the fiber is transversely isotropic in the r-θ plane and the matrix is isotropic, that is, Eq. (5.55). Thus, the stress–strain relations for the two constituents simplify, for the fiber, to:

$$\begin{aligned} \sigma_x^f &= C_{11}^f \varepsilon_x + C_{12}^f \varepsilon_\theta + C_{12}^f \varepsilon_r \\ \sigma_\theta^f &= C_{12}^f \varepsilon_x + C_{22}^f \varepsilon_\theta + C_{23}^f \varepsilon_r . \\ \sigma_r^f &= C_{12}^f \varepsilon_x + C_{23}^f \varepsilon_\theta + C_{22}^f \varepsilon_r \end{aligned} \tag{6.83}$$

And for the matrix, to:

$$\begin{aligned} \sigma_x^m &= C_{11}^m \varepsilon_x + C_{12}^m \varepsilon_\theta + C_{12}^m \varepsilon_f \\ \sigma_\theta^m &= C_{12}^m \varepsilon_x + C_{11}^m \varepsilon_\theta + C_{12}^m \varepsilon_f , \\ \sigma_r^f &= C_{12}^m \varepsilon_x + C_{12}^m \varepsilon_\theta + C_{11}^m \varepsilon_f \end{aligned} \tag{6.84}$$

where the superscripts f and m denote, respectively, the fiber and the matrix.

The elastic and thermal expansion properties of a graphite fiber are taken to be:

$$\begin{aligned} &E_x^f = 233 \text{ GPa}, \;\; E_r^f = E_\theta^f = 23.1 \text{ GPa}, \quad \nu_{xr}^f = \nu_{x\theta}^f = 0.200, \quad \nu_{r\theta}^f = 0.400 \\ &G_{xr}^f = G_{x\theta}^f = 8.96 \text{ GPa}, \quad G_{r\theta}^f = 8.27 \text{ GPa}, \\ &\alpha_x^f = -0.540 \times 10^{-6}/\text{K} \quad \alpha_r^f = \alpha_\theta^f = 10.10 \times 10^{-6}/\text{K}. \end{aligned} \tag{6.85}$$

With the above properties, the fiber is said to be transversely isotropic in the r-θ plane (i.e., in the cross section of the fiber). This means that the fiber responds the same when subjected to a stress σ_r, for example, as when it is subjected to stress σ_θ. Subjecting the material in the fiber to a stress σ_x results in a different response. Because of the transverse isotropy, when referred to the 1–2–3 coordinate system, the fiber properties can be written as:

$$\begin{aligned} &E_1^f = 233 \text{ GPa}, \quad E_2^f = E_3^f = 23.1 \text{ GPa}, \quad \nu_{12}^f = \nu_{13}^f = 0.200, \quad \nu_{23}^f = 0.400 \\ &G_{12}^f = G_{13}^f = 8.96 \text{ GPa}, \quad G_{23}^f = 8.27 \text{ GPa}, \\ &\alpha_1^f = -0.540 \times 10^{-6}/\text{K} \quad \alpha_2^f = \alpha_3^f = 10.10 \times 10^{-6}/\text{K}. \end{aligned} \tag{6.86}$$

For the polymer matrix material, the elastic and thermal expansion properties are taken to be:

$$E^m = 4.62 \text{ GPa}, \quad \nu^m = 0.360, \quad \alpha^m = 41.4 \times 10^{-6}/\text{K}, \tag{6.87}$$

where the superscript m denotes the matrix, and the matrix is assumed to be isotropic. All the properties are assumed to be independent of temperature.

Using the engineering properties of the fiber and the matrix, Eqs. (6.86) and (6.87), and the definitions of the S_{ij} and the C_{ij}, Eqs. (5.46) and (5.48), C_{11}^f, C_{12}^f, C_{22}^f, C_{23}^f, C_{11}^m, and C_{12}^m can be computed. The strains in the above equations are related to the displacements by the strain–displacement relations simplified by the assumptions of axisymmetry and independence of strain of the x coordinate. There is no shear strain response; only normal strains are given by:

$$\varepsilon_x = \frac{\partial u}{\partial x}, \quad \varepsilon_\theta = \frac{w}{r}, \quad \varepsilon_r = \frac{dw}{dr} \tag{6.88}$$

In Eq. (6.88), u is the axial displacement and w is the radial displacement. Because of the assumption of axisymmetric response, the circumferential displacement, v, is assumed to be zero. Note that the partial derivative of u with respect to x is used and that the ordinary derivative of w with respect to r is used because we can argue that w cannot be a function of x, or else the fiber would not be straight when loaded, and at this point in the development, u can be a function of both x and r.

Substituting the strain–displacement relations into the stress–strain relations and these, in turn, into Eq. (6.81), which is the third equilibrium equation, leads to an equation for the radial displacement, namely:

$$\frac{d^2 w}{dr^2} + \frac{1}{r}\frac{dw}{dr} - \frac{w}{r^2} = 0, \tag{6.89}$$

which has the solution:

$$w(r) = Ar + \frac{B}{r}. \tag{6.90}$$

The quantities A and B are the constants of integration that must be solved for by applying boundary and other conditions. The above solution is valid for both the fiber and the matrix. For a given fiber–matrix combination, then, the radial displacement is given by:

$$w^f(r) = A^f r + \frac{B^f}{r} \quad 0 \le r \le b \tag{6.91a}$$

$$w^m(r) = A^m r + \frac{B^m}{r} \quad b \le r \le c, \tag{6.91b}$$

where the range of r for each portion of the solution is given. As the strains do not vary with the axial coordinate, the solution for the axial displacement u is given by:

$$u^f(x,r) = \varepsilon_1^f x \quad 0 \le r \le b \tag{6.92a}$$

$$u^m(x,r) = \varepsilon_1^m x \quad b \le r \le c, \tag{6.92b}$$

where ε_1^f and ε_1^m are the constants.

According to Eq. (6.91a), if B^f is not zero, the radial displacement at the center of the fiber, $r = 0$, is predicted to be infinite. This is physically impossible, so the condition:

$$B^f = 0 \tag{6.93}$$

is stipulated. As a result, the strains within the fiber are given by Eq. (6.88) as:

$$\begin{aligned}
\varepsilon_x^f &= \frac{\partial u^f}{\partial x} = \varepsilon_1^f \\
\varepsilon_\theta^f &= \frac{w^f}{r} = A^f \\
\varepsilon_r^f &= \frac{dw^f}{dr} = A^f,
\end{aligned} \tag{6.94}$$

while those in the matrix are given by:

$$\begin{aligned}
\varepsilon_x^m &= \frac{\partial u^m}{\partial x} = \varepsilon_1^m \\
\varepsilon_\theta^m &= \frac{w^m}{r} = A^m + \frac{B^m}{r^2} \\
\varepsilon_r^m &= \frac{dw^m}{dr} = A^m - \frac{B^m}{r^2}.
\end{aligned} \tag{6.95}$$

With the strains defined, the stresses in the fiber and matrix, respectively, can be written as:

$$\begin{aligned}
\sigma_x^f &= C_{11}^f \varepsilon_1^f + 2C_{12}^f A^f \\
\sigma_\theta^f &= C_{12}^f \varepsilon_1^f + \left(C_{22}^f + C_{23}^f\right) A^f \\
\sigma_r^f &= C_{12}^f \varepsilon_1^f + \left(C_{22}^f + C_{23}^f\right) A^f
\end{aligned} \tag{6.96a}$$

$$\begin{aligned}
\sigma_x^m &= C_{11}^m \varepsilon_1^m + 2C_{12}^m A^m \\
\sigma_\theta^m &= C_{12}^m \varepsilon_1^m + \left(C_{11}^m + C_{12}^m\right) A^m + \left(C_{11}^m - C_{12}^m\right) \frac{B^m}{r^2}. \\
\sigma_r^m &= C_{12}^m \varepsilon_1^m + \left(C_{11}^m + C_{12}^m\right) A^m + \left(C_{12}^m - C_{11}^m\right) \frac{B^m}{r^2}
\end{aligned} \tag{6.96b}$$

When subjected to any loading, in particular an axial load *P*, the displacements at the interface between the fiber and the matrix are continuous; that is,

$$w^f(b) = w^m(b)$$
$$u^f(b) = u^m(b). \tag{6.97}$$

By substituting expressions for the displacements, Eqs. (6.91) and (6.92), this condition leads to:

$$A^f b = A^m b + \frac{B^m}{b}.$$
$$\varepsilon_1^f x = \varepsilon_1^m x. \tag{6.98}$$

The second equation leads to the conclusion that the axial strain is the same in the fiber as in the matrix, namely:

$$\varepsilon_1^f = \varepsilon_1^m = \varepsilon_1. \tag{6.99}$$

This strain is indeed the strain in the composite in the fiber direction, hence the notation ε_1, as we have been using all along. As discussed in conjunction with the finite element results, the radial stress σ_r must be the same on the fiber side of the interface as on the matrix side, or

$$\sigma_r^f(b) = \sigma_r^m(b) \tag{6.100}$$

In terms of the unknown constants, substituting for the stresses from Eqs. (6.96a) and (6.96b), Eq. (6.100) becomes:

$$C_{12}^f\varepsilon_1 + \left(C_{22}^f + C_{23}^f\right)A^f = C_{12}^m\varepsilon_1 + \left(C_{11}^m + C_{12}^m\right)A^m + \left(C_{12}^m - C_{11}^m\right)\frac{B^m}{b^2}. \tag{6.101}$$

At the outer radius of the matrix, if it is assumed that the radial stress must vanish:

$$\sigma_r^m(c) = 0. \tag{6.102}$$

Using Eq. (6.96b),

$$C_{12}^m\varepsilon_1 + \left(C_{11}^m + C_{12}^m\right)A^m + \left(C_{12}^m - C_{11}^m\right)\frac{B^m}{c^2} = 0. \tag{6.103}$$

As a final condition of the problem, the applied axial load *P* is actually the integral of the axial stresses over the cross-sectional area of the fiber–matrix combination, namely:

$$2\pi\left\{\int_0^b \sigma_x^f r\,dr + \int_b^c \sigma_x^m r\,dr\right\} = P. \tag{6.104}$$

or

$$\left\{C_{11}^{f}\varepsilon_1 + 2C_{12}^{f}A^{f}\right\}\pi b^2 + \left\{C_{11}^{m}\varepsilon_1 + 2C_{12}^{m}A^{m}\right\}\pi\left(c^2 - b^2\right) = P. \tag{6.105}$$

If we use Eqs. (6.98), (6.101), and (6.103), the constants A^f, A^m, and B^m can be solved for in terms of ε_1. As the early elasticity-based micromechanics analyses were focused on determining the overall properties of the composite, in the present case an estimate for E_1 can be obtained by substituting the expressions for A^f and A^m into Eq. (6.105), resulting in an equation of the form:

$$\sigma_1 = E_1\varepsilon_1, \tag{6.106}$$

where σ_1 is the average stress in the axial, or 1, direction. This is the stress in the composite in the principal material 1 direction and is given by:

$$\sigma_1 = \frac{P}{\pi c^2}. \tag{6.107}$$

The axial force is divided by the cross-sectional area of the fiber–matrix combination. The expression for E_1 in Eq. (6.106) is complicated, but it can be written in the form:

$$E_1 = E_1^f(1+\gamma)V^f + E^m(1+\delta)\left(1 - V^f\right). \tag{6.108}$$

The quantities γ and δ are the functions of the extensional moduli and Poisson's ratios of the fiber and matrix and the fiber volume fraction. They are given by:

$$\gamma = \frac{2\nu_{21}^{f}E^m\left(1 - \nu^f - 2\nu_{12}^{f}\nu_{21}^{f}\right)V^f\left(\nu_{12}^{f} - \nu^m\right)}{E_2^f\left(1+\nu^m\right)\left(1 + V^f\left(1 - 2\nu^m\right)\right) + E^m\left(1 - \nu^f - 2\nu_{12}^{f}\nu_{21}^{f}\right)\left(1 - V^f\right)} \tag{6.109a}$$

$$\delta = \frac{2E_2^f\nu^m V^f\left(\nu^m - \nu_{12}^{f}\right)}{E_2^f\left(1+\nu^m\right)\left(1 + V^f\left(1 - 2\nu^m\right)\right) + E^m\left(1 - \nu^f - 2\nu_{12}^{f}\nu_{21}^{f}\right)\left(1 - V^f\right)}. \tag{6.109b}$$

In the above definitions, as the fiber has been assumed to be transversely isotropic, use has been made of the fact that:

$$\begin{aligned} E_2^f &= E_3^f \\ \nu_{12}^f &= \nu_{13}^f \\ \nu_{23}^f &= \nu_{32}^f = \nu^f. \end{aligned} \tag{6.110}$$

For typical properties of the fiber and matrix, γ and δ are much less than one. Hence, to a very good first approximation, E_1 can be written as:

$$E_1 = E_1^f V^f + E^m\left(1 - V^f\right). \tag{6.111}$$

When the major Poisson's ratio of the fiber and Poisson's ratio of the matrix are equal, $\nu_{12}^f = \nu^m$, both γ and δ are identically zero and Eq. (6.111) is exact.

6.4.3.2 Axial Shear

The elasticity approach can also be used to estimate values of the shear modulus G_{12}; this modulus is often referred to as the *axial* shear modulus. The approach to this problem is based on a slightly different view of the fiber–matrix combination, namely, that of Figure 6.8. In this figure, the fiber–matrix combination is viewed by looking along the three axis toward the 1–2 plane and the boundaries of the fiber–matrix combination are deformed by shearing in the 1–2 plane. The boundaries of the portion of the fiber–matrix combination have shown the following displacements relative to the undeformed state:

$$u_1 = \frac{\gamma_{12}^0}{2} x_2$$
$$u_2 = \frac{\gamma_{12}^0}{2} x_1 \qquad (6.112)$$
$$u_3 = 0,$$

where x_1, x_2, and x_3 are used to denote the coordinates in the 1–2–3 principal material directions, and u_1, u_2, and u_3 are the displacements in those directions. The quantity γ_{12}^0 is the shear strain imposed on the boundary. Though Eq. (6.112) defines a fairly specific deformation on the boundary, in the interior of the fiber–matrix combination the displacements are assumed to be expressible as:

$$u_1 = \varnothing(x_2, x_3) - \frac{\gamma_{12}^0}{2} x_2$$
$$u_2 = \frac{\gamma_{12}^0}{2} x_1 \qquad (6.113)$$
$$u_3 = 0.$$

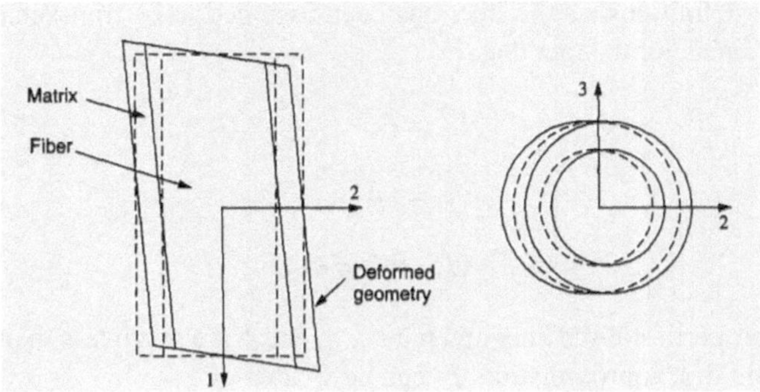

FIGURE 6.8 Concentric cylinders model: Axial shear deformation.

The function $\varnothing(x_2, x_3)$ defines the shear strains in the interior of the fiber–matrix combination, with the particular form of Eq. (6.113) leading to relatively simple expressions for the shear strains, namely:

$$\gamma_{12} = \frac{\partial \varnothing}{\partial x_2}$$
$$\gamma_{13} = \frac{\partial \varnothing}{\partial x_3}. \tag{6.114}$$

It is important to note that there is a function $\varnothing$ for the fiber region and a different function $\varnothing$ for the matrix region. As the other strains are zero, the only stresses present in the fiber–matrix combination due to the displacements given in Eq. (6.113) are:

$$\tau_{13} = G\gamma_{13}$$
$$\tau_{12} = G\gamma_{12}. \tag{6.115}$$

If we use Eq. (6.114),

$$\tau_{13} = G\frac{\partial \varnothing}{\partial x_3}$$
$$\tau_{12} = G\frac{\partial \varnothing}{\partial x_2}, \tag{6.116}$$

where it has been assumed that the material is either isotropic or transversely isotropic, namely,

$$G_{12} = G_{13} = G. \tag{6.117}$$

Because all the stresses except τ_{12} and τ_{13} are zero, the equilibrium conditions reduce to:

$$\frac{\partial \tau_{12}}{\partial x_2} + \frac{\partial \tau_{13}}{\partial x_3} = 0. \tag{6.118}$$

Substituting Eq. (6.116) in Eq. (6.118) results in:

$$\frac{\partial^2 \varnothing}{\partial x_2^{\,2}} + \frac{\partial^2 \varnothing}{\partial x_3^{\,2}} = 0. \tag{6.119}$$

At this point, it is convenient to express the problem in cylindrical coordinates; the principal material coordinate system (1–2–3) and the cylindrical coordinate system (x-r-θ) are shown in Figure 6.9. Relations between the coordinates in the two systems are given by:

$$x_2 = r\cos\theta$$
$$x_3 = r\sin\theta. \tag{6.120a}$$

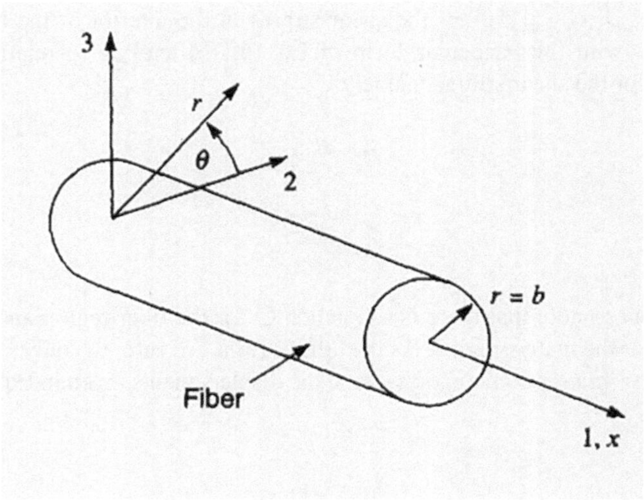

FIGURE 6.9 The x-y-z and 1–2–3 coordinate systems for the fiber–matrix combination.

The inverse results in:

$$r = \left(x_2^2 + x_3^2\right)^{\frac{1}{2}}$$
$$\theta = \tan^{-1}\left(\frac{x_3}{x_2}\right). \tag{6.120b}$$

The nonzero stresses in the cylindrical coordinate system are related to τ_{12} and τ_{13} by:

$$\tau_{xr} = \cos\theta\tau_{12} + \sin\theta\tau_{13}$$
$$\tau_{x\theta} = \cos\theta\tau_{13} - \sin\theta\tau_{12}. \tag{6.121a}$$

The inverse results in:

$$\tau_{12} = \cos\theta\tau_{xr} - \sin\theta\tau_{x\theta}$$
$$\tau_{13} = \cos\theta\tau_{x\theta} + \sin\theta\tau_{xr}. \tag{6.121b}$$

In terms of $\varnothing$, the two shear stresses in the cylindrical coordinate system are:

$$\tau_{xr} = G\frac{\partial\varnothing}{\partial r}$$
$$\tau_{x\theta} = G\frac{\partial\varnothing}{r\partial\theta} \tag{6.122}$$

Writing Eq. (6.119) in the cylindrical coordinate system results in:

$$\frac{\partial^2 \varnothing}{\partial r^2} + \frac{1}{r}\frac{\partial \varnothing}{\partial r} + \frac{1}{r^2}\frac{\partial^2 \varnothing}{\partial \theta^2} = 0. \tag{6.123}$$

Equations (6.119) and (6.123) are Laplace's equation for $\varnothing$ in the two different coordinate systems. The separation of variables technique can be used to solve Eq. (6.123); the solution is given by:

$$\varnothing(r,\theta) = a_0 + \sum_{n=1}^{\infty}\left(A_n r^n + B_n r^{-n}\right)\left(C_n \sin(n\theta) + D_n \cos(n\theta)\right). \tag{6.124}$$

For the problem here, only terms to $n = 1$ are necessary. Additionally, $a_0 = 0$ and the sin $(n\theta)$ terms are not needed. Finally, for the stresses to remain bounded at the center of the fiber, B_n for the fiber must be zero. As with the past practice, using superscripts f to denote the fiber and m to denote the matrix, the functions $\varnothing$ for these two materials in the fiber–matrix combination are:

$$\begin{aligned} \varnothing^f(r,\theta) &= a_1^f r \cos\theta \\ \varnothing^m(r,\theta) &= \left(a_1^m r + \frac{b_1^m}{r}\right)\cos\theta, \end{aligned} \tag{6.125}$$

where

$$a_1 = A_1 D_1 \text{ and } b_1 = B_1 D_1. \tag{6.126}$$

Hence,

$$\begin{aligned} \tau_{xr}^f &= G^f a_1^f \cos\theta \\ \tau_{x\theta}^f &= -G^f a_1^f \sin\theta. \end{aligned} \tag{6.127}$$

Also,

$$\begin{aligned} \tau_{xr}^m &= G^m\left(a_1^m - \frac{b_1^m}{r^2}\right)\cos\theta \\ \tau_{x\theta}^m &= -G^m\left(a_1^m + \frac{b_1^m}{r^2}\right)\sin\theta. \end{aligned} \tag{6.128}$$

Continuity of the three components of displacement at the interface between the fiber and the matrix reduces to enforcement of:

$$\varnothing^f(b,\theta) = \varnothing^m(b,\theta) \tag{6.129}$$

$$a_1^f b\cos\theta = \left(a_1^m b + \frac{b_1^m}{b}\right)\cos\theta, \tag{6.130}$$

where, recall, b is the radius of the fiber. Continuity of the stresses at the fiber–matrix interface reduces to a single condition, namely,

$$\tau_{xr}^f(b,\theta) = \tau_{xr}^m(b,\theta). \tag{6.131}$$

Using Eqs. (6.127) and (6.128), Eq. (6.131) can be written as:

$$G^f a_1^f \cos\theta = G^m\left(a_1^m - \frac{b_1^m}{r^2}\right)\cos\theta. \tag{6.132}$$

Using Eq. (6.113) to enforce the conditions of Eq. (6.112) at the boundary of the fiber–matrix combination (i.e., at $r = c$ and $\theta = 0$), provides a final condition that can be written as:

$$\varnothing^m = \gamma_{12}^0 c. \tag{6.133}$$

Substitution from Eq. (6.125) yields:

$$a_1^m c + \frac{b_1^m}{c} = \gamma_{12}^0 c. \tag{6.134}$$

Solving Eqs. (6.130), (6.132), and (6.134) leads to the solution for a_1^f, a_1^m, and b_1^m, namely,

$$a_1^f = \left\{\frac{2G^m}{\left(\dfrac{G^m - G^f}{b^2}\right) + \left(\dfrac{G^m - G^f}{c^2}\right)}\right\}\frac{\gamma_{12}^0}{b^2}$$

$$a_1^m = \left\{\frac{\left(G^m + G^f\right)}{\left(\dfrac{G^m - G^f}{b^2}\right) + \left(\dfrac{G^m - G^f}{c^2}\right)}\right\}\frac{\gamma_{12}^0}{b^2}. \tag{6.135}$$

$$b_1^m = \left\{\frac{\left(G^m - G^f\right)}{\left(\dfrac{G^m - G^f}{c^2}\right)} + \left(\frac{G^m + G^f}{b^2}\right)\right\}\gamma_{12}^0.$$

Accordingly, substituting Eq. (6.128) in the transformation relations, Eq. (6.121b), leads to:

$$\tau_{12}^m = G^m \left\{ \left(a_1^m - \frac{b_1^m}{r^2} \right) \cos\theta + \left(a_1^m + \frac{b_1^m}{r^2} \right) \sin\theta \right\}. \tag{6.136}$$

Evaluating this at $r = c$ and $\theta = 0$, and substituting for a_1^m and b_1^m from above lead to the following relation between τ_{12} and γ_{12}^0:

$$\tau_{12}^m = G^m \left\{ \frac{(G^m + G^f) - V^f (G^m - G^f)}{(G^m + G^f) + V^f (G^m - G^f)} \right\} \gamma_{12}^0. \tag{6.137}$$

This is an important expression because it provides an estimate of the composite axial shear modulus G_{12} through the relation:

$$\tau_{12}^m = G_{12} \gamma_{12}^0. \tag{6.138}$$

Specifically,

$$G_{12} = G^m \left\{ \frac{(G^m + G^f) - V^f (G^m - G^f)}{(G^m + G^f) + V^f (G^m - G^f)} \right\}. \tag{6.139}$$

REFERENCES

1. Love A.E.H. (1892), *A Treatise on the Mathematical Theory of Elasticity*, Cambridge University Press, London (also 4th ed., 1944, Dover Publications, Inc., New York).
2. Hodge P.G. Jr. (1958), "The mathematical theory of plasticity," in *Elasticity and Plasticity* (J.N. Goodier and P.G. Hodge, Jr., eds.), John Wiley & Sons, Inc., New York, p. 81.
3. Kreyszig E. (1967), *Advanced Engineering Mathematics*, 2nd ed., John Wiley & Sons, Inc., New York, p. 416.
4. Fung Y.C. (1965), *Foundations of Solid Mechanics*, Prentice-Hall, Inc., Englewood Cliffs, NJ, p. 285.
5. Green G. (1839), "On the laws of reflexion and refraction of light at the common surface of two non-crystallized media", *Transactions of the Cambridge Philosophical Society*, Vol. 7, p. 121.
6. Popov E.P. (1990), *Engineering Mechanics of Solids*, Prentice Hall, Englewood Cliffs, NJ.
7. Hyer M.W. (1998), *Stress Analysis of Fiber Reinforced Composite Materials*, McGraw-Hill, Singapore.

7 Plane Stress Assumption

In the study of fiber-reinforced composite materials, one of the important assumptions is that the properties of the fiber and the matrix are smeared into an equivalent homogeneous material, which has the properties of an orthotropic material [1]. In the previous chapters, this assumption helped in the development of the stress–strain relations and in understanding the response of fiber-reinforced material. If this assumption had not been made, then we have to consider the response of the fibers and the matrix separately. Very less progress would have been made in understanding the response of a fiber-reinforced material without this assumption.

Another equally important assumption that is made in the analysis of a fiber-reinforced composite material is the *plane stress assumption*. This assumption has its roots in the manner in which the fiber-reinforced material is used in various structural applications such as beams, plates, and cylinders. All these structural shapes have at least one characteristic dimension an order of magnitude lower than the other two dimensions. In these applications, three of the six components of stress are generally much smaller than the other three. With a plate, for example, the stresses in the plane of the plate are much larger than the stresses perpendicular to that plane. In all calculations, then, the stress components perpendicular to the plane of the structure can be set to zero, greatly simplifying the solution of many problems. In the context of fiber-reinforced plates, the stress components σ_3, τ_{23}, and τ_{13} are set to zero with the assumption that the 1–2 plane of the principal material coordinate system is in the plane of the plate. Stress components σ_1, σ_2, and τ_{12} are considered to be much larger in magnitude than components σ_3, τ_{23}, and τ_{13}. In fact, σ_1 should be the largest of all the stress components if the fibers are being utilized effectively. We use the term "plane stress" because σ_1, σ_2, and τ_{12} lie in a plane, and stresses σ_3, τ_{23}, and τ_{13} are perpendicular to this plane and are zero. Figure 7.1 shows an element of a fiber-reinforced material in a state of plane stress.

The plane stress assumption can lead to inaccuracies: some serious and some not so serious. The most serious inaccuracy occurs in the analysis of a laminate near its edge. Laminates tend to come apart in the thickness direction, or *delaminate*, at their edges, much like common plywood. An understanding of this phenomenon, illustrated in Figure 7.2, requires that all six components of stress be included in the analysis. It is exactly the stresses that are set to zero in the plane stress assumption (i.e., σ_3, τ_{23}, and τ_{13}) and that are responsible for delamination, so an analysis that ignores these stresses cannot possibly be correct for a delamination study [2].

Delaminations can also occur away from a free edge, with the layers separating in a blister fashion.

These are generally caused by the presence of imperfections between the layers. The out-of-plane stress components σ_3, τ_{23}, and τ_{13} are also important in locations where structures or components of structures are joined together. Figures 7.3 and 7.4 illustrate some examples. Figure 7.3 shows a bonded joint consisting of two laminates

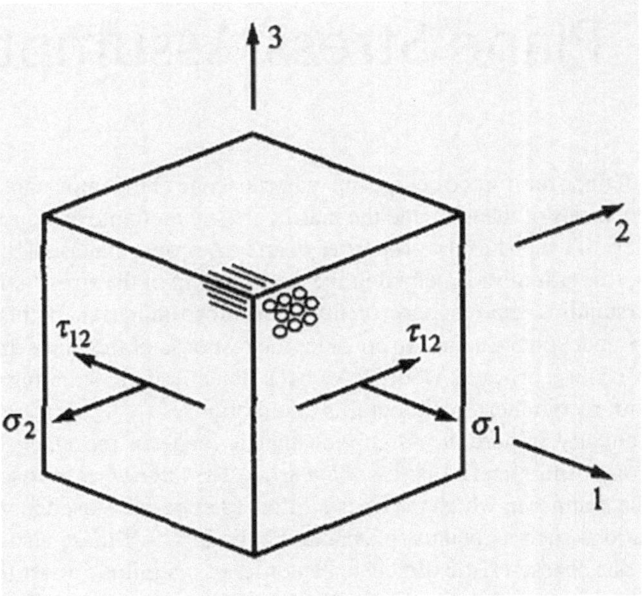

FIGURE 7.1 An element of a fiber-reinforced material in a state of plane stress.

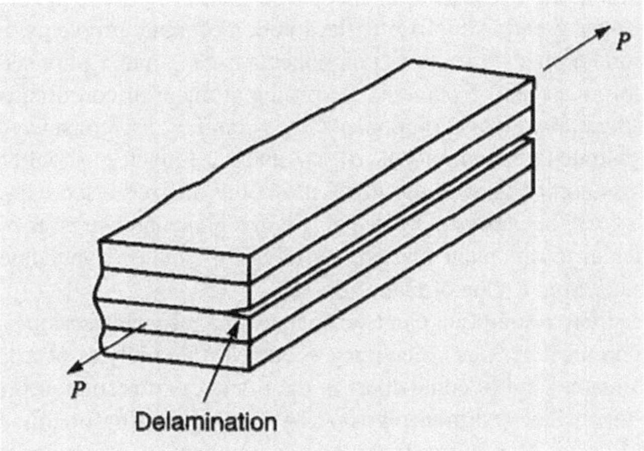

FIGURE 7.2 Region of high out-of-plane stresses: Delamination at a free edge.

subjected to tensile load P. For the load to be transferred from one laminate to the other, significant out-of-plane stresses, particularly shear, must develop in the laminates around the interface, as well as at the interface itself [3]. As another example, in many situations, stiffeners are used to increase the load capacity of plates, as shown in Figure 7.4. For the plate–stiffener combination to be effective, the plate must transfer some of the pressure load to the stiffener. Thickness direction stresses must develop in the plate and stiffener flange if load is to be transferred through

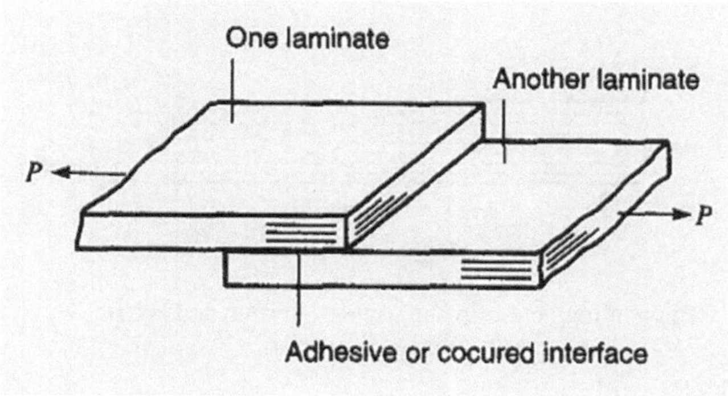

FIGURE 7.3 Region of high out-of-plane stresses: Bonded joint.

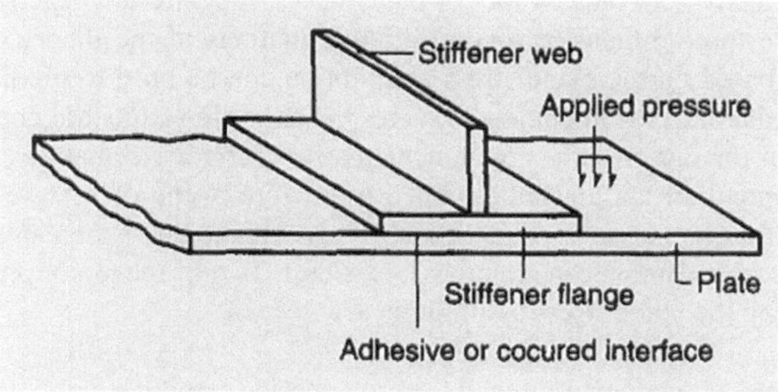

FIGURE 7.4 Region of high out-of-plane stresses: Stiffened plate.

the interface. In general, all three components of out-of-plane stress, σ_3, τ_{23}, and τ_{13}, develop in this situation. Away from the stiffener, the plate may be in a state of plane stress, so not only is there a region of the plate characterized by a fully three-dimensional stress state, there is also a transition region. In this transition region, the conditions go from truly plane stress to a fully three-dimensional stress state, making the analysis of such a problem difficult and challenging.

Figure 7.5 illustrates another area where through-the-thickness stresses are important. Often it is necessary, or desirable, to change the thickness of a laminate by gradually terminating some of the layers. Away from the terminated layer region, each portion of the laminate could well be in a state of plane stress due to the applied in-plane load P. However, the thicker region is in a different state of plane stress than the thinner region. To make the transition between the two stress states, three-dimensional effects occur. The illustrations in Figures 7.2–7.5 are prime examples of situations encountered in real composite structures. However, the plane stress assumption is accurate in so many situations that one would be remiss in not taking advantage of its simplifications [4,5]. The static, dynamic, and thermally induced deflections, and the stresses that result from these, vibration frequencies, buckling loads, and many other responses of composite structures can be accurately predicted

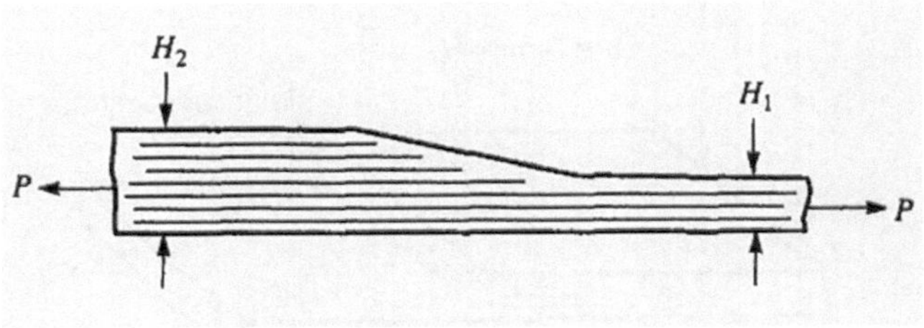

FIGURE 7.5 Region of high out-of-plane stresses: Terminated layers.

using the plane stress assumption. What is important to remember when applying the plane stress assumption is that it assumes that three stresses are *small* relative to the other three stresses and they have therefore been set to zero. They do not necessarily have to be exactly zero, and in fact in many cases, they are not exactly zero. With the aid of the three-dimensional equilibrium equations of the theory of elasticity, calculations based on the plane stress assumption can be used to predict the stress components that have been equated to zero. When these results are compared with predictions of the out-of-plane components based on rigorous analyses wherein the out-of-plane components are not assumed to be zero at the outset, we find that in many cases, the comparisons are excellent. Thus, a plane stress, or, using alternative terminology, a two-dimensional analysis, is useful. Two of the major pitfalls associated with using the plane stress assumption are:

i. The stress components equated to zero are often forgotten and no attempt is made to estimate their magnitude.
ii. It is often erroneously assumed that because the stress component σ_3 is zero and therefore ignorable, the associated strain ε_3 is also zero and ignorable.

Regarding the former point, while certain stress components may indeed be small, the material may be very weak in resisting these stresses. As was stated earlier, a fiber-reinforced material is poor in resisting all stresses except stresses in the fiber direction. Thus, several stress components may be small and so the problem conforms to the plane stress assumption. However, the out-of-plane stresses may be large enough to cause failure of the material, and therefore, they should not be completely ignored. Often they are. Regarding the second point, the stresses in the 1–2 plane of the principal material coordinate system can cause a significant strain response in the 3 direction. The assumption that ε_3 is zero just because σ_3 is negligible is wrong and, as we shall see shortly, defies the stress–strain relations that govern material behavior. It is important to keep these two points in mind as we focus our discussion in the following chapters on the plane stress condition.

7.1 STRESSES AND STRAINS UNDER PLANE STRESS CONDITION

For the plane stress assumption, σ_3, τ_{23}, and τ_{13} are set to zero in Eqs. (5.45) and (5.47). Looking at Eq. (5.45) first, we find:

$$\begin{Bmatrix} \varepsilon_1 \\ \varepsilon_2 \\ \varepsilon_3 \\ \gamma_{23} \\ \gamma_{13} \\ \gamma_{12} \end{Bmatrix} = \begin{bmatrix} S_{11} & S_{12} & S_{13} & 0 & 0 & 0 \\ S_{12} & S_{22} & S_{23} & 0 & 0 & 0 \\ S_{13} & S_{23} & S_{33} & 0 & 0 & 0 \\ 0 & 0 & 0 & S_{44} & 0 & 0 \\ 0 & 0 & 0 & 0 & S_{55} & 0 \\ 0 & 0 & 0 & 0 & 0 & S_{66} \end{bmatrix} \begin{Bmatrix} \sigma_1 \\ \sigma_2 \\ 0 \\ 0 \\ 0 \\ \tau_{12} \end{Bmatrix}. \tag{7.1}$$

From this relation, it is obvious that:

$$\gamma_{23} = \gamma_{13} = 0 \tag{7.2}$$

So, with the plane stress assumption, there can be no shear strains whatsoever in the 2–3 and 1–3 planes. That is an important ramification of the assumption. Also,

$$\varepsilon_3 = S_{13}\sigma_1 + S_{23}\sigma_2. \tag{7.3}$$

This equation indicates explicitly that for a state of plane stress, there is an extensional strain in the 3 direction. To assume that strain ε_3 is zero is *absolutely wrong.* That it is not zero is a direct result of Poisson's ratios ν_{13} and ν_{23} acting through S_{13} and S_{23}, respectively, coupling with the nonzero stress components σ_1 and σ_2. The above equation for ε_3 forms the basis for determining the thickness change of laminates subjected to in-plane loads, and for computing through-thickness, or out-of-plane, Poisson's ratios of a laminate. Despite the fact that ε_3 is not zero, the plane stress assumption leads to a relation involving only ε_1, ε_2, γ_{12} and σ_1, σ_2, and τ_{12}. By eliminating the third, fourth, and fifth equations of Eq. 7.1, we find:

$$\begin{Bmatrix} \varepsilon_1 \\ \varepsilon_2 \\ \gamma_{12} \end{Bmatrix} = \begin{bmatrix} S_{11} & S_{12} & 0 \\ S_{12} & S_{22} & 0 \\ 0 & 0 & S_{66} \end{bmatrix} \begin{Bmatrix} \sigma_1 \\ \sigma_2 \\ \tau_{12} \end{Bmatrix}. \tag{7.4}$$

In the above equation:

$$\begin{aligned} S_{11} &= \frac{1}{E_1} \quad S_{12} = \frac{-\nu_{12}}{E_1} = \frac{-\nu_{21}}{E_2} \\ S_{22} &= \frac{1}{E_2} \quad S_{66} = \frac{1}{G_{12}}. \end{aligned} \tag{7.5}$$

The 3×3 matrix of compliances is called the *reduced compliance matrix.* In matrix notation, the lower right-hand element of a 3×3 matrix is usually given the subscript 33, though in the analysis of composites, it has become conventional to retain the subscript convention from the three-dimensional formulation and maintain the subscript of the lower corner element as 66. For an isotropic material, Eq. (7.5) reduces to:

$$S_{11} = S_{22} = \frac{1}{E} \quad S_{12} = \frac{-\nu}{E} \quad S_{66} = \frac{1}{G} = \frac{2(1+\nu)}{E}. \tag{7.6}$$

If the plane stress assumption is used to simplify the inverse form of the stress–strain relation, Eq. (5.47), the result is:

$$\begin{Bmatrix} \sigma_1 \\ \sigma_2 \\ 0 \\ 0 \\ 0 \\ \tau_{12} \end{Bmatrix} = \begin{bmatrix} C_{11} & C_{12} & C_{13} & 0 & 0 & 0 \\ C_{12} & C_{22} & C_{23} & 0 & 0 & 0 \\ C_{13} & C_{23} & C_{33} & 0 & 0 & 0 \\ 0 & 0 & 0 & C_{44} & 0 & 0 \\ 0 & 0 & 0 & 0 & C_{55} & 0 \\ 0 & 0 & 0 & 0 & 0 & C_{66} \end{bmatrix} \begin{Bmatrix} \varepsilon_1 \\ \varepsilon_2 \\ \varepsilon_3 \\ \gamma_{23} \\ \gamma_{31} \\ \gamma_{12} \end{Bmatrix}. \tag{7.7}$$

From the above equation:

$$\gamma_{23} = \gamma_{13} = 0. \tag{7.8}$$

The third equation of Eq. (7.7) yields:

$$0 = C_{13}\varepsilon_1 + C_{23}\varepsilon_2 + C_{33}\varepsilon_3. \tag{7.9}$$

This gives:

$$\varepsilon_3 = -\frac{C_{13}}{C_{33}}\varepsilon_1 - \frac{C_{23}}{C_{33}}\varepsilon_2. \tag{7.10}$$

This relationship also indicates that in this state of plane stress ε_3 exists and Eq. (7.10) indicates it can be computed by knowing ε_1 and ε_2. The three-dimensional form Eq. (7.7) cannot be reduced directly to obtain a relation involving only σ_1, σ_2 and τ_{12} and ε_1, ε_2, γ_{12} by simply eliminating equations, as was done with Eq. (7.1) to obtain Eq. (7.4). However, Eq. (7.10) can be used as follows: From Eq. (7.7), the expressions for σ_1 and σ_2 are:

$$\begin{aligned} \sigma_1 &= C_{11}\varepsilon_1 + C_{12}\varepsilon_2 + C_{13}\varepsilon_3 \\ \sigma_2 &= C_{12}\varepsilon_1 + C_{22}\varepsilon_2 + C_{23}\varepsilon_3. \end{aligned} \tag{7.11}$$

Substituting Eq. (7.10) in the above equation leads to:

$$\begin{aligned} \sigma_1 &= C_{11}\varepsilon_1 + C_{12}\varepsilon_2 + C_{13}\left(-\frac{C_{13}}{C_{33}}\varepsilon_1 - \frac{C_{23}}{C_{33}}\varepsilon_2\right) \\ \sigma_2 &= C_{12}\varepsilon_1 + C_{22}\varepsilon_2 + C_{23}\left(-\frac{C_{13}}{C_{33}}\varepsilon_1 - \frac{C_{23}}{C_{33}}\varepsilon_2\right). \end{aligned} \tag{7.12}$$

Thus,

$$\sigma_1 = \left(C_{11} - \frac{C_{13}^2}{C_{33}} \right) \varepsilon_1 + \left(C_{12} - \frac{C_{13}C_{23}}{C_{33}} \right) \varepsilon_2$$
$$\sigma_2 = \left(C_{12} - \frac{C_{13}C_{23}}{C_{33}} \right) \varepsilon_1 + \left(C_{22} - \frac{C_{23}^2}{C_{33}} \right) \varepsilon_2. \tag{7.13}$$

Including the shear stress–shear strain relation, the relation between stresses and strains for the state of plane stress is written as:

$$\begin{Bmatrix} \sigma_1 \\ \sigma_2 \\ \tau_{12} \end{Bmatrix} = \begin{bmatrix} Q_{11} & Q_{12} & 0 \\ Q_{12} & Q_{22} & 0 \\ 0 & 0 & Q_{66} \end{bmatrix} \begin{Bmatrix} \varepsilon_1 \\ \varepsilon_2 \\ \gamma_{12} \end{Bmatrix} \tag{7.14}$$

The Q_{ij} are called the *reduced stiffnesses*, and from Eqs. (7.13) and (7.7):

$$Q_{11} = C_{11} - \frac{C_{13}^2}{C_{33}}$$
$$Q_{12} = C_{12} - \frac{C_{13}C_{23}}{C_{33}} \tag{7.15}$$
$$Q_{22} = C_{22} - \frac{C_{23}^2}{C_{33}}$$
$$Q_{66} = C_{66}.$$

The term "reduced" is used in relations given by Eqs. (7.4) and (7.14) because they are the result of reducing the problem from a fully three-dimensional to a two-dimensional, or plane stress, problem. However, the numerical values of the stiffnesses Q_{11}, Q_{12}, and Q_{22} are actually less than the numerical values of their respective counterparts for a fully three-dimensional problem, namely, C_{11}, C_{12}, and C_{33}, and so the stiffnesses are reduced in that sense also. It is very important to note that there is not really a numerically reduced compliance matrix. The elements in the plane stress compliance matrix, Eq. (7.5), are simply a subset of the elements from the three-dimensional compliance matrix, Eq. (7.1), and their numerical values are the identical. On the other hand, the elements of the reduced stiffness matrix, Eq. (7.15), involve a combination of elements from the three-dimensional stiffness matrix.

By inverting Eq. (7.4) and comparing it to Eq. (7.14), it is clear that:

$$Q_{11} = \frac{S_{22}}{S_{11}S_{22} - S_{12}^2}$$

$$Q_{12} = -\frac{S_{12}}{S_{11}S_{22} - S_{12}^2}$$

$$Q_{22} = \frac{S_{11}}{S_{11}S_{22} - S_{12}^2} \tag{7.16}$$

$$Q_{66} = \frac{1}{S_{66}}.$$

This provides a relationship between elements of the reduced compliance matrix and elements of the reduced stiffness matrix. A much more convenient form, and one that should be used in lieu of Eq. (7.16), can be obtained by simply writing the compliance components in Eq. (7.16) in terms of the appropriate engineering constants, namely:

$$Q_{11} = \frac{E_1}{1-\nu_{12}\nu_{21}}$$

$$Q_{12} = \frac{\nu_{12}E_2}{1-\nu_{12}\nu_{21}} = \frac{\nu_{21}E_1}{1-\nu_{12}\nu_{21}} \tag{7.17}$$

$$Q_{22} = \frac{E_2}{1-\nu_{12}\nu_{21}}$$

$$Q_{66} = G_{12}.$$

This form will be used exclusively from now on. For an isotropic material, the reduced stiffnesses become:

$$Q_{11} = Q_{22} = \frac{E}{1-\nu^2}$$

$$Q_{12} = \frac{\nu E}{1-\nu^2} \tag{7.18}$$

$$Q_{66} = G = \frac{E}{2(1+\nu)}.$$

When discussing general stress states in Chapter 5, we strongly emphasized that only one of the quantities in each of the six stress–strain pairs $\sigma_1 - \varepsilon_1$, $\sigma_2 - \varepsilon_2$, $\sigma_3 - \varepsilon_3$, $\tau_{23} - \gamma_{23}$, $\tau_{13} - \gamma_{13}$, and $\tau_{12} - \gamma_{12}$ could be specified. With the condition of plane stress, this restriction also holds. For the state of plane stress, we assume that σ_3, τ_{23}, and τ_{13} are zero. We can say nothing a priori regarding ε_3, γ_{23}, and γ_{13}. However, by using the stress–strain relations, we found, Eq. (7.2), that γ_{23} and γ_{13} are indeed zero. This is a consequence of the plane stress condition, not a stipulation. The strain ε_3 is given by

Eq. (7.3), another consequence of the plane stress condition. Of the three remaining stress–strain pairs, $\sigma_1 - \varepsilon_1$, $\sigma_2 - \varepsilon_2$, and $\tau_{12} - \gamma_{12}$, only one quantity in each of these pairs can be specified. The other quantity must be determined, as usual, by using the stress–strain relations, either Eq. (7.4) or (7.14), and the details of the specific problem being solved.

7.2 NUMERICAL RESULTS

Some of the numerical examples discussed in Section 5.5 are problems which satisfy plane stress conditions. Specifically, in Figure 5.11b, the cube of material is subjected to only one stress, as indicated by Eq. (5.64). Clearly, the conditions:

$$\sigma_3 = \tau_{23} = \tau_{13} = 0 \tag{7.19}$$

are satisfied, so the cube is in a state of plane stress. Using the plane stress stress–strain relation Eq. (7.4), we find the strains ε_1, ε_2, and γ_{12} can be directly determined as:

$$\begin{aligned} \varepsilon_1 &= S_{12}\sigma_2 \\ \varepsilon_2 &= S_{22}\sigma_2 \\ \gamma_{12} &= 0. \end{aligned} \tag{7.20}$$

The strain ε_3 must now be determined from the condition of Eq. (7.3), now not a direct part of the stress–strain relations, but rather an auxiliary equation. From Eq. (7.3):

$$\varepsilon_3 = S_{23}\sigma_2. \tag{7.21}$$

By substituting numerical values for S_{12}, S_{22}, and S_{23} from Eq. (5.56), and using the definitions of Eq. (5.67), we find the dimensional changes of the cube, as in Eq. (5.68). We can also find the dimensional change of the cube of Figure 5.11b from the plane stress stress–strain relations using Eq. (7.14), though this approach is not as direct. In particular, for the stress state of Figure 5.11b, Eq. (7.14) simplifies to:

$$\begin{aligned} 0 &= Q_{11}\varepsilon_1 + Q_{12}\varepsilon_2 \\ \sigma_2 &= Q_{12}\varepsilon_1 + Q_{22}\varepsilon_2 = -50 \text{ MPa}. \\ 0 &= Q_{66}\gamma_{12} \end{aligned} \tag{7.22}$$

Therefore,

$$\begin{aligned} \varepsilon_1 &= -\frac{Q_{12}}{Q_{11}}\varepsilon_2 \\ \gamma_{12} &= 0. \end{aligned} \tag{7.23}$$

Putting the above values in the second equation of Eq. (7.22) yields:

$$\varepsilon_2 = \left(\frac{Q_{11}}{Q_{11}Q_{22} - Q_{12}^2} \right) \sigma_2. \tag{7.24}$$

If Eq. (7.17) is used,

$$\varepsilon_2 = \frac{1}{E_2}\sigma_2 = S_{22}\sigma_2, \tag{7.25}$$

which is the second equation of Eq. (5.66). It seems that we have gone in circles – we have not! What we have shown is that the plane stress stress–strain relations yield results identical to the results obtained by using the general stress–strain relations if the problem is one of plane stress. This is an important point. If the problem is one of plane stress, then using the simpler forms, Eqs. (7.4) and (7.14), rather than the more complicated forms, Eqs. (5.45) and (5.47), gives the correct answer. The auxiliary conditions, either Eq. (7.3) or (7.10), can be used to obtain information about the strain ε_3 that is not a direct part of the plane stress stress–strain relation [6,7]. In that context, it is important to keep in mind that the out-of-plane engineering properties are still useful. For example, consider a layer of graphite-reinforced material 100 mm long, 50 mm wide, and 0.150 mm thick. As shown in Figure 7.6, this layer is subjected to a 3750 N in-plane force in the fiber direction. The through-thickness strain in the layer can be calculated from Eq. (7.3) and the numerical values in Eq. (5.56) as:

$$\varepsilon_3 = S_{13}\sigma_1 + S_{23}\sigma_2 = -\left(1.6 \times 10^{-12}\right)\frac{3750}{(0.050)\times(0.000150)} = -800\,\mu\frac{\text{mm}}{\text{mm}}. \tag{7.26}$$

7.3 EFFECTS OF FREE THERMAL AND FREE MOISTURE STRAINS

If a problem conforms to the plane stress assumption, namely, $\sigma_3 = \tau_{23} = \tau_{13} = 0$, but free thermal moisture strains are important, then, starting with Eq. (5.122) and following the same steps that led to Eq. (7.4), we can conclude that:

$$\gamma_{23} = \gamma_{13} = 0, \tag{7.27}$$

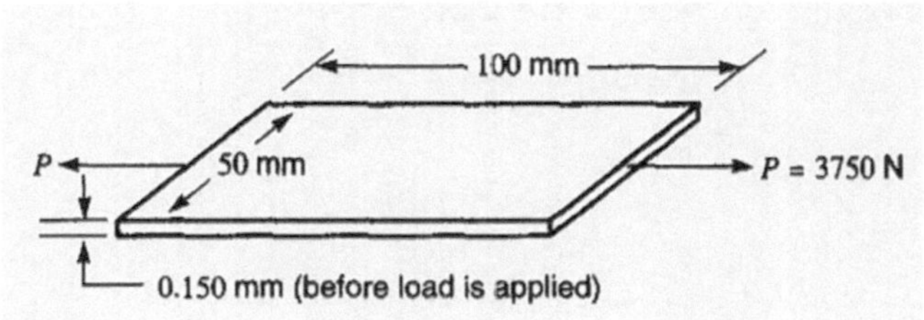

FIGURE 7.6 Layer subjected to in-plane forces.

which is identical to the case of no thermal or moisture expansion effects, as in Eq. (7.2). This is a direct consequence of the conditions $\tau_{23} = \tau_{13} = 0$. The conclusion regarding ε_3 is not exactly the same as the case with no free thermal or moisture strain effects, and in fact, the conclusion is much more far reaching. Specifically, using the condition that $\sigma_3 = 0$ in Eq. (5.122), we conclude that:

$$\varepsilon_3 = \alpha_3 \Delta T + \beta_3 \Delta M + S_{13}\sigma_1 + S_{23}\sigma_2. \tag{7.28}$$

Equation (7.28) will be the basis for determining the through-thickness, or out-of-plane, thermal or moisture expansion effects of a laminate. In this case, the through-thickness strain does not solely depend on the through-thickness expansion coefficients α_3 or β_3. The through-thickness strain involves the compliances S_{13} and S_{23}, which in turn involve the Poisson's ratios ν_{13} and ν_{23}, as well as the in-plane extensional moduli E_1 and E_2. This coupling of the through-thickness strain with the in-plane and out-of-plane elastic properties leads to important consequences. More will be said of this later. Continuing with the development, we observe that the plane stress stress–strain relations, including free thermal and moisture strain effects, become:

$$\begin{Bmatrix} \varepsilon_1 - \alpha_1 \Delta T - \beta_1 \Delta M \\ \varepsilon_2 - \alpha_2 \Delta T - \beta_2 \Delta M \\ \gamma_{12} \end{Bmatrix} = \begin{bmatrix} S_{11} & S_{12} & 0 \\ S_{12} & S_{22} & 0 \\ 0 & 0 & S_{66} \end{bmatrix} \begin{Bmatrix} \sigma_1 \\ \sigma_2 \\ \tau_{12} \end{Bmatrix}. \tag{7.29}$$

Also,

$$\begin{Bmatrix} \sigma_1 \\ \sigma_2 \\ \tau_{12} \end{Bmatrix} = \begin{bmatrix} Q_{11} & Q_{12} & 0 \\ Q_{12} & Q_{22} & 0 \\ 0 & 0 & Q_{66} \end{bmatrix} \begin{Bmatrix} \varepsilon_1 - \alpha_1 \Delta T - \beta_1 \Delta M \\ \varepsilon_2 - \alpha_2 \Delta T - \beta_2 \Delta M \\ \gamma_{12} \end{Bmatrix}. \tag{7.30}$$

Here, the mechanical strains are given by:

$$\begin{Bmatrix} \varepsilon_1^{\text{mech}} \\ \varepsilon_2^{\text{mech}} \\ \gamma_{12}^{\text{mech}} \end{Bmatrix} = \begin{Bmatrix} \varepsilon_1 - \alpha_1 \Delta T - \beta_1 \Delta M \\ \varepsilon_2 - \alpha_2 \Delta T - \beta_2 \Delta M \\ \gamma_{12} \end{Bmatrix}. \tag{7.31}$$

As in the three-dimensional stress–strain relations, Eqs. (5.122) and (5.123), $\varepsilon_1, \varepsilon_2, \gamma_{12}$ above are the total strains. The dimensional changes are therefore given by:

$$\delta\Delta_1 = \varepsilon_1 \Delta_1 \delta\Delta_2 = \varepsilon_2 \Delta_2. \tag{7.32}$$

Equally important, ε_3 in Eq. (7.28) is the total strain. This is quite obvious when one considers that the right-hand side of that equation explicitly includes free thermal strain effects, free moisture strain effects, and stress-related effects [8–10]. To continue with our examples of thermal deformations, consider a 50 mm×50 mm×50 mm element of material that is completely constrained in the 1 and 2 directions. To satisfy the plane stress condition, the element cannot be constrained in the 3 direction. The stresses in the two constrained directions and the deformation in the 3 direction due to a temperature increase of 50°C are to be determined. Moisture absorption is not an issue ($\Delta M = 0$). For this case, because of the stated conditions:

$$\varepsilon_1 = \varepsilon_2 = \gamma_{12} = 0. \tag{7.33}$$

To determine the restraining stresses, Eq. (7.30) is used to give:

$$\begin{aligned} \sigma_1 &= -(Q_{11}\alpha_1 + Q_{12}\alpha_2)\Delta T \\ \sigma_2 &= -(Q_{12}\alpha_1 + Q_{22}\alpha_2)\Delta T \\ \tau_{12} &= Q_{66} \times 0. \end{aligned} \tag{7.34}$$

To determine the deformation in the 3 direction, Eq. (7.28) is used to yield:

$$\begin{aligned} \varepsilon_3 &= \alpha_3 \Delta T + S_{13}\sigma_1 + S_{23}\sigma_2 \\ \varepsilon_3 &= \left(\alpha_3 - S_{13}(Q_{11}\alpha_1 + Q_{12}\alpha_2) - S_{23}(Q_{12}\alpha_1 + Q_{22}\alpha_2)\right)\Delta T \\ \varepsilon_3 &= \left(\alpha_3 - (S_{13}Q_{11} + S_{23}Q_{12})\alpha_1 - S_{13}(Q_{12} + S_{23}Q_{22}\alpha_2)\right)\Delta T, \end{aligned} \tag{7.35}$$

where the various terms in Eq. (7.35) are retained to show the interaction between two- and three-dimensional elastic properties, and the three coefficients of thermal expansion [11,12]. Numerically, Eqs. (7.34) and (7.35) give:

$$\begin{aligned} \sigma_1 &= -3.52 \text{ MPa} \\ \sigma_2 &= -14.77 \text{ MPa} \\ \tau_{12} &= 0 \\ \varepsilon_3 &= 1780 \times 10^{-6}. \end{aligned} \tag{7.36}$$

REFERENCES

1. Herakovich C.T. (1989), "Free edge effects in laminated composites." in *Handbook of Composites: Vol. 2, Structures and Design* (C.T. Herakovich and Y.M. Tarnopol'skii, eds.), Elsevier Science Publishing Co., New York, pp. 187–230.
2. Armanios E.A. (1989), *Interlaminar Fracture of Composites*, Trans Tech. Publications, Aedermannsdorf, Switzerland.
3. Renton W.J. and Vinson J.R. (1977), "Analysis of adhesively bonded joints between panels of composite materials", *Transactions of the ASME, Journal of Applied Mechanics*, Vol. 44, pp. 101–106.

4. Kassapoglou C. (1993), "Calculation of stresses at skin-stiffener interfaces of composite stiffened panels under shear loads", *International Journal of Solids and Structures*, Vol. 30(11), pp. 1491–1501.
5. Harrison P.N. and Johnson E.R. (1996), "A mixed variational formulation for interlaminar stresses in thickness-tapered composite laminates", *International Journal of Solids and Structures*, Vol. 33(16), pp. 2377–2399.
6. Love A.E.H. (1892), *A Treatise on the Mathematical Theory of Elasticity*, Cambridge University Press, London (also 4th ed., 1944, Dover Publications, Inc., New York).
7. Hodge P.G., Jr. (1958), "The mathematical theory of plasticity," in *Elasticity and Plasticity* (J.N. Goodier and P.G. Hodge, Jr., eds.), John Wiley & Sons, Inc., New York, p. 81.
8. Kreyszig E. (1967), *Advanced Engineering Mathematics*, 2nd ed., John Wiley & Sons, Inc., New York, p. 416.
9. Fung Y.C. (1965), *Foundations of Solid Mechanics*, Prentice-Hall, Inc., Englewood Cliffs, NJ, p. 285.
10. Green G. (1839), "On the laws of reflexion and refraction of light at the common surface of two non-crystallized media", *Transactions of the Cambridge Philosophical Society*, Vol. 7, p. 121.
11. Popov E.P. (1990), *Engineering Mechanics of Solids*, Prentice Hall, Englewood Cliffs, NJ.
12. Hyer M.W. (1998), *Stress Analysis of Fiber Reinforced Composite Materials*, McGraw-Hill, Singapore.

8 Global Coordinate System

Plane Stress Stress–Strain Relations

To this point, we have discussed the response of a fiber-reinforced material in the principal material system, where the fiber direction in a lamina is considered as the 1 direction and the other two directions are transverse to the fiber. This is suitable only for the analysis of a single lamina. Suppose we have several laminas, each having its separate fiber orientation, arranged together to form a laminate. Thus, we will be confronted with using multiple 1-2-3 coordinate systems, each with its separate orientation with respect to some global or structural coordinate system. If we are dealing with an x-y-z Cartesian coordinate system to describe the geometry of the structure, then the orientation of each principal material system must be defined with respect to the x-y-z system. If we are dealing with an x-θ-r cylindrical coordinate system to describe the structure, then the orientation of each principal material system must be defined with respect to the x-θ-r system, and so forth, for a spherical coordinate system. This leads to a large number of coordinate systems and orientations for describing the response of the fiber-reinforced structure.

As an alternative approach, we can refer the response of each layer of material to the same global system. We accomplish this by transforming the stress–strain relations from the 1-2-3 coordinate system into the global coordinate system. This will be our approach here, in particular; it will be done for a state of plane stress using the standard transformation relations for stresses and strains learned in introductory strength-of-materials courses. Transformation can also be done for a general state of stress. However, transformation here will be limited to the plane stress state because it will be useful for the development of classical lamination theory, which begins in Chapter 9. Equally important, though, is the fact that the transformation of the description of the stress–strain response of fiber-reinforced material from the principal material coordinate system to a global coordinate system results in concepts so different from what one encounters with isotropic materials that it is best to start with a simpler plane stress state and progress to the more complicated general stress state. When the concepts for the plane stress stress-state response described in a coordinate system other than the principal material coordinate system are fully understood, progression to a three-dimensional stress state is easier.

8.1 TRANSFORMATION EQUATIONS

Consider, as shown in Figure 8.1a, the familiar view of an isolated element in the principal material coordinate system. Figure 8.1b shows a similar element but one that is isolated in an *x-y-z* global coordinate system. The fibers are oriented at an angle θ with respect to the $+x$ axis of the global system. The fibers are parallel to the *x-y* plane, and the 3 and *z* axes coincide. The fibers assumed their orientation by a simple rotation of the principal material system about the 3 axis. The orientation angle θ will be considered positive when the fibers rotate counterclockwise *from* the $+x$ axis *toward* the $+y$ axis. Often the fibers not being aligned with the edges of the element are referred to as an *off-axis* condition, generally meaning the fibers are not aligned with the analysis coordinate system (i.e., off the $+x$ axis). Though we will use the notation for a rectangular Cartesian coordinate system as the global system (i.e., *x-y-z*), the global coordinate system can be considered to be any orthogonal coordinate system. The use of a Cartesian system is for convenience only, and the development is actually valid for any orthogonal coordinate system [1].

The stress–strain relations are a description of the relations between stress and strain at a point within the material. The functional form of these relations does not depend on whether the point is a point in a rectangular Cartesian coordinate system, in a cylindrical coordinate system, or in a spherical, elliptical, or parabolic coordinate system [2]. The stresses on the small volume of element are now identified in accordance with the *x-y-z* notation. The six components of stress and strain are denoted as:

$$\left\{\begin{bmatrix}\sigma_x\\ \sigma_y\\ \sigma_z\\ \tau_{yz}\\ \tau_{xz}\\ \tau_{xy}\end{bmatrix}\begin{bmatrix}\varepsilon_x\\ \varepsilon_y\\ \varepsilon_z\\ \gamma_{yz}\\ \gamma_{xz}\\ \gamma_{xy}\end{bmatrix}\right\}. \tag{8.1}$$

And the six components of stress are illustrated in Figure 8.2. Although we seem interested in describing the stress–strain relation in another coordinate system now that we have developed it and fully understand it in the 1-2-3 system, we should

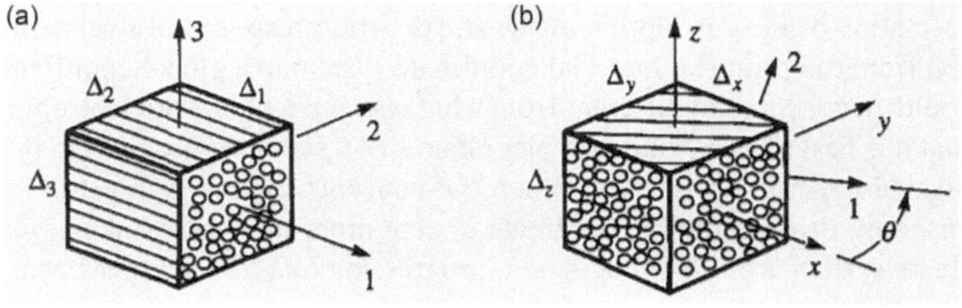

FIGURE 8.1 Element of fiber-reinforced material in (a) 1-2-3 and (b) *x-y-z* coordinate systems.

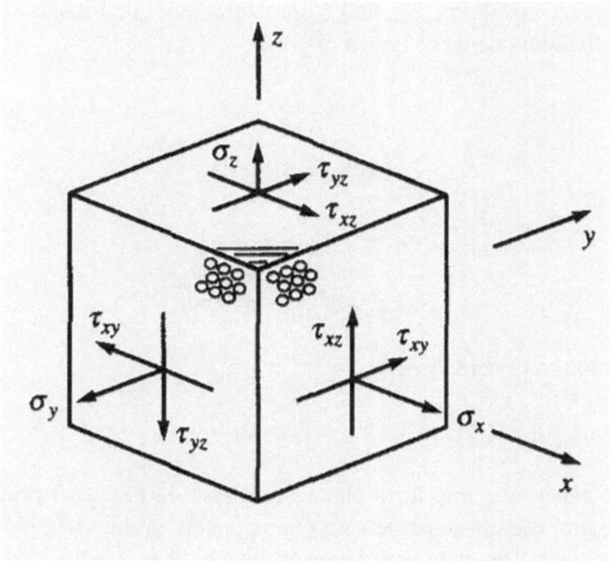

FIGURE 8.2 Components of stress in x-y-z coordinate system.

emphasize that Figure 8.2 should be interpreted quite literally. We should interpret the figure as asking, "What is the relation between the stresses and deformations denoted in Eq. (8.1) for a small volume of material whose fibers are oriented at some angle relative to the boundaries of the element rather than parallel to them?" This is the real issue! Loads will not be always applied parallel to the fibers; our intuition indicates that unusual deformations are likely to occur. The skewed orientation of the fibers must certainly cause unusual distortions of the originally cubic volume element. What are these deformations? How do they depend on fiber orientation? Are they detrimental? Are they beneficial? Fortunately, these and other questions can be answered by transforming the stress–strain relations from the 1-2-3 system to the x-y-z system. If we consider the special case as shown in Figure 8.lb, where the two coordinate systems are related to each other through a simple rotation θ about the z axis, then the stresses in the 1-2-3 system are related to the stresses in the x-y-z system by:

$$\sigma_1 = \cos^2\theta\sigma_x + \sin^2\theta\sigma_y + 2\sin\theta\cos\theta\tau_{xy}$$

$$\sigma_2 = \sin^2\theta\sigma_x + \cos^2\theta\sigma_y - 2\sin\theta\cos\theta\tau_{xy}$$

$$\sigma_3 = \sigma_z \tag{8.2}$$

$$\tau_{23} = \cos\theta\tau_{yz} - \sin\theta\tau_{xz}$$

$$\tau_{13} = \sin\theta\tau_{yz} + \cos\theta\tau_{xz}$$

$$\tau_{12} = -\sin\theta\cos\theta\sigma_x + \sin\theta\cos\theta\sigma_y + \left(\cos^2\theta - \sin^2\theta\right)\tau_{xy}.$$

For a state of plane stress, σ_3, τ_{23}, and τ_{13} are zero, and upon rearranging, the third, fourth, and fifth components of Eq. (8.2) give:

$$\sigma_z = 0$$

$$\cos\theta\tau_{yz} - \sin\theta\tau_{xz} = 0 \tag{8.3}$$

$$\sin\theta\tau_{yz} + \cos\theta\tau_{xz} = 0$$

$$\text{Since, } \cos^2\theta + \sin^2\theta = 1. \tag{8.4}$$

The only solution to Eqs. (8.3) is:

$$\tau_{yz} = \tau_{xz} = 0, \tag{8.5}$$

leading to the conclusion that for a plane stress state in the 1-2-3 principal material coordinate system, the out-of-plane stress components in the *x-y-z* global coordinate system are also zero. This may have been intuitive but here we have shown it directly. The first, second, and sixth component of Eq. (8.2) may look more familiar in the form:

$$\sigma_1 = \left(\frac{\sigma_x + \sigma_y}{2}\right) + \left(\frac{\sigma_x - \sigma_y}{2}\right)\cos 2\theta + \tau_{xy}\sin 2\theta$$

$$\sigma_2 = \left(\frac{\sigma_x + \sigma_y}{2}\right) - \left(\frac{\sigma_x - \sigma_y}{2}\right)\cos 2\theta - \tau_{xy}\sin 2\theta \tag{8.6}$$

$$\tau_{12} = -\left(\frac{\sigma_x - \sigma_y}{2}\right)\sin 2\theta + \tau_{xy}\cos 2\theta.$$

This form is derivable from Eq. (8.2) by using trigonometric identities and is the form usually found in introductory strength-of-materials courses. The form of Eq. (8.2) will be most often used as it can be put in the matrix form as:

$$\begin{Bmatrix} \sigma_1 \\ \sigma_2 \\ \tau_{12} \end{Bmatrix} = \begin{bmatrix} \cos^2\theta & \sin^2\theta & 2\sin\theta\cos\theta \\ \sin^2\theta & \cos^2\theta & -2\sin\theta\cos\theta \\ -\sin\theta\cos\theta & \sin\theta\cos\theta & \cos^2\theta - \sin^2\theta \end{bmatrix} \begin{Bmatrix} \sigma_x \\ \sigma_y \\ \tau_{xy} \end{Bmatrix}. \tag{8.7}$$

This transformation matrix of trigonometric functions will be used frequently in the plane stress analysis of fiber-reinforced composite materials, and it will be denoted by $[T]$, which is written as:

$$[T] = \begin{bmatrix} m^2 & n^2 & 2mn \\ n^2 & m^2 & -2mn \\ -mn & mn & m^2 - n^2 \end{bmatrix}. \tag{8.8}$$

In the above equation, $m = \cos\theta$, $n = \sin\theta$.

Thus, Eq. (8.7) can be written as:

$$\begin{Bmatrix} \sigma_1 \\ \sigma_2 \\ \tau_{12} \end{Bmatrix} = [T] \begin{Bmatrix} \sigma_x \\ \sigma_y \\ \tau_{xy} \end{Bmatrix} \tag{8.9}$$

or,

$$\begin{Bmatrix} \sigma_1 \\ \sigma_2 \\ \tau_{12} \end{Bmatrix} = \begin{bmatrix} m^2 & n^2 & 2mn \\ n^2 & m^2 & -2mn \\ -mn & mn & m^2 - n^2 \end{bmatrix} \begin{Bmatrix} \sigma_x \\ \sigma_y \\ \tau_{xy} \end{Bmatrix}. \tag{8.10}$$

The inverse of Eq. (8.10) is:

$$\begin{Bmatrix} \sigma_x \\ \sigma_y \\ \tau_{xy} \end{Bmatrix} = \begin{bmatrix} m^2 & n^2 & -2mn \\ n^2 & m^2 & 2mn \\ mn & -mn & m^2 - n^2 \end{bmatrix} \begin{Bmatrix} \sigma_1 \\ \sigma_2 \\ \tau_{12} \end{Bmatrix}. \tag{8.11}$$

This gives,

$$[T]^{-1} = \begin{bmatrix} m^2 & n^2 & -2mn \\ n^2 & m^2 & 2mn \\ mn & -mn & m^2 - n^2 \end{bmatrix}. \tag{8.12}$$

In a similar manner, the strains transform according to the specialized relations Eq. (8.2) as:

$$\begin{aligned} \varepsilon_1 &= \cos^2\theta\varepsilon_x + \sin^2\theta\varepsilon_y + 2\sin\theta\cos\theta\varepsilon_{xy} \\ \varepsilon_2 &= \sin^2\theta\varepsilon_x + \cos^2\theta\varepsilon_y - 2\sin\theta\cos\theta\varepsilon_{xy} \\ \varepsilon_3 &= \varepsilon_z \\ \varepsilon_{23} &= \cos\theta\varepsilon_{yz} - \sin\theta\varepsilon_{xz} \\ \varepsilon_{13} &= \sin\theta\varepsilon_{yz} + \cos\theta\varepsilon_{xz} \\ \varepsilon_{12} &= -\sin\theta\cos\theta\varepsilon_x + \sin\theta\cos\theta\varepsilon_y + \left(\cos^2\theta - \sin^2\theta\right)\varepsilon_{xy}. \end{aligned} \tag{8.13}$$

The tensor shear strains, ε, not the engineering shear strains, are being used in the above. These two measures of strain are different by a factor of two; that is,

$$\varepsilon_{23} = \frac{\gamma_{23}}{2}, \quad \varepsilon_{13} = \frac{\gamma_{13}}{2}, \quad \varepsilon_{12} = \frac{\gamma_{12}}{2}. \tag{8.14}$$

If engineering shear strain is used instead, then the transformation relations become:

$$\begin{aligned}
\varepsilon_1 &= \cos^2\theta\varepsilon_x + \sin^2\theta\varepsilon_y + 2\sin\theta\cos\theta\frac{\gamma_{xy}}{2} \\
\varepsilon_2 &= \sin^2\theta\varepsilon_x + \cos^2\theta\varepsilon_y - 2\sin\theta\cos\theta\frac{\gamma_{xy}}{2} \\
\varepsilon_3 &= \varepsilon_z \\
\gamma_{23} &= \cos\theta\gamma_{yz} - \sin\theta\gamma_{xz} \\
\gamma_{13} &= \sin\theta\gamma_{yz} + \cos\theta\gamma_{xz} \\
\frac{\gamma_{12}}{2} &= -\sin\theta\cos\theta\varepsilon_x + \sin\theta\cos\theta\varepsilon_y + \left(\cos^2\theta - \sin^2\theta\right)\frac{\gamma_{xy}}{2}.
\end{aligned} \tag{8.15}$$

As a result of the plane stress assumption, specifically by Eq. (7.2):

$$\gamma_{23} = \gamma_{13} = 0\ . \tag{8.16}$$

By analogy to Eq. (8.3), it is concluded from the fourth and fifth equation of Eq. (8.15) that:

$$\gamma_{yz} = \gamma_{xz} = 0\ . \tag{8.17}$$

Also, due to the third equation of Eq. (8.15) and Eq. (7.3):

$$\varepsilon_z = S_{13}\sigma_1 + S_{23}\sigma_2. \tag{8.18}$$

If Eq. (8.7) is used to transform the stresses, then:

$$\varepsilon_z = \left(S_{13}\cos^2\theta + S_{23}\sin^2\theta\right)\sigma_x + \left(S_{13}\sin^2\theta + S_{23}\cos^2\theta\right)\sigma_y + 2(S_{13} - S_{23})\sin\theta\cos\theta\tau_{xy}. \tag{8.19}$$

This equation is very important because it indicates that a shear stress in the *x-y* plane, τ_{xy}, produces an extensional strain, ε_z, perpendicular to that plane! For an isotropic material, $S_{13} = S_{23}$, and this simply will not happen. Shear stresses do not cause extensional strains in isotropic materials! This generation of extensional strains by shear stresses is an important characteristic of fiber-reinforced composite materials. Returning to Eq. (8.15) to focus on the strains involved in the plane stress assumption, we can write the first, second, and sixth equations as:

$$\begin{Bmatrix} \varepsilon_1 \\ \varepsilon_2 \\ \dfrac{\gamma_{12}}{2} \end{Bmatrix} = [T] \begin{Bmatrix} \varepsilon_x \\ \varepsilon_y \\ \dfrac{\gamma_{xy}}{2} \end{Bmatrix} \tag{8.20}$$

or,

$$\begin{Bmatrix} \varepsilon_1 \\ \varepsilon_2 \\ \dfrac{\gamma_{12}}{2} \end{Bmatrix} = \begin{bmatrix} m^2 & n^2 & 2mn \\ n^2 & m^2 & -2mn \\ -mn & mn & m^2 - n^2 \end{bmatrix} \begin{Bmatrix} \varepsilon_x \\ \varepsilon_y \\ \dfrac{\gamma_{xy}}{2} \end{Bmatrix}. \tag{8.21}$$

8.2 TRANSFORMED REDUCED COMPLIANCE

The stress–strain relations in the 1-2-3 principal material coordinate system, Eq. (7.4), can be written in a slightly modified form to account for the use of the tensor shear strain rather than the engineering shear strain as:

$$\begin{Bmatrix} \varepsilon_1 \\ \varepsilon_2 \\ \dfrac{\gamma_{12}}{2} \end{Bmatrix} = \begin{bmatrix} S_{11} & S_{12} & 0 \\ S_{12} & S_{22} & 0 \\ 0 & 0 & \dfrac{S_{66}}{2} \end{bmatrix} \begin{Bmatrix} \sigma_1 \\ \sigma_2 \\ \tau_{12} \end{Bmatrix}. \tag{8.22}$$

Substituting the transformations given by Eqs. (8.9) and (8.20) in Eq. (8.22) leads to:

$$[T] \begin{Bmatrix} \varepsilon_x \\ \varepsilon_y \\ \dfrac{\gamma_{xy}}{2} \end{Bmatrix} = \begin{bmatrix} S_{11} & S_{12} & 0 \\ S_{12} & S_{22} & 0 \\ 0 & 0 & \dfrac{S_{66}}{2} \end{bmatrix} [T] \begin{Bmatrix} \sigma_x \\ \sigma_y \\ \tau_{xy} \end{Bmatrix}. \tag{8.23}$$

Multiplying both sides of Eq. (8.23) by $[T]^{-1}$ results in:

$$\begin{Bmatrix} \varepsilon_x \\ \varepsilon_y \\ \dfrac{\gamma_{xy}}{2} \end{Bmatrix} = [T]^{-1} \begin{bmatrix} S_{11} & S_{12} & 0 \\ S_{12} & S_{22} & 0 \\ 0 & 0 & \dfrac{S_{66}}{2} \end{bmatrix} [T] \begin{Bmatrix} \sigma_x \\ \sigma_y \\ \tau_{xy} \end{Bmatrix}. \tag{8.24}$$

Substituting for $[T]$ and $[T]^{-1}$ from Eqs. (8.8) and (8.12), we find that multiplying these three matrices together, and multiplying the third equation by a factor of 2, yields:

$$\begin{Bmatrix} \varepsilon_x \\ \varepsilon_y \\ \gamma_{xy} \end{Bmatrix} = \begin{bmatrix} \bar{S}_{11} & \bar{S}_{12} & \bar{S}_{16} \\ \bar{S}_{12} & \bar{S}_{22} & \bar{S}_{26} \\ \bar{S}_{16} & \bar{S}_{26} & \bar{S}_{66} \end{bmatrix} \begin{Bmatrix} \sigma_x \\ \sigma_y \\ \tau_{xy} \end{Bmatrix}. \tag{8.25}$$

The $\bar{S}_{ij}$ are called the *transformed reduced compliances.* Note that the factor of ½ has been removed and the engineering shear strain reintroduced. Equation (8.25) is a fundamental equation for studying the plane stress response of fiber-reinforced composite materials. The transformed reduced compliances are defined by:

$$\bar{S}_{11} = S_{11}m^4 + (2S_{12} + S_{66})n^2m^2 + S_{22}n^4$$

$$\bar{S}_{12} = (S_{11} + S_{22} - S_{66})n^2m^2 + S_{12}\left(n^4 + m^4\right)$$

$$\bar{S}_{16} = (2S_{11} - 2S_{12} - S_{66})nm^3 - (2S_{22} - 2S_{12} - S_{66})mn^3$$

$$\bar{S}_{22} = S_{11}n^4 + (2S_{12} + S_{66})n^2m^2 + S_{22}m^4$$

$$\bar{S}_{26} = (2S_{11} - 2S_{12} - S_{66})mn^3 - (2S_{22} - 2S_{12} - S_{66})nm^3$$

$$\bar{S}_{66} = 2(2S_{11} + 2S_{22} - 4S_{12} - S_{66})n^2m^2 + S_{66}\left(n^4 + m^4\right), \tag{8.26}$$

where $m = \cos\theta$ and $n = \sin\theta$.

Equation (8.25) and the definitions Eq. (8.26) relate the strains of an element of fiber-reinforced material as measured in the *x-y-z* global coordinate system to the applied stresses measured in that coordinate system. The most profound results of Eq. (8.25) are that a normal stress σ_x will cause a shearing deformation γ_{xy} through the $\bar{S}_{16}$ term, and similarly, a normal stress σ_y will cause a shearing deformation through the $\bar{S}_{26}$ term. Equally important, because of the existence of these same $\bar{S}_{16}$ and $\bar{S}_{26}$ terms at other locations in the compliance matrix, a shear stress τ_{xy} will cause strains ε_x and ε_y. Such responses are totally different from those in metals [3,4]. In metals, normal stresses do not cause shear strains, and shear stresses do not cause extensional strains. This coupling found in fiber-reinforced composites is termed "shear–extension coupling." Shear–extension coupling is an important characteristic and is responsible for interesting and important responses of fiber-reinforced composite materials. Through a series of examples, we will examine the response of an element of fiber-reinforced material to simple stress states (i.e., σ_x only, and then τ_{xy} only) and compare the responses with the response of a similar element of metal.

Before proceeding with the examples, we should discuss two special cases of Eq. (8.26). For the first case, consider the situation when the fibers are aligned with the *x* axis, namely, $\theta = 0°$. With $\theta = 0°$, $m = 1$, and $n = 0$, Eq. (8.26) reduces to:

$$\bar{S}_{11}(0^\circ) = S_{11}$$

$$\bar{S}_{12}(0^\circ) = S_{12}$$

$$\bar{S}_{16}(0^\circ) = 0^\circ$$

$$\bar{S}_{22}(0^\circ) = S_{22}$$

$$\bar{S}_{26}(0^\circ) = 0^\circ$$

$$\bar{S}_{66}(0^\circ) = S_{66}. \tag{8.27}$$

The results of Eq. (8.27) simply state that at $\theta = 0^\circ$, the transformed reduced compliance degenerates to the reduced compliance, that is, the compliance in the principal material coordinate system. In the principal material system, there is no $\bar{S}_{16}$ or $\bar{S}_{26}$. The quantities S_{11}, S_{12}, S_{22}, and S_{66} are often referred to as the *on-axis compliances*. The barred quantities of Eq. (8.26) are frequently called the *off-axis compliances*.

For the second special case, consider isotropic materials. The compliances of Eq. (8.26) reduce to:

$$\bar{S}_{11} = \frac{1}{E},\ \bar{S}_{12} = -\frac{\nu}{E},\ \bar{S}_{16} = 0,\ \bar{S}_{22} = \frac{1}{E},\ \bar{S}_{26} = 0,\ \bar{S}_{66} = \frac{1}{G} = \frac{2(1+\nu)}{E}. \tag{8.28}$$

This can be demonstrated by using the definitions of the compliances for an isotropic material, Eq. (7.6), in Eq. (8.26). For example:

$$\bar{S}_{11} = \frac{1}{E}m^4 + \left(-\frac{2\nu}{E} + \frac{2(1+\nu)}{E}\right)n^2m^2 + \frac{1}{E}n^4 \tag{8.29a}$$

$$\bar{S}_{11} = \frac{1}{E}(m^4 + 2n^2m^2 + n^4) = \frac{1}{E}\left(m^2 + n^2\right)^2 \tag{8.29b}$$

$$\text{Using } m^2 + n^2 = 1 \tag{8.30}$$

$$\bar{S}_{11} = \frac{1}{E}. \tag{8.31}$$

Similarly,

$$\bar{S}_{22} = \frac{1}{E} \tag{8.32}$$

$$\bar{S}_{12} = \left(\frac{1}{E} + \frac{1}{E} - \frac{2(1+\nu)}{E}\right)n^2m^2 - \frac{\nu}{E}\left(n^4 + m^4\right) \tag{8.33a}$$

$$\bar{S}_{12} = -\frac{\nu}{E}\left(m^4 + 2n^2m^2 + n^4\right) = -\frac{\nu}{E}\left(m^2 + n^2\right)^2. \tag{8.33b}$$

Using Eq. (8.30):

$$\bar{S}_{12} = -\frac{\nu}{E} \tag{8.34}$$

$$\bar{S}_{16} = \left(\frac{2}{E} - 2\frac{\nu}{E} - \frac{2(1+\nu)}{E}\right)nm^3 - \left(\frac{2}{E} - 2\frac{\nu}{E} - \frac{2(1+\nu)}{E}\right)mn^3 \tag{8.35a}$$

$$\bar{S}_{16} = (0)nm^3 - (0)mn^3 = 0. \tag{8.35b}$$

Similarly,

$$\bar{S}_{26} = 0. \tag{8.36}$$

Also,

$$\bar{S}_{66} = 2\left(\frac{2}{E} + \frac{2}{E} + 4\frac{\nu}{E} - \frac{2(1+\nu)}{E}\right)n^2m^2 + \frac{2(1+\nu)}{E}\left(n^4 + m^4\right) \tag{8.37a}$$

$$\bar{S}_{66} = \frac{2(1+\nu)}{E}\left(n^4 + m^4 + 2n^2m^2\right) = \frac{2(1+\nu)}{E}\left(m^2 + n^2\right)^2. \tag{8.37b}$$

Using Eq. (8.30):

$$\bar{S}_{66} = \frac{2(1+\nu)}{E}. \tag{8.38}$$

A series of examples that illustrate the shear–extension coupling predicted by the stress–strain relations of Eq. (8.25), specifically the deformations caused by a tensile normal stress, will be discussed now. Consider a thin element of aluminum subjected to a tensile stress $\sigma_x = 155\,\text{MPa}$. As in Figure 8.3a, the aluminum element has dimensions 50 mm × 50 mm. Thickness is not important at the moment but consider the element to be thin. As σ_y and τ_{xy} are zero, the stress–strain relations of Eq. (8.25) reduce to:

$$\begin{aligned} \varepsilon_x &= \bar{S}_{11}\sigma_x \\ \varepsilon_y &= \bar{S}_{12}\sigma_x \\ \gamma_{xy} &= \bar{S}_{16}\sigma_x. \end{aligned} \tag{8.39}$$

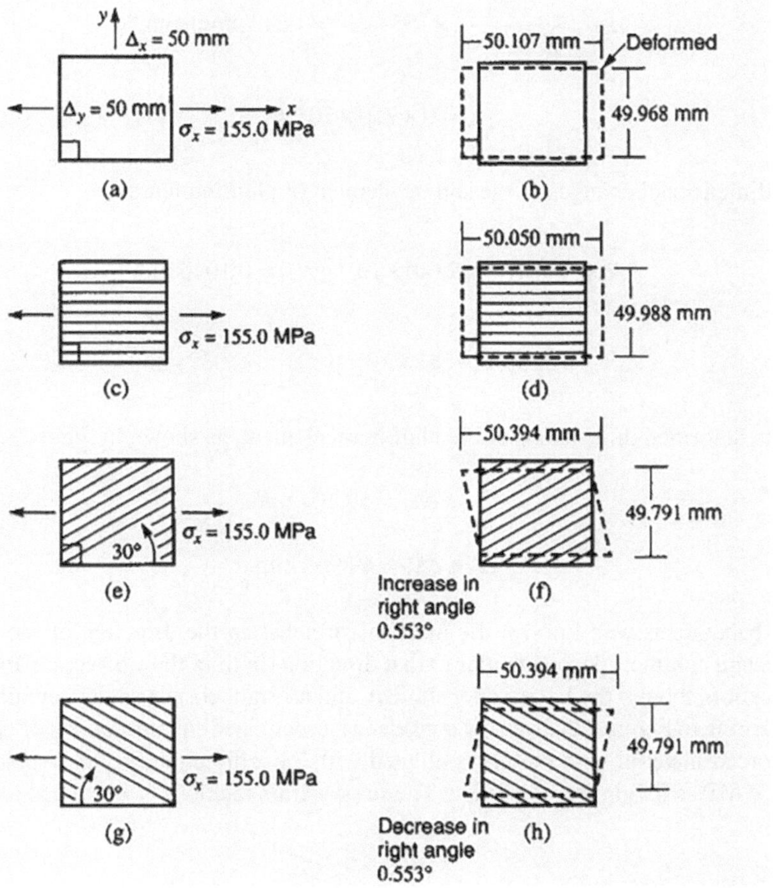

FIGURE 8.3 (a–h) Comparison of deformations of aluminum and graphite-reinforced composite under a tensile normal stress σ_x.

Using Eq. (8.28):

$$\varepsilon_x = \frac{1}{E}\sigma_x \tag{8.40}$$

$$\varepsilon_y = -\frac{\nu}{E}\sigma_x$$

$$\gamma_{xy} = 0 \times \sigma_x = 0.$$

Referring to Table 5.1, we note that the strains in the aluminum are:

$$\varepsilon_x = \frac{1}{72.4 \times 10^9} \times 155 \times 10^6 = 2140\ \mu\text{mm}\,\text{mm}^{-1}.$$

$$\varepsilon_y = \frac{-0.3}{72.4\times10^9}\times155\times10^6 = -642\ \mu\text{mm mm}^{-1} \tag{8.41}$$

$$\gamma_{xy} = 0\times250\times10^6 = 0.$$

The dimensional changes of the square element of aluminum are:

$$\delta\Delta_x = \varepsilon_x\Delta_x = \left(2140\times10^{-6}\right)(50) = 0.1070\ \text{mm}$$

$$\delta\Delta_y = \varepsilon_y\Delta_y = \left(-642\times10^{-6}\right)(50) = -0.0321\ \text{mm}. \tag{8.42}$$

So the deformed dimensions of the aluminum element, as shown in Figure 8.4b, are:

$$\Delta_x + \delta\Delta_x = 50.107\ \text{mm}$$

$$\Delta_y + \delta\Delta_y = 49.968\ \text{mm}. \tag{8.43}$$

This behavior is well known; the material stretches in the direction of the applied stress and contracts perpendicular to that direction (both in the y direction and in the z direction, though the latter is not shown), and all right corner angles remain right.

Turning to Figure 8.3c, we now consider a similar-sized square element of graphite-reinforced material with the fibers aligned with the x direction and also subjected to the 155 MPa stress in the x direction. The stress–strain relations of Eq. (8.25) reduce to:

$$\varepsilon_x = \bar{S}_{11}(0°)\sigma_x$$

$$\varepsilon_y = \bar{S}_{12}(0°)\sigma_x \tag{8.44}$$

$$\gamma_{xy} = \bar{S}_{16}(0°)\sigma_x.$$

If we use the compliances of Eq. (8.27):

$$\varepsilon_x = S_{11}\sigma_x \tag{8.45}$$

$$\varepsilon_y = S_{12}\sigma_x$$

$$\gamma_{xy} = 0\times\sigma_x = 0.$$

From Eq. (7.5) and Table 5.1, or, alternatively, directly from Eq. (5.56),

$$S_{11} = 6.45\ (\text{TPa})^{-1} \quad S_{12} = -1.6 \quad S_{13} = -1.6$$

$$S_{22} = 82.6 \quad S_{66} = 227. \tag{8.46}$$

Therefore,

$$\varepsilon_x = 6.45\times10^{-12}\times155\times10^{6} = 1000\ \mu\text{mm mm}^{-1}$$
$$\varepsilon_y = -1.6\times10^{-12}\times155\times10^{6} = -248\ \mu\text{mm mm}^{-1} \qquad (8.47)$$
$$\gamma_{xy} = 0\times227\times10^{6} = 0.$$

The dimensional changes of the graphite-reinforced element are:

$$\delta\Delta_x = \varepsilon_x\Delta_x = \left(1000\times10^{-6}\right)(50) = 0.0500\ \text{mm}$$
$$\delta\Delta_y = \varepsilon_y\Delta_y = \left(-248\times10^{-6}\right)(50) = -0.0124\ \text{mm}. \qquad (8.48)$$

The deformed dimensions are:

$$\Delta_x + \delta\Delta_x = 50.0500\ \text{mm}$$
$$\Delta_y + \delta\Delta_y = 49.988\ \text{mm}. \qquad (8.49)$$

Figure 8.3d shows the deformed shape of the graphite-reinforced element, and the deformation is similar to that of the aluminum; the element stretches in the x direction and contracts in the y direction (and in the z direction), and all right corner angles remain right.

Now consider a square element of graphite-reinforced material with the fibers oriented at $\theta = 30°$ with respect to the x axis and also subjected to a stress, as shown in Figure 8.3e, $\sigma_x = 155\,\text{MPa}$. The strains are determined by Eq. (8.25) as:

$$\varepsilon_x = \bar{S}_{11}(30°)\sigma_x$$
$$\varepsilon_y = \bar{S}_{12}(30°)\sigma_x \qquad (8.50)$$
$$\gamma_{xy} = \bar{S}_{16}(30°)\sigma_x.$$

Using $m = \sqrt{3}/2$ and $n = 1/2$, Eqs. (8.26) and (8.46) give:

$$\bar{S}_{11}(30°) = 50.8\ (\text{TPa})^{-1} \quad \bar{S}_{12}(30°) = -26.9 \quad \bar{S}_{16}(30°) = -62.2$$
$$\bar{S}_{22}(30°) = 88.9 \quad \bar{S}_{26}(30°) = -3.77 \quad \bar{S}_{66}(30°) = 126.0. \qquad (8.51)$$

From Eq. (8.50),

$$\varepsilon_x = 50.8\times10^{-12}\times155\times10^{6} = 7880\ \mu\text{mm mm}^{-1}$$

$$\varepsilon_y = -26.9\times10^{-12}\times155\times10^{6} = -4170\ \mu\text{mm mm}^{-1} \tag{8.52}$$

$$\gamma_{xy} = -62.2\times10^{-12}\times155\times10^{6} = -9640\ \mu\text{rad} = -0.553^\circ.$$

With the above numbers, dimensional changes of the graphite-reinforced element become:

$$\delta\Delta_x = \varepsilon_x\Delta_x = (0.00787)(50) = 0.394\ \text{mm} \tag{8.53}$$

$$\delta\Delta_y = \varepsilon_y\Delta_y = (-0.00417)(50) = -0.209\ \text{mm}. \tag{8.54}$$

Unlike the previous two cases, however, the original right corner angles do not remain right. The change in right angle is given by the value of γ_{xy} in Eq. (8.52), namely, −9640 μrad, or −0.553°. The deformed dimensions of the element are:

$$\Delta_x + \delta\Delta_x = 50.394\ \text{mm}$$

$$\Delta_y + \delta\Delta_y = 49.791\ \text{mm}. \tag{8.55}$$

Figure 8.3f illustrates the deformed shape of the element. It is important to properly interpret the sign of γ_{xy}. A positive γ_{xy} means that the right angle between two line segments emanating from the origin – one line segment starting from the origin and extending in the +x direction and the other line segment starting from the origin and extending in the +y direction – decreases. Because γ_{xy} in the above example is negative, the angle in the lower-left-hand corner of the element increases.

As a final example of the effects of tension normal stress in the x direction, consider an element of graphite-reinforced composite with the fibers oriented at $\theta = -30^\circ$ relative to the +x axis, as shown in Figure 8.3g. This example illustrates one of the important characteristics of the $\bar{S}_{ij}$ as regards their dependence on θ. In this situation, the stress–strain relations of Eq. (8.25) become:

$$\varepsilon_x = \bar{S}_{11}(-30^\circ)\sigma_x$$

$$\varepsilon_y = \bar{S}_{12}(-30^\circ)\sigma_x \tag{8.56}$$

$$\gamma_{xy} = \bar{S}_{16}(-30^\circ)\sigma_x.$$

An inspection of the definitions of the off-axis compliance in Eq. (8.26) reveals that $\bar{S}_{16}$ and $\bar{S}_{26}$ are odd functions of n, and hence *of* θ, while the remaining $\bar{S}_{ij}$ are even functions of θ. Therefore:

$$\bar{S}_{11}(-30^\circ) = +\bar{S}_{11}(+30^\circ) = 50.8\ (\text{TPa})^{-1}$$

$$\bar{S}_{12}(-30^\circ) = +\bar{S}_{12}(+30^\circ) = -26.9$$

$$\bar{S}_{16}(-30°) = -\bar{S}_{16}(+30°) = 62.2$$

$$\bar{S}_{22}(-30°) = +\bar{S}_{22}(+30°) = 88.9$$

$$\bar{S}_{26}(-30°) = -\bar{S}_{26}(+30°) = 3.77$$

$$\bar{S}_{66}(-30°) = +\bar{S}_{66}(+30°) = 126.0. \tag{8.57}$$

Substituting into Eq. (8.56) results in:

$$\varepsilon_x = 50.8 \times 10^{-12} \times 155 \times 10^6 = 7880\ \mu\text{mm mm}^{-1}$$

$$\varepsilon_y = -26.9 \times 10^{-12} \times 155 \times 10^6 = -4170\ \mu\text{mm mm}^{-1}$$

$$\gamma_{xy} = 62.2 \times 10^{-12} \times 155 \times 10^6 = 9640\ \mu\text{rad} = 0.553°. \tag{8.58}$$

With these numbers, dimensional changes of the −30° graphite-reinforced element become:

$$\delta\Delta_x = \varepsilon_x\Delta_x = (0.00787)(50) = 0.394\ \text{mm}$$

$$\delta\Delta_y = \varepsilon_y\Delta_y = (-0.00417)(50) = -0.209\ \text{mm}. \tag{8.59}$$

Like the +30° case, the original right corner angles do not remain right and the change in right angle is given by the value of γ_{xy} in Eq. (8.58), namely, 9640 μrad, or 0.553°. The deformed dimensions of the element are the same as the +30° case, namely:

$$\Delta_x + \delta\Delta_x = 50.394\ \text{mm}$$

$$\Delta_y + \delta\Delta_y = 49.791\ \text{mm}. \tag{8.60}$$

Figure 8.3h illustrates the deformed shape of the element. It is important to note that the change in the right corner angle for the −30° case is opposite the change for the +30° case. This ability to change the sign of the deformation by changing the fiber angle is a very important characteristic of fiber-reinforced composite materials. Here, the sign change of $\bar{S}_{16}$ was responsible for the sign of the change in right angle. Because it is also an odd function of θ, $\bar{S}_{26}$ changes sign with θ, and in other situations, it can be responsible for controlling the change in sign of a deformation. In more complicated loadings, specifically with stress components σ_x and σ_y both present, both $\bar{S}_{16}$ and $\bar{S}_{26}$ control the sign of the deformation. $\bar{S}_{16}$ and $\bar{S}_{26}$ serve "double duty" in that they couple normal stresses to shear deformation, and they couple the shear stress to extensional deformations [5,6]. Another series of examples will illustrate this latter coupling and further illustrate the influence of the sign dependence of $\bar{S}_{16}$ and $\bar{S}_{26}$ on θ. The series will again start with an element of aluminum and progress through an element of graphite-reinforced material. This progression, though adding nothing to what we already know about the behavior of aluminum, is taken specifically to show the contrasts, and in some cases the similarities, in the response of fiber-reinforced composites and isotropic materials.

Consider, as shown in Figure 8.4a, a 50 × 50 mm square of aluminum loaded by a 4.40 MPa shear stress τ_{xy}. Of interest are the deformations caused by the application of this shear stress. Because σ_x and σ_y are zero, the stress–strain relations of Eq. (8.25) reduce to:

$$\varepsilon_x = \bar{S}_{16}\tau_{xy}$$

$$\varepsilon_y = \bar{S}_{26}\tau_{xy} \tag{8.61}$$

$$\gamma_{xy} = \bar{S}_{66}\tau_{xy}.$$

For aluminum, $\bar{S}_{16}$ and $\bar{S}_{26}$ were shown to be zero, in Eq. (8.28), and as a result:

$$\varepsilon_x = 0 \times \tau_{xy}$$

$$\varepsilon_y = 0 \times \tau_{xy} \tag{8.62}$$

$$\gamma_{xy} = \bar{S}_{66}\tau_{xy} = \frac{1}{G}\tau_{xy}.$$

Using the value of shear modulus for aluminum from Table 5.1 yields:

$$\varepsilon_x = 0 \times 4.40 \times 10^6 = 0$$

$$\varepsilon_y = 0 \times 4.40 \times 10^6 = 0 \tag{8.63}$$

$$\gamma_{xy} = \bar{S}_{66}\tau_{xy} = \frac{1}{27.80 \times 10^9} \times 4.40 \times 10^6 = 158\ \mu\text{rad},$$

confirming our experience with aluminum that only a shear deformation results; the angle in the lower left corner decreases by 158.0 μrad, or 0.00905°. The lengths of the sides of the deformed element are still exactly 50 mm. Figure 8.4b shows the deformed shape.

Applying a shear stress τ_{xy} to an element of graphite-reinforced composite with the fibers aligned with the x axis, as shown in Figure 8.4c, leads to:

$$\varepsilon_x = \bar{S}_{16}(0°)\tau_{xy}$$

$$\varepsilon_y = \bar{S}_{26}(0°)\tau_{xy} \tag{8.64}$$

$$\gamma_{xy} = \bar{S}_{66}(0°)\tau_{xy}.$$

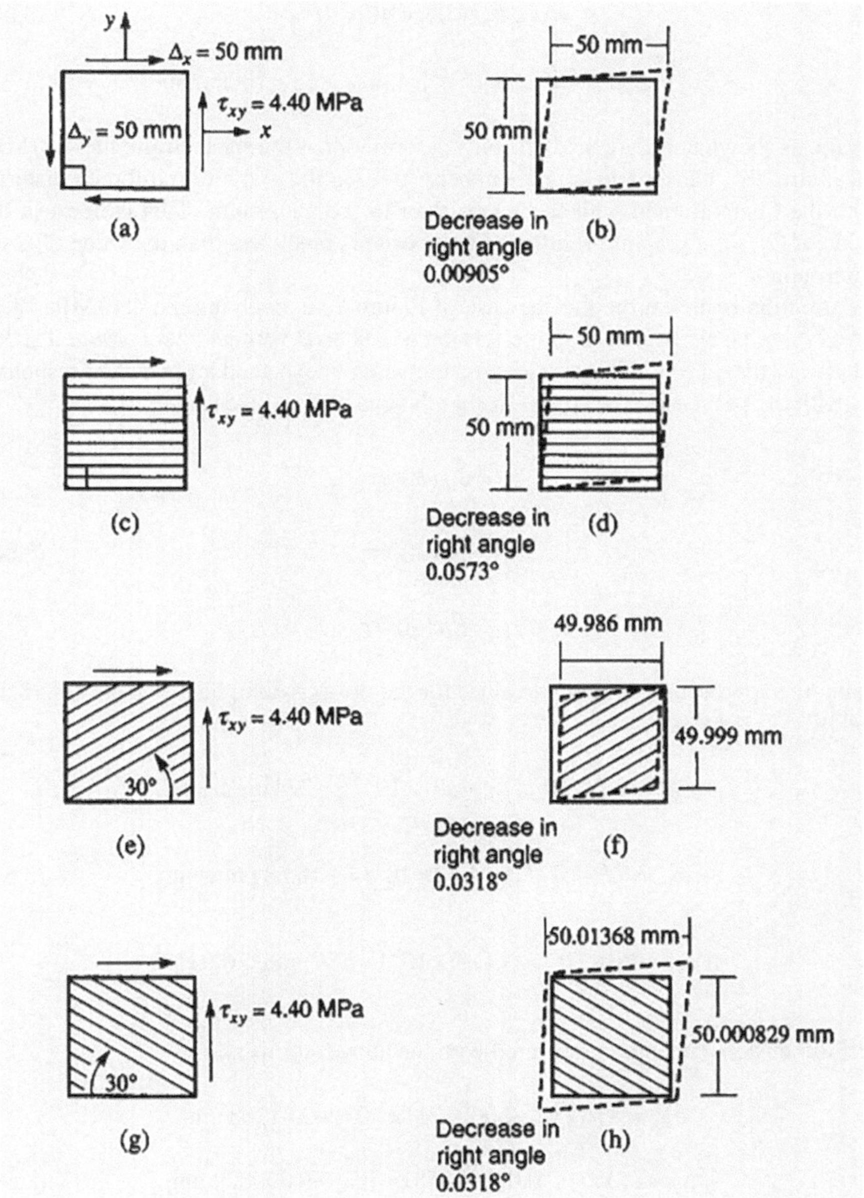

FIGURE 8.4 (a–h) Comparison of deformations of aluminum and graphite-reinforced composite under a shear stress τ_{xy}.

Using the compliances of Eq. (8.27) and the numerical values from Eq. (8.46) leads to:

$$\varepsilon_x = 0 \times \tau_{xy} = 0 \times 4.40 \times 10^6 = 0$$

$$\varepsilon_y = 0 \times \tau_{xy} = 0 \times 4.40 \times 10^6 = 0 \tag{8.65}$$

$$\gamma_{xy} = \bar{S}_{66}\tau_{xy} = \left(227 \times 10^{-12}\right)\left(4.40 \times 10^6\right) = 1000 \ \mu\text{rad}.$$

Again, as shown in Figure 8.4d, the only deformation is the shear strain; the 4.40 MPa shear stress τ_{xy} causes a much larger shear strain in the graphite-reinforced material with the fibers aligned with the x axis than in the aluminum. This is because the value of G_{12} for a graphite-reinforced composite is much less than the value of G for aluminum.

Attention is now turned to the case of Figure 8.4e, applying the 4.40 MPa shear stress τ_{xy} to an element of graphite-reinforced material with its fibers oriented at 30° relative to the $+x$ axis. This situation results in an unexpected and unusual response. As with the past cases, the stress–strain relations of Eq. (8.25) result in:

$$\varepsilon_x = \bar{S}_{16}(30°)\tau_{xy}$$

$$\varepsilon_y = \bar{S}_{26}(30°)\tau_{xy} \tag{8.66}$$

$$\gamma_{xy} = \bar{S}_{66}(30°)\tau_{xy}.$$

Using the appropriate numerical values for the off-axis compliances from Eq. (8.51) yields:

$$\varepsilon_x = \left(-62.2 \times 10^{-12}\right) \times \left(4.40 \times 10^6\right) = -274 \ \mu\text{mm mm}^{-1}$$

$$\varepsilon_y = \left(-3.77 \times 10^{-12}\right) \times \left(4.40 \times 10^6\right) = -16.58 \ \mu\text{mm mm}^{-1} \tag{8.67}$$

$$\gamma_{xy} = (126 \times 10^{-12} \times \left(4.40 \times 10^6\right) = 555 \ \mu\text{rad} = 0.0318°$$

The dimensional changes associated with the above strains are:

$$\delta\Delta_x = \varepsilon_x\Delta_x = (-0.000274) \times 50 = -0.01368 \ \text{mm}$$

$$\delta\Delta_y = \varepsilon_y\Delta_y = (-0.00001658) \times 50 = -0.000829 \ \text{mm}. \tag{8.68}$$

The shear strain is positive so the right corner angle in the lower-left-hand corner of the element decreases by 555 μrad, or 0.0318°. Figure 8.4f illustrates the deformed shape of the element; the lengths of the sides are given by:

$$\Delta_x + \delta\Delta_x = 49.986 \ \text{mm}$$

$$\Delta_y + \delta\Delta_y = 49.999 \ \text{mm}. \tag{8.69}$$

Consider the element of graphite-reinforced composite with the fibers oriented at −30° relative to the +x axis, as shown in Figure 8.4 (g). With an applied stress of $\tau_{xy} = 4.4\,\text{MPa}$, the stress–strain relations of Eq. (8.25) become:

$$\varepsilon_x = \bar{S}_{16}(-30°)\tau_{xy}$$

$$\varepsilon_y = \bar{S}_{26}(-30°)\tau_{xy} \tag{8.70}$$

$$\gamma_{xy} = \bar{S}_{66}(-30°)\tau_{xy}.$$

Using numerical values from Eq. (8.57):

$$\varepsilon_x = \left(62.2\times10^{-12}\right)\times\left(4.40\times10^{6}\right) = 274\ \mu\text{mm mm}^{-1}$$

$$\varepsilon_y = \left(3.77\times10^{-12}\right)\times\left(4.40\times10^{6}\right) = 16.58\ \mu\text{mm mm}^{-1} \tag{8.71}$$

$$\gamma_{xy} = (126\times10^{-12}\times\left(4.40\times10^{6}\right) = 555\ \mu\text{rad} = 0.0318°$$

These numbers indicate that with the fibers at $\theta = -30°$, the sides of the element increase in length. This is exactly opposite the case with θ = +30°. However, the right angle in the lower-left-hand corner decreases the same as for the $\theta = +30°$ orientation. The simple switching of the fiber angle has a significant influence on the response [7]. Figure 8.4h illustrates the deformed element; the dimensional changes are given by:

$$\delta\Delta_x = \varepsilon_x\Delta_x = (0.000274)\times50 = 0.01368\ \text{mm}$$

$$\delta\Delta_y = \varepsilon_y\Delta_y = (0.00001658)\times50 = 0.000829\ \text{mm}. \tag{8.72}$$

The new dimensions are:

$$\Delta_x + \delta\Delta_x = 50.01368\ \text{mm}$$

$$\Delta_y + \delta\Delta_y = 50.000829\ \text{mm}. \tag{8.73}$$

In the above series of examples, the stress component from each of the pairs was stipulated – two of the three stress components were zero in all the cases. In all cases, the strains in each of the stress–strain pairs were being sought. By contrast, consider again the 50 mm × 50mm square of graphite-reinforced composite, loaded in tension and with its fibers oriented at −30° with respect to the +x axis. Assume that, instead of being completely free to deform, as in the last examples, the off-axis element is constrained from any deformation in the y direction, as shown in Figure 8.5a. For this situation, σ_x is known to be 155MPa, ε_x is unknown, σ_y is unknown, ε_y is known to be zero, τ_{xy} is known to be zero, and γ_{xy} is unknown. The unknowns involve both stresses and strains, and the known involve both stresses and strains. For this particular situation, the stress–strain relations of Eq. (8.25) become:

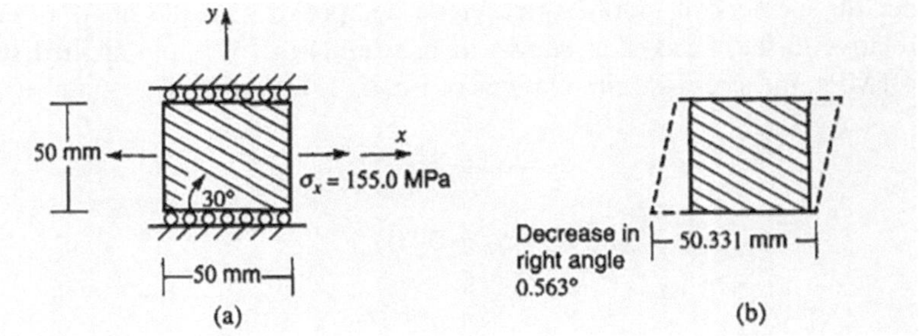

FIGURE 8.5 (a and b) Deformation of a partially constrained off-axis element of graphite-reinforced material under a tensile normal stress σ_x.

$$\varepsilon_x = \bar{S}_{11}\sigma_x + \bar{S}_{12}\sigma_y + \bar{S}_{16} \times 0 \tag{8.74}$$

$$0 = \bar{S}_{12}\sigma_x + \bar{S}_{22}\sigma_y + \bar{S}_{26} \times 0$$

$$\gamma_{xy} = \bar{S}_{16}\sigma_x + \bar{S}_{26}\sigma_y + \bar{S}_{66} \times 0.$$

From the second equation of Eq. (8.56):

$$\sigma_y = -\frac{\bar{S}_{12}}{\bar{S}_{22}}\sigma_x. \tag{8.75}$$

Substituting Eq. (8.75) in Eq. (8.74):

$$\varepsilon_x = \left(\bar{S}_{11} - \frac{\bar{S}_{12}^2}{\bar{S}_{22}}\right)\sigma_x \tag{8.76}$$

$$\gamma_{xy} = \left(\bar{S}_{16} - \frac{\bar{S}_{12}\bar{S}_{26}}{\bar{S}_{22}}\right)\sigma_x.$$

Using the numerical values for graphite-reinforced composite, from Eq. (8.57), we find:
$\varepsilon_x = 6610$ μmm mm^{-1}
$\gamma_{xy} = 9820$ μrad (0.563°)

$$\sigma_y = 46.9 \text{ MPa}. \tag{8.77}$$

According to these calculations, the applied stress in the x direction causes the x dimension to increase by:

$$\delta\Delta_x = \varepsilon_x\Delta_x = (0.00661) \times 50 = 0.331 \text{ mm}. \tag{8.78}$$

The deformed shape of the element is shown in Figure 8.5b, and compared to the case of Figure 8.3g, the addition of the restraint in the y direction decreases the change in length and increases slightly the change in right angle. Eq. (8.19), or its more fundamental form, Eq. (8.18), is the expression for the through-the-thickness strain. If we consider the situation in Figure 8.4e as an example, Eq. (8.19) becomes:

$$\varepsilon_z = 2(S_{13} - S_{23})\sin\theta\cos\theta\tau_{xy}. \tag{8.79}$$

Using numerical values of S_{13} and S_{23} from Eq. (5.56) yields:

$$\varepsilon_z = 2(-1.60 + 37.9)\times 10^{-12}\sin(30°)\cos(30°)\times 4.40\times 10^6 = 138.3\,\mu\text{mm mm}^{-1}. \tag{8.80}$$

The element of material becomes thicker due to the application of the shear stress. The change in thickness of the element of material, Δh, is given by:

$$\Delta h = \varepsilon_z h. \tag{8.81}$$

8.3 TRANSFORMED REDUCED STIFFNESSES

The inverse of the stress–strain relations of Eq. (8.25) can be derived by slightly rewriting the stress–strain relation of Eq. (7.14) to account for the factor of ½ in the shear strain as:

$$\begin{Bmatrix} \sigma_1 \\ \sigma_2 \\ \tau_{12} \end{Bmatrix} = \begin{bmatrix} Q_{11} & Q_{12} & 0 \\ Q_{12} & Q_{22} & 0 \\ 0 & 0 & 2Q_{66} \end{bmatrix} \begin{Bmatrix} \varepsilon_1 \\ \varepsilon_2 \\ \dfrac{\gamma_{12}}{2} \end{Bmatrix}. \tag{8.82}$$

Substituting Eqs. (8.9) and (8.20) into Eq. (8.82), premultiplying both sides of the resulting equation by $[T]^{-1}$, and multiplying the three matrices together give:

$$\begin{Bmatrix} \sigma_x \\ \sigma_y \\ \tau_{xy} \end{Bmatrix} = \begin{bmatrix} \bar{Q}_{11} & \bar{Q}_{12} & \bar{Q}_{16} \\ \bar{Q}_{12} & \bar{Q}_{22} & \bar{Q}_{26} \\ \bar{Q}_{16} & \bar{Q}_{26} & \bar{Q}_{66} \end{bmatrix} \begin{Bmatrix} \varepsilon_x \\ \varepsilon_y \\ \gamma_{xy} \end{Bmatrix}. \tag{8.83}$$

The $\bar{Q}_{ij}$ are called the *transformed reduced stiffnesses*, and sometimes the *off-axis reduced stiffnesses*, and they are defined by:

$$\bar{Q}_{11} = Q_{11}m^4 + 2(Q_{12} + 2Q_{66})n^2m^2 + Q_{22}n^4$$

$$\bar{Q}_{12} = (Q_{11} + Q_{22} - 4Q_{66})n^2m^2 + Q_{12}\left(n^4 + m^4\right)$$

$$\bar{Q}_{16} = (Q_{11} - Q_{12} - 2Q_{66})nm^3 + (Q_{12} - Q_{22} + 2Q_{66})mn^3$$

$$\bar{Q}_{22} = Q_{11}n^4 + 2(Q_{12} + 2Q_{66})n^2m^2 + Q_{22}m^4$$

$$\bar{Q}_{26} = (Q_{11} - Q_{12} - 2Q_{66})mn^3 + (Q_{12} - Q_{22} + 2Q_{66})nm^3$$

$$\bar{Q}_{66} = (Q_{11} + Q_{22} - 2Q_{12} - 2Q_{66})n^2m^2 + Q_{66}(n^4 + m^4). \tag{8.84}$$

As a parallel to the series of examples presented in the discussion of the reduced compliance matrix, a similar series of examples will next be discussed to illustrate the physical implications of the terms in the $\bar{Q}$ matrix. Shear–extension coupling and sign sensitivity of the $\bar{Q}_{16}$ and $\bar{Q}_{26}$ terms will again be evident. The particular examples can be considered the complement of the examples presented previously. They are termed "complementary" examples because the strain variable in each stress–strain pair is specified, whereas before, the stress variable of the pair was specified [8].

A 50 mm × 50 mm element of aluminum, as shown in Figure 8.6a, is stretched 0.050 mm in the x direction. The y dimension does not change, and the right corner angles remain right. Interest focuses on the stresses required to effect this deformation. Assume the element is in a state of plane stress. Note the complementary nature of this problem relative to the problem discussed in Figure 8.3a. In the present situation, the strains of the stress–strain pairs are known and it is the stresses that are unknown and to be solved for. In the situation of Figure 8.3a, the stresses of the stress–strain pairs were known and it was the strains that were unknown and to be solved for. Specifically, in Figure 8.3a, σ_x was the only nonzero stress.

$$\varepsilon_x = \frac{\delta\Delta_x}{\Delta_x} = \frac{0.05}{50} = 1000 \times 10^{-6}$$

$$\varepsilon_y = \frac{\delta\Delta_y}{\Delta_y} = \frac{0}{50} = 0 \tag{8.85}$$

$$\gamma_{xy} = 0.$$

To compute the stresses required to produce this deformation, the stress–strain relations of Eq. (8.83) can be used, resulting in:

$$\sigma_x = \bar{Q}_{11}\varepsilon_x$$

$$\sigma_y = \bar{Q}_{12}\varepsilon_x$$

$$\tau_{xy} = \bar{Q}_{16}\varepsilon_x. \tag{8.86}$$

Because aluminum is isotropic, Eq. (8.86) becomes:

$$\sigma_x = \frac{E}{1-\nu^2}\varepsilon_x$$

$$\sigma_y = \frac{\nu E}{1-\nu^2}\varepsilon_x$$

$$\tau_{xy} = 0 \times \varepsilon_x = 0. \tag{8.87}$$

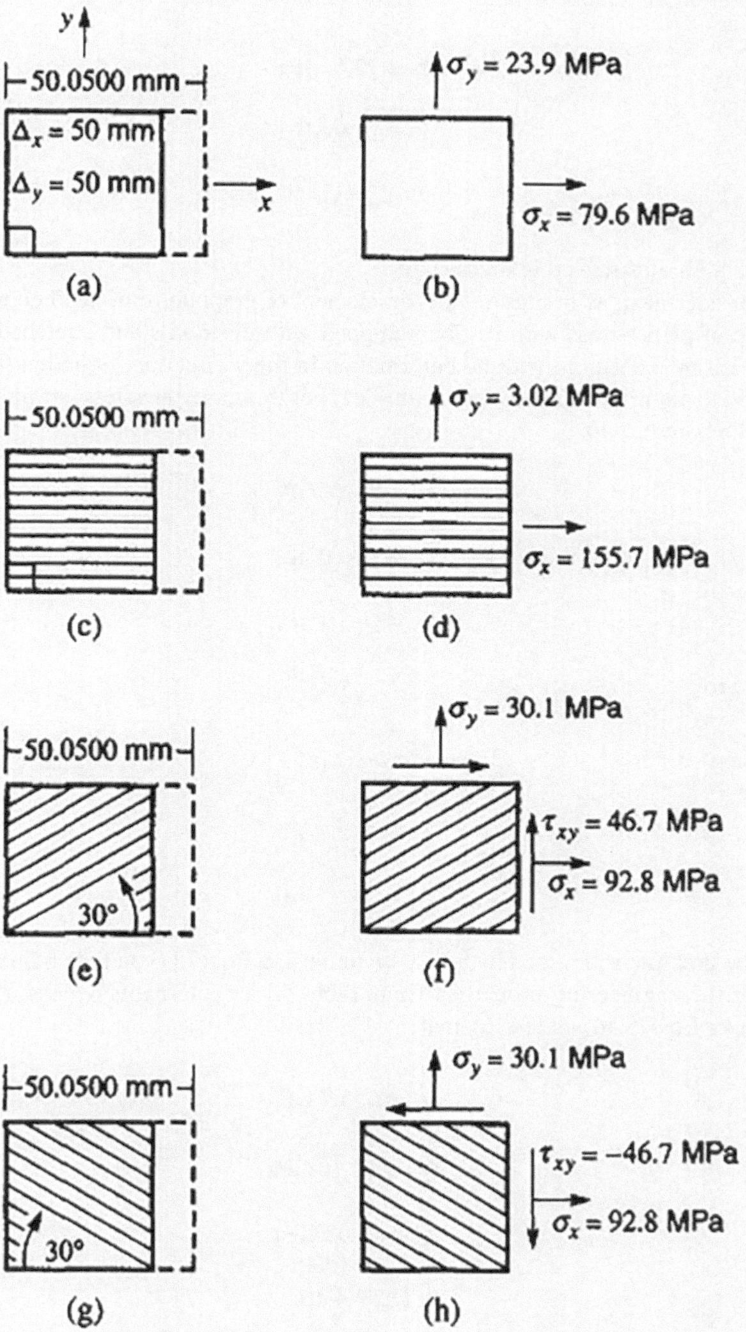

FIGURE 8.6 (a–h) Stress required to produce extensional strain in aluminum and graphite-reinforced material.

Substituting in numerical values for aluminum from Table 5.1 yields:

$$\sigma_x = 79.6 \text{ MPa}$$

$$\sigma_y = 23.9 \text{ MPa}$$

$$\tau_{xy} = 0. \tag{8.88}$$

Figure 8.6b illustrates this stress state.

Consider next, as in Figure 8.6c, an element of graphite-reinforced composite in a state of plane stress with its fibers aligned with the x axis and stretched in the x direction by 0.050 mm, with no deformation in the y direction. Equation (8.85) just applied to aluminum defines the strain state. For this case, the stress–strain relations, Eq. (8.83), reduce to:

$$\sigma_x = \bar{Q}_{11}(0°)\varepsilon_x$$

$$\sigma_y = \bar{Q}_{12}(0°)\varepsilon_x$$

$$\tau_{xy} = \bar{Q}_{16}(0°)\varepsilon_x. \tag{8.89}$$

Therefore,

$$\sigma_x = Q_{11}\varepsilon_x$$

$$\sigma_y = Q_{12}\varepsilon_x$$

$$\tau_{xy} = 0 \times \varepsilon_x = 0. \tag{8.90}$$

For the graphite-reinforced material, we find using Eq. (7.17) and the numerical values of the engineering properties from Table 5.1, or alternatively, Eqs. (5.57) and (7.15), or Eqs. (5.56) and (7.16), that:

$$Q_{11} = 155.7 \text{ GPa}$$

$$Q_{22} = 12.16 \text{ GPa}$$

$$Q_{12} = 3.02 \text{ GPa}$$

$$Q_{66} = 4.40. \tag{8.91}$$

Using the applicable stiffness from Eq. (8.91), the stresses given by Eq. (8.90) are:

$$\sigma_x = 155.7 \text{ MPa}$$

$$\sigma_y = 3.02 \text{ MPa}$$

$$\tau_{xy} = 0. \tag{8.92}$$

Consider now the situation in Figure 8.6e, a 50 mm × 50 mm element of graphite-reinforced material with the fibers oriented at +30° with respect to the +x axis and stretched 0.050 mm in the x direction. The strains are again given by Eq. (8.85), and the stresses required to produce these strains are:

$$\sigma_x = \bar{Q}_{11}(30°)\varepsilon_x$$

$$\sigma_y = \bar{Q}_{12}(30°)\varepsilon_x$$

$$\tau_{xy} = \bar{Q}_{16}(30°)\varepsilon_x. \tag{8.93}$$

Using Eqs. (8.84) and (8.91):

$$\bar{Q}_{11}(30°) = 92.8 \text{ GPa}$$

$$\bar{Q}_{12}(30°) = 30.1$$

$$\bar{Q}_{16}(30°) = 46.7$$

$$\bar{Q}_{22}(30°) = 21.0$$

$$\bar{Q}_{26}(30°) = 15.5$$

$$\bar{Q}_{66}(30°) = 31.5. \tag{8.94}$$

Using these numerical values, we find that the stresses required to produce the prescribed deformations are:

$$\sigma_x = 92.8 \text{ MPa}$$

$$\sigma_y = 30.1 \text{ MPa}$$

$$\tau_{xy} = 46.7 \text{ MPa}. \tag{8.95}$$

Figure 8.6f illustrates the stresses for this example.

As a final example of the stresses required to effect a simple elongation in the x direction, consider an element of graphite-reinforced material with the fibers oriented in the −30° direction, as shown in Figure 8.6g. For this situation:

$$\sigma_x = \bar{Q}_{11}(-30^\circ)\varepsilon_x$$

$$\sigma_y = \bar{Q}_{12}(-30^\circ)\varepsilon_x$$

$$\tau_{xy} = \bar{Q}_{16}(-30^\circ)\varepsilon_x. \tag{8.96}$$

From Eq. (8.94):

$$\bar{Q}_{11}(-30^\circ) = \bar{Q}_{11}(+30^\circ) = 92.8 \text{ GPa}$$

$$\bar{Q}_{12}(-30^\circ) = \bar{Q}_{12}(+30^\circ) = 30.1$$

$$\bar{Q}_{16}(-30^\circ) = -\bar{Q}_{16}(+30^\circ) = -46.7$$

$$\bar{Q}_{22}(-30^\circ) = \bar{Q}_{22}(+30^\circ) = 21.0$$

$$\bar{Q}_{26}(-30^\circ) = -\bar{Q}_{26}(+30^\circ) = -15.5$$

$$\bar{Q}_{66}(-30^\circ) = \bar{Q}_{66}(+30^\circ) = 31.5. \tag{8.97}$$

Substituting Eq. (8.97) into Eq. (8.96):

$$\sigma_x = 92.8 \text{ MPa}$$

$$\sigma_y = 30.1 \text{ MPa}$$

$$\tau_{xy} = -46.7 \text{ MPa}. \tag{8.98}$$

Figure 8.6h illustrates the stresses of Eq. (8.98).

Nevertheless, ε_z is not zero and it can be calculated. Consider the example of Figure 8.6e, where the element of fiber-reinforced material with its fibers oriented at +30° relative to the +x axis is stretched in the x direction at 0.050 mm. No deformation is allowed in the y direction or in shear. The resulting stresses were given by Eq. (8.95) and are illustrated in Figure 8.6f. Equation (8.19) provides us with the strain in the z direction. Specifically, using $\theta = +30^\circ$, the stresses from Eq. (8.95), and the material properties for S_{13} and S_{23} from Eq. (5.56), we find:

$$\varepsilon_z = -389 \times 10^{-6}. \tag{8.99}$$

Finally, to round out the discussion of the transformed reduced stiffness, consider the situation in Figure 8.7a, which shows a pure shearing deformation being prescribed for the 50 mm × 50 mm square element of aluminum, with the right corner angle in the lower-left-hand corner decreasing by 1000 μrad, or 0.0573°. With a pure shearing deformation, the lengths of the sides do not change.

The prescribed strain state is thus given by:

$$\varepsilon_x = 0$$

$$\varepsilon_y = 0 \tag{8.100}$$

$$\gamma_{xy} = 1000 \times 10^{-6}.$$

It is assumed the element is in a state of plane stress; that is, σ_z, τ_{xz}, and τ_{yz} are zero. With this, the stress–strain relations of Eq. (8.83) reduce to:

$$\sigma_x = \bar{Q}_{16}\gamma_{xy}$$

$$\sigma_y = \bar{Q}_{26}\gamma_{xy} \tag{8.101}$$

$$\tau_{xy} = \bar{Q}_{66}\gamma_{xy}.$$

For aluminum:

$$\sigma_x = 0 \times \gamma_{xy} = 0$$

$$\sigma_y = 0 \times \gamma_{xy} = 0 \tag{8.102}$$

$$\tau_{xy} = G\gamma_{xy}.$$

Using Table 5.1:

$$\sigma_x = 0$$

$$\sigma_y = 0 \tag{8.103}$$

$$\tau_{xy} = 27.8 \text{ MPa}.$$

For isotropic materials, only a shear stress is required to produce a prescribed shear strain, as shown in Figure 8.7b.

Prescribing the same pure shear strain state on an element of graphite-reinforced material with its fiber aligned with the x axis, as shown in Figure 8.7c, results in, from Eq. (8.83):

$$\sigma_x = \bar{Q}_{16}(0^\circ)\gamma_{xy}$$

$$\sigma_y = \bar{Q}_{26}(0^\circ)\gamma_{xy} \tag{8.104}$$

$$\tau_{xy} = \bar{Q}_{66}(0^\circ)\gamma_{xy}.$$

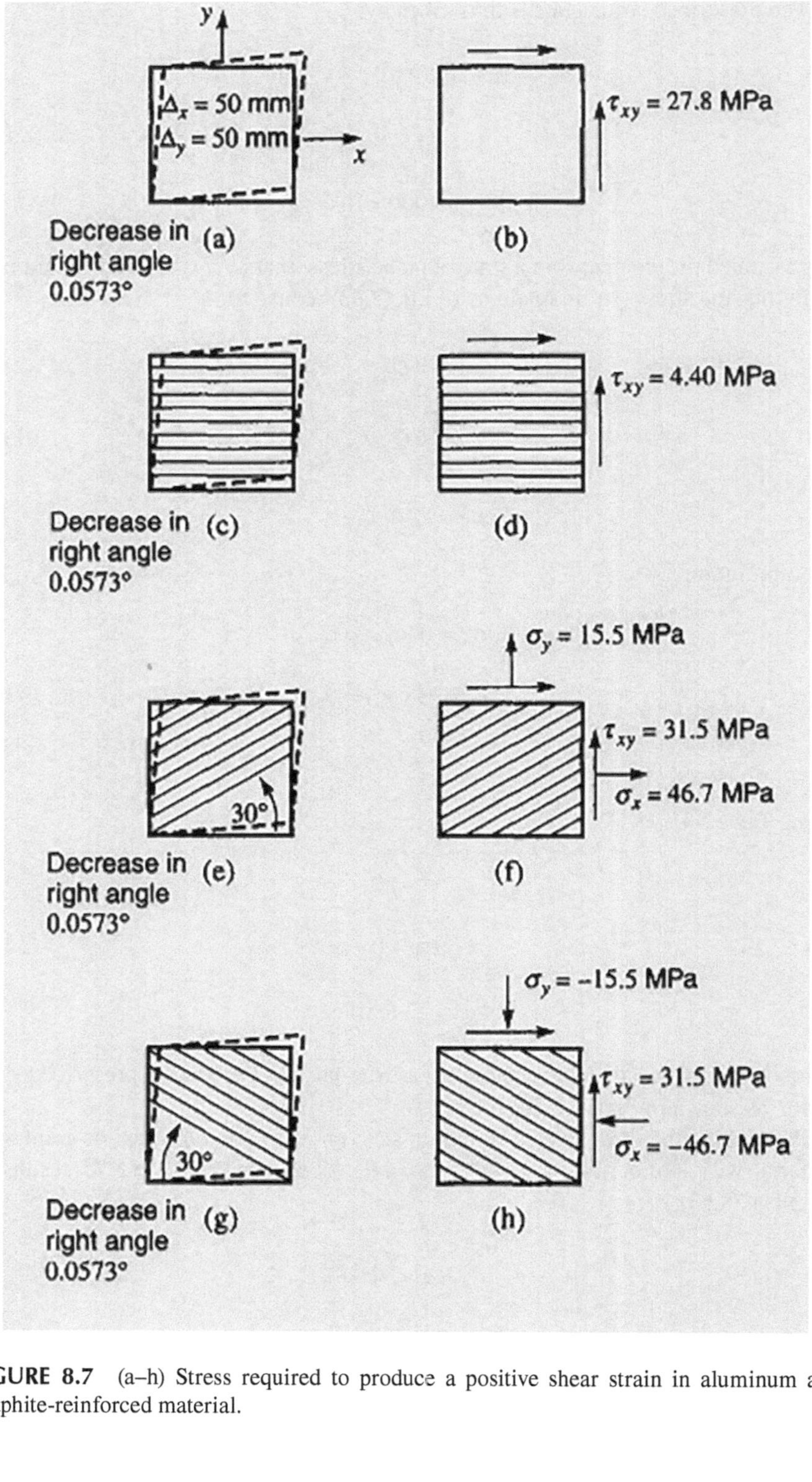

FIGURE 8.7 (a–h) Stress required to produce a positive shear strain in aluminum and graphite-reinforced material.

Using Eq. (8.91):

$$\sigma_x = 0$$
$$\sigma_y = 0 \qquad (8.105)$$
$$\tau_{xy} = 4.40 \text{ MPa}.$$

Due to the difference in values between G for aluminum and G_{12}, the level of shear stress required to produce γ_{xy} = 1000 μrad in the graphite-reinforced material with its fibers aligned with the x axis is much less than the level of shear stress required to produce that same shear strain in the aluminum, as Figure 8.7d illustrates. With the fibers oriented at + 30° relative to the + x axis, as shown in Figure 8.7e, the stresses required to maintain the pure shear deformation are given by:

$$\sigma_x = \bar{Q}_{16}(30°)\gamma_{xy}$$
$$\sigma_y = \bar{Q}_{26}(30°)\gamma_{xy} \qquad (8.106)$$
$$\tau_{xy} = \bar{Q}_{66}(30°)\gamma_{xy}.$$

Using Eq. (8.91):

$$\sigma_x = 46.7 \text{ MPa}$$
$$\sigma_y = 15.5 \text{ MPa} \qquad (8.107)$$
$$\tau_{xy} = 31.5 \text{ MPa}.$$

As might be expected by now, Eq. (8.107) leads to the conclusion that to produce a pure shear deformation with the fibers off-axis at +30° requires not only a shear stress, but also tensile stresses σ_x and σ_y, as shown in Figure 8.7f.

As the final example in this series, consider the case with the fibers oriented at –30°, as shown in Figure 8.7g. From Eq. (8.83):

$$\sigma_x = \bar{Q}_{16}(-30°)\gamma_{xy}$$
$$\sigma_y = \bar{Q}_{26}(-30°)\gamma_{xy} \qquad (8.108)$$
$$\tau_{xy} = \bar{Q}_{66}(-30°)\gamma_{xy}.$$

Using Eq. (8.97):

$$\sigma_x = -46.7 \text{ MPa}$$
$$\sigma_y = -15.5 \text{ MPa} \qquad (8.109)$$
$$\tau_{xy} = 31.5 \text{ MPa}.$$

8.4 ENGINEERING PROPERTIES IN GLOBAL COORDINATES

Engineering properties can also be defined in the *x*-*y*-*z* global coordinate system. These are often of much more use than, say, the reduced compliances or stiffnesses. The extensional moduli, Poisson's ratios, and the shear modulus may mean considerably more to many designers and engineers, because the physical interpretation of these quantities is well established and understood. The engineering properties in the *x*-*y*-*z* system are derivable directly from their definitions, just as they were in Chapter 5 for the 1-2-3 system. In this section, we shall derive the engineering properties that are important when considering a state of plane stress. They are related to the transformed reduced compliances and hence can ultimately be written in terms of the engineering properties in the 1-2-3 system.

Consider, as shown in Figure 8.8a, an off-axis element of fiber-reinforced composite material in the *x*-*y*-*z* system with its fiber oriented at some angle θ with respect to the *x* axis. The element is subjected to a normal tensile stress σ_x and all other stresses are zero, much like the situation in Figure 8.3e, though here the angle θ and the magnitude of the tensile stress are arbitrary. In response to this applied stress, the element stretches in the *x* direction and contracts in the *y* direction, and because the fibers are not aligned with the *x* axis, the right corner angles do not remain right. The extensional strain in the *x* direction is related to the stress in the *x* direction by the extensional modulus in the *x* direction, E_x; the relation between these two quantities is, by definition:

$$\varepsilon_x = \frac{\sigma_x}{E_x}. \tag{8.110}$$

Also:

$$\varepsilon_x = \bar{S}_{11}\sigma_x. \tag{8.111}$$

Comparing Eqs. (8.110) and (8.111):

$$\frac{1}{E_x} = \bar{S}_{11}. \tag{8.112}$$

The extensional modulus in the *x* direction becomes:

$$E_x = \frac{1}{\bar{S}_{11}}. \tag{8.113}$$

In terms of the principal material coordinate system engineering properties:

$$E_x = \frac{E_1}{m^4 + \left(\frac{E_1}{G_{12}} - 2\nu_{12}\right)m^2n^2 + \frac{E_1}{E_2}n^4}. \tag{8.114}$$

In addition to stretching in the *x* direction, the element contracts in the *y* direction when subjected to a tensile stress in the *x* direction (see Figure 8.8a). By definition, the relation between the contraction strain in the *y* direction and the extensional strain

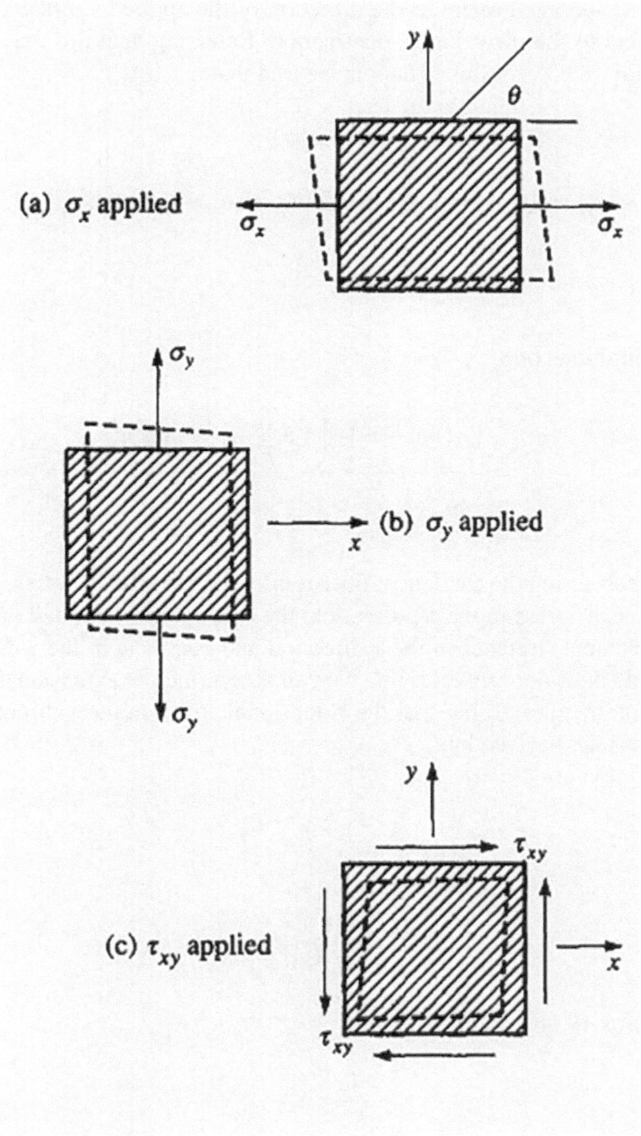

FIGURE 8.8 (a–c) Definition of engineering properties for an off-axis element with simple stress states.

in the x direction, due to simply a tensile stress in the x direction, is a Poisson's ratio. Retaining the convention established for the subscripting of the various Poisson's ratios, we define:

$$\nu_{xy} = -\frac{\varepsilon_y}{\varepsilon_x}, \tag{8.115}$$

where the first subscript refers to the direction of the applied stress and the second subscript refers to the direction of contraction. Referring again to the strain–stress relations of Eq. (8.25), for this situation, we find that:

$$\varepsilon_y = \bar{S}_{12}\sigma_x. \tag{8.116}$$

Using Eqs. (8.111) and (5.116) in the definition of ν_{xy}, we obtain:

$$\nu_{xy} = -\frac{\bar{S}_{12}}{\bar{S}_{11}}. \tag{8.117}$$

Using the definitions of $\bar{S}_{ij}$:

$$\nu_{xy} = \frac{\nu_{12}\left(m^4 + n^4\right) - \left(1 + \frac{E_1}{E_2} - \frac{E_1}{G_{12}}\right)m^2n^2}{m^4 + \left(\frac{E_1}{G_{12}} - 2\nu_{12}\right)m^2n^2 + \frac{E_1}{E_2}n^2}. \tag{8.118}$$

If instead of subjecting the element of fiber-reinforced material to a stress σ_x, it is subjected to a tensile stress in the σ_y direction, the conditions illustrated in Figure 8.8b result. The element stretches in the *y* direction and contracts in the *x* direction, and the right angle does not remain right. We can determine the extensional modulus in the *y* direction by considering that the extensional strain in the *y* direction, for this simple stress state, is given by:

$$\varepsilon_y = \frac{\sigma_y}{E_y}. \tag{8.119}$$

Also:

$$\varepsilon_y = \bar{S}_{22}\sigma_y. \tag{8.120}$$

Comparing Eqs. (8.110) and (8.111):

$$\frac{1}{E_y} = \bar{S}_{22}. \tag{8.121}$$

The extensional modulus in the *x* direction becomes:

$$E_y = \frac{1}{\bar{S}_{22}}. \tag{8.122}$$

In terms of the principal material coordinate system engineering properties:

$$E_y = \frac{E_2}{m^4 + \left(\frac{E_2}{G_{12}} - 2\nu_{21}\right)m^2n^2 + \frac{E_2}{E_1}n^4}. \tag{8.123}$$

Poisson's ratio, due to the stress in the y direction, is:

$$\nu_{yx} = -\frac{\varepsilon_x}{\varepsilon_y} \tag{8.124}$$

$$\varepsilon_x = \bar{S}_{12}\sigma_y. \tag{8.125}$$

Thus,

$$\nu_{yx} = -\frac{\bar{S}_{12}}{\bar{S}_{22}}. \tag{8.126}$$

Poisson's ratio can now be written in terms of the engineering properties in the 1-2-3 system as:

$$\nu_{yx} = \frac{\nu_{21}\left(m^4 + n^4\right) - \left(1 + \frac{E_2}{E_1} - \frac{E_2}{G_{12}}\right)m^2n^2}{m^4 + \left(\frac{E_2}{G_{12}} - 2\nu_{21}\right)m^2n^2 + \frac{E_2}{E_1}n^2}. \tag{8.127}$$

Finally, if the element is subjected to a shear stress τ_{xy}, it will deform as shown in Figure 8.8c. The change in the right corner angle in the x-y plane is denoted by γ_{xy}, and by definition, it is related to the applied shear stress by the shear modulus in the x-y plane, namely:

$$\gamma_{xy} = \frac{\tau_{xy}}{G_{xy}}. \tag{8.128}$$

Also:

$$\gamma_{xy} = \bar{S}_{661lso}; \tag{8.129}$$

Thus,

$$G_{xy} = \frac{1}{\bar{S}_{66}}. \tag{8.130}$$

In terms of engineering properties in the 1-2-3 system:

$$G_{xy} = \frac{G_{12}}{n^4 + m^4 + 2\left(2\frac{G_{12}}{E_1}(1 + 2\nu_{12}) + 2\frac{G_{12}}{E_2} - 1\right)m^2n^2}. \tag{8.131}$$

8.5 MUTUAL INFLUENCE COEFFICIENTS

Several other properties can be defined in addition to the engineering properties just defined in the previous section. These properties have their origin in the fact that when an element of fiber-reinforced material with its fiber oriented at some arbitrary angle is subjected to a normal stress, it exhibits shear strain also, and vice versa. By analogy, the *coefficient of mutual influence of the second kind* is defined as the ratio of a shear strain to an extensional strain, given that the element is subjected to only a single normal stress. The *coefficient of mutual influence of the first kind* is defined as the ratio of an extensional strain to a shear strain, given that the element is subjected to only a single shear stress. These coefficients of mutual influence can be thought of as a generalization of Poisson's ratios, as they are defined as ratios of strains.

Formally, one coefficient of mutual influence of the second kind is defined as:

$$\eta_{xy,\ x} \equiv \frac{\gamma_{xy}}{\varepsilon_x} \tag{8.132}$$

When, $\sigma_x \neq 0$ and all other stresses are zero.

Another coefficient of mutual influence of the second kind is defined as:

$$\eta_{xy,\ y} \equiv \frac{\gamma_{xy}}{\varepsilon_y} \tag{8.133}$$

When, $\sigma_y \neq 0$ and all other stresses are zero.

These coefficients relate the shear strains caused by fiber orientation effects and a normal stress to the extensional strain that is a direct result of this normal stress. This normal stress is the only stress present. In terms of the transformed reduced compliances, the coefficients of mutual influence of the second kind are given by:

$$\eta_{xy,\ x} = \frac{\bar{S}_{16}}{\bar{S}_{11}} \tag{8.134}$$

$$\eta_{xy,\ y} = \frac{\bar{S}_{26}}{\bar{S}_{22}}.$$

The coefficients of mutual influence of the first kind are defined as:

$$\eta_{\ x,xy} \equiv \frac{\varepsilon_x}{\gamma_{xy}} \tag{8.135}$$

When, $\tau_{xy} \neq 0$ and all other stresses are zero.

Another coefficient of mutual influence of the first kind is defined as:

$$\eta_{\ y,xy} \equiv \frac{\varepsilon_y}{\gamma_{xy}} \tag{8.136}$$

When, $\tau_{xy} \neq 0$ and all other stresses are zero.

These coefficients relate the extensional strains caused by fiber orientation effects and a shear stress to the shear strain that is a direct result of this shear stress. The shear stress is the only stress present. In terms of transformed reduced compliances:

$$\eta_{\ x,xy} = \frac{\bar{S}_{16}}{\bar{S}_{66}} \tag{8.137}$$

$$\eta_{y,xy} = \frac{\bar{S}_{26}}{\bar{S}_{66}}.$$

8.6 FREE THERMAL AND MOISTURE STRAINS

Using the transformations of Eq. (8.15), the thermally induced or moisture-induced strains can be transformed from the 1-2-3 principal material coordinate system to the x-y-z global coordinate system.

Considering the free thermal strains, the transformations become:

$$\begin{aligned}
\varepsilon_x^T(T, T_{ref}) &= \cos^2\theta\varepsilon_1^T(T, T_{\text{ref}}) + \sin^2\theta\varepsilon_2^T(T, T_{\text{ref}}) \\
\varepsilon_y^T(T, T_{ref}) &= \sin^2\theta\varepsilon_1^T(T, T_{\text{ref}}) + \cos^2\theta\varepsilon_2^T(T, T_{\text{ref}}) \\
\varepsilon_z^T(T, T_{\text{ref}}) &= \varepsilon_3^T(T, T_{\text{ref}}) \\
\gamma_{yz}^T(T, T_{\text{ref}}) &= 0 \\
\gamma_{xz}^T(T, T_{\text{ref}}) &= 0 \\
\frac{1}{2}\gamma_{xy}^T(T, T_{\text{ref}}) &= \left(\varepsilon_1^T(T, T_{\text{ref}}) - \varepsilon_2^T(T, T_{\text{ref}})\right)\sin\theta\cos\theta.
\end{aligned} \tag{8.138}$$

If the strains are assumed to be linearly dependent on the difference between a particular temperature, T, and the reference temperature, T_{ref}, then, if we use Eqs. (5.96) and (5.97), Eq. (8.138) becomes:

$$\begin{aligned}
\varepsilon_x^T(T, T_{\text{ref}}) &= \left(\cos^2\theta\alpha_1 + \sin^2\theta\alpha_2\right)\Delta T \\
\varepsilon_y^T(T, T_{\text{ref}}) &= \left(\sin^2\theta\alpha_1 + \cos^2\theta\alpha_2\right)\Delta T \\
\varepsilon_z^T(T, T_{\text{ref}}) &= \alpha_3\Delta T \\
\gamma_{xy}^T(T, T_{\text{ref}}) &= 2(\alpha_1 - \alpha_2)\sin\theta\cos\theta\,\Delta T.
\end{aligned} \tag{8.139}$$

If the *coefficients of thermal deformation* (CTD) in the x-y-z system are defined to be such that, due to a temperature change ΔT,

$$\begin{aligned}
\varepsilon_x^T(T, T_{\text{ref}}) &= \alpha_x\ \Delta T \\
\varepsilon_y^T(T, T_{\text{ref}}) &= \alpha_y\ \Delta T \\
\varepsilon_z^T(T, T_{\text{ref}}) &= \alpha_z\Delta T \\
\gamma_{xy}^T(T, T_{\text{ref}}) &= \alpha_{xy}\ \Delta T.
\end{aligned} \tag{8.140}$$

The CTD in the *x-y-z* system become:

$$\alpha_x = \cos^2\theta\alpha_1 + \sin^2\theta\alpha_2$$

$$\alpha_y = \sin^2\theta\alpha_1 + \cos^2\theta\alpha_2$$

$$\alpha_{xy} = 2(\alpha_1 - \alpha_2)\sin\theta\cos\theta$$

$$\alpha_z = \alpha_3. \tag{8.141}$$

Though there are no free thermal shear strains in the 1-2-3 system, this is not the case for the *x-y-z* system. Heating or cooling a small element of material with its fibers not aligned with the *x* or *y* axis results in a change in the right angle of the corners in the *x-y* plane. A simple example will serve to underscore the importance of fiber orientation effects on free thermal strains. Consider, as shown in Figure 8.9a, an unconstrained 50 mm × 50 mm × 50 mm off-axis element of graphite-reinforced material with its fibers oriented at 45° relative to the *x* axis. Assume the temperature of the material is increased by 50°C, and the deformed shape of the material and the lengths of the original 50 mm sides are of interest. From Table 5.1:

$$\alpha_1 = -\ 0.01800 \times 10^{-6}\ /\ ^\circ\text{C}$$

$$\alpha_2 = 24.30 \times 10^{-6}\ /\ ^\circ\text{C}. \tag{8.142}$$

The material shrinks in the fiber direction when heated; that is, the diagonal *AC* contracts. On the other hand, the diagonal *BD* expands. Intuitively, then, the corners *A*, *B*, C, and *D* cannot remain orthogonal when the material is heated. As Figure 8.9b shows, corners *A* and *C* must open, while corners *D* and *B* must close. Quantitative information regarding the shape changes can be obtained by using Eq. (8.141) as:

$$\alpha_x = \left(-0.01800\cos^2 45^\circ + 24.30\sin^2 45^\circ\right) \times 10^{-6}$$

$$\alpha_y = (-0.01800\sin^2 45^\circ + 24.30\cos^2 45^\circ) \times 10^{-6}$$

$$\alpha_{xy} = 2(-0.01800 - 24.30)\sin 45^\circ \cos 45^\circ \times 10^{-6}. \tag{8.143}$$

Thus,

$$\alpha_x = \alpha_y = 12.14 \times 10^{-6}\ /\ ^\circ\text{C} \tag{8.144}$$

$$\alpha_{xy} = -24.30 \times 10^{-6}\ /\ ^\circ\text{C}.$$

For $\Delta T = 50^\circ\text{C}$, from Eq. (8.140):

$$\varepsilon_x^T = 607 \times 10^{-6} = \varepsilon_y^T$$

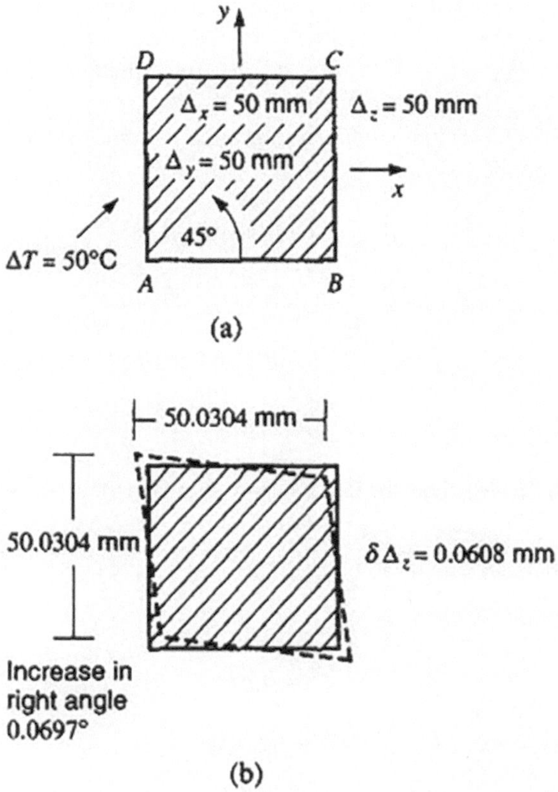

FIGURE 8.9 Thermal deformations of an unconstrained off-axis element. (a) Initial shape, (b) deformed shape.

$$\gamma_{xy}^T = -1216 \times 10^{-6}. \tag{8.145}$$

Using the definition of free thermal strain:

$$\delta\Delta_x^T = \varepsilon_x^T \Delta_x = \left(607 \times 10^{-6}\right) \times 50 = 0.0304 \text{ mm}$$

$$\delta\Delta_y^T = \varepsilon_y^T \Delta_y = \left(607 \times 10^{-6}\right) \times 50 = 0.0304 \text{ mm}. \tag{8.146}$$

The change in the right corner angle in the *x*-*y* plane is:

$$\gamma_{xy}^T = -1216 \times 10^{-6} \text{ rad} = 0.0697°. \tag{8.147}$$

The free thermal strain in the *z* direction is, by Eqs. (5.140) and (5.141):

$$\varepsilon_z = \varepsilon_3 = \alpha_3 \Delta T = \left(24.3 \times 10^{-6}\right) \times 50 = 1215 \times 10^{-6}. \tag{8.148}$$

Thus,

$$\delta\Delta_z = \varepsilon_z \Delta_z = \left(1215 \times 10^{-6}\right) \times 50 = 0.0608 \text{ mm}. \tag{8.149}$$

By analogy to Eq. (8.141), the *coefficients of moisture deformation*, or CMD, in the *x-y-z* system are defined as:

$$\begin{aligned}
\beta_x &= \cos^2\theta\beta_1 + \sin^2\theta\beta_2 \\
\beta_y &= \sin^2\theta\beta_1 + \cos^2\theta\beta_2 \\
\beta_{xy} &= 2(\beta_1 - \beta_2)\sin\theta\cos\theta \\
\beta_z &= \beta_3.
\end{aligned} \tag{8.150}$$

By analogy to Eq. (8.140), then, the free moisture strains in the *x-y-z* system are given by:

$$\begin{aligned}
\varepsilon_x^M &= \beta_x \ \Delta M \\
\varepsilon_y^M &= \beta_y \ \Delta M \\
\varepsilon_z^M &= \beta_3 \Delta M \\
\gamma_{xy}^M &= \beta_{xy} \ \Delta M.
\end{aligned} \tag{8.151}$$

8.7 EFFECTS OF FREE THERMAL AND MOISTURE STRAINS ON PLANE STRESS STRESS–STRAIN RELATIONS IN GLOBAL COORDINATE SYSTEM

Considering Eq. (7.29) first, including the factor of 2 with the shear strain:

$$\begin{Bmatrix} \varepsilon_1 - \alpha_1\Delta T - \beta_1\Delta M \\ \varepsilon_2 - \alpha_2\Delta T - \beta_2\Delta M \\ \dfrac{\gamma_{12}}{2} \end{Bmatrix} = \begin{bmatrix} S_{11} & S_{12} & 0 \\ S_{12} & S_{22} & 0 \\ 0 & 0 & \dfrac{S_{66}}{2} \end{bmatrix} \begin{Bmatrix} \sigma_1 \\ \sigma_2 \\ \tau_{12} \end{Bmatrix}. \tag{8.152}$$

This can be expanded to the form:

$$\begin{Bmatrix} \varepsilon_1 \\ \varepsilon_2 \\ \dfrac{\gamma_{12}}{2} \end{Bmatrix} - \begin{Bmatrix} \alpha_1\Delta T \\ \alpha_2\Delta T \\ \dfrac{0}{2} \end{Bmatrix} - \begin{Bmatrix} \beta_1\Delta M \\ \beta_2\Delta M \\ \dfrac{0}{2} \end{Bmatrix} = \begin{bmatrix} S_{11} & S_{12} & 0 \\ S_{12} & S_{22} & 0 \\ 0 & 0 & \dfrac{S_{66}}{2} \end{bmatrix} \begin{Bmatrix} \sigma_1 \\ \sigma_2 \\ \tau_{12} \end{Bmatrix}. \tag{8.153}$$

The second term on the left-hand side can be rewritten with the aid of Eq. (8.141) in the form:

$$\begin{Bmatrix} \alpha_x \Delta T \\ \alpha_y \Delta T \\ \frac{1}{2}\alpha_{xy}\Delta T \end{Bmatrix} = \begin{bmatrix} \cos^2\theta & \sin^2\theta & -2\sin\theta\cos\theta \\ \sin^2\theta & \cos^2\theta & -2\sin\theta\cos\theta \\ \sin\theta\cos\theta & -\sin\theta\cos\theta & \cos^2\theta - \sin^2\theta \end{bmatrix} \begin{Bmatrix} \alpha_1 \Delta T \\ \alpha_2 \Delta T \\ \frac{0}{2} \end{Bmatrix}. \tag{8.154}$$

Taking the inversion:

$$\begin{Bmatrix} \alpha_1 \Delta T \\ \alpha_2 \Delta T \\ \frac{0}{2} \end{Bmatrix} = [T] \begin{Bmatrix} \alpha_x \Delta T \\ \alpha_y \Delta T \\ \frac{1}{2}\alpha_{xy}\Delta T \end{Bmatrix}. \tag{8.155}$$

Similarly,

$$\begin{Bmatrix} \beta_1 \Delta M \\ \beta_2 \Delta M \\ \frac{0}{2} \end{Bmatrix} = [T] \begin{Bmatrix} \beta_x \Delta M \\ \beta_y \Delta M \\ \frac{1}{2}\beta_{xy}\Delta M \end{Bmatrix}. \tag{8.156}$$

Using Eqs. (8.9), (8.20), (8.155), and (8.156), Eq. (8.153) becomes:

$$[T]\begin{Bmatrix} \varepsilon_x \\ \varepsilon_y \\ \frac{\gamma_{xy}}{2} \end{Bmatrix} - [T]\begin{Bmatrix} \alpha_x \Delta T \\ \alpha_y \Delta T \\ \frac{1}{2}\alpha_{xy}\Delta T \end{Bmatrix} - [T]\begin{Bmatrix} \beta_x \Delta M \\ \beta_y \Delta M \\ \frac{1}{2}\beta_{xy}\Delta M \end{Bmatrix} = \begin{Bmatrix} S_{11} & S_{12} & 0 \\ S_{12} & S_{22} & 0 \\ 0 & 0 & \frac{S_{66}}{2} \end{Bmatrix} \times [T]\begin{Bmatrix} \sigma_x \\ \sigma_y \\ \tau_{xy} \end{Bmatrix}. \tag{8.157}$$

Multiplying both sides by $[T]^{-1}$, combining the terms on the left-hand side, and multiplying the three matrices together on the right side, accounting for the factors of 1/2, lead to:

$$\begin{Bmatrix} \varepsilon_x - \alpha_x \Delta T - \beta_x \Delta M \\ \varepsilon_y - \alpha_y \Delta T - \beta_y \Delta M \\ \gamma_{xy} - \alpha_{xy} \Delta T - \beta_{xy} \Delta M \end{Bmatrix} = \begin{bmatrix} \bar{S}_{11} & \bar{S}_{12} & \bar{S}_{16} \\ S_{12} & \bar{S}_{22} & \bar{S}_{26} \\ \bar{S}_{16} & \bar{S}_{26} & \bar{S}_{66} \end{bmatrix} \begin{Bmatrix} \sigma_x \\ \sigma_y \\ \tau_{xy} \end{Bmatrix}. \tag{8.158}$$

Eq. (8.158) is the off-axis counterpart to Eq. (7.29), where the mechanical strains are given by:

$$\begin{Bmatrix} \varepsilon_x^{\text{mech}} \\ \varepsilon_y^{\text{mech}} \\ \gamma_{xy}^{\text{mech}} \end{Bmatrix} = \begin{Bmatrix} \varepsilon_x - \alpha_x \Delta T - \beta_x \Delta M \\ \varepsilon_y - \alpha_y \Delta T - \beta_y \Delta M \\ \gamma_{xy} - \alpha_{xy} \Delta T - \beta_{xy} \Delta M \end{Bmatrix}. \tag{8.159}$$

The inverse of Eq. (8.158) can be derived as:

$$\begin{Bmatrix} \sigma_x \\ \sigma_y \\ \tau_{xy} \end{Bmatrix} = \begin{bmatrix} \bar{Q}_{11} & \bar{Q}_{12} & \bar{Q}_{16} \\ \bar{Q}_{12} & \bar{Q}_{22} & \bar{Q}_{26} \\ \bar{Q}_{16} & \bar{Q}_{26} & \bar{Q}_{66} \end{bmatrix} \begin{Bmatrix} \varepsilon_x - \alpha_x \Delta T - \beta_x \Delta M \\ \varepsilon_y - \alpha_y \Delta T - \beta_y \Delta M \\ \gamma_{xy} - \alpha_{xy} \Delta T - \beta_{xy} \Delta M \end{Bmatrix}. \tag{8.160}$$

ε_z can be obtained directly from Eq. (7.28), namely:

$$\varepsilon_z = \varepsilon_3 = \alpha_3 \Delta T + \beta_3 \Delta M + S_{13}\sigma_1 + S_{23}\sigma_2. \tag{8.161}$$

If we use the stress transformation Eq. (8.7), Eq. (8.161) becomes:

$$\varepsilon_z = \varepsilon_3 = \alpha_3 \Delta T + \beta_3 \Delta M + \left(S_{13}\cos^2\theta + S_{23}\sin^2\theta\right)\sigma_x + \left(S_{13}\sin^2\theta + S_{23}\cos^2\theta\right)\sigma_y$$
$$+ 2(S_{13} - S_{23})\sin\theta\cos\theta\tau_{xy}. \tag{8.162}$$

The mechanical extensional strain in the z direction is given by:

$$\varepsilon_z^{\text{mech}} = \varepsilon_z - \alpha_z \Delta T + \beta_z \Delta M = \left(S_{13}\cos^2\theta + S_{23}\sin^2\theta\right)\sigma_x + \left(S_{13}\sin^2\theta + S_{23}\cos^2\theta\right)\sigma_y$$
$$+ 2(S_{13} - S_{23})\sin\theta\cos\theta\tau_{xy}. \tag{8.163}$$

Even in the presence of free thermal and free moisture strain effects, for the condition of plane stress there are no shear strains whatsoever in the y-z and x-z planes:

$$\gamma_{yz} = \gamma_{xz} = \gamma_{yz}^T = \gamma_{xz}^T = \gamma_{yz}^{\text{mech}} = \gamma_{xz}^{\text{mech}} = 0 \tag{8.164}$$

and similarly for free moisture strain effects.

As an example, assume that instead of being completely free to deform, the off-axis element in the example of Figure 8.9 is completely restrained in the x-y plane. There is no restraint in the z direction, and thus, the problem is one of plane stress. Because of the constraints:

$$\delta\Delta_x = \delta\Delta_y = 0. \tag{8.165}$$

Because the total strains are given by:

$$\varepsilon_x = \frac{\delta\Delta_x}{\Delta_x}$$

$$\varepsilon_y = \frac{\delta\Delta_y}{\Delta_y}. \tag{8.166}$$

Thus,

$$\varepsilon_x = \varepsilon_y = 0. \tag{8.167}$$

Also,

$$\gamma_{xy} = 0. \tag{8.168}$$

As moisture effects are not present, incorporating the restraint effects of Eqs. (8.167) and (8.168), Eq. (8.160) becomes:

$$\begin{Bmatrix} \sigma_x \\ \sigma_y \\ \tau_{xy} \end{Bmatrix} = \begin{bmatrix} \bar{Q}_{11} & \bar{Q}_{12} & \bar{Q}_{16} \\ \bar{Q}_{12} & \bar{Q}_{22} & \bar{Q}_{26} \\ \bar{Q}_{16} & \bar{Q}_{26} & \bar{Q}_{66} \end{bmatrix} \begin{Bmatrix} -\alpha_x \Delta T \\ -\alpha_y \Delta T \\ -\alpha_{xy} \Delta T \end{Bmatrix}. \tag{8.169}$$

For $\theta = 45°$and for the graphite-reinforced material:

$$\bar{Q}_{11} = \bar{Q}_{22} = 47.9 \text{ GPa}$$

$$\bar{Q}_{12} = 39.1 \text{ GPa}$$

$$\bar{Q}_{16} = \bar{Q}_{26} = 35.9 \text{ GPa}$$

$$\bar{Q}_{66} = 40.5 \text{ GPa}. \tag{8.170}$$

Using the numerical values of α_x, α_y, and α_{xy} from Eq. (8.144), with $\Delta T = 50°C$, we find that Eq. (8.169) yields:

$$\sigma_x = \sigma_y = -9.15 \text{ MPa}$$

$$\tau_{xy} = 5.62 \text{ MPa}. \tag{8.171}$$

For the present constrained problem, using $\theta = 45°$, the values of the stresses from Eq. (8.171), and the values of S_{13} and S_{23} from Eq. (5.56), we compute ε_z from Eq. (8.162) to be:

$$\varepsilon_z = 1780 \times 10^{-6}. \tag{8.172}$$

For this problem, the nonzero mechanical strains are given by:

$$\begin{Bmatrix} \varepsilon_x^{\text{mech}} \\ \varepsilon_y^{\text{mech}} \\ \varepsilon_z^{\text{mech}} \\ \gamma_{xy}^{\text{mech}} \end{Bmatrix} = \begin{Bmatrix} -\alpha_x \Delta T \\ -\alpha_y \Delta T \\ \varepsilon_z - \alpha_z \Delta T \\ \alpha_{xy} \Delta T \end{Bmatrix} = \begin{Bmatrix} -607 \\ -607 \\ 565 \\ -1216 \end{Bmatrix} \times 10^{-6}. \tag{8.173}$$

As a closing example, let us examine another variant of the problem of Figure 8.9. As in Figure 8.10a, let us assume that the 50 mm × 50 mm × 50mm off-axis element of graphite-reinforced material is not completely constrained; rather, it is partially constrained by frictionless rollers from deformation in the y direction, but is otherwise free to deform. The problem is one of plane stress, and because of the constraints:

$$\varepsilon_y = 0. \tag{8.174}$$

Because of the rollers and the lack of contact on the edges perpendicular to the x axis:

$$\tau_{xy} = \sigma_x = 0. \tag{8.175}$$

Using Eq. (8.158) with $\Delta M = 0$ gives:

$$\begin{Bmatrix} \varepsilon_x - \alpha_x \Delta T \\ 0 - \alpha_y \Delta T \\ \gamma_{xy} - \alpha_{xy} \Delta T \end{Bmatrix} = \begin{bmatrix} \bar{S}_{11} & \bar{S}_{12} & \bar{S}_{16} \\ \bar{S}_{12} & \bar{S}_{22} & \bar{S}_{26} \\ \bar{S}_{16} & \bar{S}_{26} & \bar{S}_{66} \end{bmatrix} \begin{Bmatrix} 0 \\ \sigma_y \\ 0 \end{Bmatrix}. \tag{8.176}$$

Thus,

$$\varepsilon_x - \alpha_x \Delta T = \bar{S}_{12} \sigma_y$$

$$-\alpha_y \Delta T = \bar{S}_{22} \sigma_y$$

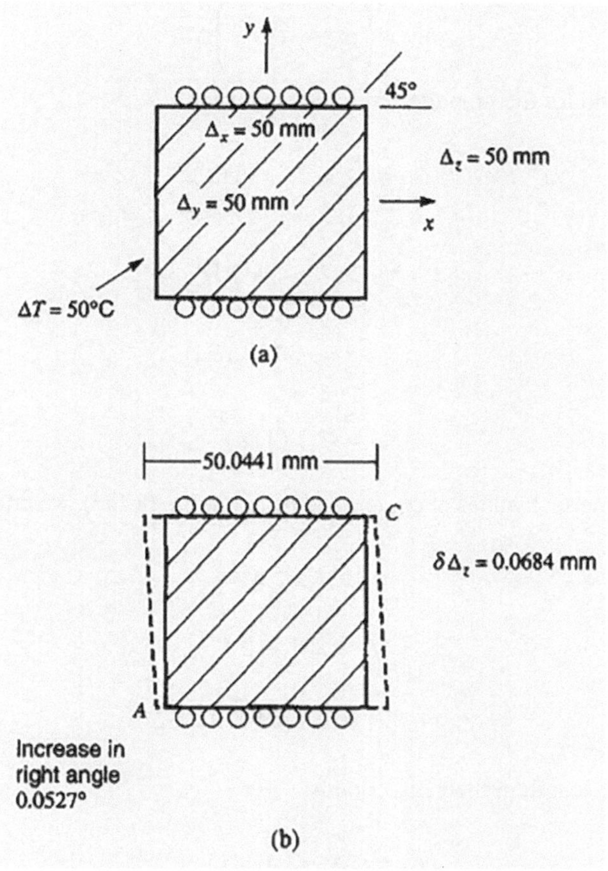

FIGURE 8.10 Thermal deformations of a partially constrained off-axis element of fiber-reinforced material. (a) Initial shape, (b) deformed shape.

$$\gamma_{xy} - \alpha_{xy}\Delta T = \bar{S}_{26}\sigma_y. \tag{8.177}$$

The stress required to constrain the deformation in the y direction is given by the second equation, namely:

$$\sigma_y = \frac{-\alpha_y \Delta T}{\bar{S}_{22}}. \tag{8.178}$$

Substituting Eq. (8.178) in Eq. (8.177):

$$\varepsilon_x = \left(\alpha_x - \frac{\bar{S}_{12}}{\bar{S}_{22}}\alpha_y\right)\Delta T$$

$$\gamma_{xy} = \left(\alpha_{xy} - \frac{\bar{S}_{26}}{\bar{S}_{22}} \alpha_y \right) \Delta T. \tag{8.179}$$

For $\theta = 45°$ and for the graphite-reinforced material:

$$\bar{S}_{11} = \bar{S}_{22} = 78.3\ (\text{TPa})^{-1}$$

$$\bar{S}_{12} = -35.3\ (\text{TPa})^{-1}$$

$$\bar{S}_{16} = \bar{S}_{26} = -38.1\ (\text{TPa})^{-1}$$

$$\bar{S}_{66} = 92.3\ (\text{TPa})^{-1}. \tag{8.180}$$

Using the numerical values of α_x, α_y, and α_{xy} from Eq. (8.144), we find that:

$$\sigma_y = -7.75\ \text{MPa}$$

$$\varepsilon_x = 881 \times 10^{-6}$$

$$\gamma_{xy} = -921 \times 10^{-6}\ \text{rad} = -0.0527°. \tag{8.181}$$

The change in length in the x direction is:

$$\delta\Delta_x = \varepsilon_x \Delta_x = 0.0441\ \text{mm}. \tag{8.182}$$

The right angles at corners A and C increase by 0.0527°. Because $\tau_{xy} = \sigma_x = 0$, the strain in the z direction for the situation in Figure 8.10, from Eq. (8.162), simplifies to:

$$\varepsilon_z = \alpha_3 \Delta T + \left(S_{13} \sin^2\theta + S_{23} \cos^2\theta \right) \sigma_y \tag{8.183}$$

or,

$$\varepsilon_z = 1368 \times 10^{-6}. \tag{8.184}$$

If we use the value $\Delta_z = 50\,\text{mm}$, the change in thickness of the element is given by:

$$\delta\Delta_z = 0.0684\ \text{mm}. \tag{8.185}$$

Figure 8.10b illustrates the deformations due to a 50°C temperature change. For this problem, the nonzero mechanical strains are:

$$\begin{Bmatrix} \varepsilon_x^{\text{mech}} \\ \varepsilon_y^{\text{mech}} \\ \varepsilon_z^{\text{mech}} \\ \gamma_{xy}^{\text{mech}} \end{Bmatrix} = \begin{Bmatrix} \varepsilon_x - \alpha_x \Delta T \\ -\alpha_y \Delta T \\ \varepsilon_z - \alpha_z \Delta T \\ \gamma_{xy} - \alpha_{xy} \Delta T \end{Bmatrix} = \begin{Bmatrix} 274 \\ -607 \\ 153 \\ 295 \end{Bmatrix} \times 10^{-6}. \tag{8.186}$$

REFERENCES

1. B. St. de Venant (1864), in his edition of Navier's *Resume des sur l'application de la Mecanique*, app. 3, Carilian-Goeury, Paris.
2. Hooke R. (1678), *De Potentia Restitutiva*, London.
3. Love A.E.H. (1892), *A Treatise on the Mathematical Theory of Elasticity*, Cambridge University Press, London (also 4th ed., 1944, Dover Publications, Inc., New York).
4. Hodge P.G. Jr. (1958), "The mathematical theory of plasticity," in *Elasticity and Plasticity* (J. N. Goodier and P. G. Hodge, Jr., eds.), John Wiley & Sons, Inc., New York, p. 81.
5. Kreyszig E. (1967), *Advanced Engineering Mathematics*, 2nd ed., John Wiley & Sons, Inc., New York, p. 416.
6. Fung Y.C. (1965), *Foundations of Solid Mechanics*, Prentice-Hall, Inc., Englewood Cliffs, NJ, p. 285.
7. Green G. (1839), "On the laws of reflexion and refraction of light at the common surface of two non-crystallized media", *Transactions of the Cambridge Philosophical Society*, Vol. 7, p. 121.
8. Popov E.P. (1990), *Engineering Mechanics of Solids*, Prentice Hall, Englewood Cliffs, NJ.

9 Classical Lamination Theory

In the previous chapters, the tools needed to find the elastic response of a fiber-reinforced material were developed on the assumption that the fiber and the matrix material were smeared into an equivalent homogenous material. As explained previously, the fiber-reinforced materials are made of multiple layers of material to form a laminate. Each layer may have its own fiber orientation and thickness. Some of the layers may contain fibers of carbon, while others may use glass fibers. Two laminates involving the same number of layers and having the same set of fiber orientations may show entirely different behavior due to the arrangement of the layers. As an example, consider a four-layer laminate with the two outer layers having fibers oriented at 0° and the two inner layers having fibers oriented at 90°. Now consider another four-layer laminate with the fibers oriented at 90° in the two outer layers and those in the inner layers at 0°. As Figure 9.1 shows, when subjected to the same level of bending moment, *M*, the first laminate will deform much less than the second. The nomenclature $[0/90)]_s$ and $[90/0]_s$ of Figure 9.1 will be dealt with shortly.

In the first case, the fibers that resist bending are further apart, resulting in a larger bending stiffness. On the other hand, as shown in Figure 9.2, both laminates will stretch the same amount in the x direction when an in-plane load is applied in that direction. Both laminates have two layers with their fibers parallel to the load and two layers with their fibers perpendicular to the load. The real issue, then, is to understand how laminates respond to loads, how the fiber angles of the individual layers influence laminate response, how the stacking arrangement of the layers influences the response, how changing material properties in a group of layers changes response, and so forth. Furthermore, the stresses within each layer, in addition to depending on the magnitude and character of the loading, must also depend on the arrangement of the layers, the fiber orientation in each layer, the material properties of each layer, and the like [1,2]. How are the stresses influenced by these parameters? The number of variables that can be changed in a laminated fiber-reinforced composite structure, as well as the number of responses that can be studied, is immense. It is of prime concern to understand how changing these variables influences laminate response and ultimately structural response. As an end product, it is important to be able to design laminates so that structures have a specific response, so that deformations are within certain limits and stress levels are below a given level.

We can thus evaluate the influence of fiber directions, stacking arrangements, material properties, and so forth, on laminate and structural response. However, before we introduce the simplified theory, commonly called the *classical lamination theory* (CLT), we will briefly discuss the nomenclature associated with laminates, including the manner in which fiber angles are specified and the stacking arrangement is identified.

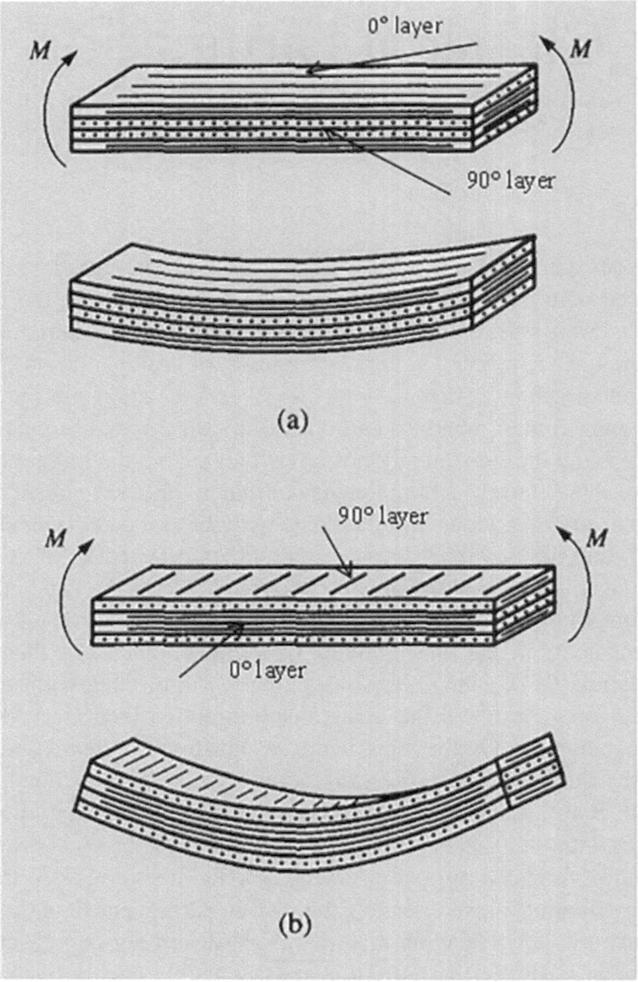

FIGURE 9.1 Bending deformations of (a) $[0/90]_s$ and (b) $[90/0]_s$ laminates.

9.1 LAMINATE NOMENCLATURE

Figure 9.3 illustrates a global Cartesian coordinate system and a general laminate consisting of N layers. The upper portion of the figure is a cross-sectional view in the x-z plane ($y = 0$ plane), and the lower portion is a plan-form view. The laminate thickness is denoted by H, and the thickness of an individual layer by h. Not all layers necessarily have the same thickness, so the thickness of the kth layer is denoted as h_k. The geometric mid-plane may be within a particular layer or at an interface between layers. What is important is that a geometric mid-plane can be defined. Herein, the $+z$ axis will be downward and the laminate extends in the z direction from $-H/2$ to $+H/2$. We refer to the layer at the negative-most z location as layer 1, the next layer in as layer 2, the layer at an arbitrary location as layer k, and the layer at the

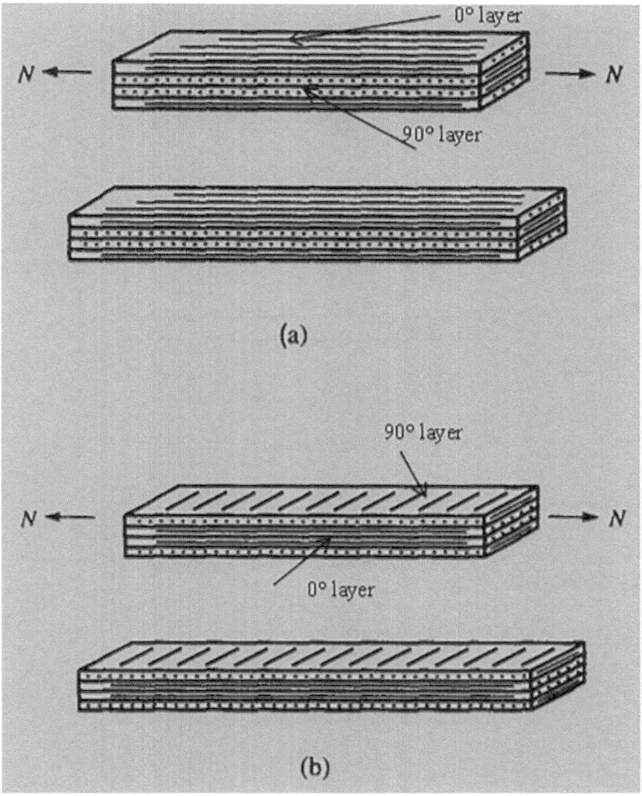

FIGURE 9.2 Similar extensions of (a) $[0/90]_s$ and (b) $[90/0]_s$ laminates.

positive-most z position as layer N. In the plan-form view, as shown in Figure 9.3b, layer N is closest to the reader. The locations of the layer interfaces are denoted by a subscripted z; the first layer is bounded by locations z_0 and z_1, the second layer by z_1 and z_2, the kth layer by z_{k-1} and z_k, and the Nth layer by z_{N-1} and z_N.

To identify the fiber angles of the various layers, the fiber angle relative to the $+x$ axis of each layer is specified. The specification starts with layer 1, the layer at the negative-most z location. For example, we denote the laminate in the upper portion of Figure 9.1 as a [0/90/90/0] laminate. We denote the laminate in the lower portion as a [90/0/0/90] laminate, where in each case we assume the x axis is oriented in the lengthwise direction of each laminate. The leftmost entry in the laminate notation refers to the orientation of layer 1. In cases where the stacking sequence to the one side of the $z = 0$ plane, the laminate geometric mid-plane, is a mirror image of the stacking sequence on the other side of the $z = 0$ plane, the stacking notation can be abbreviated by referring to only one-half of the laminate and subscripting the stacking notation with an S, which means symmetric. The upper and lower laminates of Figure 9.1 can thus be denoted by $[0/90]_s$ and $[90/0]_s$, respectively. With this notation, the leftmost entry in stacking specification is either layer 1 or layer N; that is, the stacking specification starts with the outer layer on each side of the laminate.

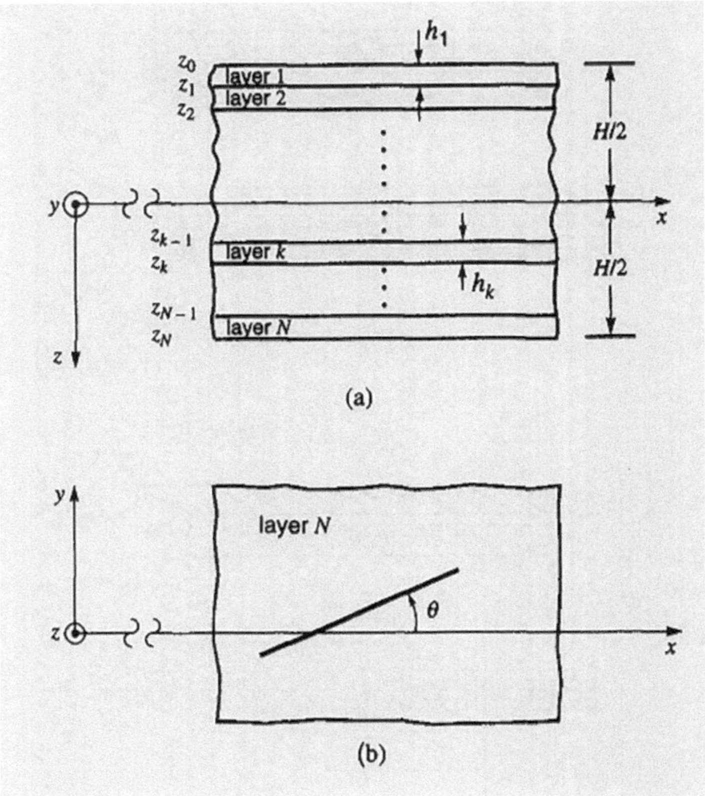

FIGURE 9.3 Nomenclature of a laminate. (a) View in x-z plane, (b) plan-form view.

To categorize a laminate as symmetric, it is necessary that the material properties, fiber orientation, and thickness of the layer at a specific location to one side of the geometric mid-plane be identical to the material properties, fiber orientation, and thickness of the layer at the mirror image location on the other side. Otherwise, the laminate is not truly symmetric. The influence of symmetry, or its lack, will be discussed later. Sufficient it to say that many aspects of the response of a laminate strongly depend on whether or not it is symmetric. In fact, symmetric laminates have been emphasized to such an extent that if one encounters the notation [0/90/90/0] in a discussion of composites, one usually assumes that an eight-layer laminate is being discussed, with stacking sequence [0/90/90/0/0/90/90/0]. To emphasize that indeed the complete laminate is being specified, the subscript T for total is sometimes used. Thus, the upper and lower laminates of Figure 9.1 could be denoted as $[0/90/90/0]_T$ and $[90/0/0/90]_T$, respectively.

When the stacking sequence involves adjacent layers of opposite orientation, as is often the case, shorthand notation is used. For example, if a six-layer laminate has the stacking sequence $[+45/-45/0/0/-45/+45]_T$, it would be abbreviated as $[\pm 45/0]_s$. Here, the $\pm$ is used to contract the notation and indicates there is a layer with its fibers oriented at +45° with respect to the $+x$ axis and adjacent to it another

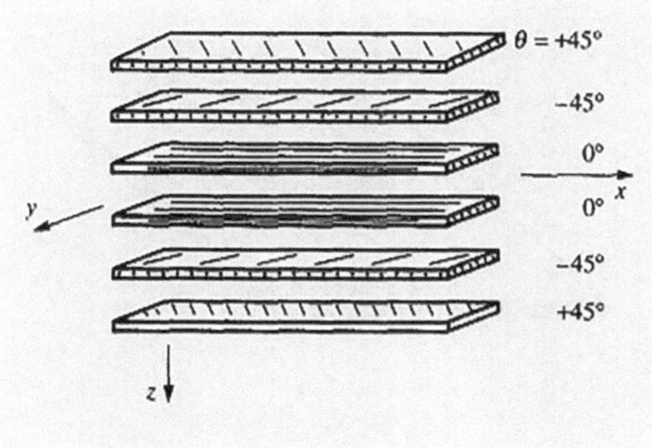

FIGURE 9.4 A $[\pm 45/0]_s$ laminate.

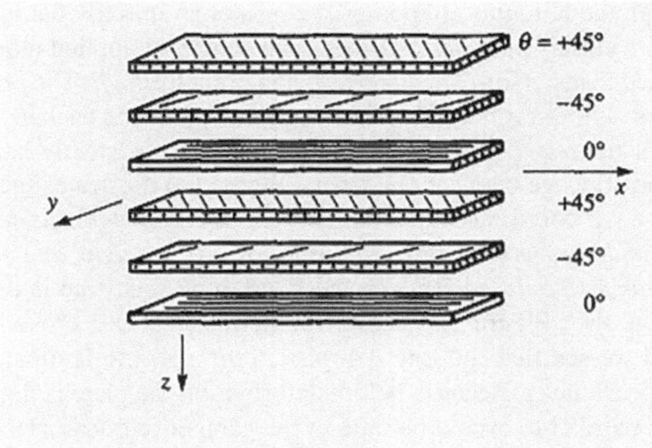

FIGURE 9.5 A $[(\pm 45/0)_2]_T$ laminate.

layer with its fibers oriented at −45° with relative to the +x axis. Next to the −45° layer is a layer with its fibers aligned with the x axis, as shown in Figure 9.4. When a stacking sequence of a subset consisting of several layers is repeated within a laminate, further shorthand notation is often used. If a 12-layer laminate has a stacking arrangement of $[+45/-45/0/+45/-45/0/0/-45/+45/0/-45/+45]_T$, it can be contracted to read $[(\pm 45/0)_2]_s$. Accordingly, a laminate denoted by $[(\pm 45/0)_2]_T$ would represent a six-layer laminate with an unsymmetric stacking arrangement of $[+45/-45/0/+45/-45/0]$, as shown in Figure 9.5.

9.2 THE KIRCHHOFF HYPOTHESIS

An important assumption was made by Kirchhoff in the mid-1800s regarding the analysis of structures and also of materials within the structures [3,4]. The assumption has been widely used for predicting the response of beams, plates, and shells.

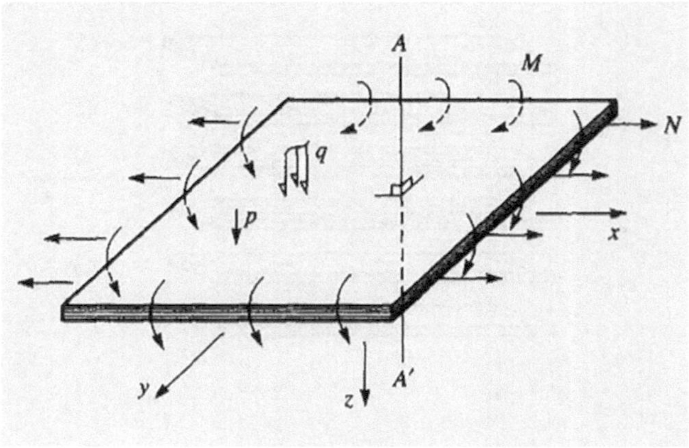

FIGURE 9.6 Laminated plate with undeformed normal AA' acted upon by loads.

Beginning with the Kirchhoff hypothesis, consider an initially flat laminated plate acted upon by a variety of loads. The loads can consist of applied moments, M; distributed applied loads, q; in-plane loads, N; and point loads, P. The plate consists of multiple layers of fiber-reinforced materials, and the fibers in each layer are parallel to the plane of the plate. We assume that all layers are perfectly bonded together and there is no slippage between the layers. Figure 9.6 illustrates the loaded plate, showing the x-y-z coordinate system. *The Kirchhoff hypothesis focuses on the deformation of lines which before deformation are straight and normal to the laminate's geometric mid-surface.* In the figure, one such line is drawn and it is denoted as line AA'. Figure 9.7a shows the detail of an x-z cross section of the laminate, and we see that the line AA' passes through the laminate and specifically through each layer. Because before deformation the plate is flat and the layer interfaces are parallel to each other and to the geometric mid-surface of the plate, the line AA' is normal to each interface. Figure 9.7b shows the details of this line viewed in a y-z cross section. The Kirchhoff hypothesis is simple: *It assumes that despite the deformations caused by the applied loads, line AA' remains straight and normal to the deformed geometric mid-plane and does not change length.* Specifically, the normal line does not deform; it simply translates and rotates as a consequence of the deformation, a very simple but yet very far-reaching assumption. Figure 9.8 shows the normal of Figure 9.6 having simply rotated and translated due to the deformations caused by the applied loads. That the line remains straight and normal to the geometric mid-plane after deformation is an important part of the assumption.

Figure 9.9 illustrates the consequences of normal remaining straight by again showing the details of a laminate x-z cross section. Because the interfaces remain parallel with each other and with the geometric mid-plane, the line AA' remains normal to each interface. The line is continuously straight through the thickness of the laminate, as shown in Figure 9.9b. The line is not a series of straight-line segments, as shown in Figure 9.9c; rather, it is a single straight line. That the line does not

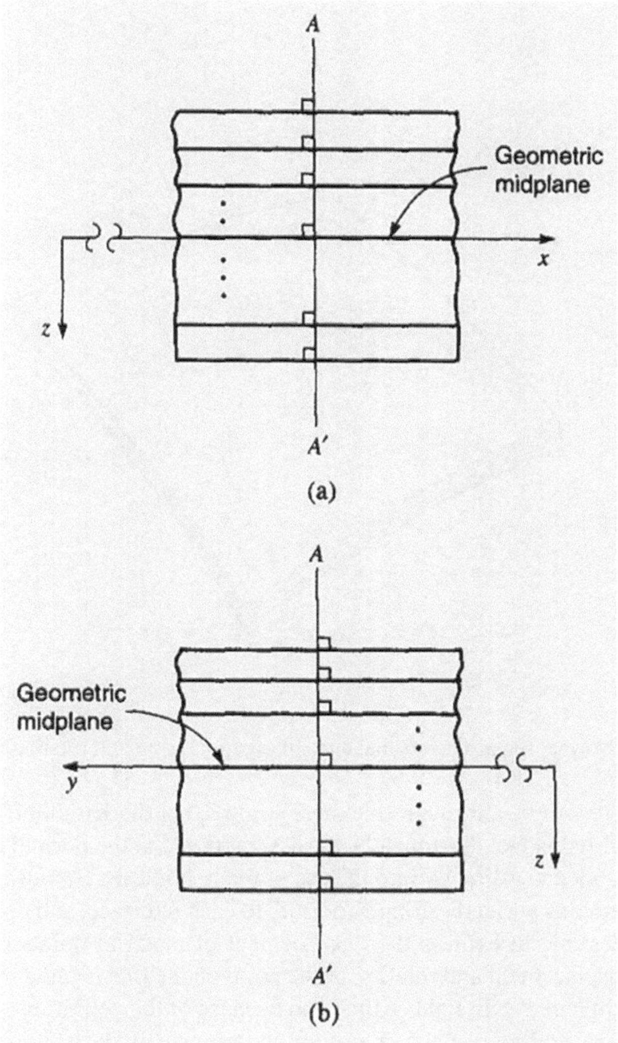

FIGURE 9.7 Cross sections showing normal before deformations: (a) x-z cross section and (b) y-z cross section.

change length is another important part of the assumption. For the length of the line to remain unchanged, the top and bottom surfaces of the laminate must remain the same distance apart in the thickness direction of the laminate. The distance between points t and t' as shown in Figure 9.9b, then, is the same as the distance between t and t' as shown in Figure 9.9a.

According to the hypothesis, there is no through-thickness strain ε_z along the line AA'. This is counter to what we know the stress–strain relations predict, namely, stresses in the x direction causing strains in the z direction, as shown in Eq. (8.19).

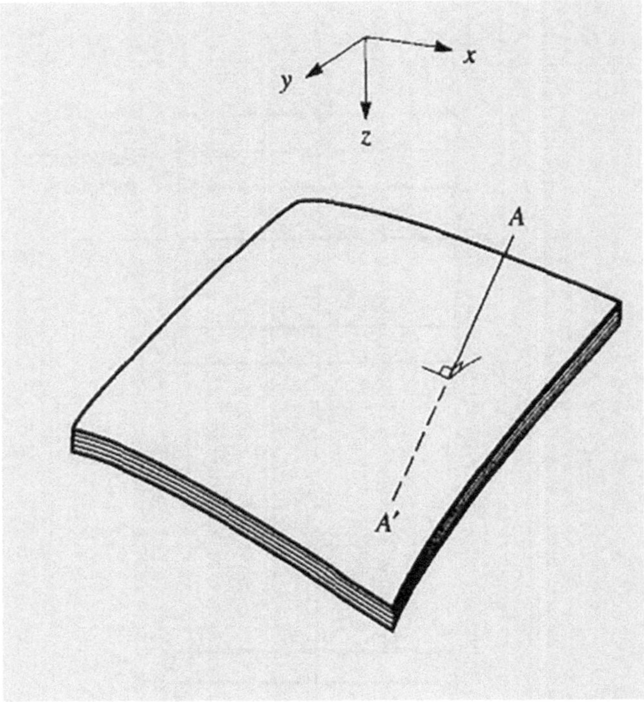

FIGURE 9.8 Normal remaining normal and simply translating and rotating.

Generally, there will be a through-thickness strain ε_z but the Kirchhoff hypothesis is inconsistent with this fact. Fortunately, the assumption that the normal remains fixed in length does not enter directly into the use of the hypothesis. Because it is assumed that the line remains perfectly straight, normal to each interface, and does not change length, it is possible to express the displacement of material points on the line in terms of the displacement and rotation of the point on the line located at the laminate geometric mid-plane [5]. In light of this, and because of the definition of stress resultants that will naturally arise at a later point in laminate analysis, it is convenient to think of the laminate geometric mid-plane as a *reference surface*. The mechanics of a laminate will be expressed in terms of what is happening at the reference surface. If the response of the reference surface is known, the strains, displacements, and stresses at each point along the normal line through the thickness of the laminate can be determined. This is an important advantage. Rather than treating a laminate as a three-dimensional domain and having to analyze it as such, the analysis of laminates degenerates to studying what is happening to the reference surface, a two-dimensional domain. With cylinders, for example, analysis degenerates to having to know what is happening to the surface of the cylinder located at the mean cylinder radius. As will be seen, understanding what is happening at the reference surface can be complicated enough. However, treating a plate or cylinder as a three-dimensional domain can become intractable [6]. Hereafter, the laminate geometric mid-surface will be referred to as the reference surface of the laminate.

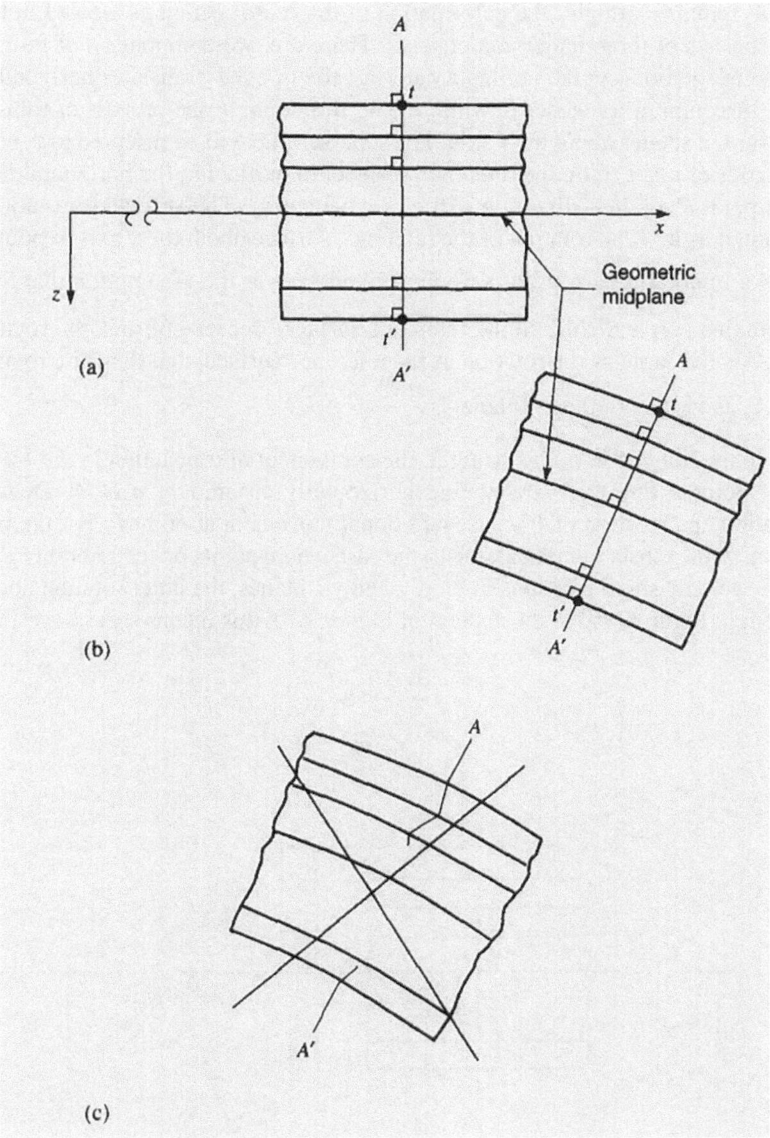

FIGURE 9.9 Implications of the Kirchhoff hypothesis: (a) undeformed, (b) deformed, and (c) alternate deformation.

9.3 EFFECTS OF THE KIRCHHOFF HYPOTHESIS

To study the implications of the Kirchhoff hypothesis, and to take advantage of it, let us examine the deformation of an x-z cross section of the plate being discussed. Figure 9.10 details the deformation of a cross section, and in particular the displacements of point P, a point located at an arbitrary distance z below point P^0, a point on the reference surface, points P and P^0 both being on the line AA'. Because the line AA' remains straight, the deformation of the cross section as viewed in the x-z plane consists of three major components. There are two components of translation and one of rotation. As the laminate deforms, the line AA' translates horizontally in the $+x$ direction; it translates down in the $+z$ direction; in the process of translating downward, it rotates about the y axis. The superscript 0 will be reserved to denote the kinematics of point P^0 on the reference surface. In particular, the horizontal translation of point P^0 in the x direction will be denoted as u^0. The vertical translation will be denoted as w^0. The rotation of the reference surface about the y axis at point P^0 is $\frac{\partial w^0}{\partial x}$. An important part of the Kirchhoff hypothesis is the assumption that the line AA' remains perpendicular to the reference surface. Because of this, the rotation of line AA' is the same as the rotation of the reference surface, and thus, the rotation of line AA', as viewed in the x-z plane, is $\frac{\partial w^0}{\partial x}$.

Since the line AA' remains straight, the component of translation in the $+x$ direction of point P due to P^0 translating horizontally an amount u is u^0. Downward translation and rotation of P^0 cause additional movement at point P. For the present case, we shall restrict our discussion to the case where points on the reference surface experience only small rotations in the x-z and y-z planes, the latter rotation not being apparent in Figure 9.10. In the context of Figure 9.10, this means

$$\frac{\partial w^0}{\partial x} < 1. \tag{9.1}$$

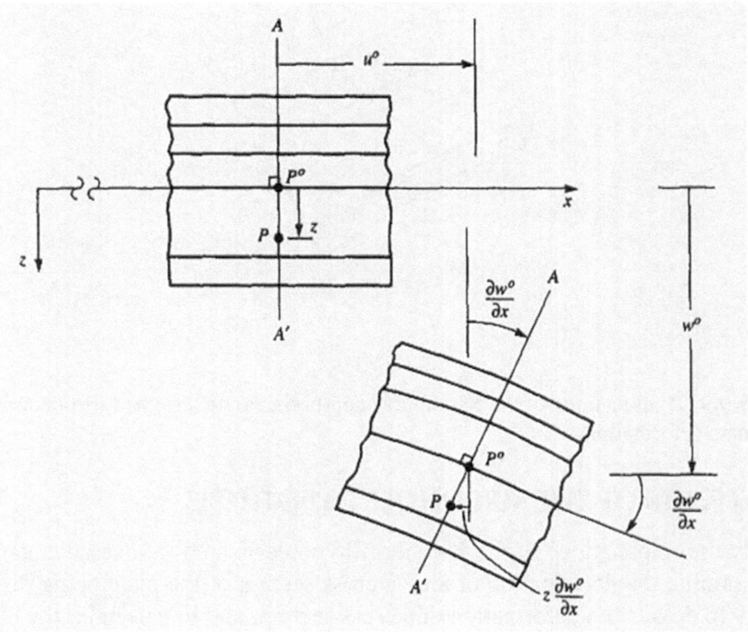

FIGURE 9.10 Deformation as viewed in the x-z plane.

By less than unity is meant that sines and tangents of angles of rotation are replaced by the rotations themselves, and cosines of the angles of rotation are replaced by 1. With this approximation, then, the rotation of point P^0 causes point P to translate horizontally in the minus x direction by an amount:

$$z\frac{\partial w^0}{\partial x}. \tag{9.2}$$

This negative horizontal translation is denoted in Figure 9.10, and the total translation of point P in the x direction, denoted as $u(x, y, z)$, is thus the sum of two effects, namely:

$$u(x, y, z) = u^0(x, y) - z\frac{\partial w^0(x, y)}{\partial x}. \tag{9.3}$$

Point P is located at (x, y, z), an arbitrary position within the laminate. The displacement of that point in the x direction is a function of all three coordinates, and thus the notation $u(x, y, z)$. The displacements and rotations of point P^0 on the reference surface, however, depend only on x and y, and hence the notation $u^0(x, y)$ and $\frac{\partial w^0(x, y)}{\partial x}$. Clearly, due to the kinematics of the Kirchhoff hypothesis, the displacement of point (x, y, z) depends *linearly* on z, the distance the point is away from the reference surface. Completing the picture of displacements of the x-z cross section, we see that as a result of the small rotation assumption, the vertical translation of point P is the same as the vertical translation of point P^0; that is, the vertical translation of point P is independent of z. As we shall see shortly, and as discussed just a few paragraphs ago, this leads to contradictory results for through-thickness extensional strains. With this independence of z:

$$w(x, y, z) = w^0(x, y), \tag{9.4}$$

where again the notation $w(x, y, z)$ indicates that the vertical displacement of point P at location (x, y, z) is, in general, a function of x, y, and z, but the hypothesis renders the vertical displacement independent of z and exactly equal to the reference surface displacement. Another interpretation of the independence of z is that all points on the line AA' move vertically the same amount. A similar picture emerges if the deformation is viewed in the y-z plane. As shown in Figure 9.11, the translation of point P in the $+y$ direction is:

$$v(x, y, z) = v^0(x, y) - z\frac{\partial w^0(x, y)}{\partial y}. \tag{9.5}$$

In the above, v^0 is the translation of point P^0 on the reference surface in the $+y$ direction and $\frac{\partial w^0(x, y)}{\partial y}$ is the rotation of that point about the x axis. In summary, then, the displacement of an arbitrary point P with coordinates (x, y, z) is given by:

$$u(x, y, z) = u^0(x, y) - z\frac{\partial w^0(x, y)}{\partial x}$$

$$v(x, y, z) = v^0(x, y) - z\frac{\partial w^0(x, y)}{\partial y} \tag{9.6}$$

$$w(x, y, z) = w^0(x,y).$$

The important points to note from the Kirchhoff hypothesis are that the in-plane displacements $u(x, y, z)$ and $v(x, y, z)$ everywhere within the laminate vary linearly with z, and the out-of-plane displacement $w(x, y, z)$ is independent of z.

9.4 LAMINATE STRAINS

From the strain–displacement relations and Eq. (9.6), the extensional strain in the x direction, ε_x, is given by:

$$\varepsilon_x(x, y, z) \equiv \frac{\partial u(x, y, z)}{\partial x} = \frac{\partial u^0(x, y)}{\partial x} - z\frac{\partial^2 w^0(x, y)}{\partial x^2}, \tag{9.7}$$

where the triple horizontal bars are to be interpreted as "is defined as." We can see that the strain ε_x is composed of two parts. The first term, $\frac{\partial u^0(x, y)}{\partial x}$, is the extensional strain of the reference surface in the x direction. Because we are restricting our discussion to small rotations of the reference surface, the second term, $\frac{\partial^2 w^0(x, y)}{\partial x^2}$, is the curvature of the reference surface in the x direction. In general, the curvature, which is the inverse of the radius of curvature, involves more than just the second derivative of w. However, for the case of small rotations, the curvature and the second derivative are identical [7,8]. Accordingly, the strain ε_x is written as:

$$\varepsilon_x(x, y, z) = \varepsilon_x^0(x, y) + zk_x^0(x, y), \tag{9.8}$$

where we use the notation:

$$\varepsilon_x^0(x, y) = \frac{\partial u^0(x, y)}{\partial x}$$

$$k_x^0(x, y) = -\frac{\partial^2 w^0(x, y)}{\partial x^2}. \tag{9.9}$$

The quantity ε_x^0 is referred to as the extensional strain of the reference surface in the x direction, and k_x^0 is referred to as the curvature of the reference surface in the x direction.

The other five strain components are given by:

$$\varepsilon_y(x, y, z) \equiv \frac{\partial v(x, y, z)}{\partial y} = \varepsilon_y^0(x, y) + zk_y^0(x, y)$$

$$\varepsilon_z(x, y, z) \equiv \frac{\partial w(x, y, z)}{\partial y} = \frac{\partial w^0(x, y)}{\partial z} = 0 \tag{9.10}$$

$$\gamma_{yz}(x, y, z) \equiv \frac{\partial w(x, y, z)}{\partial y} + \frac{\partial v(x, y, z)}{\partial z} = \frac{\partial w^0(x, y)}{\partial y} - \frac{\partial w^0(x, y)}{\partial y} = 0$$

$$\gamma_{xz}(x, y, z) \equiv \frac{\partial w(x, y, z)}{\partial x} + \frac{\partial u(x, y, z)}{\partial z} = \frac{\partial w^0(x, y)}{\partial x} - \frac{\partial w^0(x, y)}{\partial x} = 0$$

$$\gamma_{xy}(x, y, z) \equiv \frac{\partial v(x, y, z)}{\partial x} + \frac{\partial u(x, y, z)}{\partial y} = \gamma_{xy}^0 + zk_{xy}^0,$$

where we can define:

$$\varepsilon_y^0(x, y) = \frac{\partial v^0(x, y)}{\partial y}$$

$$k_y^0(x, y) = -\frac{\partial^2 w^0(x, y)}{\partial y^2} \tag{9.11}$$

$$\gamma_{xy}^0(x, y) = \frac{\partial v^0(x, y)}{\partial x} + \frac{\partial u^0(x, y)}{\partial y}$$

$$k_{xy}^0 = -2\frac{\partial^2 w^0(x, y)}{\partial x \partial y}.$$

The quantities ε_y^0, k_y^0, γ_{xy}^0, and k_{xy}^0 are referred to as the reference surface extensional strain in the y direction, the reference surface curvature in the y direction, the reference surface in-plane shear strain, and the reference surface twisting curvature, respectively. The notation used emphasizes again the fact that the reference surface strains and curvatures are functions only of x and y, while the strains are, in general, functions of x, y, and z. The term "in-plane shear strain" in connection with γ_{xy}^0 is used to indicate that it is related to changes in right angles between perpendicular line segments lying in the reference surface. Three of the six strain components are exactly zero. The two shear strains through the thickness are zero because the Kirchhoff hypothesis assumes that lines perpendicular to the reference surface before deformation remain perpendicular after the deformation; right angles in the thickness direction do not change when the laminate deforms (see Figures 9.10 and 9.11). By definition, then, the through-thickness shear strains must be exactly zero. If they were not computed to be zero in Eq. (9.10), there would be an inconsistency. The fact that the through-thickness extensional strain ε_z is predicted to be zero is the inconsistency we referred to previously that is due to the assumption that the length of the normal AA' does not change when the laminate deforms. Eqs. (5.45) and (5.47) indicate that extensional strains in all three directions are an integral part of the stress–strain relations.

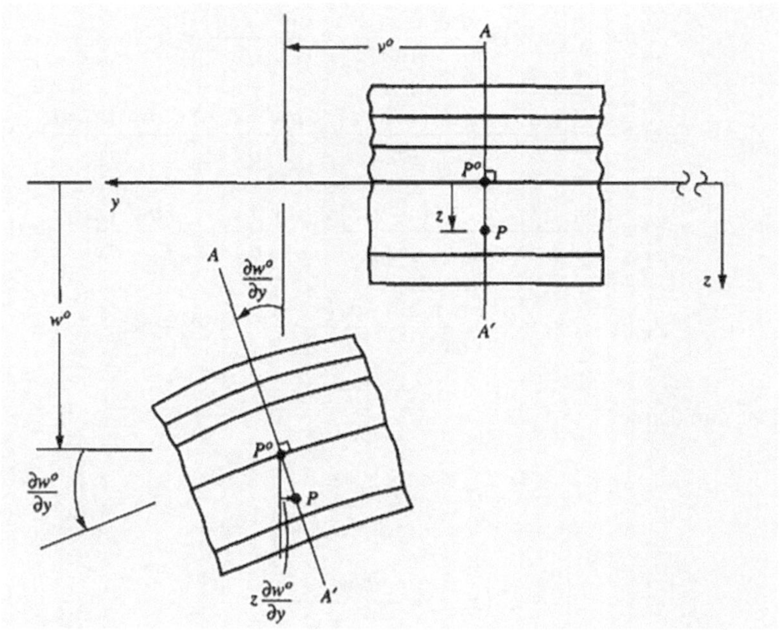

FIGURE 9.11 Deformation as viewed in the y-z plane.

For a state of plane stress, Eqs. (7.3) and (7.10) indicate that the extensional strain ε_z is not zero for this situation either. Hence, the third equation of Eq. (9.6) leads to an inconsistency. This is inherent in the Kirchhoff hypothesis and cannot be resolved within the context of the theory. The issue can be resolved by using the third term of Eq. (9.6) when the vertical displacement of a point in the cross section is needed, and using the stress–strain relations, such as Eq. (7.3) or (7.10), when the through-thickness extensional strain component ε_z is needed. The just stated equations imply that if the reference surface strains and curvatures are known at every point within a laminate, then the strains at every point within the three-dimensional volume are known.

As an example, consider the situation in Figure 9.12a, which illustrates a small segment of a four-layer laminate deformed such that at a point P^0 on the reference surface, the extensional strain in the x direction is 1000×10^{-6}. The radius of curvature of the reference surface, R^0, is 0.2 m – the latter corresponding to a curvature of 5 m^{-1}. The x-direction extensional strain as a function of the z coordinate through the thickness of the laminate above and below point P^0 on the reference surface is given by:

$$\varepsilon_x = 0.001 + (z)(5). \tag{9.12}$$

Note the sign of 0.300 mm is:

$$\varepsilon_x = 0.001 + \left(-300 \times 10^{-6}\right)(5) = -500\ \mu\text{mm mm}^{-1}. \tag{9.13}$$

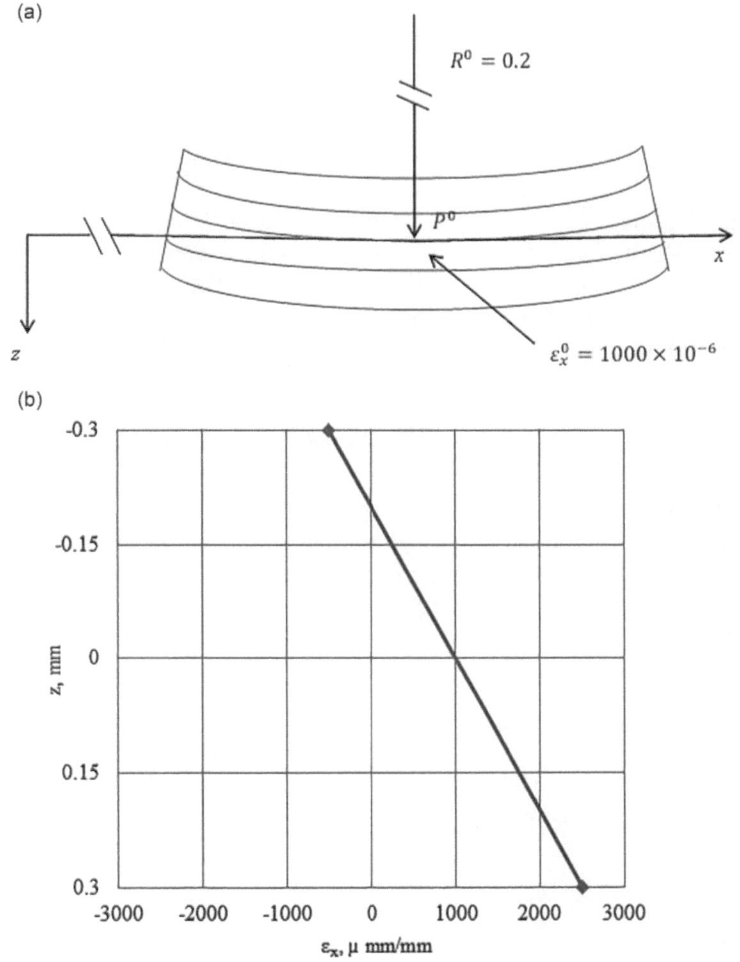

FIGURE 9.12 (a) ε_x^0 and k_x^0 are specified at a point on the reference surface of a laminate, (b) linear variation of ε_x with z through the laminate thickness.

The nondimensional character of the strain becomes apparent. At the bottom of the laminate, where $z = +0.300\,\text{mm}$,

$$\varepsilon_x = 0.001 + \left(300 \times 10^{-6}\right)(5) = 2500\ \mu\text{mm}\ \text{mm}^{-1}. \tag{9.14}$$

Figure 9.12b shows the linear variation of strain with z through the laminate thickness.

9.5 LAMINATE STRESSES

The second important assumption of CLT is that *each point within the volume of a laminate is in a state of plane stress.* Using the strains that result from the Kirchhoff hypothesis, Eq. (9.10), we find that the stress–strain relations for a laminate become:

$$\begin{Bmatrix} \sigma_x \\ \sigma_y \\ \tau_{xy} \end{Bmatrix} = \begin{bmatrix} \bar{Q}_{11} & \bar{Q}_{12} & \bar{Q}_{16} \\ \bar{Q}_{12} & \bar{Q}_{22} & \bar{Q}_{26} \\ \bar{Q}_{16} & \bar{Q}_{26} & \bar{Q}_{66} \end{bmatrix} \begin{Bmatrix} \varepsilon_x^0 + zk_x^0 \\ \varepsilon_y^0 + zk_y^0 \\ \gamma_{xy}^0 + zk_{xy}^0 \end{Bmatrix}. \tag{9.15}$$

As can be seen, because the strains are functions of z, the stresses are functions of z. For a laminate, however, the stresses depend on z for another very important reason: The material properties in one layer, represented in Eq. (9.15) by the reduced stiffnesses $\bar{Q}_{ij}$, are generally different than the material properties of another layer, due both to varying fiber orientation from layer to layer, and to the fact that different materials can be used in different layers [9]. Consequently, the transformed reduced stiffnesses are functions of location through the thickness and thus functions of z. The stresses therefore vary with z not only because the strains vary linearly with z but also because the reduced stiffnesses vary with z. In fact, the reduced stiffnesses vary in a piecewise-constant fashion, having one set of values through the thickness of one layer, another set of values through another layer, a third set of values through yet another layer, and so forth [10]. The net result is that, in general, the variation of the stresses through the thickness of the laminate is discontinuous and very much unlike the variation through the thickness of an isotropic material.

9.6 STRESS DISTRIBUTIONS

In this section, some examples related to the implications of Kirchhoff hypothesis have been discussed. This will also allow us to introduce other important concepts and generally illustrate the implications of CLT.

9.6.1 $[0/90]_s$ Laminate Subjected to Known ε_x^0

Consider a four-layer graphite–epoxy laminate, similar to the one shown in Figure 9.1, subjected to loads such that at a particular point (x, y) on the reference surface:

$$\begin{aligned} \varepsilon_x^0(x, y) &= 1000 \times 10^{-6} \\ \varepsilon_y^0(x, y) &= 0 \\ \gamma_{xy}^0(x, y) &= 0 \\ k_x^0(x, y) &= 0 \\ k_y^0(x, y) &= 0 \\ k_{xy}^0(x, y) &= 0. \end{aligned} \tag{9.16}$$

Using Eq. (9.10), we find that the strain distribution through the thickness of the laminate is given by:

$$\varepsilon_x(x,\ y,\ z) = \varepsilon_x^0(x,\ y) + zk_x^0(x,\ y) = 1000 \times 10^{-6} \tag{9.17}$$
$$\varepsilon_y(x,\ y,\ z) = \varepsilon_y^0(x,\ y) + zk_y^0(x,\ y) = 0$$
$$\gamma_{xy}(x,\ y,\ z) = \gamma_{xy}^0 + zk_{xy}^0 = 0.$$

The thickness of a layer, h, will be assumed to be 0.150 mm. From Eq. (9.17), through the entire thickness of the laminate, above and below the point (x, y) on the reference surface, the only nonzero strain is the extensional strain in the x direction, as shown in Figure 9.13. This represents a laminate stretched, or extended, in the x direction

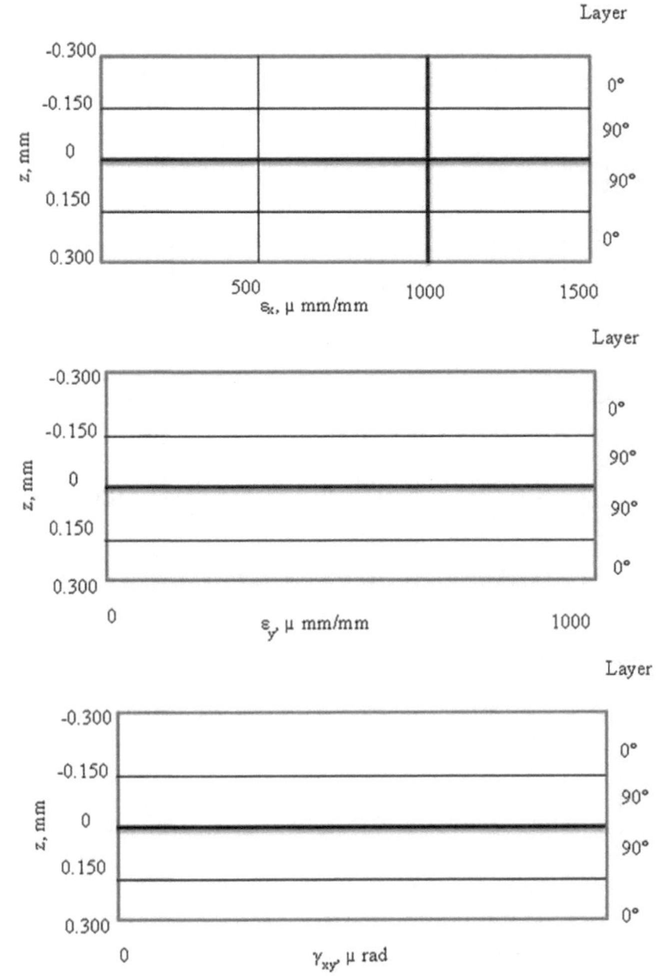

FIGURE 9.13 Strain distribution through the thickness of $[0/90]_s$ laminate subjected to $\varepsilon_x^0 = 1000 \times 10^{-6}$.

but with no associated Poisson contraction in the y direction, and with no in-plane shear strain and no curvature effects. For this $[0/90]_s$ laminate, we find, referring to the nomenclature of Figure 9.3, the layer interface locations are:

$$z_0 = -0.300 \text{ mm}, \quad z_1 = -0.150 \text{ mm}, \quad z_2 = 0$$

$$z_3 = 0.150 \text{ mm}, \quad z_4 = 0.300 \text{ mm}. \tag{9.18}$$

The stresses at each point through the thickness of the laminate are determined by using the above reference surface strains and curvatures in the stress–strain relation, Eq. (9.15) specifically:

$$\begin{Bmatrix} \sigma_x \\ \sigma_y \\ \tau_{xy} \end{Bmatrix} = \begin{bmatrix} \bar{Q}_{11} & \bar{Q}_{12} & \bar{Q}_{16} \\ \bar{Q}_{12} & \bar{Q}_{22} & \bar{Q}_{26} \\ \bar{Q}_{16} & \bar{Q}_{26} & \bar{Q}_{66} \end{bmatrix} \begin{Bmatrix} 1000 \times 10^{-6} \\ 0 \\ 0 \end{Bmatrix}. \tag{9.19}$$

For the two 0° layers:

$$\begin{Bmatrix} \sigma_x \\ \sigma_y \\ \tau_{xy} \end{Bmatrix} = \begin{bmatrix} \bar{Q}_{11}(0^\circ) & \bar{Q}_{12}(0^\circ) & \bar{Q}_{16}(0^\circ) \\ \bar{Q}_{12}(0^\circ) & \bar{Q}_{22}(0^\circ) & \bar{Q}_{26}(0^\circ) \\ \bar{Q}_{16}(0^\circ) & \bar{Q}_{26}(0^\circ) & \bar{Q}_{66}(0^\circ) \end{bmatrix} \begin{Bmatrix} 1000 \times 10^{-6} \\ 0 \\ 0 \end{Bmatrix}. \tag{9.20}$$

For the two 90° layers:

$$\begin{Bmatrix} \sigma_x \\ \sigma_y \\ \tau_{xy} \end{Bmatrix} = \begin{bmatrix} \bar{Q}_{11}(90^\circ) & \bar{Q}_{12}(90^\circ) & \bar{Q}_{16}(90^\circ) \\ \bar{Q}_{12}(90^\circ) & \bar{Q}_{22}(90^\circ) & \bar{Q}_{26}(90^\circ) \\ \bar{Q}_{16}(90^\circ) & \bar{Q}_{26}(90^\circ) & \bar{Q}_{66}(90^\circ) \end{bmatrix} \begin{Bmatrix} 1000 \times 10^{-6} \\ 0 \\ 0 \end{Bmatrix}. \tag{9.21}$$

From Eq. (8.84), the stresses in the 0° layers are:

$$\begin{Bmatrix} \sigma_x \\ \sigma_y \\ \tau_{xy} \end{Bmatrix} = \begin{bmatrix} Q_{11} & Q_{12} & 0 \\ Q_{12} & Q_{22} & 0 \\ 0 & 0 & Q_{66} \end{bmatrix} \begin{Bmatrix} 1000 \times 10^{-6} \\ 0 \\ 0 \end{Bmatrix}. \tag{9.22}$$

From Eqs. (9.21) and (8.84), the stresses in the 90° layers are:

$$\begin{Bmatrix} \sigma_x \\ \sigma_y \\ \tau_{xy} \end{Bmatrix} = \begin{bmatrix} Q_{22} & Q_{12} & 0 \\ Q_{12} & Q_{11} & 0 \\ 0 & 0 & Q_{66} \end{bmatrix} \begin{Bmatrix} 1000 \times 10^{-6} \\ 0 \\ 0 \end{Bmatrix}. \tag{9.23}$$

Using numerical values for the reduced stiffnesses, Eq. (8.91), we find that for the 0° layers:

$$\sigma_x = 155.7 \text{ MPa}$$
$$\sigma_y = 3.02 \text{ MPa}$$
$$\tau_{xy} = 0. \tag{9.24}$$

And for the 90° layers:

$$\sigma_x = 12.16 \text{ MPa}$$
$$\sigma_y = 3.02 \text{ MPa}$$
$$\tau_{xy} = 0. \tag{9.25}$$

Figure 9.14 illustrates the distribution of the stresses through the thickness of the laminate. For the problem here, the stress component *ax* is constant within each layer, but varies from layer to layer. Formally, σ_x is said to be *piecewise constant* through the thickness of the laminate [11,12]. On the other hand, the stress component σ_y is constant through the thickness of the laminate. The constancy of σ_y is a very special condition that can occur in cross-ply laminates or other special lamination arrangements. It can occur because $\bar{Q}_{12}(0°) = \bar{Q}_{12}(90°) = Q_{12}$. Note that a tensile value for σ_y is needed to overcome the natural tendency of the laminate to contract in the y direction. Due to the Poisson effect, stretching in the x direction causes contraction in the y direction. By the statement of the problem in Eq. (9.19), contraction in the y direction is here stipulated to be zero ($\varepsilon_y = 0$), so a tensile stress is required to enforce this. Also special is the fact that the shear stress is zero at all z locations.

If we consider a four-layer aluminum laminate made by perfecting bonding together four layers of aluminum, the distribution of strain through the thickness would be as shown in Figure 9.13. As we have emphasized, the reference surface strains and curvatures dictate the distribution of the strains through the thickness. Specified values of reference surface strains and curvatures produce the same distributions of ε_x, ε_y, and γ_{xy} through the thickness of aluminum as through the thickness of a composite. The stresses for each layer in the aluminum laminate are given by Eq. (9.19); the reduced stiffnesses for aluminum are used instead of the reduced stiffnesses of the composite [13]. From our previous examples with aluminum and the numerical values for aluminum from Table 5.1:

$$\sigma_x = 79.6 \text{ MPa}$$
$$\sigma_y = 23.9 \text{ MPa}$$
$$\tau_{xy} = 0. \tag{9.26}$$

Due to the assumption of perfection bonding between layers, the four layers of aluminum act as one. Figure 9.15 shows the distribution of the stresses through the thickness of the aluminum. The characteristics of this figure should be compared with that of Figure 9.14, particularly the continuity of the stress component σ_x and

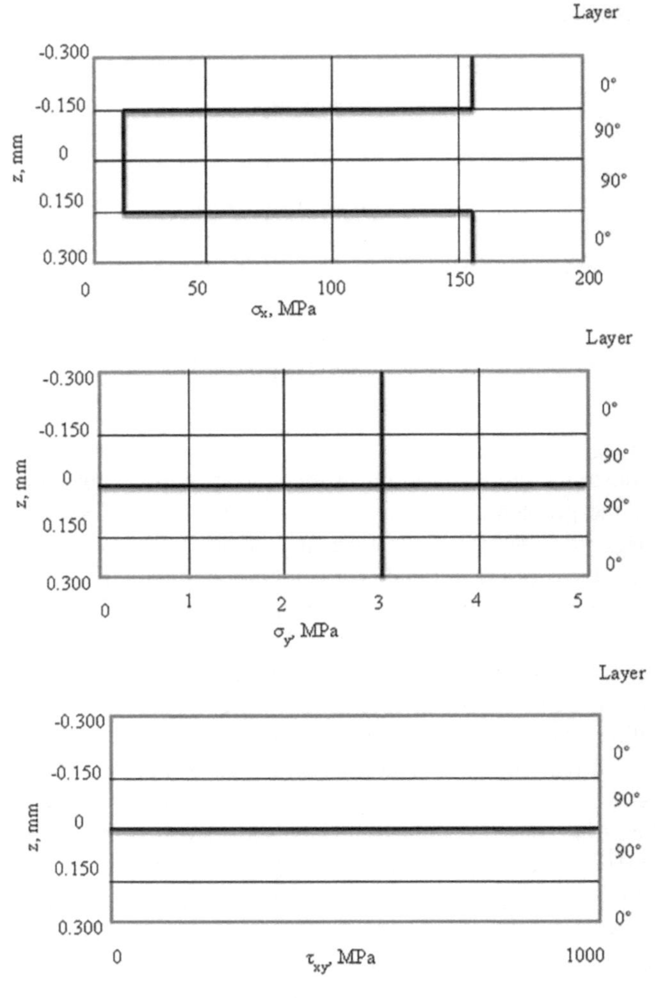

FIGURE 9.14 Stress distribution through the thickness of $[0/90]_s$ laminate subjected to $\varepsilon_x^0 = 1000 \times 10^{-6}$.

the magnitude of σ_y relative to σ_x. We introduced an *x-y-z* global coordinate system simply for convenience, specifically so we would not have to deal with a number of coordinate systems. However, we should address fundamental issues in the principal material system. To do this, we employ the transformation relations for stress and strain of Chapter 8, specifically Eqs. (8.10) and (8.21). As they will be used frequently in this chapter, we here reproduce those relations:

$$\begin{Bmatrix} \sigma_1 \\ \sigma_2 \\ \tau_{12} \end{Bmatrix} = [T] \begin{Bmatrix} \sigma_x \\ \sigma_y \\ \tau_{xy} \end{Bmatrix} = \begin{bmatrix} m^2 & n^2 & 2mn \\ n^2 & m^2 & -2mn \\ -mn & mn & m^2 - n^2 \end{bmatrix} \begin{Bmatrix} \sigma_x \\ \sigma_y \\ \tau_{xy} \end{Bmatrix} \tag{9.27}$$

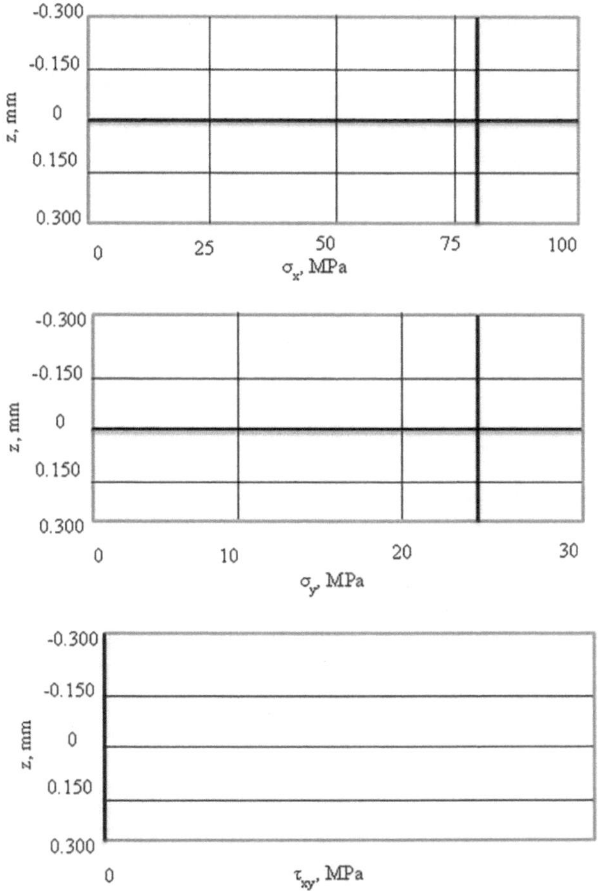

FIGURE 9.15 Stress distribution through the thickness of a four-layer aluminum laminate subjected to $\varepsilon_x^0 = 1000 \times 10^{-6}$.

$$\begin{Bmatrix} \varepsilon_1 \\ \varepsilon_2 \\ \dfrac{\gamma_{12}}{2} \end{Bmatrix} = [T] \begin{Bmatrix} \varepsilon_x \\ \varepsilon_y \\ \dfrac{\gamma_{xy}}{2} \end{Bmatrix} = \begin{bmatrix} m^2 & n^2 & 2mn \\ n^2 & m^2 & -2mn \\ -mn & mn & m^2 - n^2 \end{bmatrix} \begin{Bmatrix} \varepsilon_x \\ \varepsilon_y \\ \dfrac{\gamma_{xy}}{2} \end{Bmatrix}. \quad (9.28)$$

For the particular problem here, no transformation is needed for the top and bottom layers. As their fibers are oriented at $\theta = 0°$, $\varepsilon_1 = \varepsilon_x$, $\varepsilon_2 = \varepsilon_y$, and $\gamma_{xy} = \gamma_{12}$. Nevertheless, to remain formal, for the 0° layers, since $m = 1$ and $n = 0$, Eq. (9.28) results in:

$$\begin{Bmatrix} \varepsilon_1 \\ \varepsilon_2 \\ \dfrac{\gamma_{12}}{2} \end{Bmatrix} = [T(0°)] \begin{Bmatrix} \varepsilon_x \\ \varepsilon_y \\ \dfrac{\gamma_{xy}}{2} \end{Bmatrix} = \begin{bmatrix} 1 & 0 & 0 \\ 0 & 1 & 0 \\ 0 & 0 & 1 \end{bmatrix} \begin{Bmatrix} \varepsilon_x \\ \varepsilon_y \\ \dfrac{\gamma_{xy}}{2} \end{Bmatrix}. \quad (9.29)$$

Substituting for the values of strain in the *x*-*y*-*z* system for the 0° layers, we find:

$$\begin{Bmatrix} \varepsilon_1 \\ \varepsilon_2 \\ \dfrac{\gamma_{12}}{2} \end{Bmatrix} = \begin{bmatrix} 1 & 0 & 0 \\ 0 & 1 & 0 \\ 0 & 0 & 1 \end{bmatrix} \begin{Bmatrix} 1000 \times 10^{-6} \\ 0 \\ 0 \end{Bmatrix}. \tag{9.30}$$

In the 0° layers:

$$\begin{aligned} \varepsilon_1 &= 1000 \times 10^{-6} \\ \varepsilon_2 &= 0 \\ \gamma_{12} &= 0. \end{aligned} \tag{9.31}$$

For the 90° layers, $m = 0$ and $n = 1$ and Eq. (9.28) becomes:

$$\begin{Bmatrix} \varepsilon_1 \\ \varepsilon_2 \\ \dfrac{\gamma_{12}}{2} \end{Bmatrix} = [T(90°)] \begin{Bmatrix} \varepsilon_x \\ \varepsilon_y \\ \dfrac{\gamma_{xy}}{2} \end{Bmatrix} = \begin{bmatrix} 0 & 1 & 0 \\ 1 & 0 & 0 \\ 0 & 0 & -1 \end{bmatrix} \begin{Bmatrix} 1000 \times 10^{-6} \\ 0 \\ 0 \end{Bmatrix}. \tag{9.32}$$

In the 90° layers:

$$\begin{aligned} \varepsilon_1 &= 0 \\ \varepsilon_2 &= 1000 \times 10^{-6} \\ \gamma_{12} &= 0. \end{aligned} \tag{9.33}$$

Figure 9.16 shows the distribution of these principal material system strains ε_1, ε_2, and γ_{12}. The stresses in the principal material system are computed using Eq. (9.27). As with the strains, the stresses σ_1, σ_2, and τ_{12} for the 0° layers are σ_x, σ_y, and τ_{xy}, respectively; so for the 0° layers:

$$\begin{aligned} \sigma_1 &= 155.7 \text{ MPa} \\ \sigma_2 &= 3.02 \text{ MPa} \\ \tau_{12} &= 0. \end{aligned} \tag{9.34}$$

For the 90° layers:

$$\begin{Bmatrix} \sigma_1 \\ \sigma_2 \\ \tau_{12} \end{Bmatrix} = [T(90°)] \begin{Bmatrix} \sigma_x \\ \sigma_y \\ \tau_{xy} \end{Bmatrix} = \begin{bmatrix} 0 & 1 & 0 \\ 1 & 0 & 0 \\ 0 & 0 & -1 \end{bmatrix} \begin{Bmatrix} 12.16 \\ 3.02 \\ 0 \end{Bmatrix} \tag{9.35}$$

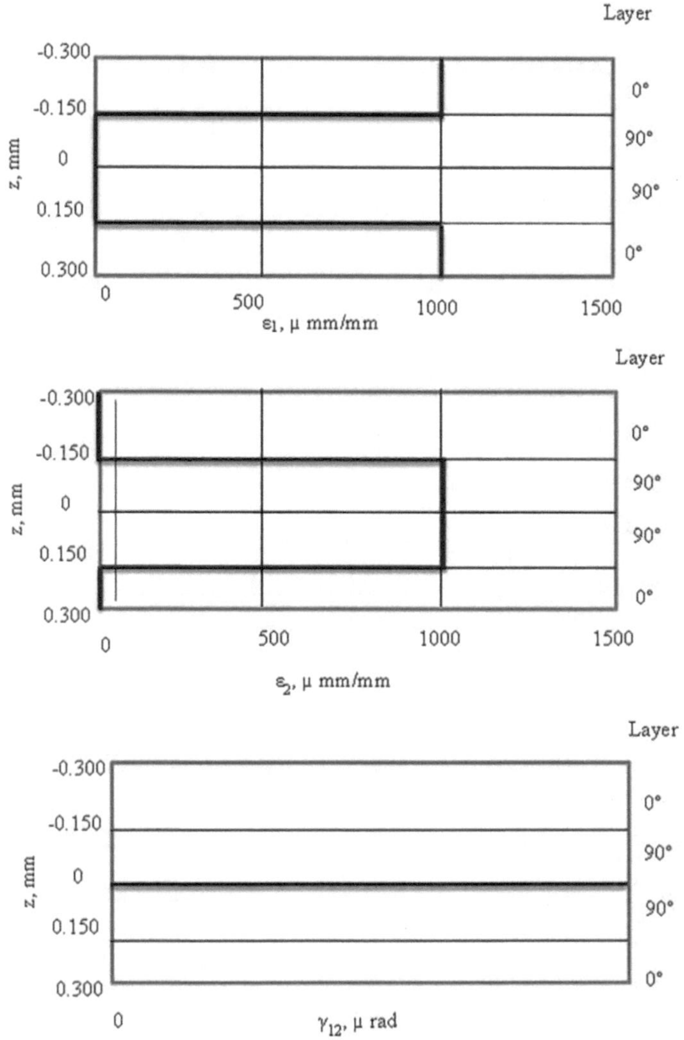

FIGURE 9.16 Principal material system strain distribution through the thickness of $[0/90]_s$ laminate subjected to $\varepsilon_x^0 = 1000 \times 10^{-6}$.

or,

$$\sigma_1 = 3.02 \text{ MPa}$$

$$\sigma_2 = 12.16 \text{ MPa}$$

$$\tau_{12} = 0. \tag{9.36}$$

The thickness distribution of the principal material system stresses, as shown in Figure 9.17, like the distribution of the laminate system stresses, is discontinuous with z. The discontinuity results not only because of the way the abrupt changes in

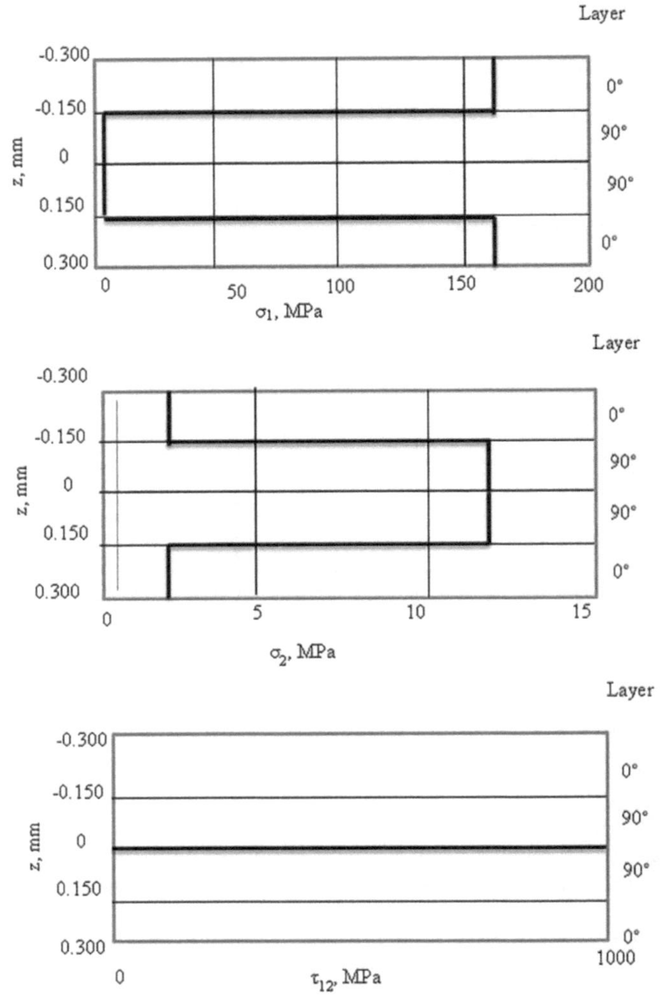

FIGURE 9.17 Principal material system stress distribution through the thickness of $[0/90]_s$ laminate subjected to $\varepsilon_x^0 = 1000 \times 10^{-6}$.

the material properties from layer to layer influence σ_x, σ_y, and τ_{xy} but also, like the principal material system strains, because the direction in which these stresses act changes abruptly from layer to layer.

9.6.2 $[0/90]_s$ Laminate Subjected to Known k_x^0

Consider the same four-layer $[0/90]_s$ laminate subjected to loads such that at a particular point (x, y) on the reference surface:

$$\varepsilon_x^0(x, y) = 0$$

$$\begin{aligned}
\varepsilon_y^0(x, y) &= 0 \\
\gamma_{xy}^0(x, y) &= 0 \\
k_x^0(x, y) &= 3.33\ \text{m}^{-1} \\
k_y^0(x, y) &= 0 \\
k_{xy}^0(x, y) &= 0.
\end{aligned} \tag{9.37}$$

We find that the strain distribution through the thickness of the laminate is not independent of z; rather, it is linear in z and is given by:

$$\begin{aligned}
\varepsilon_x(x, y, z) &= \varepsilon_x^0(x, y) + zk_x^0(x, y) = 3.33\ z \\
\varepsilon_y(x, y, z) &= \varepsilon_y^0(x, y) + zk_y^0(x, y) = 0 \\
\gamma_{xy}(x, y, z) &= \gamma_{xy}^0 + zk_{xy}^0 = 0.
\end{aligned} \tag{9.38}$$

Through the entire thickness of the laminate, above and below the point on the reference surface, the only nonzero strain is the extensional strain in the x direction. This nonzero strain is linear in z, as shown in Figure 9.18, and is zero on the reference surface, where $z = 0$; -1000×10^{-6} on the top surface, where $z = -0.300\,\text{mm}$; and $+1000 \times 10^{-6}$ on the bottom surface, where $z = +0.300\,\text{mm}$. By the convention being used here, if the reference surface is deformed such that there is a negative second derivative of w^0 with respect to x (i.e., positive k_x^0), then there is the tendency toward a compressive strain in the upper surface of the laminate. The stresses at each point through the thickness of the laminate are determined by using the strains from Eq. (9.38) in the stress–strain relations, Eq. (9.19), namely:

$$\begin{Bmatrix} \sigma_x \\ \sigma_y \\ \tau_{xy} \end{Bmatrix} = \begin{bmatrix} \bar{Q}_{11} & \bar{Q}_{12} & \bar{Q}_{16} \\ \bar{Q}_{12} & \bar{Q}_{22} & \bar{Q}_{26} \\ \bar{Q}_{16} & \bar{Q}_{26} & \bar{Q}_{66} \end{bmatrix} \begin{Bmatrix} 3.33z \\ 0 \\ 0 \end{Bmatrix}. \tag{9.39}$$

For the two 0° layers:

$$\begin{Bmatrix} \sigma_x \\ \sigma_y \\ \tau_{xy} \end{Bmatrix} = \begin{bmatrix} \bar{Q}_{11}(0^\circ) & \bar{Q}_{12}(0^\circ) & \bar{Q}_{16}(0^\circ) \\ \bar{Q}_{12}(0^\circ) & \bar{Q}_{22}(0^\circ) & \bar{Q}_{26}(0^\circ) \\ \bar{Q}_{16}(0^\circ) & \bar{Q}_{26}(0^\circ) & \bar{Q}_{66}(0^\circ) \end{bmatrix} \begin{Bmatrix} 3.33z \\ 0 \\ 0 \end{Bmatrix}. \tag{9.40}$$

For the two 90° layers:

$$\begin{Bmatrix} \sigma_x \\ \sigma_y \\ \tau_{xy} \end{Bmatrix} = \begin{bmatrix} \bar{Q}_{11}(90^\circ) & \bar{Q}_{12}(90^\circ) & \bar{Q}_{16}(90^\circ) \\ \bar{Q}_{12}(90^\circ) & \bar{Q}_{22}(90^\circ) & \bar{Q}_{26}(90^\circ) \\ \bar{Q}_{16}(90^\circ) & \bar{Q}_{26}(90^\circ) & \bar{Q}_{66}(90^\circ) \end{bmatrix} \begin{Bmatrix} 3.33z \\ 0 \\ 0 \end{Bmatrix}. \tag{9.41}$$

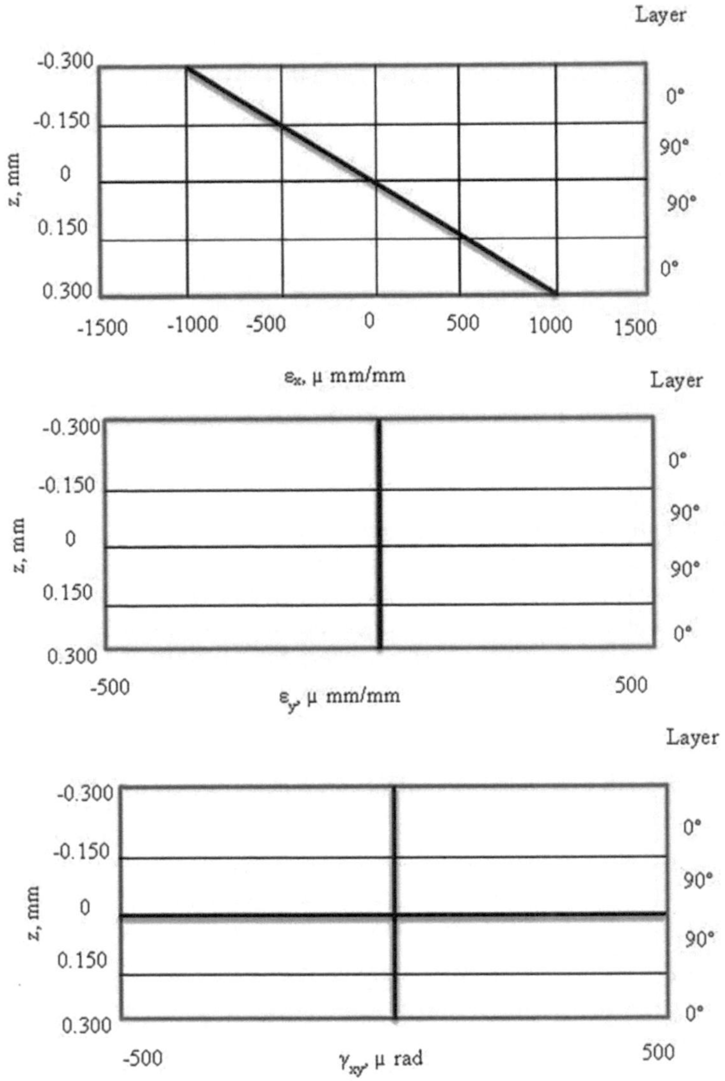

FIGURE 9.18 Strain distribution through the thickness of $[0/90]_s$ laminate subjected to $k_x^0 = 3.33\ \text{m}^{-1}$.

From Eq. (8.84), the stresses in the 0° layers are:

$$\begin{Bmatrix} \sigma_x \\ \sigma_y \\ \tau_{xy} \end{Bmatrix} = \begin{bmatrix} Q_{11} & Q_{12} & 0 \\ Q_{12} & Q_{22} & 0 \\ 0 & 0 & Q_{66} \end{bmatrix} \begin{Bmatrix} 3.33z \\ 0 \\ 0 \end{Bmatrix}. \tag{9.42}$$

From Eqs. (9.21) and (8.84), the stresses in the 90° layers are:

$$\left\{\begin{array}{c}\sigma_x\\ \sigma_y\\ \tau_{xy}\end{array}\right\}=\left[\begin{array}{ccc}Q_{22} & Q_{12} & 0\\ Q_{12} & Q_{11} & 0\\ 0 & 0 & Q_{66}\end{array}\right]\left\{\begin{array}{c}3.33z\\ 0\\ 0\end{array}\right\}. \tag{9.43}$$

Using numerical values for the reduced stiffnesses, Eq. (8.91), we find that for the 0° layers:

$$\begin{aligned}\sigma_x &= 519{,}000z \text{ MPa}\\ \sigma_y &= 10{,}060z \text{ MPa}\\ \tau_{xy} &= 0.\end{aligned} \tag{9.44}$$

These relations are valid for z in the range -0.300 mm $\leq z \leq -0.150$ mm and $+0.150$ mm $\leq z \leq +0.300$ mm.

And for the 90° layers:

$$\begin{aligned}\sigma_x &= 40{,}500z \text{ MPa}\\ \sigma_y &= 10{,}060z \text{ MPa}\\ \tau_{xy} &= 0.\end{aligned} \tag{9.45}$$

These relations are valid for z in the range -0.150 mm $\leq z \leq +0.150$ mm.

Figure 9.19 illustrates the stress distribution; the maximum tensile stress of $\sigma_x =$ 155.7 MPa occurs in the 0° layer at the bottom of the laminate where $z = +0.300$ mm, and the maximum compressive stress of $\sigma_x = -155.7$ MPa occurs at the top of the laminate where $z = -0.300$ mm. Whatever the layer arrangement, the deformations specified by Eq. (9.37) require that there be a value for the stress component σ_y. This is because Eq. (9.37) specifies that at the particular point on the reference surface, where Eq. (9.37) is valid, there is no curvature in the y direction, k_y^0. If the deformations specified in Eq. (9.37) were assumed to be valid at every point on the entire reference surface of a laminated plate, not just at a particular point, then the plate would appear as shown in Figure 9.20a, namely, deformed into a cylindrical surface. The cylindrical shape would be such that there would be curvature in the x direction, but not the y direction. With this deformation, the upper surface of the plate would experience compressive strains in the x direction and the lower surface would experience tensile strains. If the curvature were not specified to be zero in the y direction, the compressive strains on the top surface in the x direction, through the natural tendency of the Poisson effect, would produce tensile strains in the y direction, and the tensile strains on the lower surface in the x direction would produce contraction strains in the y direction. As a result, the plate would be *saddle-shaped*, as shown in Figure 9.20b. This curvature in the y direction due to Poisson effects is referred to as *anticlastic* curvature [14]. This is the natural tendency of the plate if it is given a curvature in the x direction and nothing is specified about the curvature in the y direction. However, here we have specified the curvature in the y direction to be zero; the

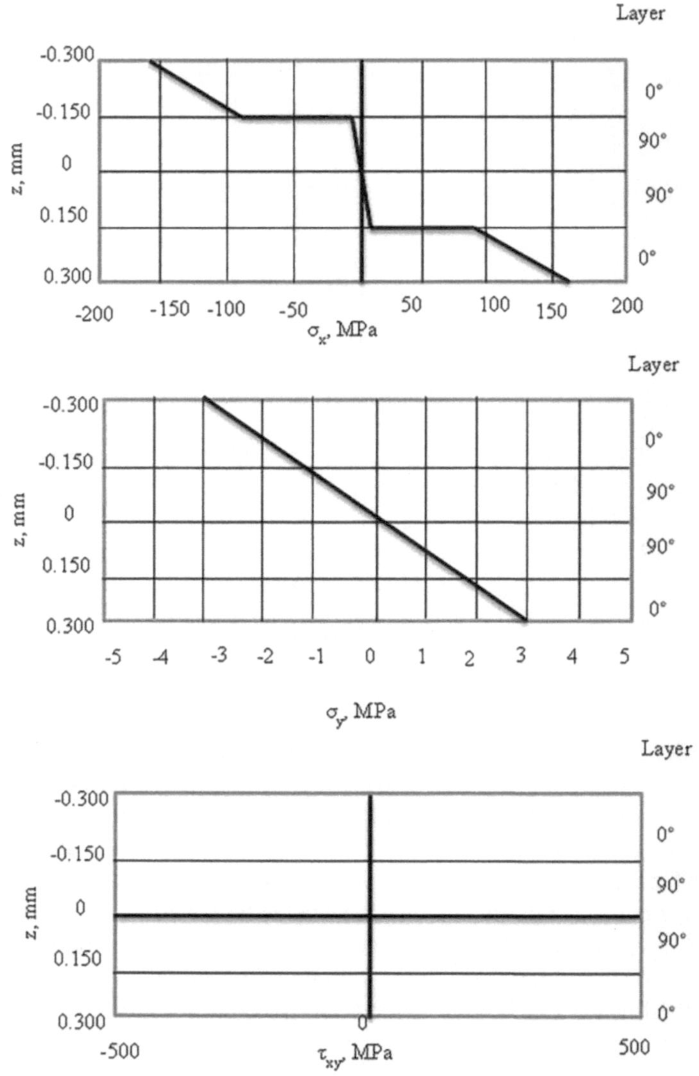

FIGURE 9.19 Stress distribution through the thickness of $[0/90]_s$ laminate subjected to $k_x^0 = 3.33\ \text{m}^{-1}$.

anticlastic curvature is suppressed. To overcome the tendency to develop anticlastic curvature in the y direction, a stress component, σ_y, is required.

The distribution of σ_y as shown in Figure 9.19 is the distribution required to have this particular laminate remain flat in the y direction, yet cylindrical in the x direction.

The strain distribution in the principal material coordinate system can be determined using the transformation equations as discussed earlier. In the 0° layers, using the transformation matrix $[T]$ for the 0° layers yields:

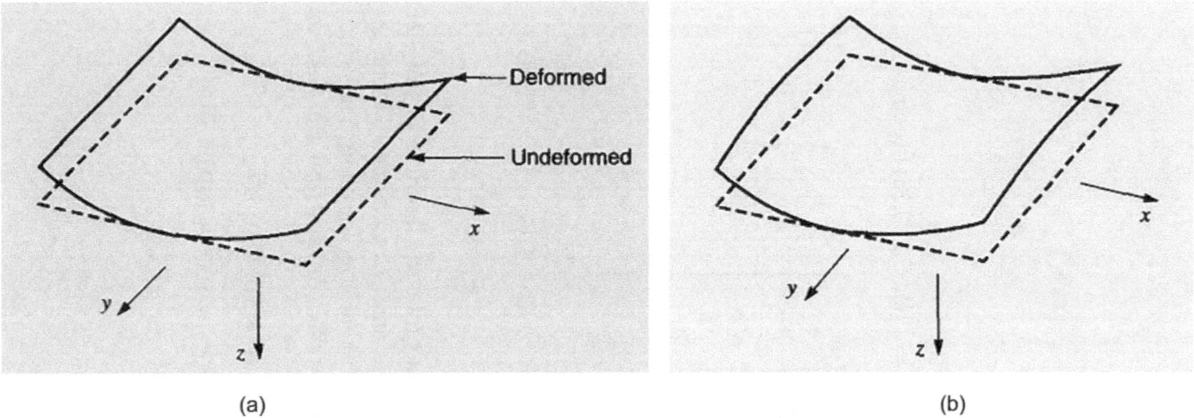

FIGURE 9.20 Deformations of laminated plates. (a) Cylindrical shape defined by k_x^0, with $k_y^0 = 0$, and (b) anticlastic curvature, $k_y^0 \neq 0$.

$$\begin{Bmatrix} \varepsilon_1 \\ \varepsilon_2 \\ \dfrac{\gamma_{12}}{2} \end{Bmatrix} = \begin{bmatrix} 1 & 0 & 0 \\ 0 & 1 & 0 \\ 0 & 0 & 1 \end{bmatrix} \begin{Bmatrix} 3.33z \\ 0 \\ 0 \end{Bmatrix}. \tag{9.46}$$

In the 0° layers:

$$\begin{aligned} \varepsilon_1 &= 3.33z \\ \varepsilon_2 &= 0 \\ \gamma_{12} &= 0. \end{aligned} \tag{9.47}$$

For the 90° layers, $m = 0$ and $n = 1$ and Eq. (9.28) becomes:

$$\begin{Bmatrix} \varepsilon_1 \\ \varepsilon_2 \\ \dfrac{\gamma_{12}}{2} \end{Bmatrix} = [T(90°)] \begin{Bmatrix} \varepsilon_x \\ \varepsilon_y \\ \dfrac{\gamma_{xy}}{2} \end{Bmatrix} = \begin{bmatrix} 0 & 1 & 0 \\ 1 & 0 & 0 \\ 0 & 0 & -1 \end{bmatrix} \begin{Bmatrix} 3.33z \\ 0 \\ 0 \end{Bmatrix}. \tag{9.48}$$

In the 90° layers:

$$\begin{aligned} \varepsilon_1 &= 0 \\ \varepsilon_2 &= 3.33z \\ \gamma_{12} &= 0. \end{aligned} \tag{9.49}$$

These principal material system strains, as shown in Figure 9.21, are piecewise linear; the discontinuities result because, as mentioned before, they represent strains in different directions at the different layer locations. The principal material system stresses in the 0° layers can be written directly from Eq. (9.42) as:

$$\begin{aligned} \sigma_1 &= 519{,}000z \text{ MPa} \\ \sigma_2 &= 10{,}060z \text{ MPa} \\ \tau_{12} &= 0. \end{aligned} \tag{9.50}$$

For the 90° layers:

$$\begin{Bmatrix} \sigma_1 \\ \sigma_2 \\ \tau_{12} \end{Bmatrix} = [T(90°)] \begin{Bmatrix} \sigma_x \\ \sigma_y \\ \tau_{xy} \end{Bmatrix} = \begin{bmatrix} 0 & 1 & 0 \\ 1 & 0 & 0 \\ 0 & 0 & -1 \end{bmatrix} \begin{Bmatrix} 40{,}500z \\ 10{,}060z \\ 0 \end{Bmatrix} \tag{9.51}$$

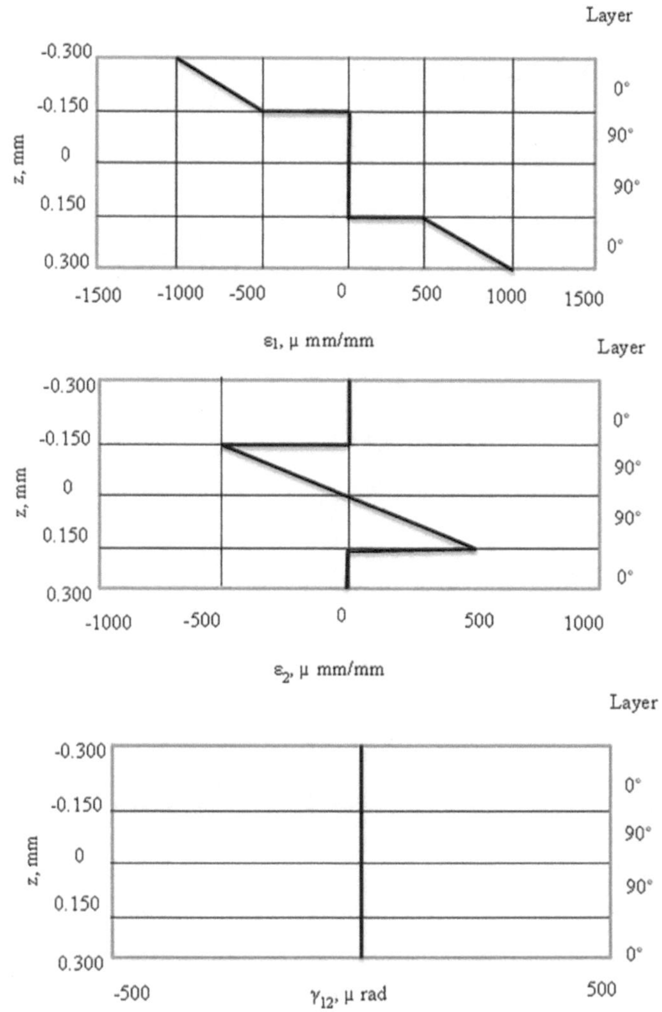

FIGURE 9.21 Principal material system strain distribution through the thickness of $[0/90]_s$ laminate subjected to $k_x^0 = 3.33\ \text{m}^{-1}$.

or,

$$\sigma_1 = 10{,}060z \text{ MPa}$$

$$\sigma_2 = 40{,}500z \text{ MPa}$$

$$\tau_{12} = 0. \tag{9.52}$$

Figure 9.22 shows the piecewise linear variations of the principal material system stresses, and the distributions are very dissimilar to anyone would see with metals. Again, consider a laminate constructed of four aluminum layers and subjected to the

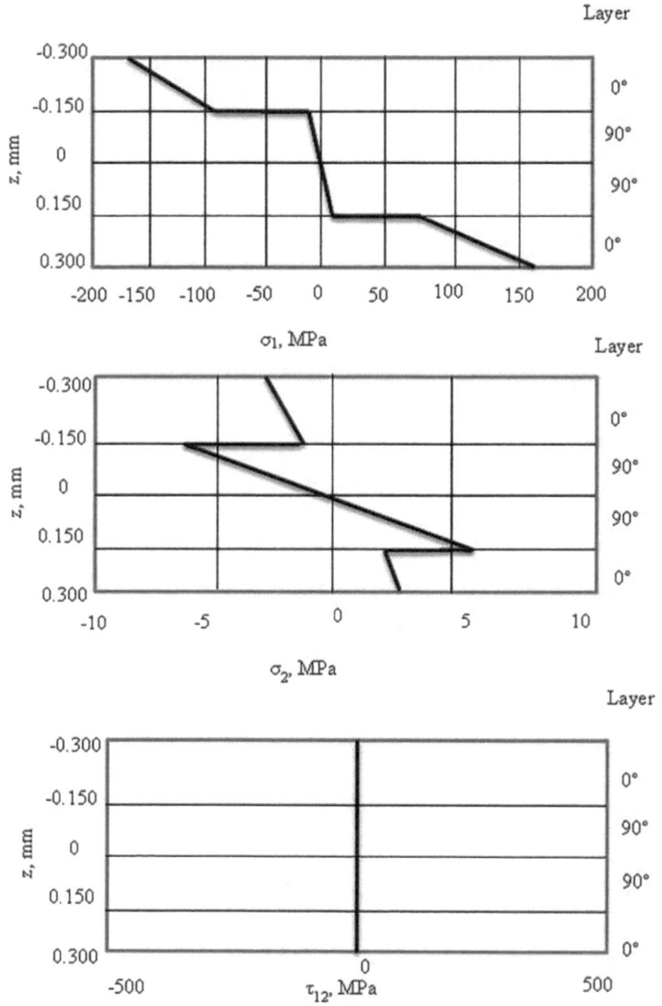

FIGURE 9.22 Principal material system stress distribution through the thickness of $[0/90]_s$ laminate subjected to $k_x^0 = 3.33\ \text{m}^{-1}$.

reference surface deformations given by Eq. (9.37). Equation (9.38) remains valid, and the strain distribution for this case is that shown by Figure 9.18. The four layers of aluminum act as one, and thus, Eq. (9.15) with the reduced stiffnesses of aluminum is valid for the entire thickness, $-0.300\ \text{mm} \le z \le +0.300\,\text{mm}$; as shown in Figure 9.23, that equation leads to:

$$\sigma_x = 265{,}000z\ \text{MPa}$$
$$\sigma_y = 79{,}600z\ \text{MPa}$$
$$\tau_{xy} = 0. \tag{9.53}$$

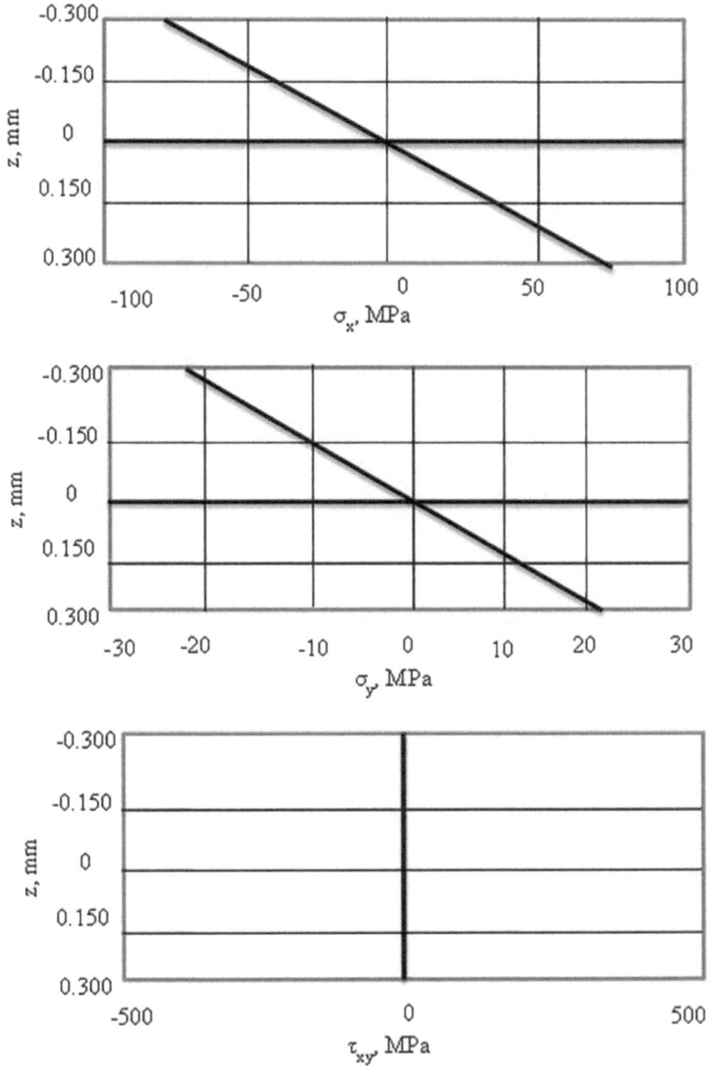

FIGURE 9.23 Stress distribution through the thickness of a four-layer aluminum laminate subjected to $k_x^0 = 3.33\ \text{m}^{-1}$.

The simple linear variation of σ_x for the aluminum is a contrast to the piecewise linear nature of the distribution of σ_x for the laminate, as shown in Figure 9.19, which is caused in the latter case by the discontinuous nature of the material properties.

9.7 FORCE AND MOMENT RESULTANTS

To keep the reference surface of a laminated plate from exhibiting anticlastic curvature at the point (x, y), stresses σ_y are required, as shown in Figure 9.19. Equally important, though not discussed specifically, to produce the specified curvature in the x direction, σ_x stresses are required, which are also illustrated in Figure 9.19. From the perspective of an entire laminated plate, as shown in Figure 9.24, bending moments are required along the edge of the plate to produce the deformations of Figure 9.20a. Similarly, to keep a laminated plate deformed as specified by the first example, Eq. (9.16), Figure 9.14 indicates that in-plane stresses in the x and y directions are required. These stresses are manifested as in-plane loads along the edges of the plate, as shown in Figure 9.25. The loads and moments required to produce the specified mid-plane deformations in any particular problem are actually *integrals* through the laminate thickness of the stresses. We refer to these integrals through the

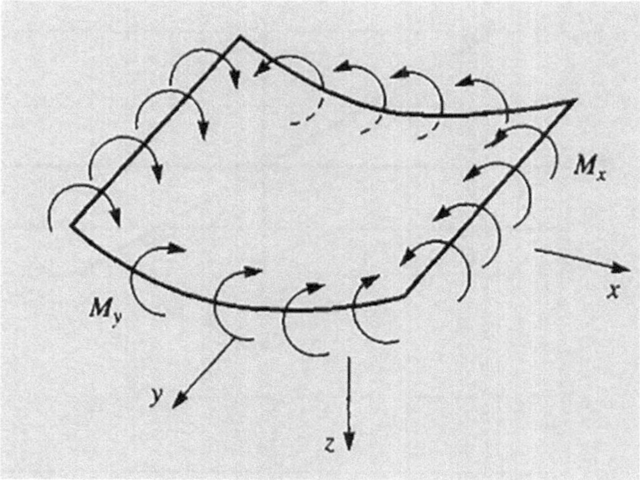

FIGURE 9.24 Moments required for deforming a laminated plate into a cylindrical shape.

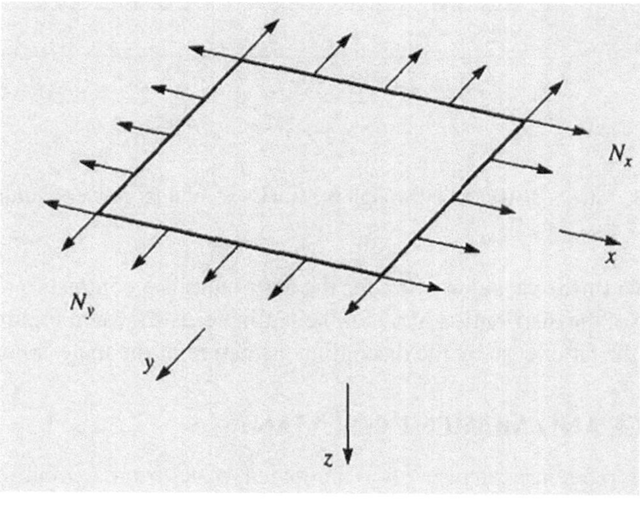

FIGURE 9.25 Forces required for stretching a laminated plate in the x direction.

thickness as *stress resultants*. Specifically, the in-plane force resultant in the x direction, N_x, is defined as:

$$N_x = \int_{-\frac{H}{2}}^{+\frac{H}{2}} \sigma_x \, dz, \qquad (9.54)$$

where H is the thickness of the laminate. We define the bending moment resultant due to σ_x as:

$$M_x = \int_{-\frac{H}{2}}^{+\frac{H}{2}} \sigma_x z \, dz. \qquad (9.55)$$

The force resultant has the units of force per unit length, and the moment resultant has the units of moment per unit length. The unit of length is length in the y direction. For the problem discussed in Section 9.6.1:

$$N_x = \int_{-\frac{H}{2}}^{+\frac{H}{2}} \sigma_x \, dz = \int_{z_0}^{z_1} \sigma_x \, dz + \int_{z_1}^{z_2} \sigma_x \, dz + \int_{z_2}^{z_3} \sigma_x \, dz + \int_{z_3}^{z_4} \sigma_x \, dz. \qquad (9.56)$$

Substituting from Eqs. (9.24) and (9.25), the numerical values for the stresses that are valid over the appropriate range of z, and from Eq. (9.17) substituting for the values of the interface locations z_k, we find:

$$\begin{aligned} N_x = &\int_{-300\times10^{-6}}^{-150\times10^{-6}} 155.7\times10^6 \, dz + \int_{-150\times10^{-6}}^{0} 12.16\times10^6 \, dz \\ &+ \int_{0}^{150\times10^{-6}} 12.16\times10^6 \, dz + \int_{150\times10^{-6}}^{300\times10^{-6}} 155.7\times10^6 \, dz \end{aligned}. \qquad (9.57)$$

Because the stresses in each integrand are constant, they can be taken outside the integrals, resulting in:

$$\begin{aligned} N_x = &\,155.7\times10^6 \int_{-300\times10^{-6}}^{-150\times10^{-6}} dz + 12.16\times10^6 \int_{-150\times10^{-6}}^{0} dz \\ &+ 12.16\times10^6 \int_{0}^{150\times10^{-6}} dz + 155.7\times10^6 \int_{150\times10^{-6}}^{300\times10^{-6}} dz \end{aligned}. \qquad (9.58)$$

The integrals on z are just the thickness of each layer, here 0.150 mm, and so:

$$N_x = \left[(155.7 + 12.16 + 12.16 + 155.7)\times10^6\right]\left(150\times10^{-6}\right) = 50{,}400 \text{ N m}^{-1}. \qquad (9.59)$$

Force resultants based on σ_y and τ_{xy} can be defined in a similar manner, specifically:

$$N_y = \int_{-\frac{H}{2}}^{+\frac{H}{2}} \sigma_y \, dz \qquad (9.60)$$

$$N_{xy} = \int_{-\frac{H}{2}}^{+\frac{H}{2}} \tau_{xy}\, dz\ , \tag{9.61}$$

where the latter is referred to as the shear force resultant. For the problem discussed in Section 9.6.1, from Eqs. (9.24) and (9.25):

$$N_y = (4)\left(3.02 \times 10^6\right)\left(150 \times 10^{-6}\right) = 1809\ \text{N}\,\text{m}^{-1} \tag{9.62}$$
$$N_{xy} = 0.$$

Because the definitions of the stress resultants are integrals with respect to z, and because z is measured from the reference surface, the stress resultants should be considered forces per unit in-plane length acting at the reference surface. The proper interpretation is that if at a point (x, y) on the reference surface of a $[0/90]_s$ graphite–epoxy laminate, the force resultants are:

$$N_x = 50{,}400\ \text{N}\,\text{m}^{-1},\quad N_y = 1809\ \text{N}\,\text{m}^{-1},\quad N_{xy} = 0, \tag{9.63}$$

Then that point on the reference surface will deform in the fashion given by Eq. (9.16), namely:

$$\begin{aligned}
\varepsilon_x^0(x, y) &= 1000 \times 10^{-6} \\
\varepsilon_y^0(x, y) &= 0 \\
\gamma_{xy}^0(x, y) &= 0 \\
k_x^0(x, y) &= 0 \\
k_y^0(x, y) &= 0 \\
k_{xy}^0(x, y) &= 0.
\end{aligned} \tag{9.64}$$

The stress and strain distributions of Figures 9.13, 9.14, 9.16, and 9.17 result from the application of these stress resultants. To further clarify the physical meaning of Eqs. (9.63) and (9.64), consider, for example, a $[0/90]_s$ graphite-reinforced laminate with a length L_x in the x direction of 0.250 m and a width L_y in the y direction of 0.125 m. To have every point on the entire 0.0312 m^2 reference surface subjected to the deformations of Eq. (9.64) requires a load:

$$N_x \times L_y = 50{,}400\ \text{N}\,\text{m}^{-1} \times 0.125\ \text{m} = 6300\ \text{N} \tag{9.65a}$$

in the x direction and a load:

$$N_x \times L_x = 1809\ \text{N}\,\text{m}^{-1} \times 0.250\ \text{m} = 452\ \text{N} \tag{9.65b}$$

in the y direction.

These loads, as shown in Figure 9.26, should be uniformly distributed along the appropriate edges. The displacements of the reference surface for this case can be determined by considering Eqs. (9.9–9.11) and (9.64), namely:

$$\varepsilon_x^0(x, y) = \frac{\partial u^0(x, y)}{\partial x} = 1000 \times 10^{-6}$$

$$\varepsilon_y^0(x, y) = \frac{\partial v^0(x, y)}{\partial y} = 0. \tag{9.66}$$

Integrating these two equations results in:

$$u^0(x, y) = 0.001x + g(y)$$

$$v^0(x, y) = h(x), \tag{9.67}$$

where $g(y)$ and $h(x)$ are the functions of integration resulting from integrating partial derivatives. As a result, because the shear strain $\gamma_{xy}^0(x, y)$ is zero, that is,

$$\gamma_{xy}^0(x, y) = \frac{\partial v^0(x, y)}{\partial x} + \frac{\partial u^0(x, y)}{\partial y} = \frac{dh(x)}{dx} + \frac{dg(y)}{dy} = 0. \tag{9.68}$$

The quantity $\frac{dg(y)}{dy}$ is a function of y only, and the quantity $\frac{dh(x)}{dx}$ is a function of x only, and Eq. (9.68) specifies that they add to yield a constant, namely, zero. The only way that a function of x can add to a function of y to produce a constant is if

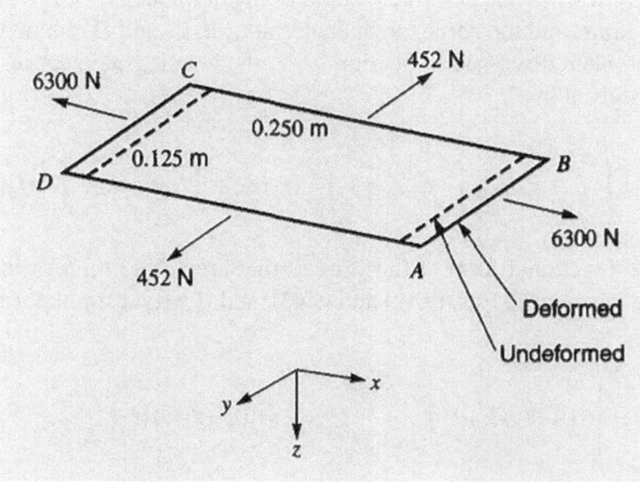

FIGURE 9.26 Forces required to produce a state of deformation: $\varepsilon_x^0 = 1000 \times 10^{-6}$ in the $[0/90]_s$ laminate.

the two functions are constants. Here, they must be the same constant, differing by a sign. Specifically:

$$\frac{dg(y)}{dy} = C_1, \quad \frac{dh(x)}{dx} = -C_1. \tag{9.69}$$

The result of integrating these two equations is:

$$\begin{aligned} g(y) &= C_1 y + C_2 \\ h(x) &= -C_1 x + C_3. \end{aligned} \tag{9.70}$$

The constants C_2 and C_3 represent the rigid body translations of the laminate in the x and y directions, respectively, and C_1 represents the rigid body rotation about the z axis. Setting u^0 and v^0 to zero at the origin of coordinate system, $x = 0$ and $y = 0$, thereby suppressing the rigid body translations, and arbitrarily suppressing the rigid body rotation about the z axis by setting C_1 to zero, results in expressions for the displacements of the laminate reference surface, namely:

$$\begin{aligned} u^0(x, y) &= 0.001x \\ v^0(x, y) &= 0 \\ w^0(x, y) &= 0. \end{aligned} \tag{9.71}$$

The third equation results from the fact that curvatures do not develop due to in-plane loads for this laminate. Though it is not needed for the determination of Eq. (9.71), we assume here that the origin of the x-y-z global coordinate system is at the geometric center of the laminate. With the displacements of Eq. (9.71), we can see that u^0 varies linearly with x. The deformed length of the laminate is 250.25 mm, the width remains the same, and the corner right angles at A, B, C, and D remain right.

For the problem discussed in Section 9.6.2, the bending moment resultant in the x direction is given by:

$$M_x = \int_{-\frac{H}{2}}^{+\frac{H}{2}} \sigma_x z\,dz = \int_{z_0}^{z_1} \sigma_x z\,dz + \int_{z_1}^{z_2} \sigma_x z\,dz + \int_{z_2}^{z_3} \sigma_x z\,dz + \int_{z_3}^{z_4} \sigma_x z\,dz. \tag{9.72}$$

If we use the functional form of the stresses that are valid in the various ranges of the interface locations z, Eqs. (9.42) and (9.43), and if we substitute for the values of z_k, then:

$$\begin{aligned} M_x = \Bigg\{ & 519{,}000 \times 10^6 \int_{-300\times10^{-6}}^{-150\times10^{-6}} z^2\,dz + 40{,}500 \times 10^6 \int_{-150\times10^{-6}}^{0} z^2\,dz \\ & + 40{,}500 \times 10^6 \int_{0}^{150\times10^{-6}} z^2\,dz + 519{,}000 \times 10^6 \int_{150\times10^{-6}}^{300\times10^{-6}} z^2\,dz \Bigg\}. \end{aligned} \tag{9.73}$$

This results in:

$$M_x = 8.27 \text{ Nm m}^{-1}. \tag{9.74}$$

The sign is consistent with the sense of the moments as shown in Figure 9.24; the moments there are shown in a positive sense. Moment resultants associated with the other two stresses can be defined as:

$$M_y = \int_{-\frac{H}{2}}^{+\frac{H}{2}} \sigma_y z\, dz \tag{9.75}$$

$$M_{xy} = \int_{-\frac{H}{2}}^{+\frac{H}{2}} \tau_{xy} z\, dz. \tag{9.76}$$

The latter is referred to as the twisting moment resultant. For this problem:

$$\begin{aligned} M_y &= \int_{-\frac{H}{2}}^{+\frac{H}{2}} \sigma_y z\, dz = \int_{z_0}^{z_1} \sigma_y z\, dz + \int_{z_1}^{z_2} \sigma_y z\, dz + \int_{z_2}^{z_3} \sigma_y z\, dz + \int_{z_3}^{z_4} \sigma_y z\, dz \\ &= \left(10{,}060 \times 10^6\right)\left\{\int_{-300\times10^{-6}}^{300\times10^{-6}} z^2\, dz\right\} = 0.1809 \text{ Nm m}^{-1} \end{aligned} \tag{9.77}$$

Using Eqs. (9.42) and (9.43):

$$M_{xy} = 0. \tag{9.78}$$

Take care to note the sign of the bending moment resultant M_y, as it is consistent with the sense as shown in Figure 9.24. Finally, the definitions of the moment resultants imply that the moments are taken about the point $z = 0$, that is, the reference surface. The proper interpretation is that if at a point (x, y) on the reference surface of a $[0/90]_s$ graphite–epoxy laminate, the moment resultants are:

$$M_x = 8.27 \text{ Nm m}^{-1}, \quad M_y = 0.1809 \text{ Nm m}^{-1}, \quad M_{xy} = 0. \tag{9.79}$$

Then, that point on the reference surface will deform in the fashion given by Eq. (9.37), namely:

$$\begin{aligned} \varepsilon_x^0(x, y) &= 0 \\ \varepsilon_y^0(x, y) &= 0 \\ \gamma_{xy}^0(x, y) &= 0 \\ k_x^0(x, y) &= 3.33 \text{ m}^{-1} \end{aligned} \tag{9.80}$$

$$k_y^0(x, y) = 0$$

$$k_{xy}^0(x, y) = 0.$$

The stress and strain distributions of Figures 9.18, 9.19, 9.21, and 9.22 result from the application of these stress resultants. Consider again the 0.250 m long by 0.125 m wide $[0/90]_s$ laminate; if it is to have the deformations of Eq. (9.80) at every point on its entire 0.0312 m^2 reference surface, then along the 0.125 m widthwise side, there must be a total bending moment of:

$$M_x \times L_y = 8.27 \times 0.1255 \text{ m} = 1.033 \text{ Nm} \tag{9.81a}$$

and along the 0.250 m lengthwise edge, there must be a total bending moment of:

$$M_y \times L_x = 0.1809 \times 0.250 \text{ m} = 0.0452 \text{ Nm}. \tag{9.81b}$$

Figure 9.27 illustrates these moments, and it is assumed they are uniformly distributed along the edges. Here, the double-headed arrows are used to indicate moments, as opposed to the curved arrows in Figure 9.24. We are introducing the double-headed arrow notation to avoid confusion in later figures. The sense of the moments along the 0.250 m lengthwise edges is such as to counter anticlastic curvature effects. The out-of-plane displacements of the reference surface are obtained from the definitions of the curvatures and Eq. (9.80) as follows:

$$k_x^0(x, y) = -\frac{\partial^2 w^0(x, y)}{\partial x^2} = 3.33 \tag{9.82}$$

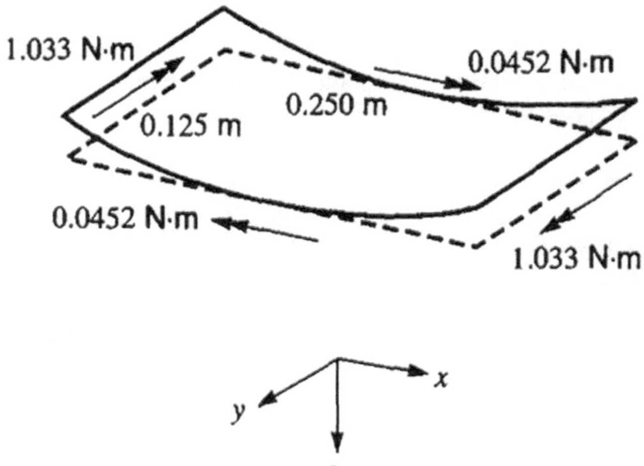

FIGURE 9.27 Moments required to produce a state of deformation: $k_x^0 = 3.33\text{ m}^{-1}$ in the $[0/90]_s$ laminate.

$$k_y^0(x, y) = -\frac{\partial^2 w^0(x, y)}{\partial y^2} = 0 \tag{9.83}$$

$$k_{xy}^0 = -2\frac{\partial^2 w^0(x, y)}{\partial x \partial y} = 0. \tag{9.84}$$

Integrating the first two equations results in two different expressions for $w^0(x, y)$:

$$w^0(x, y) = -\frac{1}{2}3.33x^2 + q(y)x + r(y)$$

$$w^0(x, y) = s(x)y + t(x), \tag{9.85}$$

where $q(y)$, $r(y)$, $s(x)$, and $t(x)$ are the arbitrary functions of integration. Computing the reference surface twisting curvature, which is equal to zero, from both expressions leads to:

$$k_{xy}^0 = -2\frac{\partial^2 w^0(x, y)}{\partial x \partial y} = -2\frac{dq(y)}{dy} = 0 \tag{9.86}$$

$$k_{xy}^0 = -2\frac{\partial^2 w^0(x, y)}{\partial x \partial y} = -2\frac{ds(x)}{dx} = 0.$$

From these two equations, we can conclude that $q(y)$ and $s(x)$ are the constants; that is:

$$q(y) = K_1$$

$$s(x) = K_2. \tag{9.87}$$

Thus, the two expressions for $w^0(x, y)$ become:

$$w^0(x, y) = -\frac{1}{2}3.33x^2 + K_1x + r(y)$$

$$w^0(x, y) = K_2y + t(x). \tag{9.88}$$

As there can be only one expression for $w^0(x, y)$, we conclude that:

$$r(y) = K_2y + K_3 \tag{9.89}$$

$$t(x) = -\frac{1}{2}3.33x^2 + K_1x + K_3,$$

with K_3 being a constant that is common to both functions. The final expression for $w^0(x, y)$ is:

$$w^0(x, y) = -\frac{1}{2}3.33x^2 + K_1x + K_2y + K_3. \tag{9.90}$$

The constants K_1 and K_2 represent the rigid body rotations of the laminate about the y and x axes, respectively, and K_3 represents the rigid body translation in the z direction. If w^0, $\frac{\partial w^0}{\partial x}$, and $\frac{\partial w^0}{\partial y}$ are arbitrarily set equal to zero at the origin of the coordinate system, $x = 0$ and $y = 0$, then to suppress rigid body motion results in requiring K_1, K_2, and K_3 to be zero. The deformed shape of the laminate is thus given by:

$$\begin{aligned} u^0(x, y) &= 0 \\ v^0(x, y) &= 0 \\ w^0(x, y) &= -\frac{1}{2}3.33x^2. \end{aligned} \tag{9.91}$$

The following couplings are very important:

i. The coupling of the deformations of the reference surface with the distribution of strains through the thickness
ii. The coupling of the distribution of strains through the thickness with the distribution of stresses through the thickness
iii. The coupling of the distribution of stresses through the thickness with the stress resultants that act at the reference surface.

The couplings begin to tie together the analysis of laminates in a cause-and-effect relation. Generally, the loads on a laminate are specified and we want to know the resulting stresses. The effect is the reference surface deformations and stresses, while the cause is the stress resultants, or loads. If we know the reference surface deformations, we can determine the stresses. At this point, we cannot determine the reference surface deformations from the stress resultants; we can only compute the stress resultants, given that we know the reference surface deformations. Later, we will be able to compute the reference surface deformations from the stress resultants, and thus determine the thickness distribution of the strains and stresses from the given loads.

REFERENCES

1. Herakovich C.T. (1989), "Free edge effects in laminated composites." in *Handbook of Composites:* vol. 2, *Structures and Design* (C. T. Herakovich and Y. M. Tarnopol'skii, eds.), Elsevier Science Publishing Co, New York.
2. Armanios E.A. (1989), *Interlaminar Fracture of Composites*, Trans Tech. Publications, Aedermannsdorf, Switzerland.
3. Renton W.J. and Vinson J.R. (1977), "Analysis of adhesively bonded joints between panels of composite materials", *Transactions of the ASME, Journal of Applied Mechanics*, Vol. 44, pp. 101–106.
4. Kassapoglou C. (1993), "Calculation of stresses at skin-stiffener interfaces of composite stiffened panels under shear loads", *International Journal of Solids and Structures*, Vol. 30(11), pp. 1491–1501.
5. Harrison P.N. and Johnson E.R. (1996), "A mixed variational formulation for interlaminar stresses in thickness-tapered composite laminates", *International Journal of Solids and Structures*, Vol. 33(16), pp. 2377–2399.

6. Love A.E.H. (1892), *A Treatise on the Mathematical Theory of Elasticity*, Cambridge University Press, London (also 4th ed., 1944, Dover Publications, Inc., New York).
7. Hodge P.G. Jr. (1958), "The mathematical theory of plasticity," in *Elasticity and Plasticity* (J. N. Goodier and P. G. Hodge, Jr., eds.), John Wiley & Sons, Inc., New York, p. 81.
8. Kreyszig E. (1967), *Advanced Engineering Mathematics*, 2nd ed., John Wiley & Sons, Inc., New York, p. 416.
9. Fung Y.C. (1965), *Foundations of Solid Mechanics*, Prentice-Hall, Inc., Englewood Cliffs, NJ, p. 285.
10. Green G. (1839), "On the laws of reflexion and refraction of light at the common surface of two non-crystallized media", *Transactions of the Cambridge Philosophical Society*, Vol. 7, p. 121.
11. Popov E.P. (1990), *Engineering Mechanics of Solids*, Prentice Hall, Englewood Cliffs, NJ.
12. Hyer M.W. (1998), *Stress Analysis of Fiber Reinforced Composite Materials*, McGraw-Hill, Singapore.
13. Ashton J.E. and Whitney J.M. (1970), *Theory of Laminated Plates*, Technomic Publishing Co., Lancaster, PA.
14. Jones R. M. (1975), *Mechanics of Composite Materials*, McGraw-Hill, New York.

10 The *ABD* Matrix

In the previous chapters, the classical lamination theory was developed enabling us to calculate the strain distributions through the thickness of the laminate in terms of the strains and curvatures at a point (x, y) on the reference surface. Using the stress–strain relations, the stress distributions were obtained. These stresses and strains were then determined for the principal material coordinate system by making use of the transformation equations. The stress resultants, namely, the force and moment resultants, were then predicted from the given strains and curvatures. All these steps are the result of the plane stress assumption and the Kirchhoff hypothesis. Figure 10.1 illustrates the connection between these steps. What remains in the development of classical lamination theory is to be able to specify the force and moment resultants acting at a point (x, y) on the reference surface, and then to be able to compute the strains and curvatures of that point on the reference surface that these resultants cause. We want to fill in the missing link in the upper-left portion of Figure 10.1; namely, compute the reference surface strains and curvatures, knowing the stress resultants. With these computed reference surface strains and curvatures, we can then, as in the previous examples, compute the strain and stress distributions through the thickness of the laminate. Relating the stress resultants to the reference surface strains and curvatures is an important step [1,2]. In the application of composite materials to structures, we are often given the forces and moments that act on a laminate, and we want to know the stresses and strains that are caused by these loads. Having a relation between the force and moment resultants, and the strains and curvatures at a point (x, y) on the reference surface will allow us to do this. In the following sections, this relationship has been developed.

10.1 FORCE AND MOMENT RESULTANTS

The stress resultants in the x direction, N_x, in the y direction, N_y, and in shear, N_{xy}, are defined as:

$$N_x = \int_{-\frac{H}{2}}^{+\frac{H}{2}} \sigma_x \, dz$$

$$N_y = \int_{-\frac{H}{2}}^{+\frac{H}{2}} \sigma_y \, dz \tag{10.1}$$

$$N_{xy} = \int_{-\frac{H}{2}}^{+\frac{H}{2}} \tau_{xy} \, dz.$$

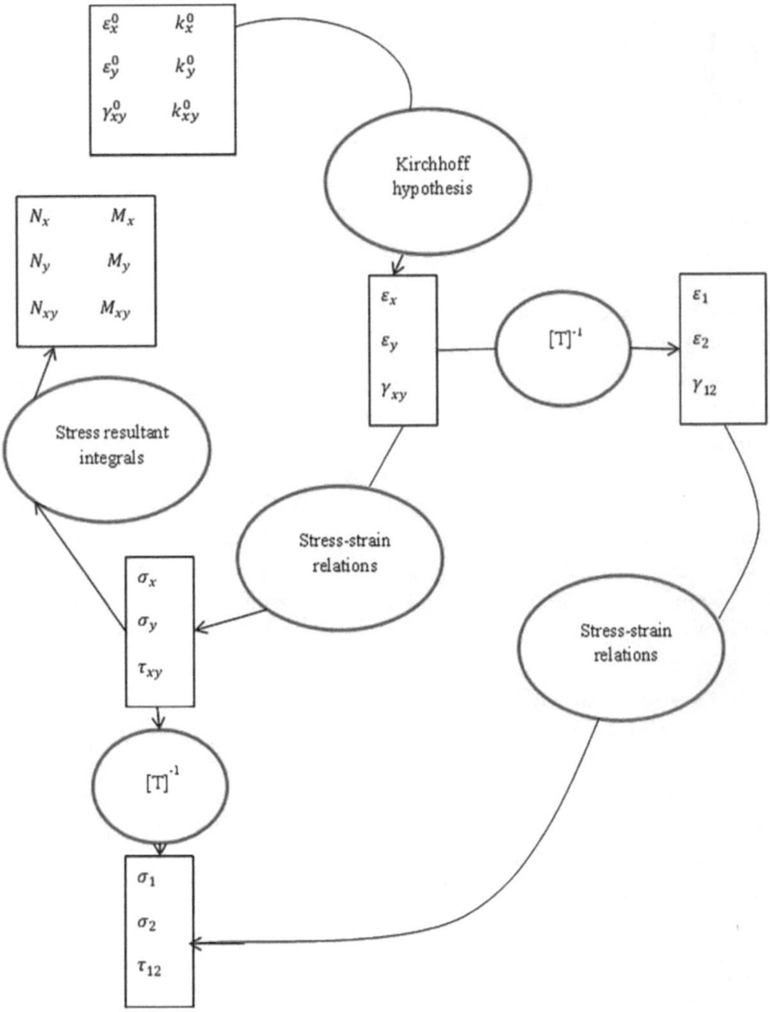

FIGURE 10.1 Interrelations among the concepts developed so far showing the missing link in the upper-left-hand corner.

The resultants N_x and N_y will be referred to as the *normal* force resultants, and N_{xy} will be referred to as the *shear* force resultant. Figure 10.2 illustrates a small element of laminate surrounding a point (x, y) on the geometric mid-plane, and indicates the directions of these three stress resultants [3–6]. The units of the stress resultants are force per unit length; the unit of length is a unit of length in the x direction or in the y direction. If the small element of laminate has dimensions dx by dy, then the force in the x direction due to N_x, the force in the x direction due to N_{xy}, the force in the y direction due to N_y, and the force in the y direction due to N_{xy} are, respectively:

$$N_x dy \quad N_{xy} dx \quad N_y dx \quad N_{xy} dy. \tag{10.2}$$

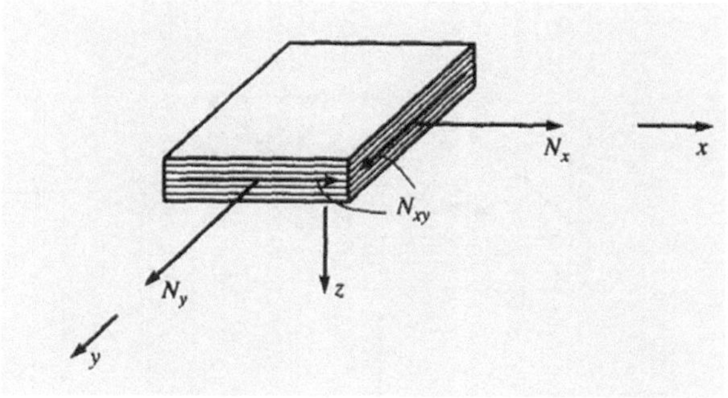

FIGURE 10.2 Definitions of force resultants N_x, N_y, and N_{xy}.

The moment resultants M_x, M_y, and M_{xy} are defined as:

$$M_x = \int_{-\frac{H}{2}}^{+\frac{H}{2}} \sigma_x z\, dz$$

$$M_y = \int_{-\frac{H}{2}}^{+\frac{H}{2}} \sigma_y z dz \tag{10.3}$$

$$M_{xy} = \int_{-\frac{H}{2}}^{+\frac{H}{2}} \tau_{xy} z\, dz.$$

The resultants M_x and M_y will be referred to as the bending moment resultants, and M_{xy} will be referred to as the twisting moment resultant. In Figure 10.3, the sense of these three moment resultants is illustrated, and the senses shown are important. The sense illustrated in Figure 10.3 is consistent with the cross-product definition of moments learned in elementary mechanics courses (i.e., $\vec{M} = \vec{r} \times \vec{F}$). The units of the moment resultants are moment per unit length, again the unit of length being a unit of length in either the x direction or the y direction. If a small element of the laminate has dimensions dx by dy, then the *moments* about the positive y and the positive x axis due to M_x, M_y, and M_{xy} are, respectively, written as:

$$M_x dy \quad M_{xy} dx \quad -M_y dx \quad -M_{xy} dy. \tag{10.4}$$

As the definitions of the moment resultants involve integrals with respect to the coordinate z and because z is measured relative to the reference surface, the moment resultants have to be thought of as producing moments on the reference surface. The force resultants can be considered to act anywhere through the thickness of the laminate. But when force resultants are used in conjunction with moment resultants, it is most consistent to think of the force resultants as producing forces on the reference surface.

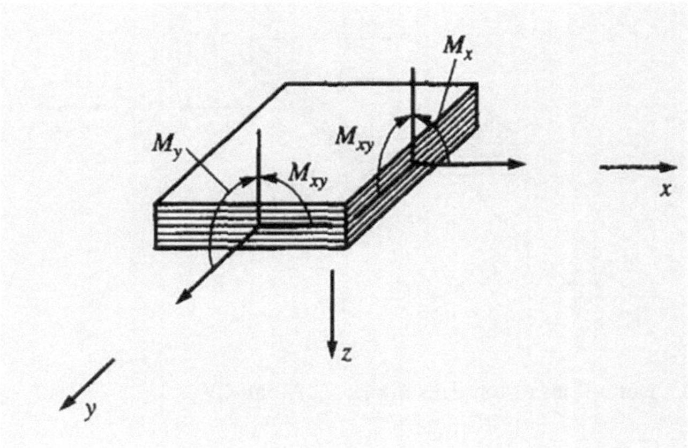

FIGURE 10.3 Definitions of moment resultants M_x, M_y, and M_{xy}.

10.2 THE ABD MATRIX

From the stress–strain relations, the stresses can be written in terms of the strains by the following relation:

$$\begin{Bmatrix} \sigma_x \\ \sigma_y \\ \tau_{xy} \end{Bmatrix} = \begin{bmatrix} \bar{Q}_{11} & \bar{Q}_{12} & \bar{Q}_{16} \\ \bar{Q}_{12} & \bar{Q}_{22} & \bar{Q}_{26} \\ \bar{Q}_{16} & \bar{Q}_{26} & \bar{Q}_{66} \end{bmatrix} \begin{Bmatrix} \varepsilon_x \\ \varepsilon_y \\ \gamma_{xy} \end{Bmatrix}. \tag{10.5}$$

The strains can, in turn, be written in terms of the strains and curvatures of the reference surface to produce another relation as follows:

$$\begin{Bmatrix} \sigma_x \\ \sigma_y \\ \tau_{xy} \end{Bmatrix} = \begin{bmatrix} \bar{Q}_{11} & \bar{Q}_{12} & \bar{Q}_{16} \\ \bar{Q}_{12} & \bar{Q}_{22} & \bar{Q}_{26} \\ \bar{Q}_{16} & \bar{Q}_{26} & \bar{Q}_{66} \end{bmatrix} \begin{Bmatrix} \varepsilon_x^0 + zk_x^0 \\ \varepsilon_y^0 + zk_y^0 \\ \gamma_{xy}^0 + zk_{xy}^0 \end{Bmatrix}. \tag{10.6}$$

The reference surface strains and curvatures, by definition, do not depend on z. Let us examine the expression for the normal force resultant N_x by substituting for σ_x from Eq. (10.6) into the integrand of the first equation of Eq. (10.1), specifically:

$$N_x = \int_{-\frac{H}{2}}^{+\frac{H}{2}} \sigma_x \, dz = \int_{-\frac{H}{2}}^{+\frac{H}{2}} \left\{ \bar{Q}_{11}\left(\varepsilon_x^0 + zk_x^0\right) + \bar{Q}_{12}\left(\varepsilon_y^0 + zk_y^0\right) + \bar{Q}_{16}\left(\gamma_{xy}^0 + zk_{xy}^0\right) \right\} dz. \tag{10.7}$$

Expanding the integrand leads to:

$$N_x = \int_{-\frac{H}{2}}^{+\frac{H}{2}} \left\{ \bar{Q}_{11}\varepsilon_x^0 + \bar{Q}_{11}zk_x^0 + \bar{Q}_{12}\varepsilon_y^0 + \bar{Q}_{12}zk_y^0 + \bar{Q}_{16}\gamma_{xy}^0 + \bar{Q}_{16}zk_{xy}^0 \right\} dz. \tag{10.8}$$

Because the reference surface deformations are not functions of *z*, they can be taken outside of the integrals, resulting in:

$$N_x = \left\{\int_{-\frac{H}{2}}^{+\frac{H}{2}} \bar{Q}_{11}\,dz\right\}\varepsilon_x^0 + \left\{\int_{-\frac{H}{2}}^{+\frac{H}{2}} \bar{Q}_{11} z\,dz\right\}k_x^0 + \left\{\int_{-\frac{H}{2}}^{+\frac{H}{2}} \bar{Q}_{12}\,dz\right\}\varepsilon_y^0 + \left\{\int_{-\frac{H}{2}}^{+\frac{H}{2}} \bar{Q}_{12} z\,dz\right\}k_y^0$$
$$+\left\{\int_{-\frac{H}{2}}^{+\frac{H}{2}} \bar{Q}_{16}\,dz\right\}\gamma_{xy}^0 + \left\{\int_{-\frac{H}{2}}^{+\frac{H}{2}} \bar{Q}_{16} z\,dz\right\}k_{xy}^0 + \left\{\int_{-\frac{H}{2}}^{+\frac{H}{2}} \bar{Q}_{16}\,dz\right\}\gamma_{xy}^0 + \left\{\int_{-\frac{H}{2}}^{+\frac{H}{2}} \bar{Q}_{16} z\,dz\right\}k_{xy}^0. \quad (10.9)$$

The first integral in Eq. (10.9) is:

$$\int_{-\frac{H}{2}}^{+\frac{H}{2}} \bar{Q}_{11}\,dz. \quad (10.10)$$

Because the reduced stiffnesses are materials properties, they are constant within a layer. However, from layer to layer, the values of the reduced stiffnesses change [7–10]. Because of this piecewise-constant property of the reduced stiffnesses, the integral through the thickness can be expanded to give:

$$\int_{-\frac{H}{2}}^{+\frac{H}{2}} \bar{Q}_{11}\,dz = \int_{z_0}^{z_1} \bar{Q}_{11_1}\,dz + \int_{z_1}^{z_2} \bar{Q}_{11_2}\,dz + \int_{z_2}^{z_3} \bar{Q}_{11_3}\,dz + \cdots \int_{z_{k-1}}^{z_k} \bar{Q}_{11_k}\,dz + \cdots \int_{z_{N-1}}^{z_N} \bar{Q}_{11_N}\,dz. \quad (10.11)$$

The subscript on $\bar{Q}_{11}$ indicates the layer number. Because within each layer, $\bar{Q}_{11}$ is a constant; in the above series of integrals, it can be removed from each integral and the series of integrals written as:

$$\int_{-\frac{H}{2}}^{+\frac{H}{2}} \bar{Q}_{11}\,dz = \bar{Q}_{11_1}\int_{z_0}^{z_1} dz + \bar{Q}_{11_2}\int_{z_1}^{z_2} dz + \bar{Q}_{11_3}\int_{z_2}^{z_3} dz + \cdots \bar{Q}_{11_k}\int_{z_{k-1}}^{z_k} dz + \cdots \bar{Q}_{11_N}\int_{z_{N-1}}^{z_N} dz. \quad (10.12)$$

After integration:

$$\int_{-\frac{H}{2}}^{+\frac{H}{2}} \bar{Q}_{11}\,dz = \bar{Q}_{11_1}(z_1 - z_0) + \bar{Q}_{11_2}(z_2 - z_1) + \bar{Q}_{11_3}(z_3 - z_2)$$
$$+ \cdots + \bar{Q}_{11_k}(z_k - z_{k-1}) + \cdots + \bar{Q}_{11_N}(z_N - z_{N-1}) \quad (10.13)$$

or

$$\int_{-\frac{H}{2}}^{+\frac{H}{2}} \bar{Q}_{11}\,dz = \sum_{k=1}^{N} \bar{Q}_{11_k}\left(z_k - z_{k-1}\right). \tag{10.14}$$

The difference in the z values is recognized as simply the thickness of the kth layer, h_k, so Eq. (10.14) becomes:

$$\int_{-\frac{H}{2}}^{+\frac{H}{2}} \bar{Q}_{11}\,dz = \sum_{k=1}^{N} \bar{Q}_{11_k} h_k. \tag{10.15}$$

The integral of $\bar{Q}_{11}$ through the thickness is denoted by A_{11}; that is:

$$\int_{-\frac{H}{2}}^{+\frac{H}{2}} \bar{Q}_{11}\,dz = A_{11} = \sum_{k=1}^{N} \bar{Q}_{11_k} h_k. \tag{10.16}$$

In a similar manner:

$$\int_{-\frac{H}{2}}^{+\frac{H}{2}} \bar{Q}_{12}\,dz = A_{12} = \sum_{k=1}^{N} \bar{Q}_{12_k} h_k \tag{10.17}$$

$$\int_{-\frac{H}{2}}^{+\frac{H}{2}} \bar{Q}_{16}\,dz = A_{16} = \sum_{k=1}^{N} \bar{Q}_{16_k} h_k. \tag{10.18}$$

Thus, the first, third, and fifth terms on the right-hand side of Eq. (10.9) become:

$$A_{11}\varepsilon_x^0 + A_{12}\varepsilon_y^0 + A_{16}\ \gamma_{xy}^0. \tag{10.19}$$

The second, fourth, and sixth terms on the right-hand side of Eq. (10.9) involve integrals of $\bar{Q}_{11} \times z$, $\bar{Q}_{12} \times z$, and $\bar{Q}_{16} \times z$, rather than just $\bar{Q}_{11}$, $\bar{Q}_{12}$, and $\bar{Q}_{16}$. Specifically, the second term is:

$$\int_{-\frac{H}{2}}^{+\frac{H}{2}} \bar{Q}_{11} z\,dz \tag{10.20}$$

$$\Rightarrow \int_{-\frac{H}{2}}^{+\frac{H}{2}} \bar{Q}_{11} z\,dz = \bar{Q}_{11_1}\int_{z0}^{z1} z\,dz + \bar{Q}_{11_2}\int_{z1}^{z2} z\,dz + \bar{Q}_{11_3}$$
$$\int_{z2}^{z3} z\,dz + \ldots \bar{Q}_{11_k}\int_{z_{k-1}}^{z_k} z\,dz + \cdots \bar{Q}_{11_N}\int_{z_{N-1}}^{z_N} z\,dz. \tag{10.21}$$

After integration:

$$\int_{-\frac{H}{2}}^{+\frac{H}{2}} \bar{Q}_{11}\,dz = \frac{1}{2}\Big\{\bar{Q}_{11_1}\left(z_1^2 - z_0^2\right) + \bar{Q}_{11_2}\left(z_2^2 - z_1^2\right) + \bar{Q}_{11_3}\left(z_3^2 - z_2^2\right)$$

$$+ \cdots + \bar{Q}_{11_k}\left(z_k^2 - z_{k-1}^2\right) + \cdots + \bar{Q}_{11_N}\left(z_N^2 - z_{N-1}^2\right)\Big\} \tag{10.22}$$

or

$$\int_{-\frac{H}{2}}^{+\frac{H}{2}} \bar{Q}_{11} z\,dz = \frac{1}{2}\sum_{k=1}^{N} \bar{Q}_{11_k}\left(z_k^2 - z_{k-1}^2\right). \tag{10.23}$$

The integral of $\bar{Q}_{11}z$ through the thickness is denoted by B_{11}; that is:

$$\int_{-\frac{H}{2}}^{+\frac{H}{2}} \bar{Q}_{11} z\,dz = B_{11} = \frac{1}{2}\sum_{k=1}^{N} \bar{Q}_{11_k}\left(z_k^2 - z_{k-1}^2\right). \tag{10.24}$$

In a similar manner:

$$\int_{-\frac{H}{2}}^{+\frac{H}{2}} \bar{Q}_{12} z\,dz = B_{12} = \frac{1}{2}\sum_{k=1}^{N} \bar{Q}_{12_k}\left(z_k^2 - z_{k-1}^2\right) \tag{10.25}$$

$$\int_{-\frac{H}{2}}^{+\frac{H}{2}} \bar{Q}_{16} z\,dz = B_{16} = \frac{1}{2}\sum_{k=1}^{N} \bar{Q}_{16_k}\left(z_k^2 - z_{k-1}^2\right). \tag{10.26}$$

The expression for N_x in Eq. (10.9) becomes, after slight rearrangement:

$$N_x = A_{11}\varepsilon_x^0 + A_{12}\varepsilon_y^0 + A_{16}\gamma_{xy}^0 + B_{11}k_x^0 + B_{12}k_y^0 + B_{16}k_{xy}^0. \tag{10.27}$$

Because the A_{ij} and B_{ij} are the integrals, they represent *smeared* or *integrated* properties of the laminate. Eq. (10.27) represents a relation between the normal force resultant N_x and the six reference surface deformations. This is part of the missing link in Figure 10.1. Five other equations will be developed that will lead to a total of six relations between the six stress resultants and the six reference surface deformations. The missing link will then be complete [11–13]. However, before proceeding to complete the link, let us compute A_{11}, A_{12}, A_{16}, B_{11}, B_{12}, B_{16} for the laminates discussed in Sections 9.6.1 and 9.6.2.

For the four-layer $[0/90]_s$ laminate of Sections 9.6.1 and 9.6.2, as each layer is having the same thickness h, by Eq. (10.16):

$$A_{11} = \sum_{k=1}^{N} \bar{Q}_{11_k} h_k = \left\{\bar{Q}_{11_1} + \bar{Q}_{11_2} + \bar{Q}_{11_3} + \bar{Q}_{11_4}\right\} \times h \tag{10.28}$$

or

$$A_{11}=\sum_{k=1}^{N}\bar{Q}_{11_k}h_k=\left\{\bar{Q}_{11}(0°)+\bar{Q}_{11}(90°)+\bar{Q}_{11}(90°)+\bar{Q}_{11}(0°)\right\}\times h. \tag{10.29}$$

Substituting the numerical values from Sections 9.6.1 and 9.6.2 for the appropriate $\bar{Q}_{ij}$:

$$A_{11}=\{155.7+12.16+12.16+155.7\}\times 10^9\times 150\times 10^{-6} \tag{10.30}$$

$$A_{11}=50.4\times 10^6\ \mathrm{N\,m^{-1}}. \tag{10.31}$$

Similarly,

$$A_{12}=\left\{\bar{Q}_{12}(0°)+\bar{Q}_{12}(90°)+\bar{Q}_{12}(90°)+\bar{Q}_{12}(0°)\right\}\times h$$

$$A_{12}=\{3.02+3.02+3.02+3.02\}\times 10^9\times 150\times 10^{-6}$$

$$A_{12}=1.809\times 10^6\ \mathrm{N\,m^{-1}} \tag{10.32}$$

$$A_{16}=\left\{\bar{Q}_{16}(0°)+\bar{Q}_{16}(90°)+\bar{Q}_{16}(90°)+\bar{Q}_{16}(0°)\right\}\times h$$

$$A_{16}=0.$$

For all cross-ply laminates, A_{16} will be zero because for each layer, $\bar{Q}_{16}=0$. The B_{ij} can be computed from Eqs. (10.24)–(10.26) as follows:

$$B_{11}=\frac{1}{2}\sum_{k=1}^{N}\bar{Q}_{11_k}\left(z_k^2-z_{k-1}^2\right)=\frac{1}{2}\left\{\begin{array}{c}\bar{Q}_{11_1}\left(z_1^2-z_0^2\right)+\bar{Q}_{11_2}\left(z_2^2-z_1^2\right)+\\ \bar{Q}_{11_3}\left(z_3^2-z_2^2\right)+\bar{Q}_{11_4}\left(z_4^2-z_3^2\right)\end{array}\right\} \tag{10.33}$$

$$B_{11}=\frac{1}{2}\left\{\begin{array}{c}\bar{Q}_{11_1}(0°)\left(z_1^2-z_0^2\right)+\bar{Q}_{11_2}(90°)\left(z_2^2-z_1^2\right)+\\ \bar{Q}_{11_3}(90°)\left(z_3^2-z_2^2\right)+\bar{Q}_{11_4}(0°)\left(z_4^2-z_3^2\right)\end{array}\right\}. \tag{10.34}$$

Using the numerical values from Sections 9.6.1 and 9.6.2 for the appropriate $\bar{Q}_{ij}$:

$$B_{11}=\frac{1}{2}\left\{\begin{array}{c}155.7\left[(-150)^2-(300)^2\right]+12.16\left[(0)^2-(-150)^2\right]+\\ 12.16\left[(150)^2-(0)^2\right]+155.7\left[(300)^2-(150)^2\right]\end{array}\right\}\times 10^9\times\left(10^{-6}\right)^2. \tag{10.35}$$

This gives:

$$B_{11}=0. \tag{10.36}$$

For symmetric laminates, there is no need to go through the algebra to compute the values of the B_{ij}. They are *all* exactly zero; thus,

$$B_{12} = 0 \quad B_{16} = 0. \tag{10.37}$$

To continue with the development of the laminate stiffness matrix, we can substitute the expressions for σ_y and τ_{xy} into the definitions in Eq. (10.1) for N_y and N_{xy}, respectively, to yield:

$$N_y = \int_{-\frac{H}{2}}^{+\frac{H}{2}} \left\{ \bar{Q}_{12}\varepsilon_x^0 + \bar{Q}_{12}zk_x^0 + \bar{Q}_{22}\varepsilon_y^0 + \bar{Q}_{22}zk_y^0 + \bar{Q}_{26}\gamma_{xy}^0 + \bar{Q}_{26}zk_{xy}^0 \right\} dz \tag{10.38}$$

$$N_{xy} = \int_{-\frac{H}{2}}^{+\frac{H}{2}} \left\{ \bar{Q}_{16}\varepsilon_x^0 + \bar{Q}_{16}zk_x^0 + \bar{Q}_{26}\varepsilon_y^0 + \bar{Q}_{26}zk_y^0 + \bar{Q}_{66}\gamma_{xy}^0 + \bar{Q}_{66}zk_{xy}^0 \right\} dz\,. \tag{10.39}$$

This gives:

$$N_y = A_{12}\varepsilon_x^0 + A_{22}\varepsilon_y^0 + A_{26}\gamma_{xy}^0 + B_{12}k_x^0 + B_{22}k_y^0 + B_{26}k_{xy}^0 \tag{10.40}$$

$$N_{xy} = A_{16}\varepsilon_x^0 + A_{26}\varepsilon_y^0 + A_{66}\gamma_{xy}^0 + B_{16}k_x^0 + B_{26}k_y^0 + B_{66}k_{xy}^0. \tag{10.41}$$

Writing the results in matrix notation:

$$\begin{Bmatrix} N_x \\ N_y \\ N_{xy} \end{Bmatrix} = \begin{bmatrix} A_{11} & A_{12} & A_{16} \\ A_{12} & A_{22} & A_{26} \\ A_{16} & A_{26} & A_{66} \end{bmatrix} \begin{Bmatrix} \varepsilon_x^0 \\ \varepsilon_y^0 \\ \gamma_{xy}^0 \end{Bmatrix} + \begin{bmatrix} B_{11} & B_{12} & B_{16} \\ B_{12} & B_{22} & B_{26} \\ B_{16} & B_{26} & B_{66} \end{bmatrix} \begin{Bmatrix} k_x^0 \\ k_y^0 \\ k_{xy}^0 \end{Bmatrix}. \tag{10.42}$$

The above matrix equation constitutes three of the six relations between the six stress resultants and the six reference surface deformations. Three equations have yet to be derived, but with the above notation, it is clear that the A_{ij} and B_{ij} are the components of three-by-three matrices. We can compute the numerical values of the additional A_{ij} and B_{ij} for the laminates of CLT Examples discussed in Sections 9.6.1 and 9.6.2 in a manner similar to the previous calculations of A_{11}, A_{12}, A_{16}, B_{11}, B_{12}, and B_{16}.

For deriving the remaining three equations, we use the moment expressions, Eq. (10.3), by substituting for σ_x from Eq. (10.6) into the first equation of Eq. (10.3) to yield:

$$M_x = \int_{-\frac{H}{2}}^{+\frac{H}{2}} \left\{ \bar{Q}_{11}\varepsilon_x^0 + \bar{Q}_{11}zk_x^0 + \bar{Q}_{12}\varepsilon_y^0 + \bar{Q}_{12}zk_y^0 + \bar{Q}_{16}\gamma_{xy}^0 + \bar{Q}_{16}zk_{xy}^0 \right\} z\,dz \tag{10.43}$$

$$M_x = \left\{ \int_{-\frac{H}{2}}^{+\frac{H}{2}} \bar{Q}_{11} z\, dz \right\} \varepsilon_x^0 + \left\{ \int_{-\frac{H}{2}}^{+\frac{H}{2}} \bar{Q}_{11} z^2\, dz \right\} k_x^0 + \left\{ \int_{-\frac{H}{2}}^{+\frac{H}{2}} \bar{Q}_{12} z\, dz \right\} \varepsilon_y^0$$
$$+ \left\{ \int_{-\frac{H}{2}}^{+\frac{H}{2}} \bar{Q}_{12} z^2\, dz \right\} k_y^0 + \left\{ \int_{-\frac{H}{2}}^{+\frac{H}{2}} \bar{Q}_{16} z\, dz \right\} \gamma_{xy}^0 + \left\{ \int_{-\frac{H}{2}}^{+\frac{H}{2}} \bar{Q}_{16} z^2\, dz \right\} k_{xy}^0. \tag{10.44}$$

The first integral on the right-hand side of Eq. (10.44) is:

$$\int_{-\frac{H}{2}}^{+\frac{H}{2}} \bar{Q}_{11} z\, dz. \tag{10.45}$$

The above term from Eq. (10.24) is B_{11}. The first, third, and fifth terms on the right-hand side of Eq. (10.44) become:

$$B_{11}\varepsilon_x^0 + B_{12}\varepsilon_y^0 + B_{16}\gamma_{xy}^0. \tag{10.46}$$

The second, fourth, and sixth terms on the right-hand side of Eq. (10.24) are new; the second term is:

$$\int_{-\frac{H}{2}}^{+\frac{H}{2}} \bar{Q}_{11} z^2\, dz. \tag{10.47}$$

This can be expanded to:

$$\int_{-\frac{H}{2}}^{+\frac{H}{2}} \bar{Q}_{11} z^2\, dz = \bar{Q}_{11_1} \int_{z_0}^{z_1} z^2\, dz + \bar{Q}_{11_2} \int_{z_1}^{z_2} z^2\, dz + \bar{Q}_{11_3} \int_{z_2}^{z_3} z^2\, dz$$
$$+ \cdots + \bar{Q}_{11_k} \int_{z_{k-1}}^{z_k} z^2\, dz + \cdots \bar{Q}_{11_N} \int_{z_{N-1}}^{z_N} z^2\, dz. \tag{10.48}$$

Because of the constancy of the reduced stiffnesses within layers:

$$\int_{-\frac{H}{2}}^{+\frac{H}{2}} \bar{Q}_{11} z^2\, dz = \frac{1}{3} \left\{ \begin{array}{c} \bar{Q}_{11_1}\left(z_1^3 - z_0^3\right) + \bar{Q}_{11_2}\left(z_2^3 - z_1^3\right) + \\ \bar{Q}_{11_3}\left(z_3^3 - z_2^3\right) \cdots + \bar{Q}_{11_k}\left(z_k^3 - z_{k-1}^3\right) + \cdots + \bar{Q}_{11_N}\left(z_N^3 - z_{N-1}^3\right) \end{array} \right\} \tag{10.49}$$

or

$$\int_{-\frac{H}{2}}^{+\frac{H}{2}} \bar{Q}_{11} z^2\, dz = \frac{1}{3} \sum_{k=1}^{N} \bar{Q}_{11_k}\left(z_k^3 - z_{k-1}^3\right). \tag{10.50}$$

This integral is denoted as D_{11}; that is:

$$\int_{-\frac{H}{2}}^{+\frac{H}{2}} \bar{Q}_{11} z^2\, dz = D_{11} = \frac{1}{3}\sum_{k=1}^{N} \bar{Q}_{11_k}\left(z_k^3 - z_{k-1}^3\right). \tag{10.51}$$

Referring to the fourth and sixth terms in Eq. (10.44), we find that:

$$\int_{-\frac{H}{2}}^{+\frac{H}{2}} \bar{Q}_{12} z^2\, dz = D_{12} = \frac{1}{3}\sum_{k=1}^{N} \bar{Q}_{12_k}\left(z_k^3 - z_{k-1}^3\right) \tag{10.52}$$

$$\int_{-\frac{H}{2}}^{+\frac{H}{2}} \bar{Q}_{16} z^2\, dz = D_{16} = \frac{1}{3}\sum_{k=1}^{N} \bar{Q}_{16_k}\left(z_k^3 - z_{k-1}^3\right). \tag{10.53}$$

The expression for the bending moment resultant M_x in Eq. (10.44) becomes:

$$M_x = B_{11}\varepsilon_x^0 + B_{12}\varepsilon_y^0 + B_{16}\gamma_{xy}^0 + D_{11}k_x^0 + D_{12}k_y^0 + D_{16}k_{xy}^0. \tag{10.54}$$

For the four-layer $[0/90]_s$ laminate of Sections 9.6.1 and 9.6.2:

$$\begin{aligned} D_{11} &= \frac{1}{3}\sum_{k=1}^{N} \bar{Q}_{11_k}\left(z_k^3 - z_{k-1}^3\right) \\ &= \frac{1}{3}\left\{\bar{Q}_{11_1}\left(z_1^3 - z_0^3\right) + \bar{Q}_{11_2}\left(z_2^3 - z_1^3\right) + \bar{Q}_{11_3}\left(z_3^3 - z_2^3\right) + \bar{Q}_{11_4}\left(z_4^3 - z_3^3\right)\right\} \end{aligned} \tag{10.55}$$

or

$$\begin{aligned} D_{11} = \frac{1}{3}\Big\{&\bar{Q}_{11}(0^\circ)\left(z_1^3 - z_0^3\right) + \bar{Q}_{11}(90^\circ)\left(z_2^3 - z_1^3\right) + \bar{Q}_{11}(90^\circ)\left(z_3^3 - z_2^3\right) \\ &+ \bar{Q}_{11}(0^\circ)\left(z_4^3 - z_3^3\right)\Big\}. \end{aligned} \tag{10.56}$$

Thus,

$$D_{11} = 2.48 \text{ Nm} \tag{10.57}$$

$$\begin{aligned} D_{12} &= \frac{1}{3}\sum_{k=1}^{N} \bar{Q}_{12_k}\left(z_k^3 - z_{k-1}^3\right) \\ &= \frac{1}{3}\left\{\bar{Q}_{12_1}\left(z_1^3 - z_0^3\right) + \bar{Q}_{12_2}\left(z_2^3 - z_1^3\right) + \bar{Q}_{12_3}\left(z_3^3 - z_2^3\right) + \bar{Q}_{12_4}\left(z_4^3 - z_3^3\right)\right\} \end{aligned} \tag{10.58}$$

$$\begin{aligned} D_{12} = \frac{1}{3}\Big\{&\bar{Q}_{12}(0^\circ)\left(z_1^3 - z_0^3\right) + \bar{Q}_{12}(90^\circ)\left(z_2^3 - z_1^3\right) \\ &+ \bar{Q}_{12}(90^\circ)\left(z_3^3 - z_2^3\right) + \bar{Q}_{12}(0^\circ)\left(z_4^3 - z_3^3\right)\Big\} \end{aligned} \tag{10.59}$$

$$D_{12} = 0.0543 \text{ Nm}. \tag{10.60}$$

Because Q_{16} is zero for each layer:

$$D_{16} = 0. \tag{10.61}$$

To finish the development of the laminate stiffness matrix, we can substitute the expressions for σ_y and τ_{xy} into the definitions in Eq. (10.3) for M_y and M_{xy}. The result is:

$$M_y = \int_{-\frac{H}{2}}^{+\frac{H}{2}} \left\{ \bar{Q}_{12}\varepsilon_x^0 + \bar{Q}_{12}zk_x^0 + \bar{Q}_{22}\varepsilon_y^0 + \bar{Q}_{22}zk_y^0 + \bar{Q}_{26}\gamma_{xy}^0 + \bar{Q}_{26}zk_{xy}^0 \right\} z\,dz \tag{10.62}$$

$$M_{xy} = \int_{-\frac{H}{2}}^{+\frac{H}{2}} \left\{ \bar{Q}_{16}\varepsilon_x^0 + \bar{Q}_{16}zk_x^0 + \bar{Q}_{26}\varepsilon_y^0 + \bar{Q}_{26}zk_y^0 + \bar{Q}_{66}\gamma_{xy}^0 + \bar{Q}_{66}zk_{xy}^0 \right\} z\,dz. \tag{10.63}$$

Thus:

$$M_y = B_{12}\varepsilon_x^0 + B_{22}\varepsilon_y^0 + B_{26}\gamma_{xy}^0 + D_{12}k_x^0 + D_{22}k_y^0 + D_{26}k_{xy}^0 \tag{10.64}$$

$$M_{xy} = B_{16}\varepsilon_x^0 + B_{26}\varepsilon_y^0 + B_{66}\gamma_{xy}^0 + D_{16}k_x^0 + D_{26}k_y^0 + D_{66}k_{xy}^0. \tag{10.65}$$

These three moments can be written in the matrix notation as:

$$\begin{Bmatrix} M_x \\ M_y \\ M_{xy} \end{Bmatrix} = \begin{bmatrix} B_{11} & B_{12} & B_{16} \\ B_{12} & B_{22} & B_{26} \\ B_{16} & B_{26} & B_{66} \end{bmatrix} \begin{Bmatrix} \varepsilon_x^0 \\ \varepsilon_y^0 \\ \gamma_{xy}^0 \end{Bmatrix} + \begin{bmatrix} D_{11} & D_{12} & D_{16} \\ D_{12} & D_{22} & D_{26} \\ D_{16} & D_{26} & D_{66} \end{bmatrix} \begin{Bmatrix} k_x^0 \\ k_y^0 \\ k_{xy}^0 \end{Bmatrix}. \tag{10.66}$$

The combined matrix relation is:

$$\begin{Bmatrix} N_x \\ N_y \\ N_{xy} \\ M_x \\ M_y \\ M_{xy} \end{Bmatrix} = \begin{bmatrix} A_{11} & A_{12} & A_{16} & B_{11} & B_{12} & B_{16} \\ A_{12} & A_{22} & A_{26} & B_{12} & B_{22} & B_{26} \\ A_{16} & A_{26} & A_{66} & B_{16} & B_{26} & B_{66} \\ B_{11} & B_{12} & B_{16} & D_{11} & D_{12} & D_{16} \\ B_{12} & B_{22} & B_{26} & D_{12} & D_{12} & D_{26} \\ B_{16} & B_{26} & B_{66} & D_{16} & D_{26} & D_{66} \end{bmatrix} \begin{Bmatrix} \varepsilon_x^0 \\ \varepsilon_y^0 \\ \gamma_{xy}^0 \\ k_x^0 \\ k_y^0 \\ k_{xy}^0 \end{Bmatrix}. \tag{10.67}$$

Equation (10.67) is called the *laminate stiffness matrix* or the *ABD* matrix [14]. The *ABD* matrix defines a relationship between the stress resultants (i.e., loads) applied to a laminate, and the reference surface strains and curvatures (i.e., deformations). This form is a direct result of the Kirchhoff hypothesis, the plane stress assumption, and the definition of the stress resultants. The laminate stiffness matrix involves everything that is used to define the laminate-layer material properties, fiber orientation, thickness, and location.

Completing the calculations for the $[0/90]_s$ laminate:

$$[A]=\begin{bmatrix} 50.4 & 1.809 & 0 \\ 1.809 & 50.4 & 0 \\ 0 & 0 & 2.64 \end{bmatrix}\times 10^6\ \mathrm{N\,m^{-1}} \tag{10.68}$$

$$[B]=\begin{bmatrix} 0 & 0 & 0 \\ 0 & 0 & 0 \\ 0 & 0 & 0 \end{bmatrix} N \tag{10.69}$$

$$[D]=\begin{bmatrix} 2.48 & 0.0543 & 0 \\ 0.0543 & 0.542 & 0 \\ 0 & 0 & 0.0792 \end{bmatrix}\mathrm{Nm}\ . \tag{10.70}$$

10.3 CLASSIFICATION OF LAMINATES

The form of the *ABD* matrix strongly depends on whether the laminate is what we referred to as symmetric, whether it consists of just 0° and 90° layers, or whether for every layer at orientation $+\theta$ there is a layer at $-\theta$. In this section, we would like to formally classify laminates as to their stacking arrangement, and indicate the effects of the various classifications on the *ABD* matrix [15].

10.3.1 Symmetric Laminates

A laminate is said to be *symmetric* if for every layer to one side of the laminate reference surface with a specific thickness, specific material properties, and specific fiber orientation, there is another layer with the identical distance on the *opposite* side of the reference surface with the identical thickness, material properties, and fiber orientation. If a laminate is not symmetric, then it is referred to as an *unsymmetric* laminate. Note that this pairing on opposite sides of the reference surface must occur for *every* layer. Although we have restricted our examples to laminates with layers of the same material and same thickness, this does not have to be the case for a laminate to be symmetric. A four-layer laminate with the two outer layers made of glass-reinforced material oriented at +45° and the two inner layers made of graphite-reinforced material oriented at −30° is a symmetric laminate. For a symmetric laminate, all the components of the matrix are identically zero. Consequently, the full six-by-six set of equations in Eq. (10.67) decouples into two three-by-three sets of equations, namely:

$$\begin{Bmatrix} N_x \\ N_y \\ N_{xy} \end{Bmatrix} = \begin{bmatrix} A_{11} & A_{12} & A_{16} \\ A_{12} & A_{22} & A_{26} \\ A_{16} & A_{26} & A_{66} \end{bmatrix} \begin{Bmatrix} \varepsilon_x^0 \\ \varepsilon_y^0 \\ \gamma_{xy}^0 \end{Bmatrix} \tag{10.71}$$

$$\begin{Bmatrix} M_x \\ M_y \\ M_{xy} \end{Bmatrix} = \begin{bmatrix} D_{11} & D_{12} & D_{16} \\ D_{12} & D_{22} & D_{26} \\ D_{16} & D_{26} & D_{66} \end{bmatrix} \begin{Bmatrix} k_x^0 \\ k_y^0 \\ k_{xy}^0 \end{Bmatrix}. \tag{10.72}$$

10.3.2 Balanced Laminates

A laminate is said to be *balanced* if for every layer with a specified thickness, specific material properties, and specific fiber orientation, there is another layer with the identical thickness, material properties, but opposite fiber orientation somewhere in the laminate. The layer with opposite fiber orientation does not have to be on the opposite side of the reference surface, nor immediately adjacent to the other layer, nor anywhere in particular. The other layer can be anywhere within the thickness. A laminate does not have to be symmetric to be balanced. The symmetric $[\pm30/0]_s$ laminate and the unsymmetric $[\pm30/0]_T$ laminate are both balanced laminates. If a laminate is balanced, then the stiffness matrix components A_{16} and A_{26} are always zero because the $\bar{Q}_{16}$ and $\bar{Q}_{26}$ from the layer pairs with opposite orientation are of opposite sign, and the net contribution to A_{16} and A_{26} from these layer pairs is then zero. To classify as balanced, all off-axis layers must occur in pairs. As a special case, all laminates consisting entirely of layers with their fibers oriented at either 0 or 90°, to be discussed shortly, are balanced, as $\bar{Q}_{16}$ and $\bar{Q}_{26}$ are zero for every layer, resulting in A_{16} and A_{26} being zero. The *ABD* matrix of a balanced but otherwise general laminate is not that much simpler than the *ABD* matrix of a general unsymmetric, unbalanced laminate. The full six-by-six form, Eq. (10.67), applies but with the A_{16} and A_{26} components set to zero.

10.3.3 Symmetric Balanced Laminates

A laminate is said to be a *symmetric balanced* laminate if it meets both the criterion for being symmetric and the criterion for being balanced. If this is the case, then Eq. (10.67) takes the decoupled form:

$$\begin{Bmatrix} N_x \\ N_y \end{Bmatrix} = \begin{bmatrix} A_{11} & A_{12} \\ A_{12} & A_{22} \end{bmatrix} \begin{Bmatrix} \varepsilon_x^0 \\ \varepsilon_y^0 \end{Bmatrix} \tag{10.73}$$

$$N_{xy} = A_{66}\gamma_{xy}^0$$

$$\begin{Bmatrix} M_x \\ M_y \\ M_{xy} \end{Bmatrix} = \begin{bmatrix} D_{11} & D_{12} & D_{16} \\ D_{12} & D_{22} & D_{26} \\ D_{16} & D_{26} & D_{66} \end{bmatrix} \begin{Bmatrix} k_x^0 \\ k_y^0 \\ k_{xy}^0 \end{Bmatrix}. \tag{10.74}$$

10.3.4 Cross-Ply Laminates

A laminate is said to be a *cross-ply* laminate if every layer has its fibers oriented at either 0° or 90°. If this is the case, because $\bar{Q}_{16}$ and $\bar{Q}_{26}$ are zero for every layer, then $A_{16}, A_{26}, B_{16}, B_{26}, D_{16}$, and D_{26} are zero, and there is some decoupling of the six equations. In particular, the six equations decouple to a set of four equations and a set of two equations, namely:

$$\begin{Bmatrix} N_x \\ N_y \\ M_x \\ M_y \end{Bmatrix} = \begin{bmatrix} A_{11} & A_{12} & B_{11} & B_{12} \\ A_{12} & A_{22} & B_{12} & B_{22} \\ B_{11} & B_{12} & D_{11} & D_{12} \\ B_{12} & B_{22} & D_{12} & D_{22} \end{bmatrix} \begin{Bmatrix} \varepsilon_x^0 \\ \varepsilon_y^0 \\ k_x^0 \\ k_y^0 \end{Bmatrix} \tag{10.75}$$

$$\begin{Bmatrix} N_{xy} \\ M_{xy} \end{Bmatrix} = \begin{bmatrix} A_{66} & B_{66} \\ B_{66} & D_{66} \end{bmatrix} \begin{Bmatrix} \gamma_{xy}^0 \\ k_{xy}^0 \end{Bmatrix}. \tag{10.76}$$

10.3.5 Symmetric Cross-Ply Laminates

A laminate is said to be a *symmetric cross-ply* laminate if it meets the criterion for being symmetric and the criterion for being cross-ply. This results in the simplest form of the *ABD* matrix. All the B_{ij} are zero, and A_{16}, A_{26}, D_{16}, and D_{26} are zero. Specifically:

$$\begin{Bmatrix} N_x \\ N_y \end{Bmatrix} = \begin{bmatrix} A_{11} & A_{12} \\ A_{12} & A_{22} \end{bmatrix} \begin{Bmatrix} \varepsilon_x^0 \\ \varepsilon_y^0 \end{Bmatrix} \tag{10.77}$$

$$N_{xy} = A_{66}\gamma_{xy}^0$$

and

$$\begin{Bmatrix} M_x \\ M_y \end{Bmatrix} = \begin{bmatrix} D_{11} & D_{12} \\ D_{12} & D_{22} \end{bmatrix} \begin{Bmatrix} k_x^0 \\ k_y^0 \end{Bmatrix} \tag{10.78}$$

$$M_{xy} = D_{66}k_{xy}^0.$$

REFERENCES

1. Herakovich C.T. (1989), "Free edge effects in laminated composites." in *Handbook of Composites:* vol. 2, *Structures and Design* (C. T. Herakovich and Y. M. Tarnopol'skii, eds.), Elsevier Science Publishing Co, New York.
2. Armanios E.A. (1989), *Interlaminar Fracture of Composites*, Trans Tech. Publications, Aedermannsdorf, Switzerland.
3. Renton W.J. and Vinson J.R. (1977), "Analysis of adhesively bonded joints between panels of composite materials", *Transactions of the ASME, Journal of Applied Mechanics*, Vol. 44, pp. 101–106.
4. Kassapoglou C. (1993), "Calculation of stresses at skin-stiffener interfaces of composite stiffened panels under shear loads", *International Journal of Solids and Structures*, Vol. 30(11), pp. 1491–1501.
5. Harrison P.N. and Johnson E.R. (1996), "A mixed variational formulation for interlaminar stresses in thickness-tapered composite laminates", *International Journal of Solids and Structures*, Vol. 33(16), pp. 2377–2399.
6. Love A.E.H. (1892), *A Treatise on the Mathematical Theory of Elasticity*, Cambridge University Press, London (also 4th ed., 1944, Dover Publications, Inc., New York).
7. Hodge P.G. Jr. (1958), "The mathematical theory of plasticity," in *Elasticity and Plasticity* (J. N. Goodier and P. G. Hodge, Jr., eds.), John Wiley & Sons, Inc., New York, p. 81.
8. Kreyszig E. (1967), *Advanced Engineering Mathematics*, 2nd ed., John Wiley & Sons, Inc., New York, p. 416.
9. Fung Y.C. (1965), *Foundations of Solid Mechanics*, Prentice-Hall, Inc., Englewood Cliffs, NJ, p. 285.
10. Green G. (1839), "On the laws of reflexion and refraction of light at the common surface of two non-crystallized media", *Transactions of the Cambridge Philosophical Society*, Vol. 7, p. 121.
11. Popov E.P. (1990), *Engineering Mechanics of Solids*, Prentice Hall, Englewood Cliffs, NJ.
12. Hyer M.W. (1998), *Stress Analysis of Fiber Reinforced Composite Materials*, McGraw-Hill, Singapore.
13. Ashton J.E. and Whitney J.M. (1970), *Theory of Laminated Plates*, Technomic Publishing Co., Lancaster, PA.
14. Jones R. M. (1975), *Mechanics of Composite Materials*, McGraw-Hill, New York.
15. Nemeth M.P. (1986), "The importance of anisotropy on buckling of compression-loaded symmetric composite plates", *AIAA Journal*, Vol. 24(11), pp. 1831–1835.

11 Failure Theories for Composite Materials

The field of study of the mechanical behavior of materials such as metals, ceramics, and polymers is quiet mature in comparison with the composite materials such as carbon/epoxy or glass/polyester that have revolutionized the automobile and space industry. Several researchers have been trying to apply the concepts developed for the monolithic materials to composites. Significant success has been achieved with respect to the effective elastic properties, laminate plate theory, and homogenization, but the development of theories of failure for the composite materials is still lagging behind that of monolithic materials. Several uncertainties and controversies still exist in predicting composite failure.

Yielding and fracture are the main concepts related to the failure of monolithic materials, especially metals. Yielding phenomenon is analyzed using the plasticity approach, whereas the fracture phenomenon is studied using the fracture mechanics approach. For polymer matrix composites (PMCs), which are the most common class of composite materials, concepts in both these fields must be modified significantly. As far as yielding as the initiation of plastic flow is concerned, polymers as well as fibers of glass or carbon, which are the common constituents of PMCs, do not have shear-driven thresholds for inelastic behavior as in metals. Furthermore, instability of crack growth, which is the basis of brittle fracture in metals, is not the main concern in laminated PMCs, where unstable matrix cracks seldom lead to failure as they are in most cases arrested at interfaces within the composite volume. Most of the failure processes in composites then consist of the following:

i. Multiple crack formation
ii. Linkage of crack surfaces
iii. Extensive fiber failure.

These events may eventually separate a composite part in two or more pieces, but prior to that may cause significant changes in the deformational response that induces failure to perform the design function. These considerations for composites question the so-called strength, which in monolithic materials is synonymous with "failure." More relevant to composites is the concept of "criticality" associated with the function for which the composite part was designed and manufactured. Thus, failure theories for composites should aim at describing and predicting appropriate criticalities. Over the years, numerous failure theories for composites have emerged. The common industry practice is to use only the theory or model that has been validated by experimental evidence. Because of the uncertainty of validation of the failure theories, an ambitious exercise named "worldwide failure exercise" (WWFE) was conducted to objectively test how different theories fare against experimental data. Although useful, the validation against test data cannot fully explain why a particular theory works, or does not do so. As it has turned out, not a single theory has the ability to agree with test data in all cases examined. The effort made here is to complement the WWFE in trying to understand the reasons for the lack of validation.

To begin, historical development of failure theories for composites is reviewed, examining the foundation of each theory. Attempt is made to unfold the underlying assumptions – explicit or implicit – with the aim to gain insight into the limitations and relevance in each case. Following that, a comprehensive scheme for failure assessment is proposed and its ingredients are illustrated by examples. It is noted that the objective here is not to provide a comprehensive review of all failure theories proposed so far. For this, the reader is referred to WWFE [1,2]. It is believed that this close examination suffices to reveal the nature of the limitations inherent in the approach to which these theories, and other later ones, belong.

11.1 THEORIES OF FAILURE

The serious work on composite materials analysis began in the early 1960s. It was primarily Zvi Hashin who initiated it. Necessarily, the study of the elastic properties of composites came before anything else. The seminal work on bounds for heterogeneous elasticity was given by Hashin [3]. Many years of work followed immediately by many people engaged in related activities. The only exact, three-dimensional, non-dilute treatment of the related elastic properties problem was given by Christensen and Lo [4], and eventually, it came to be known as the *generalized self-consistent method*. While all this was occurring, the failure problem was incubating and ominously looming. With work progressing well on the elastic properties, Tsai and Wu [5] initiated the first significant published work on the failure of composite materials. This was later followed by an entirely different and contradictory approach taken by Hashin [6]. After that, the floodgates opened to an all-enveloping surge of further work on the failure of fiber composite materials. Puck and colleagues were among those involved. The objective was to develop the analytical forms for the failure criteria of fiber composites. Everyone seemed to want their own personalized criterion. None of the proposed forms fulfilled the promise. What exactly are failure criteria? These are targeted to be reasonable, manageable, logical analytical forms that encompass the failure states in terms of the stresses. The enormous success of the *Mises criterion* with isotropic materials provided the motivation if not the methodology.

11.2 HILL'S THEORY OF FAILURE

Hill noted that at large plastic strains and in processes such as rolling, drawing, and extrusion, metals develop a preferred orientation due to crystalline plane alignment. A metal sheet can then be viewed as having three orthogonal planes of symmetry as illustrated in Figure 11.1. For such orthotropic materials, Hill proposed a straightforward mathematical generalization of the von Mises yield criterion for isotropic

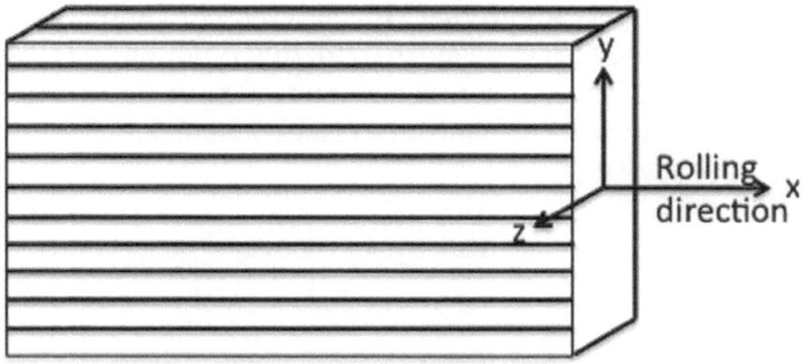

FIGURE 11.1 A metal sheet with preferred orientation resulting in three planes of symmetry denoted by the rectangular coordinate axes *x*, *y*, *z*.

metals as follows. Starting with the von Mises criterion expressed in the *x*–*y*–*z* Cartesian coordinates:

$$(\sigma_{xx}-\sigma_{yy})^2+(\sigma_{yy}-\sigma_{zz})^2+(\sigma_{zz}-\sigma_{xx})^2+6(\sigma_{xy}^2+\sigma_{yz}^2+\sigma_{zx}^2)=2A^2, \quad (11.1)$$

where *A* is the yield stress in uniaxial loading in any direction, and making two assumptions, namely, that there is no Bauschinger effect – that is, the yield stress is the same in tension and compression, and that the plastic potential is not affected by superposition of hydrostatic pressure, Hill generalized the isotropic yield criterion to the following form:

$$F(\sigma_{yy}-\sigma_{zz})^2+G(\sigma_{zz}-\sigma_{xx})^2+H(\sigma_{xx}-\sigma_{yy})^2+2(L\tau_{yz}^2+M\tau_{zx}^2+N\tau_{xy}^2)=1, \quad (11.2)$$

where *F*, *G*, *H*, *L*, *M*, and *N* are the material constants representing the direction-dependent yielding. Denoting by *X*, *Y*, and *Z*, the yield strength values in the principal orthotropic directions, *x*, *y*, and *z*, respectively, one obtains:

$$\frac{1}{X^2}=G+H$$

$$\frac{1}{Y^2}=H+F$$

$$\frac{1}{Z^2}=F+G$$

$$2F=\frac{1}{Y^2}+\frac{1}{Z^2}-\frac{1}{X^2}$$

$$2G=\frac{1}{Z^2}+\frac{1}{X^2}-\frac{1}{Y^2}$$

$$2H=\frac{1}{X^2}+\frac{1}{Y^2}-\frac{1}{Z^2}. \quad (11.3)$$

While in isotropic yielding with no Bauschinger effect the initiation of yielding in pure shear is related to the tensile yield strength, this is not the case for *anisotropic yielding*. Hill [7] assumed three independent yield strength values corresponding to pure shear in the three orthogonal symmetry planes, and denoting these by R, S, and T, obtained:

$$\begin{aligned} 2L &= \frac{1}{R^2} \\ 2M &= \frac{1}{S^2} \\ 2N &= \frac{1}{T^2}. \end{aligned} \tag{11.4}$$

Hill [7] further showed that if rotational symmetry about one of the symmetry axes exists in a plane normal to the axis, then the six yield strength constants will reduce to four independent constants. The two relationships between the yield strength constants in case of rotational symmetry about the x axis are:

$$H = G,\ L = G + 2F. \tag{11.5}$$

11.3 TSAI–HILL THEORY OF FAILURE

If the x axis is aligned with the fiber direction in a unidirectional composite, then the y–z plane would be the plane of isotropy; that is, the composite will be transversely isotropic. Azzi and Tsai [8] assumed this to be the case. They further recognized that the unidirectional composites used commonly as plies in a laminate are thin, and therefore, placing the z axis in the thickness direction, they neglected the stress components σ_{zz}, τ_{zx}, and τ_{zy}. Using this condition in Eq. (11.2) gives:

$$(G + H)\sigma_{xx}^2 + (F + H)\sigma_{yy}^2 - 2H\sigma_{xx}\sigma_{yy} + 2N\tau_{xy}^2 = 1. \tag{11.6}$$

From the first of Eq. (11.5), that is, $H = G$, it can be seen that using Eq. (11.3) gives $Z = Y$. Thus, this is the consequence of transverse isotropy, and it is obvious on physical grounds. Using this condition in Eq. (11.6), and further using Eqs. (11.3) and (11.4), gives the familiar form of the so-called Tsai–Hill failure criterion, stated below:

$$\frac{\sigma_{xx}^2}{X^2} + \frac{\sigma_{yy}^2}{Y^2} - \frac{\sigma_{xx}\sigma_{yy}}{X^2} + \frac{\tau_{xy}^2}{T^2} = 1. \tag{11.7}$$

It is clear from the review above that Eq. (7) is in all essence the yield criterion for thin metal sheets with one preferred orientation lying in the plane of the sheet. The adaptation of this criterion to composite materials by Azzi and Tsai [8] consists of regarding the preferred orientation as the fiber direction in a thin layer of unidirectional composite. The various failure modes as per Eq. (11.7) are shown in Figure 11.2.

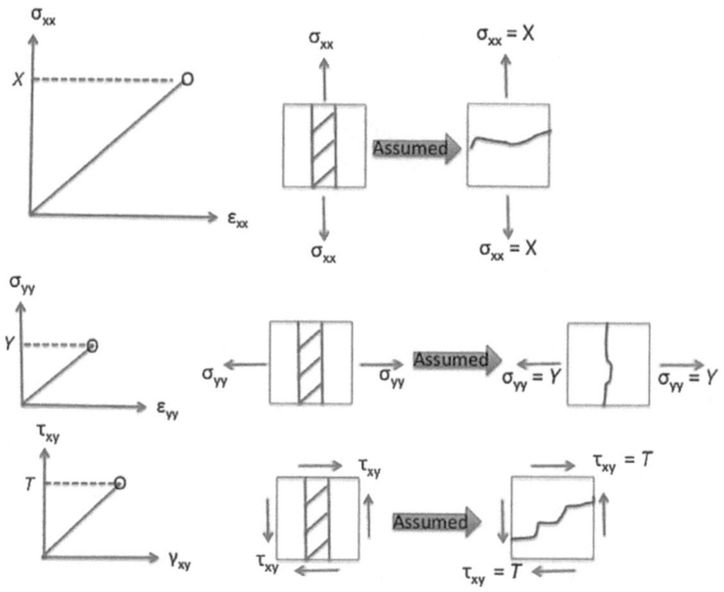

FIGURE 11.2 Basic failure modes as per Eq. (11.7).

11.4 HOFFMAN THEORY OF FAILURE

The next failure theory of interest for unidirectional composites appeared in 1967 by Hoffman [9]. This author assumed, without physical evidence, that all failure modes in composites were brittle fracture. This implied that the linear elastic response in each stress component terminated abruptly at a limiting value (fracture strength) for that stress component and that the fracture strength in combined stresses could be formally represented by a "yield condition" in spite of the assumed brittle fracture. Although Azzi and Tsai [8] did not explicitly state these assumptions, they essentially implied them by renaming yield stresses in Hill [7] as the composite strength values. The essential difference between Hoffman [9] and Azzi and Tsai [8] in this respect is that the former assigned different strength in tension and compression in each principal composite direction (i.e., along fibers and normal to fibers). In the context of Hill [7], this meant that the Bauschinger effect, which Hill neglected, was now included. Accordingly, Hoffman [9] added three linear terms to Eq. (11.2), increasing the number of material constants from six in Hill's version to nine. For a lamina (a unidirectional fiber-reinforced sheet in plane stress), the modified criterion led to the graphical representation in the form of an ellipsoid with its center shifted from the origin of the orthogonal axes σ_{xx}, σ_{yy}, and τ_{xy}.

The expression for Hoffman failure criterion is:

$$F(\sigma_{yy} - \sigma_{zz})^2 + G(\sigma_{zz} - \sigma_{xx})^2 + H(\sigma_{xx} - \sigma_{yy})^2$$
$$+ O\sigma_{xx} + P\sigma_{yy} + Q\sigma_{zz} + R\tau_{yz}^2 + S\tau_{zx}^2 + T\tau_{xy}^2 = 1, \tag{11.8}$$

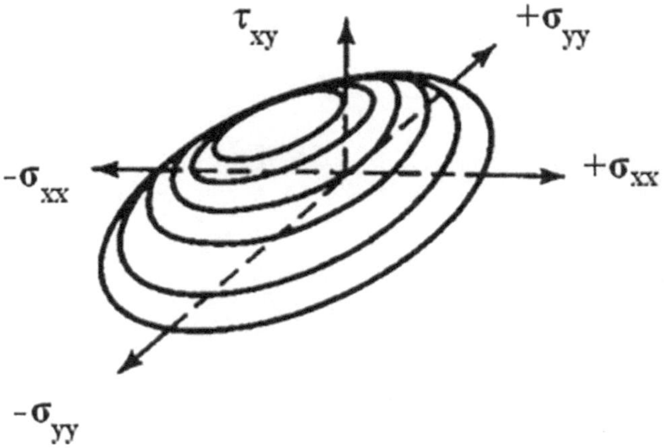

FIGURE 11.3 Hoffman failure surface.

where the nine constants are determined from the nine strengths in the principal material coordinates: X_t, X_c, Y_t, Y_c, Z_t, Z_c, S_{yz}, S_{zx}, and S_{xy}. For plane stress in the *x-y* plane ($\sigma_{zz} = \tau_{zx} = \tau_{yz} = 0$) and transverse isotropy in the *y-z* plane ($Z_t = Y_t$, $Z_c = Y_c$, $S_{zx} = S_{xy}$), Eq. (11.8) reduces to:

$$-\frac{\sigma_{xx}^2}{X_c X_T} - \frac{\sigma_{yy}^2}{Y_c Y_T} + \frac{\sigma_{xx}\sigma_{yy}}{X_c X_T} + \frac{\tau_{xy}^2}{S_{xy}^2} + \frac{X_c + X_T}{X_c X_T}\sigma_{xx} + \frac{Y_c + Y_T}{Y_c Y_T}\sigma_{yy} = 1. \tag{11.9}$$

Here, X_c is a negative number. For equal strengths in tension and compression ($X_c = -X_T = -X$ and $Y_c = -Y_T = -Y$), the Hoffman failure criterion reduces to the Tsai–Hill criterion. Both the criteria are ellipsoids in σ_{xx}, σ_{yy}, and τ_{xy} space as shown in Figure 11.3.

11.5 MAXIMUM STRESS FAILURE THEORY

In the maximum stress failure criterion, each and every one of the stresses in principal material coordinates must be less than the respective strengths; otherwise, failure is said to have occurred. Thus for tensile stresses:

$$\begin{aligned} \sigma_{xx} &< X_T \\ \sigma_{yy} &< Y_T. \end{aligned} \tag{11.10}$$

However, for compressive stresses,

$$\begin{aligned} \sigma_{xx} &> X_C \\ \sigma_{yy} &> Y_C \end{aligned} \tag{11.11}$$

and

$$\left|\tau_{xy}\right| < S. \tag{11.12}$$

If any of the inequalities is not satisfied, the material is said to have failed by the failure mechanism associated with X_t, X_c, Y_t, Y_c, or S, respectively. There is no interaction between the modes of failure in this criterion. There are in reality five subcriteria.

11.6 MAXIMUM STRAIN THEORY

This theory is very similar to the maximum stress failure theory. However, the strains are limited in this theory instead of the stresses. The material is said to have failed if one or more of the following inequalities are not satisfied:

$$\begin{aligned} &\varepsilon_{xx} < X_{\varepsilon_t} \\ &\varepsilon_{yy} < Y_{\varepsilon_t} \\ &\left|\gamma_{xy}\right| < S_{\varepsilon}. \end{aligned} \tag{11.13}$$

Including for materials with different strength in tension and compression:

$$\begin{aligned} &\varepsilon_{xx} > X_{\varepsilon_c} \\ &\varepsilon_{yy} > Y_{\varepsilon_c}, \end{aligned} \tag{11.14}$$

where

$X_{\varepsilon_t}\left(X_{\varepsilon_c}\right)$ = the maximum tensile (compressive) normal strain in the x direction
$Y_{\varepsilon_t}\left(Y_{\varepsilon_c}\right)$ = the maximum tensile (compressive) normal strain in the y direction
S_{ε} = maximum shear strain in the x-y coordinates.

11.7 THE TSAI–WU FAILURE CRITERION

A commonly used failure criterion is the Tsai–Wu failure criterion [5]. This criterion states that the failure condition is reached when the combined applied stresses satisfy the following equality:

$$F_{11}^{*}\sigma_1^{*2} + 2F_{12}^{*}\sigma_1^{*}\sigma_2^{*} + F_{22}^{*}\sigma_2^{*2} + F_{66}^{*}\tau_{12}^{*2} + F_1^{*}\sigma_1^{*} + F_2^{*}\sigma_2^{*} = 1, \tag{11.15}$$

where the coefficients F_{11}^{*}, F_{12}^{*}, F_{22}^{*}, F_{66}^{*}, F_1^{*}, and F_2^{*} must be determined experimentally. Note that the stress components appearing in the criterion are expressed in the fiber-aligned triad. Consider first the case where a single stress component σ_1^{*} is applied. At failure in tension and in compression, the above equality must be satisfied, implying:

$$F_{11}^{*}\sigma_{1t}^{*f2} + F_1^{*}\sigma_{1t}^{*f} = 1$$

$$F_{11}^{*}\sigma_{1c}^{*f2} - F_{1}^{*}\sigma_{1c}^{*f} = 1. \tag{11.16}$$

The second test involves stress component σ_2^* only and yields:

$$\begin{aligned} F_{22}^{*}\sigma_{2t}^{*f2} + F_{2}^{*}\sigma_{2t}^{*f} &= 1 \\ F_{22}^{*}\sigma_{2c}^{*f2} - F_{2}^{*}\sigma_{2c}^{*f} &= 1. \end{aligned} \tag{11.17}$$

Finally, the last test involves τ_{12}^* only and implies $F_{66}^*\tau_{12}^{*f2} = 1$. These five equations can be solved for five of the coefficients appearing in Eq. (11.15) to find:

$$\begin{aligned} F_{11}^{*} &= \frac{1}{\sigma_{1t}^{*f}\sigma_{1c}^{*f}} \\ F_{22}^{*} &= \frac{1}{\sigma_{2t}^{*f}\sigma_{2c}^{*f}} \\ F_{66}^{*} &= \frac{1}{\tau_{12}^{*f2}} \\ F_{1}^{*} &= \frac{\sigma_{1c}^{*f} - \sigma_{1t}^{*f}}{\sigma_{1t}^{*f}\sigma_{1c}^{*f}} \\ F_{2}^{*} &= \frac{\sigma_{2c}^{*f} - \sigma_{2t}^{*f}}{\sigma_{2t}^{*f}\sigma_{2c}^{*f}}. \end{aligned} \tag{11.18}$$

These results are introduced in the initial statement of the failure criterion, Eq. (11.15), to yield:

$$\bar{\sigma}_{11}^{*2} + 2\bar{F}_{12}^{*}\bar{\sigma}_{11}^{*}\bar{\sigma}_{22}^{*} + \bar{\sigma}_{22}^{*2} + \bar{\tau}_{12}^{*2} + \bar{F}_{1}^{*}\bar{\sigma}_{11}^{*} + \bar{F}_{2}^{*}\bar{\sigma}_{22}^{*} = 1, \tag{11.19}$$

where the following nondimensional stress components are defined:

$$\begin{aligned} \bar{\sigma}_{11}^{*} &= \frac{\sigma_{1}^{*}}{\sqrt{\sigma_{1t}^{*f}\sigma_{1c}^{*f}}} \\ \bar{\sigma}_{22}^{*} &= \frac{\sigma_{2}^{*}}{\sqrt{\sigma_{2t}^{*f}\sigma_{2c}^{*f}}} \\ \bar{\tau}_{12}^{*} &= \frac{\tau_{12}^{*}}{\tau_{12}^{*f}} \\ \bar{F}_{1}^{*} &= \frac{\sigma_{1c}^{*f} - \sigma_{1t}^{*f}}{\sqrt{\sigma_{1t}^{*f}\sigma_{1c}^{*f}}} \end{aligned}$$

$$\bar{F}_2^* = \frac{\sigma_{2c}^{*f} - \sigma_{2t}^{*f}}{\sqrt{\sigma_{2t}^{*f}\sigma_{2c}^{*f}}}. \tag{11.20}$$

Coefficient $\bar{F}_{12}^*$ is as yet undetermined. Clearly, an additional test involving a biaxial state of applied stress (i.e., a test where both σ_1^* and σ_2^* are applied simultaneously) is required to determine this coefficient. Because such a biaxial test is very difficult to perform, coefficient $\bar{F}_{12}^*$ is often selected by fitting the prediction of the criterion to available experimental data. $\bar{F}_{12}^* = -1/2$ has been found to provide the best fit. The final statement of the Tsai–Wu criterion becomes:

$$\bar{\sigma}_{11}^{*2} - \bar{\sigma}_{11}^*\bar{\sigma}_{22}^* + \bar{\sigma}_{22}^{*2} + \bar{\tau}_{12}^{*2} + \bar{F}_1^*\bar{\sigma}_{11}^* + \bar{F}_2^*\bar{\sigma}_{22}^* = 1. \tag{11.21}$$

As an example, consider the simple test as shown in Figure 11.4. A single stress component, σ_1, is applied to a lamina with fibers running at an angle θ. The stress rotation formula yields the applied stresses in the fiber-aligned triad as $\sigma_1^* = \sigma_1 \cos^2\theta$, $\sigma_2^* = \sigma_2 \sin^2\theta$, and $\tau_{12}^* = -\sigma_1 \cos\theta \sin\theta$. The level of applied stress that corresponds to failure satisfies the failure criterion of Eq. (11.21), that is,

$$\sigma_1^2\left[\frac{\cos^4\theta}{\sigma_{1t}^{*f}\sigma_{1c}^{*f}} - \frac{\sin^2\theta\cos^2\theta}{\sqrt{\sigma_{1t}^{*f}\sigma_{1c}^{*f}\sigma_{2t}^{*f}\sigma_{2c}^{*f}}} + \frac{\sin^4\theta}{\sigma_{2t}^{*f}\sigma_{2c}^{*f}} + \frac{\sin^2\theta\cos^2\theta}{\tau_{12}^{*f2}}\right]$$

$$+\sigma_1\left[\frac{\bar{F}_1^*\cos^2\theta}{\sqrt{\sigma_{1t}^{*f}\sigma_{1c}^{*f}}} - + \frac{\bar{F}_2^*\sin^2\theta}{\sqrt{\sigma_{2t}^{*f}\sigma_{2c}^{*f}}}\right] - 1 = 0. \tag{11.22}$$

This second-order equation can be solved to find the failure load.

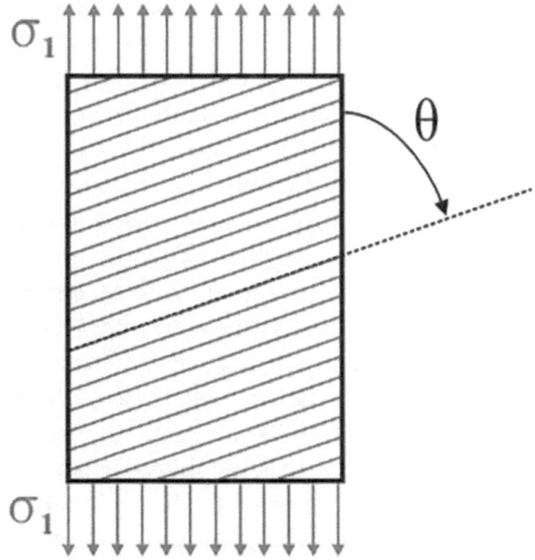

FIGURE 11.4 A single stress component, σ_1, being applied to a lamina with fibers running at an angle θ.

11.8 HASHIN THEORY

Hashin [6] failure criteria distinguish among the various different failure modes of the unidirectional lamina: tensile and compressive fiber failures and tensile and compressive matrix failure, and they are given separately for each of these failure modes as follows:

Tensile fiber mode: $\sigma_1 > 0$

$$\left(\frac{\sigma_1}{X^t}\right)^2 + \left(\frac{\tau_{12}}{S}\right)^2 = 1. \tag{11.23}$$

Compressive fiber mode: $\sigma_1 < 0$

$$\sigma_1 = X^c. \tag{11.24}$$

Tensile matrix mode: $\sigma_2 > 0$

$$\left(\frac{\sigma_2}{Y^t}\right)^2 + \left(\frac{\tau_{12}}{S}\right)^2 = 1. \tag{11.25}$$

Compressive matrix mode: $\sigma_2 < 0$

$$\left(\frac{\sigma_2}{2S^T}\right)^2 + \left[\left(\frac{Y^t}{2S^T}\right)^2 - 1\right]\frac{\sigma_2}{Y^c} + \left(\frac{\tau_{12}}{S}\right)^2 = 1, \tag{11.26}$$

where S^T is the transverse or out-of-plane shear strength, while S is the longitudinal or in-plane shear strength, and the rest are defined previously. Similar to Tsai–Hill failure criterion, Hashin failure criteria are also quadratic polynomials of stresses. However, they are derived from general polynomials of the first four transversely isotropic stress invariants with the highest order term for each of these four invariants being selected in such a way that results in a highest order of stress to be quadratic. Furthermore, Hashin failure criteria assume that two distinct fiber and matrix failure modes occur in two different fracture planes and only relevant stress components on the associated fracture plane of each mode will contribute to the failure criteria for that failure mode. As a consequence, the failure envelope described by Hashin failure criteria is only piecewise smooth, with each smooth branch modeling a distinct failure mode. Hashin criteria provide an improvement for the prediction of intra-lamina failures. However, as pointed out by Hashin, the quadratic failure criteria for matrix mode imply that the fracture plane is the maximum transverse shear plane and this may not be always true. A more general approach based on the Mohr–Coulomb failure theory for matrix failure is also suggested by Hashin to overcome this limitation. Puck and Schurmann [10] and more recently Davila et al. [11] have taken some form of this general approach in their development of matrix failure criteria.

REFERENCES

1. Hinton M.J., Kaddour A.S., Soden P.D. (2004), *Failure Criteria in Fibre Reinforced Polymer Composites: The World-Wide Failure Exercise*, Elsevier, Amsterdam.
2. Soden P.D., Kaddour A.S., Hinton M.J. (2004), "Recommendations for designers and researchers resulting from the world-wide failure exercise", *Composite Science and Technology*, Vol. 64, pp. 589–604.
3. Hashin Z. (1965), "On elastic behavior of fire reinforced materials of arbitrary transverse phase geometry", *Journal of Mechanics and Physics of Solids*, Vol. 13, pp. 119–134.
4. Christensen R.M. and Lo K.H. (1979), "Solutions for the effective shear properties in three phase sphere and cylinder models", *Journal of Mechanics and Physics of Solids*, Vol. 27 pp. 315–330.
5. Tsai S.W. and Wu E.M. (1971), "A general theory of strength for anisotropic materials", *Journal of Composite Materials*, Vol. 5, pp. 58–80.
6. Hashin Z. (1980), "Failure criteria for unidirectional fiber composites", *Journal of Applied Mechanics*, Vol. 47, pp. 329–334.
7. Hill R. (1948), "A theory of the yielding and plastic flow of anisotropic materials", *Proceedings of the Royal Society A*, Vol. 193, pp. 281–297.
8. Azzi V.D. and Tsai S. (1965), "Anisotropic strength of composites", *Experimental Mechanics*, Vol. 5, pp. 283–328.
9. Hoffman O (1967), "The brittle strength of orthotropic materials", *Journal of Composite Materials*, Vol. 1, pp. 200–206.
10. Puck A. and Schurmann H. (1998), "Failure analysis of FRP laminates by means of physically based phenomenological models", *Composite Science and Technology*, Vol. 58, pp. 1045–1068.
11. Davila C., Camanho P., Rose C. (2005), "Failure criteria for FRP laminates", *Journal of Composite Materials*, Vol. 39, pp. 323–345.

12 Mechanics of Short-Fiber-Reinforced Composites

Short-fiber composites have become increasingly popular in recent years because of their easy adaptability to conventional techniques of manufacturing and low fabrication cost. The increasing number of applications of short-fiber composites makes them more important to understand and predict their mechanical and thermal properties, which strongly depend on fiber orientation and volume fraction. Methods of the analysis of relationships between microstructures, and overall properties and strength of composites have been developed from the simplest relationships between averaged values (e.g., Voigt/Reuss estimations) to the complex multiscale numerical models, which take into account the nonlinear behavior of components, evolving microstructures, and real inhomogeneous phase arrangements. In this chapter, several approaches and methods of the analysis of the interrelations between the microstructures, and the mechanical behavior and strength of materials in particular have been discussed. This chapter reviews and evaluates models that predict the stiffness of short-fiber composites. The polymer processing community has made substantial progress in modeling process-induced fiber orientation, particularly in injection molding, and these results are now routinely used to predict mechanical properties. Real injection-molded composites invariably have misoriented fibers of highly variable length, but aligned fiber properties are always determined as a prelude to modeling the more realistic situation. Hence, the focus here will be on composites having aligned fibers with uniform length and mechanical properties. In selecting models for consideration, the general requirements are that each model must include the effects of fiber and matrix properties and the fiber volume fraction, include the effect of fiber aspect ratio, and predict a complete set of elastic constants for the composite. Any model not meeting these criteria has been excluded from consideration. All of the models use the same basic assumptions:

a. Fibers and the matrix are linearly elastic.
b. Matrix is isotropic, and fibers are either isotropic or transversely isotropic.
c. Fibers are axisymmetric, identical in shape and size, and can be characterized by an aspect ratio (l/d).
d. Fibers and matrix are well bonded at their interface, and remain that way during deformation.
e. Interfacial slip, fiber–matrix debonding, or matrix microcracking is not considered.

12.1 NOTATION

Vectors will be denoted by lowercase Roman letters, the second-order tensors by lowercase Greek letters, and the fourth-order tensors by capital Roman letters. Whenever possible, vectors and tensors are written as boldface characters; indicial notation is used where necessary. A subscript or superscript f indicates a quantity associated with the fibers, and m denotes a matrix quantity. Thus, the fibers have Young's modulus E_f and Poisson's ratio ν_f, while the corresponding matrix properties are E_m and ν_m. The symbol I represents the fourth-order unit tensor. $\boldsymbol{C}$ and $\boldsymbol{S}$ denote the stiffness and compliance tensors, respectively, and σ and ε are the total stress and infinitesimal strain tensors. Hence, the constitutive equations for the fiber and matrix materials are:

$$\sigma^f = C^f \, \varepsilon^f \tag{12.1}$$

$$\sigma^m = C^m \, \varepsilon^m. \tag{12.2}$$

Let x denote the position vector. When a composite material is loaded, the pointwise stress field $\sigma(x)$ and the corresponding strain field $\varepsilon(x)$ will be nonuniform on the microscale. The solution of these nonuniform fields is a formidable problem. However, many useful results can be obtained in terms of the average stress and strain. Consider a representative averaging volume V. Choose V large enough to contain many fibers, but small compared to any length scale over which the average loading or deformation of the composite varies. The volume-average stress $\bar{\sigma}$ is defined as the average of the pointwise stress $\sigma(x)$ over the volume V.

$$\bar{\sigma} \equiv \frac{1}{V}\int_V \sigma(x)dV. \tag{12.3}$$

The average strain $\bar{\varepsilon}$ is defined similarly. It is also convenient to define volume-average stresses and strains for the fiber and matrix phases. To obtain these, first partition the averaging volume V into the volume occupied by the fibers V_f and the volume occupied by the matrix V_m. We consider only two-phase composites, so that:

$$V = V_f + V_m. \tag{12.4}$$

The fiber and matrix volume fractions are simply $v_f = V_f/V$ and $v_m = V_m/V$ and, since only fibers and matrix are present, $v_f + v_m = 1$. The average fiber and the matrix stresses are averages over the corresponding volumes:

$$\bar{\sigma}^f \equiv \frac{1}{V_f}\int_{V_f} \sigma(x)dV \text{ and } \bar{\sigma}^m \equiv \frac{1}{V_m}\int_{V_m} \sigma(x)dV. \tag{12.5}$$

Similarly, average strains for the fiber and matrix are defined. The relationships between the fiber and matrix averages, and the overall averages can be derived from the preceding definitions; they are:

$$\bar{\sigma} = V_f \bar{\sigma}^f + V_m \bar{\sigma}^m \tag{12.6}$$

$$\bar{\varepsilon} = V_f \bar{\varepsilon}^f + V_m \bar{\varepsilon}^m. \tag{12.7}$$

An important related result is the average strain theorem. Let the averaging volume V be subjected to surface displacements $u^0(x)$ consistent with a uniform strain ε^0. Then, the average strain within the region is:

$$\bar{\varepsilon} = \varepsilon^0. \tag{12.8}$$

This theorem is proved by substituting the definition of the strain tensor ε in terms of the displacement vector u into the definition of average strain $\bar{\varepsilon}$, and applying Gauss's theorem. The result is:

$$\bar{\varepsilon}_{ij} = \frac{1}{V}\int_S \left(u_i^0 n_j + n_i u_j^0\right) dS, \tag{12.9}$$

where S denotes the surface of V and n is a unit vector normal to dS.

The average strain within a volume V is completely determined by the displacements on the surface of the volume, so displacements consistent with a uniform strain must produce the identical value of average strain. A corollary of this principle is that if we define a perturbation strain $\varepsilon^C(x)$ as the difference between the local strain and the average,

$$\varepsilon^C(x) \equiv \varepsilon(X) - \bar{\varepsilon} \tag{12.10}$$

Then the volume average of $\varepsilon^C(x)$ must equal zero:

$$\bar{\varepsilon}^C = \frac{1}{V}\int_V \varepsilon^C(x)\, dV = 0. \tag{12.11}$$

The corresponding theorem for average stress also holds. Thus, if surface tractions consistent with a σ^0 are exerted on S, then the average stress is:

$$\bar{\sigma} = \sigma^0. \tag{12.12}$$

12.2 AVERAGE PROPERTIES

The goal of micromechanics models is to predict the average elastic properties of the composite, but even these need careful definition. Direct approach given by Hashin [1] is followed here. Subject the representative volume V to surface displacements consistent with a uniform strain ε^0; the average stiffness of the composite is the tensor C that maps this uniform strain to the average stress. Using Eq. (12.6), we have:

$$\bar{\sigma} = C \cdot \bar{\varepsilon}. \tag{12.13}$$

Average compliance S is defined in the same way, applying tractions consistent with a uniform stress σ^0 on the surface of the averaging volume. Then, using Eq. (12.10):

$$\bar{\varepsilon} = S\bar{\sigma}. \tag{12.14}$$

It should be clear that $S = C^{-1}$. Many authors have defined the average stiffness and compliance through the integral of the strain energy over V; this is equivalent to the direct approach. An important related concept, first introduced by Hill [2], is the idea of strain and stress concentration tensors A and B. These are essentially the ratios between the average fiber strain (or stress) and the corresponding average in the composite. More precisely:

$$\bar{\varepsilon}^f = A\bar{\varepsilon} \tag{12.15}$$

$$\bar{\sigma}^f = B\bar{\sigma}. \tag{12.16}$$

A and **B** are the fourth-order tensors, and in general, they must be found from a solution of the microscopic stress or strain fields. Different micromechanics models provide different ways to approximate **A** or **B**. Note that **A** and **B** have both the minor symmetries of a stiffness or compliance tensor, but lack the major symmetry. That is,

$$A_{ijkl} = A_{jikl} = A_{ijlk}. \tag{12.17}$$

But in general:

$$A_{ijkl} \neq A_{klij}. \tag{12.18}$$

For later use, it will be convenient to have an alternate strain concentration tensor $\hat{A}$ that relates the average fiber strain to the average matrix strain:

$$\bar{\varepsilon}^f = \hat{A}\bar{\varepsilon}^m. \tag{12.19}$$

This is related to **A** by:

$$\mathbf{A} = \hat{A}\left[(1 - V_f)\mathbf{I} + V_f\hat{A}\right]^{-1}. \tag{12.20}$$

So, the two forms are easily interchanged. Using equations now in hand, one can express the average composite stiffness in terms of the strain concentration tensor **A** and the fiber and matrix properties. Combining Eqs. (12.1), (12.2), (12.6), (12.7), (12.13), and (12.15), the resulting equation is:

$$\mathbf{C} = \mathbf{C}^m + V_f\left(\mathbf{C}^f - \mathbf{C}^m\right)\mathbf{A}. \tag{12.21}$$

The dual equation for the compliance is:

$$\mathbf{S} = \mathbf{S}^m + V_f\left(\mathbf{S}^f - \mathbf{S}^m\right)\mathbf{B}. \tag{12.22}$$

Equations (12.21) and (12.22) are not independent since $S = C^{-1}$. Hence, the strain concentration tensor **A** and the stress concentration tensor **B** are not independent either. The choice of which one to use in any instance is a matter of convenience. To illustrate the use of the strain concentration and stress concentration tensors, it can be noted that the Voigt average corresponds to the assumption that the fiber and the matrix both experience the same, uniform strain. Then, $\bar{\varepsilon}^{f} = \bar{\varepsilon}$, $\mathbf{A} = \mathbf{I}$, and from Eq. (12.21), the composite modulus is:

$$\mathbf{C}^{\text{Voigt}} = \mathbf{C}^{\text{m}} + V_{\text{f}}\left(\mathbf{C}^{\text{f}} - \mathbf{C}^{\text{m}}\right) = V_{\text{f}}\mathbf{C}^{\text{f}} + V_{\text{m}}\mathbf{C}^{\text{m}}. \tag{12.23}$$

Voigt average is an upper bound on the composite modulus. Reuss average assumes that the fiber and matrix both experience the same, uniform stress. This means that the stress concentration tensor B equals the unit tensor I, and from (12.22), the compliance is:

$$\mathbf{S}^{\text{Reuss}} = \mathbf{S}^{\text{m}} + V_{\text{f}}\left(\mathbf{S}^{\text{f}} - \mathbf{S}^{\text{m}}\right) = V_{\text{f}}\mathbf{S}^{\text{f}} + V_{\text{m}}\mathbf{S}^{\text{m}}. \tag{12.24}$$

This represents a lower bound on the stiffness of the composite. In the next section, various models that can be used for predicting the modulus and damping properties of the composite have been reviewed.

12.3 THEORETICAL MODELS

Some major micromechanical models that have been employed for this comparative study are discussed below.

12.3.1 Cox Shear-Lag Model

Shear-lag model is widely used to analyze the fiber–matrix stress transfer in undamaged and damaged composites. This model was developed initially by Cox [3] and then expanded and modified by many authors. The so-called shear-lag method is often used for the analysis of stress transfer problems in composites. The term "shear lag" can be traced, prior to its use in composites, to the analysis of bending of I beams and T beams with wide flanges and to box beams. Simple beam theory predicts that the axial displacements in the flanges of such beams are only a function of the distance from the neutral axis and independent of the distance from the web. This simple theory also predicts zero shear stress and zero shear strain in the flange. In reality, the true axial displacements "lag" behind the beam theory predictions. This "lag" is caused by load diffusion which can be viewed (using equilibrium arguments) as a consequence of nonzero shear stresses in the flange – hence the term "shear lag." In these beam analyses, "shear lag" is an effect and not an analysis method. Many possible analysis methods can evaluate the "shear-lag" effect. These methods generally result in defining an effective flange width that is less than the actual flange width. If effective width is not considered when designing beams, the resulting beams can be seriously under designed. Assuming that the load transfer from matrix to fiber occurs via shear stresses on the interface between them, Cox considered the force balance in

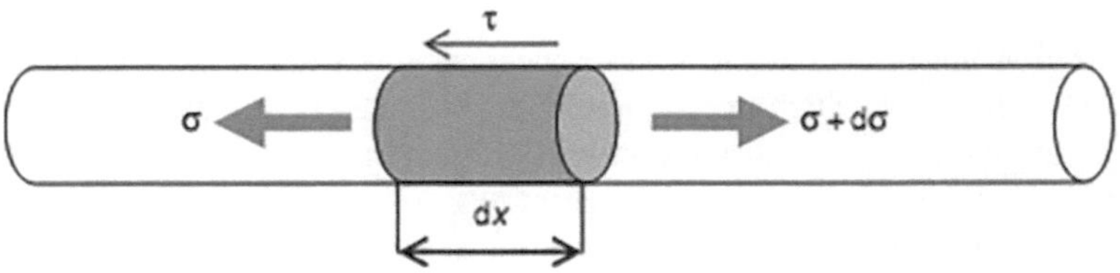

FIGURE 12.1 Cox shear-lag model.

a section of the fiber, and derived the formula which relates the rate of change of the stress along the fiber length, and the interfacial shear stress (Figure 12.1):

$$\frac{d\sigma_x}{dx} = -\frac{\tau}{r}, \tag{12.25}$$

where r is the radius of the fiber, τ is the interface shear stress, and x is the coordinate along the fiber length. This formula is referred to as the basic shear-lag equation. After some manipulations, this model leads to the following second-order differential equation:

$$\frac{d^2\sigma}{dx^2} - \beta^2\sigma = -\beta^2\sigma_\infty, \tag{12.26}$$

where β is the so-called shear-lag parameter, σ_∞ is the far-field fiber stress, and σ is the fiber stress.

Solving this equation, one determines the stress distribution along the x axis of the fiber:

$$\sigma = E\varepsilon\left[1 - \cosh(\beta x)\,\mathrm{sech}(\beta xr)\right], \tag{12.27}$$

where E is the Young's modulus of the fiber and ε is the strain in the composite. Cox derived a formula for the shear-lag parameter, β for the case of a cylindrical fiber of radius r, embedded into a cylindrical layer of matrix:

$$\beta = -\frac{1}{r}\sqrt{\frac{2G_m}{E\ln\left(\frac{s}{r}\right)}}, \tag{12.28}$$

where G_m is the matrix shear modulus, s is the average distance between the fiber axes, r is the fiber radius, and E is the axial modulus of the fiber. Using Eqs. (12.1)–(12.3), one can determine the critical fiber length l_{cr}, at which both the matrix and the fiber fail at the same strain. Assuming constant shear stress τ in Eq. (12.1), Kelly and Tyson [4] derived the following formula for the critical fiber length:

$$l_c = \frac{\sigma_c r}{\tau_c}, \tag{12.29}$$

where σ_c is the fiber failure stress and τ_c is the matrix–fiber interfacial shear strength. Figure 12.2 shows the effect of the fiber length on the stress distribution in the fiber under tensile loading. If $l < l_c$ (the case of discontinuous reinforcement), the stress in the fiber is below the critical level, and the fiber is not utilized fully. If $l > l_c$ (longer fibers), a large part of the fiber is overloaded, and multiple cracking can be observed in the fiber.

If the fiber length is equal to the critical size $l = l_c$, both the matrix and fiber fail at the same load, ensuring the most efficient reinforcement of the composite. Kelly and Tyson used this model to explain the experimental observation that the fiber breaking strength is a linear function of the wire content in different composites. Nairn [5] carried out a rigorous theoretical analysis of a model, for which the shear assumptions are exact. Using axisymmetric elasticity equations, he demonstrated that the rigorous analysis leads to Eq. (12.2) as well. Nairn demonstrated further that the shear-lag method gives reasonably good estimations of average axial stress in the fiber and total strain energy in the specimen, yet the method is not applicable for low fiber volume fractions. For the case of a broken fiber embedded into the ductile matrix, the shear-lag model was generalized by Landis and McMeeking [6], who derived the shear-lag equation for this case, and verified the model by comparing it with finite element analysis. The longitudinal modulus E_{11}, which depends on the fiber aspect ratio s, is obtained from Cox's shear-lag model as:

$$E_{cox} = (1 - V_f)E_m + \alpha\left(1 - \frac{\tanh\beta}{\beta}\right)V_f E_f, \tag{12.30}$$

where

$$\beta = \frac{1}{d}\sqrt{\frac{E_m}{(1+\nu)E_f \cdot \ln\left(\frac{\pi}{4V_f}\right)}} \tag{12.31}$$

$\alpha = 1$, for aligned fibers.

For the in-plane shear modulus G_{12}, we used the Wilczynski equation [7]:

$$G_{12} = G_m \frac{G_f(1+V_f) + G_m(1-V_f)}{G_f(1-V_f) + G_m(1+V_f)}, \tag{12.32}$$

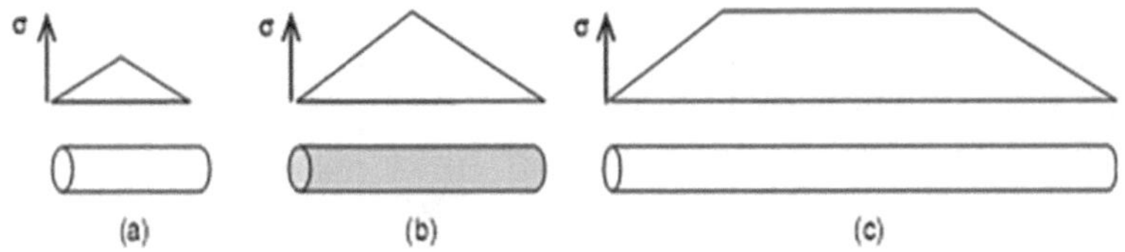

FIGURE 12.2 Stress distribution in the fiber (subject to the tensile loading) along the length: (a) Short fiber ($l < l_c$); (b) critical fiber length $l = l_c$; (c) longer fiber $l > l_c$.

where G_f and G_m are the shear modulus of the fiber and matrix, respectively. Historically, shear-lag models were the first micromechanics models for short-fiber composites, as well as the first to examine behavior near the ends of broken fibers in a continuous-fiber composite. Despite some serious theoretical flaws, shear-lag models have enjoyed enduring popularity, perhaps due to their algebraic simplicity and their physical appeal. Classical shear-lag models only predict the longitudinal modulus E_{11}, so they do not meet the criterion of predicting a complete set of elastic constants. However, they have been included here because of their historical importance and their widespread use. One could obtain a complete stiffness model by using the shear-lag prediction for E_{11} and some continuous-fiber model for the remaining elastic constants. If the fiber is anisotropic, then its axial modulus should be used in the shear-lag equations. Following Cox, the shear-lag analysis focuses on a single fiber of length l and radius r_f, which is encased in a concentric cylindrical shell of matrix having radius R. The fiber is aligned parallel to the z axis, that is, 1 direction, as shown in Figure 12.3.

Only the axial stress σ_{11} and axial strain ε_{11} are of interest, and Poisson effects are neglected, so that $\sigma_{11}^f = E_f\,\varepsilon_{11}^f$. The outer cylindrical surface of the matrix is subjected to displacement boundary conditions consistent with an average axial strain, and one solves for the fiber stress σ_{11}^f (z), that is, the average stress over the fiber cross section at z. Axial equilibrium of the fiber requires that:

$$d\sigma_{11}^f/dz = -2(\tau_{12}/r_f), \tag{12.33}$$

where τ_{12} is the axial shear stress at the fiber surface. The key assumption of shear-lag theory is that τ_{12} is proportional to the difference in displacement (w) between the fiber surface and the outer matrix surface:

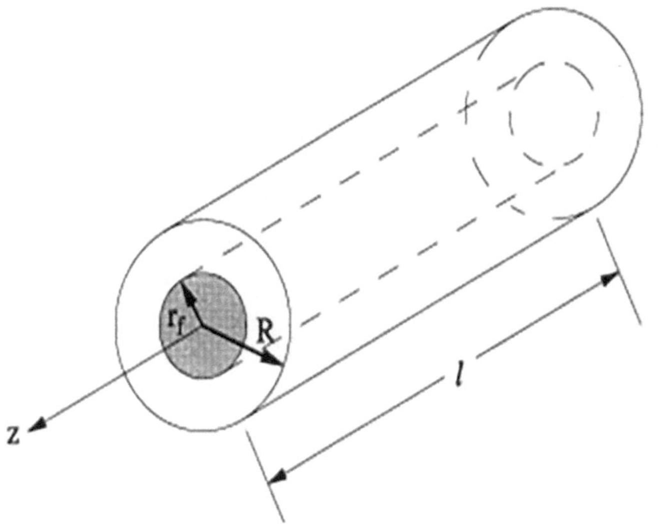

FIGURE 12.3 Idealized fiber and matrix geometry used in shear-lag models.

$$\tau_{12} = (H/2\pi r_f)\left[w(R,z) - w(r_f,z)\right], \tag{12.34}$$

where H is a constant that depends on matrix properties and fiber volume fraction. Solving Eq. (12.33) for σ_{11}^{f} (z) and applying boundary conditions of zero stress at the fiber ends give an average fiber stress of:

$$\bar{\sigma}_{11}^{f} = E_m \bar{\varepsilon}_{11}\left[1 - \frac{\tanh(\beta l/2)}{(\beta l/2)}\right], \tag{12.35}$$

where

$$\beta^2 = \frac{H}{\pi r_f^2 E_f}. \tag{12.36}$$

The above expression may be rewritten for the average fiber strain as:

$$\bar{\varepsilon}_{11}^{f} = \eta_l \bar{\varepsilon}_{11}, \tag{12.37}$$

Where η_l is a length-dependent "efficiency factor," given as:

$$\eta_l = \left[1 - \frac{\tanh\left(\frac{\beta l}{2}\right)}{\left(\frac{\beta l}{2}\right)}\right]. \tag{12.38}$$

Note that η_l is a scalar analog of the strain concentration tensor **A** defined in Eq. (12.15), and $1/\beta$ is a characteristic length for stress transfer between the fiber and the matrix. Cox found the coefficient H by solving a second idealized problem. The concentric cylinder geometry is maintained, but the outer cylindrical surface of the matrix is held stationary and the inner cylinder, which is now rigid, is subjected to a uniform axial displacement. An elasticity solution for the matrix layer then gives:

$$H = \frac{2\pi G_m}{\ln(R/r_f)}. \tag{12.39}$$

Rosen [8] simplified this part of the problem by assuming that the matrix shell was thin compared to the fiber radius, $(R - r_f) \ll r_f$, obtaining:

$$H = \frac{2\pi G_m}{\left(\frac{R}{r_f} - 1\right)}. \tag{12.40}$$

Rosen's approximation gives much larger errors at lower volume fractions, so will not be considered further. It remains to choose the radius R of the matrix cylinder, and the exact choice is important. Several choices have been used, all of which can be written in the form:

$$\frac{R}{r_f} = \sqrt{\frac{K_R}{V_f}}, \tag{12.41}$$

where K_R is a constant that depends on the assumption used to find R. Table 12.1 summarizes the choices for K_R. Cox assumed a hexagonal packing, and chose R as the distance between centers of nearest-neighbor fibers as shown in Figure 12.4a. It seems more realistic to let R equal half of the distance between nearest neighbors (Figure 12.4b), a choice labeled "hexagonal" in Table 12.1. Rosen and later Carman and Reifsnider [9] chose $r_f^2/R^2 = V_f$, so that the concentric cylinder model in Figure 12.3 would have the same fiber volume fraction as the composite. This is the same R as the composite cylinders model of Hashin and Rosen [10]. Robinson and Robinson [11] assumed a square array of fibers, and chose R as half the distance between centers of nearest neighbors as shown in Figure 12.4c. Each of these choices gives a somewhat different dependence of η_l on fiber volume fraction, with larger values of K_R producing lower values of E_{11}. Shear-lag models are usually completed by combining the average fiber stress in Eq. (12.35) with an average matrix stress to produce a modified rule of mixtures for the axial modulus.

TABLE 12.1
Values of K_R Used in Shear-Lag Models

Fiber Packing	**K_R**
Cox	$\frac{2\pi}{\sqrt{3}} = 3.628$
Composite cylinder	1.000
Hexagonal	$\frac{\pi}{2\sqrt{3}} = 0.907$
Square	$\frac{\pi}{4} = 0.785$

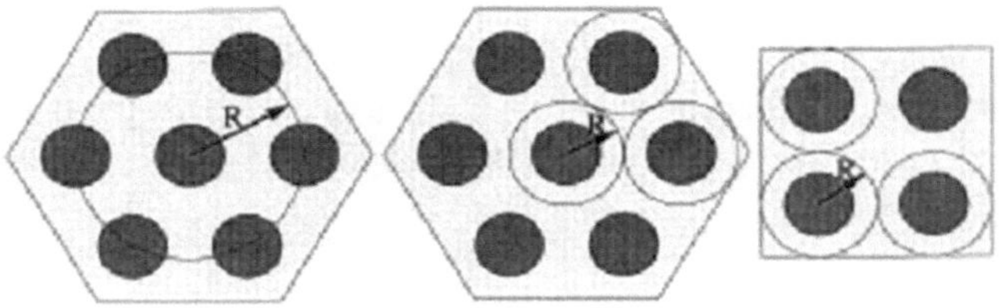

(a) (b) (c)

FIGURE 12.4 Fiber packing arrangements: (a) Cox, (b) hexagonal, and (c) square, used to find R in shear-lag models.

$$E_{11} = E_f V_f \eta_l + (1 - V_f) E_m = E_m + V_f (E_f - E_m) \eta_l. \qquad (12.42)$$

This equation is an exact scalar analog of the general tensorial stiffness formula, Eq. (12.21). A model by Fukuda and Kawata [12] for the axial stiffness of aligned short-fiber composites is closely related to shear-lag theory. They begin with a 2-D elasticity solution for the shear stress around a single slender fiber in an infinite matrix. The usual shear-lag relation, Eq. (12.33), is used to transform this into an equation for the fiber stress distribution, which is then approximated by a Fourier series. The coefficients of a truncated series are evaluated analytically using Galerkin's method. This is a dilute theory in which modulus varies linearly with fiber volume fraction. Like any shear-lag theory, Fukuda and Kawata's theory predicts that E_{11} approaches the rule-of-mixtures result as the fiber aspect ratio approaches infinity. But for short fibers, Fukuda and Kawata's theory gives much lower E_{11} values than shear-lag theory.

In Fukuda and Kawata's theory, the ratio of fiber strain to matrix strain is governed by the parameter $l/d\left(E_m/E_f\right)$. In contrast, for shear-lag theory, Eq. (12.38), the governing parameter is $\beta l/2$, which is proportional to $l/d\sqrt{\left(E_m/E_f\right)}$. Thus, for high modulus ratio and low aspect ratio, Fukuda and Kawata's theory tends to underpredict E_{11}. For this reason, this theory is not pursued further.

12.3.2 Eshelby's Equivalent Inclusion

A fundamental result used in several different models is Eshelby's equivalent inclusion given by Eshelby [13]. Eshelby solved for the elastic stress field in and around an ellipsoidal particle in an infinite matrix. By letting the particle be a prolate ellipsoid of revolution, one can use Eshelby's result to model the stress and strain fields around a cylindrical fiber. Eshelby first posed and solved a different problem, that of a homogeneous inclusion as shown in Figure 12.5. Consider an infinite solid body with stiffness $\mathbf{C}^m$ that is initially stress-free. All subsequent strains will be measured from this state. A particular small region of the body will be called the inclusion, and

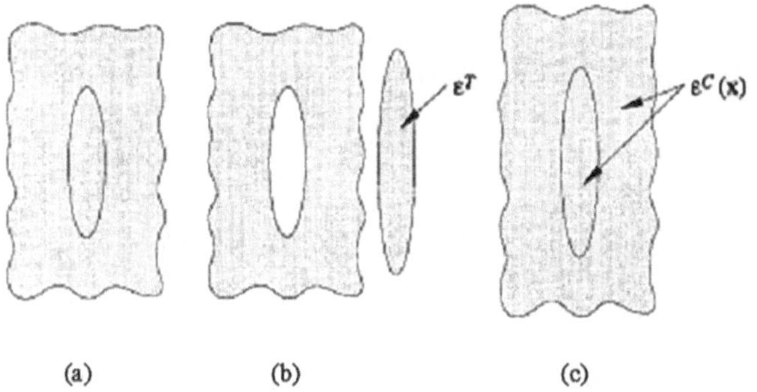

FIGURE 12.5 Eshelby's equivalent inclusion problem. Starting from the stress-free state (a), the inclusion undergoes a stress-free transformation strain ε^T (b). Fitting the inclusion and matrix back together (c) produces the strain state ε^C (x) in both the inclusion and the matrix.

the rest of the body will be called the matrix. Suppose that the inclusion undergoes some type of transformation such that, if it were a separate body, it would acquire a uniform strain ε^T with no surface traction or stress. ε^T is called the transformation strain, or the eigen strain. This strain might be acquired through a phase transformation, or by a combination of a temperature change and a different thermal expansion coefficient in the inclusion. In fact, the inclusion is bonded to the matrix, so when the transformation occur, the whole body develops some complicated strain field $\boldsymbol{\varepsilon}^C(x)$ relative to its shape before the transformation. Within the matrix, the stress $\boldsymbol{\sigma}^m$ is simply the stiffness times this strain:

$$\boldsymbol{\sigma}^m(x) = \mathbf{C}^m \varepsilon^C(x). \tag{12.43}$$

But within the inclusion, the transformation strain does not contribute to the stress, so the inclusion stress is:

$$\sigma^I(x) = C^m\left(\varepsilon^C - \varepsilon^T\right). \tag{12.44}$$

The key result of Eshelby was to show that within an ellipsoidal inclusion, the strain ε^C is uniform and is related to the transformation strain by:

$$\varepsilon^C = \mathbf{E}\varepsilon^T, \tag{12.45}$$

where E is called Eshelby's tensor and it depends only on inclusion aspect ratio and the matrix elastic constants. A detailed derivation and applications are given by Mura [14], and analytical expressions for Eshelby's tensor for an ellipsoid of revolution in an isotropic matrix appear in many papers. The strain field $\varepsilon^C(x)$ in the matrix is highly nonuniform, but this more complicated part of the solution can often be ignored. The second step in Eshelby's approach is to demonstrate equivalence between the homogeneous inclusion problem and an inhomogeneous inclusion of the same shape. Consider two infinite bodies of matrix, as shown in Figure 12.6. One has a homogeneous inclusion with some transformation strain ε^T; the other has an inclusion with a different stiffness C^f, but no transformation strain. Subject both the bodies to a uniform applied strain ε^A at infinity. We wish to find the transformation strain ε^T that gives the two problems the same stress and strain distributions.

For the first problem, the inclusion stress is Eq. (12.44) with the applied strain added:

$$\sigma^I = C^m\left(\varepsilon^A + \varepsilon^C - \varepsilon^T\right). \tag{12.46}$$

However, the second problem has no ε^T but a different stiffness, giving a stress of:

$$\sigma^I = C^f\left(\varepsilon^A + \varepsilon^C\right). \tag{12.47}$$

Equating these two expressions gives the transformation strain that makes two problems equivalent. Using Eq. (12.45) and some rearrangement, the result is:

$$-\left[\mathbf{C}^m + \left(\mathbf{C}^f - \mathbf{C}^m\right)\mathbf{E}\right]\varepsilon^T = \left(\mathbf{C}^f - \mathbf{C}^m\right)\varepsilon^A. \tag{12.48}$$

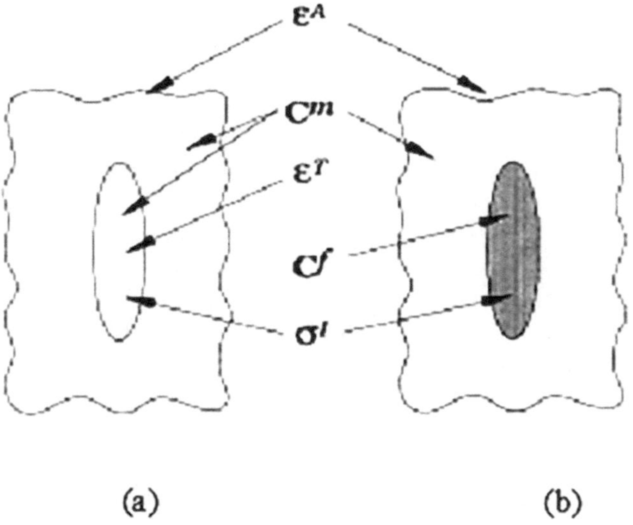

FIGURE 12.6 Eshelby's equivalent inclusion problem. The inclusion (a), with transformation strain ε^T has the same stress σ^I and strain as the inhomogeneity. (b). When both the bodies are subjected to a far-field strain ε^A.

Note that ε^T is proportional to ε^A, which makes the stress in the equivalent inhomogeneity proportional to the applied strain.

12.3.3 Dilute Eshelby's Model

If the volume fraction of inclusions is very small, the interaction between reinforcing elements can be neglected. In this case, the dilute distribution model can be used to analyze the effective elastic properties of composites. The dilute distribution model, as well as many other micromechanical approaches, is based on the theory developed by Eshelby. Eshelby considered the stress and strain fields in a medium with an elliptical region which undergoes a transformation and changes its shape or size. He has shown that uniform stress and strain states are induced in the transformed elastic homogeneous inclusion ("elliptical region"), embedded into an infinite matrix subject to uniform strain. Eshelby introduced a so-called Eshelby tensor **E**, which relates the strains in an inclusion in the infinite elastic matrix ε_{constr} to the strain of the same inclusion, placed outside the matrix and free of the stresses imposed by the matrix $\varepsilon_{unconstr.}$ One can use Eshelby's result to find the stiffness of a composite with ellipsoidal fibers at dilute concentrations. From Eq. (12.21), we know that to find the stiffness, one only has to find the strain concentration tensor A. To do this, first note that for a dilute composite, the average strain is identical to the applied strain since this is the strain at infinity:

$$\bar{\varepsilon} = \varepsilon^A. \tag{12.49}$$

Also, from Eshelby, the fiber strain is uniform and is given by:

$$\bar{\varepsilon}^{f} = \varepsilon^{A} + \varepsilon^{c}, \tag{12.50}$$

where the right-hand side is evaluated within the fiber. Now writing the equivalence between the stresses in the homogeneous and the inhomogeneous inclusions, Eqs. (12.46) and (12.47):

$$C^{f}\left(\varepsilon^{A} + \varepsilon^{C}\right) = C^{m}\left(\varepsilon^{A} + \varepsilon^{C} - \varepsilon^{T}\right). \tag{12.51}$$

And then using Eqs. (12.45), (12.49), and (12.50) to eliminate ε^{T}, ε^{A}, and ε^{C} from this equation, giving:

$$\left[\mathbf{I} + \mathbf{E}\mathbf{S}^{m}\left(\mathbf{C}^{f} - \mathbf{C}^{m}\right)\bar{\varepsilon}^{f} = \bar{\varepsilon}\right]. \tag{12.52}$$

Comparing this equation with Eq. (12.15) shows that the strain concentration tensor for Eshelby's equivalent inclusion is:

$$\mathbf{A}^{\text{Eshelby}} = \left[\mathbf{I} + \mathbf{E}\mathbf{S}^{m}\left(\mathbf{C}^{f} - \mathbf{C}^{m}\right)\right]^{-1}. \tag{12.53}$$

This can be used in Eq. (12.21) to predict the moduli of aligned fiber composites, a result first developed by Russel [15]. Calculations using this model to explore the effects of particle aspect ratio on stiffness are presented by Chow [16]. While Eshelby's solution treats only ellipsoidal fibers, the fibers in most short-fiber composites are much better approximated as right circular cylinders. The relationship between ellipsoidal and cylindrical particles was considered by Steif and Hoysan [17], who developed a very accurate finite element technique for determining the stiffening effect of a single fiber of given shape. For very short particles, $l/d = 4$, they found reasonable agreement for E_{11} by letting the cylinder and the ellipsoid have the same l/d. The ellipsoidal particle gave a slightly stiffer composite, with the difference between the two results increasing as the modulus ratio, E_f/E_m, increased. Henceforth, the cylinder aspect ratio will be used in place of the ellipsoid aspect ratio in Eshelby-type models. Because Eshelby's solution only applies to a single particle surrounded by an infinite matrix, $\mathbf{A}^{\text{Eshelby}}$ is independent of fiber volume fraction and the stiffness predicted by this model increases linearly with fiber volume fraction. Modulus predictions based on Eqs. (12.53) and (12.21) should be accurate only at low volume fractions, say up to v_f of 1%. The more difficult problem is to find some way to include interactions between fibers in the model, and to produce accurate results at higher volume fractions. The next section deals with the approaches for doing that.

12.3.4 Mori–Tanaka Model

a. In the approach, suggested by Mori and Tanaka [18], each inclusion behaves as an isolated inclusion, subject to the averaged stress fields acting on it from all the other inclusions as shown in Figure 12.7. The stresses, acting on an

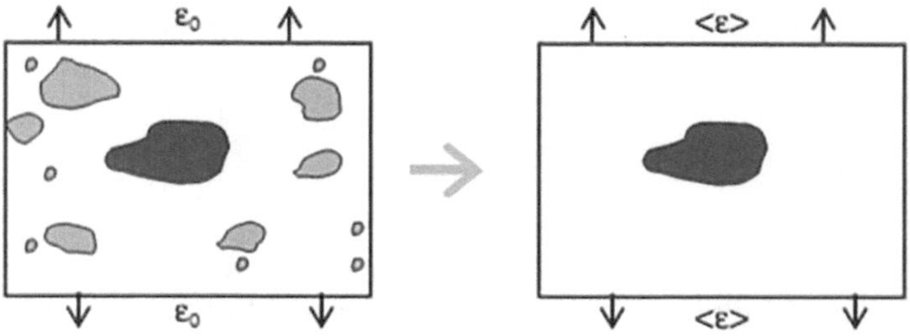

FIGURE 12.7 Mori–Tanaka model. The interaction between the inclusions is taken into account by averaging the strain fields acting on a given inclusion from all the other inclusions.

inclusion and caused by the presence of other inclusions, are superimposed on the applied stress. The idea of Mori and Tanaka was to combine the Eshelby approach and the effective field concept. This is done by defining the strain concentration tensor, which relates the strain in the inclusion to the unknown strain in the matrix instead of the applied strain, as in the case of the dilute distribution model. Benveniste [19] expanded the relations suggested by Mori–Tanaka and provided a general method for determining the effective properties, based on this theory.

Mori and Tanaka approach belongs to the group of effective field methods. This model leads to the same formulae as the lower Hashin–Shtrikman bound (i.e., when the matrix is the softer phase) for spheres and many other inclusion shapes. A family of models for non-dilute composite materials has evolved from a proposal originally made by Mori and Tanaka [18]. Benveniste [19] has provided a particularly simple and clear explanation of the Mori–Tanaka approach, which is used here to introduce the approach. The strain concentration tensor has already been introduced in Eq. (12.15). Suppose that a composite is to be made of a certain type of reinforcing particle, and that, for a single particle in an infinite matrix, we know the dilute strain concentration tensor $\mathbf{A}^{\text{Eshelby}}$:

$$\bar{\varepsilon}^{\text{f}} = \mathbf{A}^{\text{Eshelby}}\bar{\varepsilon}. \tag{12.54}$$

Mori–Tanaka assumption is that, when many identical particles are introduced in the composite, the average fiber strain is given by:

$$\bar{\varepsilon}^{\text{f}} = \mathbf{A}^{\text{Eshelby}}\bar{\varepsilon}^{\text{m}}. \tag{12.55}$$

That is, within a concentrated composite, each particle "sees" a far-field strain equal to the average strain in the matrix. Using the alternate strain concentrator defined in Eq. (12.19), the Mori–Tanaka assumption can be re-stated as:

$$\hat{\text{A}}^{\text{MT}} = \mathbf{A}^{\text{Eshelby}}. \tag{12.56}$$

Equation (12.20) then gives the Mori–Tanaka strain concentrator as:

$$\mathbf{A}^{MT} = \mathbf{A}^{Eshelby}\left[(1 - v_f)\mathbf{I} + v_f\mathbf{A}^{Eshelby}\right]^{-1}. \tag{12.57}$$

This is the basic equation for implementing a Mori–Tanaka model. Mori–Tanaka approach for modeling composites was first introduced by Wakashima et al. [20] for modeling thermal expansions of composites with aligned ellipsoidal inclusions. (Mori and Tanaka's paper treats only the homogeneous inclusion problem, and says nothing about composites). Mori–Tanaka predictions for the longitudinal modulus of a short-fiber composite were first developed by Taya and Mura [21] and Taya and Chou [22], whose work also included the effects of cracks and of a second type of reinforcement. Weng [23] generalized their method, and Tandon and Weng [24] used the Mori–Tanaka approach to develop equations for the complete set of elastic constants of a short-fiber composite. Tandon and Weng's equations for the plane-strain bulk modulus k_{23} and the major Poisson ratio ν_{12} must be solved iteratively. The usual development of the Mori–Tanaka model differs somewhat from Benveniste's explanation. For an average applied stress $\bar{\sigma}$, the reference strain ε^0 is defined as the strain in a homogeneous body of matrix at this stress:

$$\bar{\sigma} = \mathbf{C}^{\mathbf{m}}\varepsilon^0. \tag{12.58}$$

Within the composite, the average matrix strain differs from the reference strain by some perturbation $\widetilde{\varepsilon^m}$:

$$\bar{\varepsilon}^m = \varepsilon^0 + \widetilde{\varepsilon^m}. \tag{12.59}$$

A fiber in the composite will have an additional strain perturbation, $\widetilde{\varepsilon^f}$, such that:

$$\bar{\varepsilon}^f = \varepsilon^0 + \widetilde{\varepsilon^m} + \widetilde{\varepsilon^f}. \tag{12.60}$$

However, the equivalent inclusion will have this strain plus the transformation strain ε^T. The stress equivalence between the inclusion and the fiber then becomes:

$$\mathbf{C}^{\mathbf{f}}\left(\varepsilon^0 + \widetilde{\varepsilon^m} + \widetilde{\varepsilon^f}\right) = \mathbf{C}^{\mathbf{m}}\left(\varepsilon^0 + \widetilde{\varepsilon^m} + \widetilde{\varepsilon^f} - \varepsilon^T\right). \tag{12.61}$$

Comparing this to the dilute version, Eq. (12.51), and noting that ε^A in the dilute problem is equivalent to $\left(\varepsilon^0 + \widetilde{\varepsilon^m}\right)$ here. The development is completed by assuming that the extra fiber perturbation is related to the transformation strain by Eshelby's tensor:

$$\widetilde{\varepsilon^f} = \mathbf{E}\varepsilon^T. \tag{12.62}$$

Combining this with Eqs. (12.59) and (12.60) reveals that Eq. (12.62) contains the essential Mori–Tanaka assumption: The fiber in a concentrated composite sees the average strain of the matrix.

b. Tandon and Weng [24] derived explicit expressions for the elastic constants of a short-fiber composite using the Mori–Tanaka approach. Their formulae for the plane-strain bulk modulus k_{23} and the major Poisson ratio ν_{12} are coupled, and must be solved iteratively. Tandon and Weng calculations are given as:

$$E_{11} = \frac{E_m}{1 + \frac{V_f(A_1 + 2V_m A_2)}{A}} \tag{12.63}$$

$$E_{22} = \frac{E_m}{1 + V_f\left[(1 - V_m)A_4 - 2V_m A_3 + (1 + V_m)A_5 A\right]/2A} \tag{12.64}$$

$$G_{12} = \frac{G_m\,(1 + V_f\,)}{2(1 - V_f)S_{1212} + \frac{G_m}{G_f - G_m}} \tag{12.65}$$

$$G_{23} = \frac{G_m(1 + V_f\,)}{2(1 - V_f)S_{2323} + \frac{G_m}{G_f - G_m}} \tag{12.66}$$

$$\nu_{12} = \frac{\nu_m A - V_f(A_3 - \nu_m A_4)}{A + V_f(A_1 + 2\nu_m A_2)} \tag{12.67}$$

$$\nu_{23} = \frac{E_{22}}{2G_{23}} - 1. \tag{12.68}$$

The equation for ν_{12} shown here was derived by Tucker and Liang [25] and provides a non-iterative formula to the iterative equation of ν_{12} presented by Tandon and Weng. The constants are found using Eq. (12.69).

$$\begin{aligned}
A_1 &= D_1(B_4 + B_5) - 2B_2 \\
A_2 &= (1 + D_1)B_2 - (B_4 + B_5) \\
A_3 &= B_1 - D_1 B_3 \\
A_4 &= (1 + D_1)B_1 - 2B_3 \\
A_5 &= (1 - D_1)/(B_4 - B_5) \\
A &= 2B_2 B_3 - B_1(B_4 + B_5).
\end{aligned} \tag{12.69}$$

The constants B_i and D_j are found from the following:

$$B_1 = V_f D_1 + D_2 + (1 - V_f)(D_1 S_{1111} + 2S_{2211}) \tag{12.70}$$

$$B_2 = V_f + D_3 + (1 - V_f)(D_1 S_{1122} + S_{2222} + S_{2233}) \tag{12.71}$$

$$B_3 = V_f + D_3 + (1 - V_f)(S_{1111} + (1 + D_1) S_{2211}) \tag{12.72}$$

$$B_4 = V_f D_1 + D_2 + (1 - V_f)(S_{1122} + D_1 S_{2222} + S_{2233}) \tag{12.73}$$

$$B_5 = V_f + D_3 + (1 - V_f)(S_{1122} + S_{2222} + D_1 S_{2233}) \tag{12.74}$$

$$D_1 = 1 + 2(\mu_f - \mu_m)/(\lambda_f - \lambda_m) \tag{12.75}$$

$$D_2 = (\lambda_m + 2\mu_m)/(\lambda_f - \lambda_m), \tag{12.76}$$

where μ_m, μ_f, λ_m, and λ_f are Lame's constants for the matrix and fiber materials. Lame's constants are related to the Young's modulus E and Poisson's ratio ν by:

$$\lambda = E\nu/(1+\nu)(1-2\nu) \tag{12.77}$$

$$\mu = E/2(1+\nu). \tag{12.78}$$

The S_{ijkl} in Eqs. (12.70) to (12.74) are the Eshelby tensor components for a spheroidal inclusion defined as:

$$S_{1111} = \frac{1}{2(1-\nu_m)}\left\{(1-2\nu_m) + \frac{3\alpha^2 - 1}{\alpha^2 - 1} - \left[1 - 2\nu_m + \frac{3\alpha^2}{\alpha^2 - 1}\right]g\right\} \tag{12.79}$$

$$S_{2222} = S_{3333} = \frac{3}{8(1-\nu_m)}\frac{\alpha^2}{\alpha^2 - 1} + \frac{1}{4(1-\nu_m)}\left[1 - 2\nu_m - \frac{9}{4(\alpha^2 - 1)}\right]g \tag{12.80}$$

$$S_{2233} = S_{3322} = \frac{1}{4(1-\nu_m)}\left\{\frac{\alpha^2}{2(\alpha^2 - 1)} - \left[1 - 2\nu_m + \frac{3}{4(\alpha^2 - 1)}\right]g\right\} \tag{12.81}$$

$$S_{2211} = S_{3311} = -\frac{1}{2(1-\nu_m)}\frac{\alpha^2}{\alpha^2 - 1} + \frac{1}{4(1-\nu_m)}\left\{\frac{3\alpha^2}{\alpha^2 - 1} - (1 - 2\nu_m)\right\}g \tag{12.82}$$

$$S_{1122} = S_{1133} = -\frac{1}{2(1-\nu_m)}\left[1 - 2\nu_m + \frac{1}{\alpha^2 - 1}\right] + \frac{1}{2(1-\nu_m)}\left[1 - 2\nu_m + \frac{3}{2(\alpha^2 - 1)}\right]g \tag{12.83}$$

$$S_{2323} = S_{3232} = \frac{1}{4(1-\nu_m)}\left\{\frac{\alpha^2}{2(\alpha^2-1)} + \left[1-2\nu_m - \frac{3}{4(\alpha^2-1)}\right]g\right\} \tag{12.84}$$

$$S_{1212} = S_{1313} = \frac{1}{4(1-\nu_m)}\left\{(1-2\nu_m) - \frac{\alpha^2+1}{\alpha^2-1} - \frac{1}{2}\left[1-2\nu_m - \frac{3(\alpha^2+1)}{\alpha^2-1}\right]g\right\}, \tag{12.85}$$

where α is the aspect ratio of the fiber, and g is a parameter defined as:

$$g = \frac{\alpha}{(\alpha^2-1)3/2}\left\{\alpha(\sqrt{\alpha^2-1} - \cosh^{-1}(\alpha)\right\}. \tag{12.86}$$

Using Eqs. (12.79–12.86), Eshelby's tensor **E** is evaluated and further used in Eq. (12.53) to find the strain concentration tensor **A**. This Eshelby's tensor **E** has been used in all the models to find the strain concentration tensor. The strain concentration tensor is then used in Eq. (12.21) to find the stiffness and damping of the composite.

12.3.5 Chow Model

Some other micromechanics models are equivalent to the Mori–Tanaka approach, though this equivalence has not always been recognized. Chow [26] considered Eshelby's inclusion problem and concluded that in a concentrated composite, the inclusion strain would be the sum of two terms: the dilute result given by Eshelby and the average strain in the matrix:

$$\left(\varepsilon^C\right)^f = \mathbf{E}\varepsilon^T + \left(\bar{\varepsilon}^C\right)^m. \tag{12.87}$$

This can be combined with the definition of the average strain from Eq. (12.7) to relate the inclusion strain $\left(\varepsilon^C\right)f$ to the transformation strain ε^T:

$$\left(\varepsilon^C\right)f = (1-V_f)\mathbf{E}\varepsilon^T. \tag{12.88}$$

Chow then extended this result to an inhomogeneity following the usual arguments, Eqs. (12.46)–(12.53), which produces a strain concentration tensor:

$$\mathbf{A}^{Chow} = \left[\mathbf{I} + (1-V_f)\mathbf{E}\mathbf{S}^m\left(\mathbf{C}^f - \mathbf{C}^m\right)\right]^{-1}, \tag{12.89}$$

which is equivalent to the Mori–Tanaka result (Eq. 12.57). Chow was apparently unaware of the connection between his approach and the Mori–Tanaka scheme, but he seems to have been the first to apply the Mori–Tanaka approach to predict the stiffness of short-fiber composites. A more recent development is the equivalent poly-inclusion model of Ferrari [27]. Rather than using the strain concentration tensor A,

Ferrari used an effective Eshelby tensor $\hat{E}$, defined as the tensor that relates inclusion strain to transformation strain at the finite volume fraction:

$$\left(\varepsilon^{C}\right)f = \hat{E}\,\varepsilon^{T}. \tag{12.90}$$

Once $\hat{E}$ has been defined, it is straightforward to derive a strain concentration tensor **A** and a composite modulus. Ferrari considered admissible forms for $\hat{E}$, given the requirements that $\hat{E}$ must (a) produce a symmetric stiffness tensor C, (b) approach Eshelby's tensor **E** as volume fraction approaches zero, and (c) give a composite stiffness that is independent of the matrix stiffness as volume fraction approaches unity. He proposed a simple form that satisfies these criteria:

$$\hat{E} = (1 - V_f)\mathbf{E}. \tag{12.91}$$

Combination of Eqs. (12.90) and (12.91) is identical to Chow's assumption as in Eq. (12.88) and for aligned fibers of uniform length, Ferrari's equivalent polyinclusion model, Chow's model, and the Mori–Tanaka model are identical. Important differences between the equivalent polyinclusion model and the Mori–Tanaka model arise when the fibers are misoriented or have different lengths.

12.3.6 Modified Halpin–Tsai or Finegan Model

Dynamic properties of carbon nanofiber/polypropylene composites can be expressed analytically from the theory of short-fiber composites. The elastic–viscoelastic correspondence principle – "A linear elastostatic analysis can be converted to a vibratory linear viscoelastic analysis by replacing the elastic moduli with the corresponding complex moduli" – is used to define the dynamic properties. The complex modulus has a real and an imaginary part. The damping loss factor can be expressed as the ratio of imaginary to the real part of the complex modulus.

$$\eta = \frac{C''}{C'}, \tag{12.92}$$

where $C^* = C' + iC''$ = complex modulus of the composite
C' = storage modulus of the composite
C'' = loss modulus of the composite
η = damping loss factor of the composite.

A composite modulus that takes into account the fiber orientation in the direction of injection has been expressed by Advani and Tucker [28] by using tensorial notation. The most general description of fiber orientation is the probability distribution function for orientation Ψ. Some of the assumptions made are as follows:

a. Fibers are rigid cylinders, uniform in length and diameter.
b. Number of fibers per unit volume is uniform, though the orientation of these fibers may not be uniform.

With these assumptions, the orientation of a single fiber can be described by the angles (θ, Φ) as shown in Figure 12.8. The figure presents the definition of the θ, Φ angles used in expressing the orientation of a single fiber by using three Cartesian components (p_1, p_2, p_3) of a unit vector oriented along the fiber axis:

$$p_1 = \sin\theta\cos\Phi$$

$$p_2 = \sin\theta\sin\Phi \tag{12.93}$$

$$p_3 = \cos\theta.$$

a_{ij} and a_{ijkl} give information on how well the fibers are aligned to 1, 2, 3 axes (shown in Figure 12.8).

The orientation state at a point in space can be described by a probability distribution function Ψ (θ, Φ). This function defines the probability of finding a fiber between angles θ_1 & (θ_1 + dθ) & Φ_1 & (Φ_1 + dΦ_1) in the way as:

$$P(\theta_1 \le \theta \le \theta_1 + d\theta,\ \Phi_1 \le \Phi \le \Phi_1 + d\Phi) = \Psi(\theta_1, \Phi_1)\sin\theta_1 d\theta d\Phi. \tag{12.94}$$

Another way is to associate a unit vector p with the fiber. Since the length of the vector is fixed, hence

$$p_i p_i = 1. \tag{12.95}$$

The set of all possible directions of p corresponds to unit sphere. The integral over the surface of the unit sphere is denoted by:

$$\oint Dp = \int_{\Phi=0}^{2\pi} \int_{\theta=0}^{\pi} \sin\theta\ d\theta\ d\Phi. \tag{12.96}$$

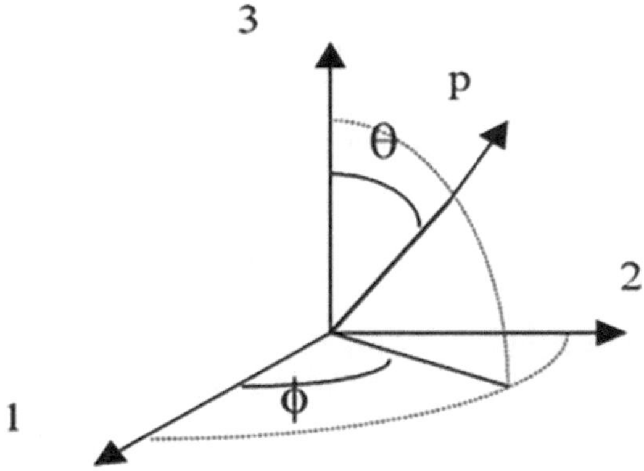

FIGURE 12.8 Coordinate system and definitions of θ, Φ, p.

The function Ψ must satisfy the following conditions:

a. A fiber oriented at an angle (θ, Φ) is indistinguishable from a fiber oriented at angles $(\pi - \theta, \Phi + \pi)$, so Ψ should be periodic. Therefore,

$$\Psi(\theta,\Phi) = \Psi(\pi - \theta, \Phi + \pi) \tag{12.97}$$

$$\text{or } \Psi(p) = \Psi(-p).$$

b. Since every fiber has some orientation, so Ψ must be normalized, that is,

$$\oint \Psi(p)\,dp = \int_{\theta=0}^{\pi}\int_{\Phi=0}^{2\pi} \Psi(\theta, \Phi)\sin\theta \; d\theta \; d\Phi = 1. \tag{12.98}$$

c. The third condition is continuity condition which describes the change in Ψ with time when fibers are changing orientation. If fibers move with bulk motion of fluid, then Ψ may be regarded as a convected quantity. One set of orientation tensors can be defined by forming products of vector p and integrating the product of these vectors with distribution function over all possible directions. Therefore,

$$\frac{D\Psi}{Dt} = -\frac{\alpha}{\alpha p}(\Psi \dot{p}\ ), \tag{12.99}$$

where $\frac{\alpha}{\alpha p}$ = gradient operator on the surface of unit sphere.

Because $\Psi(p) = \Psi(-p)$, the odd order integrals are zero. Thus, only even order tensors are of interest. The second- and fourth-order orientation tensors are:

$$a_{ij} = \oint p_i p_j \Psi(p)\,dp$$

$$a_{ijkl} = \oint p_i p_j p_k p_l \Psi(p)\,dp. \tag{12.100}$$

One can also form another version of orientation tensors:

$$b_{ij} = a_{ij} - \frac{1}{3}\delta_{ij} \tag{12.101}$$

$$b_{ijkl} = a_{ijkl} - \frac{1}{7}\left(\delta_{ij}a_{kl} + \delta_{ik}a_{jl} + \delta_{il}a_{jk} + \delta_{jk}a_{il} + \delta_{jl}a_{ik} + \delta_{kl}a_{ij}\right)$$

$$+ \frac{1}{35}\left(\delta_{ij}\delta_{kl} + \delta_{ik}\delta_{jl} + \delta_{il}\delta_{jk}\right), \tag{12.102}$$

where δ_{ij} is the unit tensor.

If we define tensor basis functions of p as:

$$f_{ij}(p) = p_i p_j - \frac{1}{3}\delta_{ij}$$

$$f_{ijkl}(p) = p_i p_j p_k p_l - \frac{1}{7}\left(\delta_{ij}p_k p_l + \delta_{ik}p_j p_l + \delta_{il}p_j p_k + \delta_{jk}p_i p_l + \delta_{jl}p_i p_k + \delta_{kl}p_i p_j\right)$$

$$+\frac{1}{35}\left(\delta_{ij}\delta_{kl} + \delta_{ik}\delta_{jl} + \delta_{il}\delta_{jk}\right) \quad (12.103)$$

Then the distribution function is

$$\Psi(p) = \frac{1}{4\pi} + \frac{15}{8\pi}b_{ij}f_{ij}(p) + \frac{315}{32\pi}b_{ijkl}f_{ijkl}(p) + \cdots. \quad (12.104)$$

A planar orientation state is one in which all the fibers lie in a single plane. Taking 1–2 plane as the plane of orientation:

$$\Psi(\theta,\Phi) = \delta\left(\theta - \frac{\pi}{2}\right)\Psi(\Phi), \quad (12.105)$$

where
δ = Dirac delta function
$\Psi(\Phi)$ = Planar distribution function which satisfies the normalization condition

$$\int_0^{2\pi} \Psi(\Phi)\,d\Phi = 1. \quad (12.106)$$

Therefore, planar orientation tensors are:

$$a_{ij} = \int_0^{2\pi} \Psi(\Phi)p_i p_j\,d\Phi$$

$$a_{ijkl} = \int_0^{2\pi} \Psi(\Phi)p_i p_j p_k p_l\,d\Phi \quad (12.107)$$

$$b'_{ij}(p) = a_{ij} - \frac{1}{2}\delta_{ij}$$

$$b'_{ijkl} = a_{ijkl} - \frac{1}{6}\left(\delta_{ij}a_{kl} + \delta_{ik}a_{jl} + \delta_{il}a_{jk} + \delta_{jk}a_{il} + \delta_{jl}a_{ik} + \delta_{kl}a_{ij}\right)$$

$$+\frac{1}{24}\left(\delta_{ij}\delta_{kl} + \delta_{ik}\delta_{jl} + \delta_{il}\delta_{jk}\right),$$

where i and j can take values = 1 and 2.

Basis functions are:

$$f'_{ij}(p) = p_i p_j - \frac{1}{2}\delta_{ij}$$

$$f'_{ijkl}(p) = p_i p_j p_k p_l - \frac{1}{6}\left(\delta_{ij}p_k p_l + \delta_{ik}p_j p_l + \delta_{il}p_j p_k + \delta_{jk}p_i p_l + \delta_{jl}p_i p_k + \delta_{kl}p_i p_j\right)$$

$$+ \frac{1}{24}\left(\delta_{ij}\delta_{kl} + \delta_{ik}\delta_{jl} + \delta_{il}\delta_{jk}\right). \tag{12.108}$$

Therefore,

$$\Psi(\Phi) = \frac{1}{2\pi} + \frac{2}{\pi} b'_{ij} f'_{ij}\,(p) + \frac{8}{\pi}\, b'_{ijkl}\; f'_{ijkl}\,(p) + \cdots. \tag{12.109}$$

Now, considering a tensor property T(p) associated with a unidirectional microstructure aligned in the direction of p, for p to be a complete description of direction of T, T must be unchanged by any coordinate rotation that either preserves or reverses the direction of p; that is, T must be transversely isotropic, with p as its axis of symmetry. For a second-order tensor to be transversely isotropic, it must have the form:

$$T_{ij}(p) = A_1 p_i p_j + A_2\delta_{ij}, \tag{12.110}$$

where A_1 and A_2 are the scalar constants. If we take orientation average, then:

$$\langle T\rangle_{ij} = A_1 a_{ij} + A_2\delta_{ij}. \tag{12.111}$$

That is, orientation average of second-order tensor is completely determined by the second-order orientation tensor and by the unidirectional property tensor. Similarly, for a fourth-order tensor:

$$T_{ijkl}(p) = B_1\left(p_i p_j p_k p_l\right) + B_2\left(p_i p_j \delta_{kl} + p_k p_l \delta_{ij}\right) + B_3\left(p_i p_k \delta_{jl} + p_i p_l \delta_{jk} + p_j p_l \delta_{ik} + p_j p_k \delta_{il}\right)$$

$$+ B_4\left(\delta_{ij}\delta_{kl}\right) + B_5\left(\delta_{ik}\delta_{jl} + \delta_{il}\delta_{jk}\right). \tag{12.112}$$

These B's are five scalar constants related to five independent components of a transversely isotropic elasticity tensor. Therefore:

$$\langle T\rangle_{ijkl} = B_1\left(a_{ijkl}\right) + B_2\left(a_{ij}\delta_{kl} + a_{kl}\delta_{ij}\right) + B_3\left(a_{ik}\delta_{jl} + a_{il}\delta_{jk} + a_{jl}\delta_{ik} + a_{jk}\delta_{il}\right)$$

$$+ B_4\left(\delta_{ij}\delta_{kl}\right) + B_5\left(\delta_{ik}\delta_{jl} + \delta_{il}\delta_{jk}\right). \tag{12.113}$$

For longitudinal direction:

$$a_{11} = \int_{\Phi=0}^{2\pi}\int_{\theta=0}^{\pi} p_1(\theta, \Phi)p_1(\theta, \Phi)\,\Psi(\theta)\sin\theta\, d\theta\, d\Phi \tag{12.114}$$

$$a_{1111} = \int_{\Phi=0}^{2\pi} \int_{\theta=0}^{\pi} p_1(\theta, \Phi)\, p_1(\theta, \Phi) p_1(\theta, \Phi) p_1(\theta, \Phi) \Psi(\theta) \sin\theta \, d\theta \, d\Phi,$$

where $p_1(\theta, \Phi)$ is given by components of p and weighting factor $\Psi(\theta)$ is determined experimentally as:

$$\Psi(\theta) = I(\theta) \times \text{Norm}$$

$$I(\theta) = I_1 \left[\exp\{-\theta/\theta_{fwhm}\}^2 \right] + I_0.$$

$I_1 = 44.3$, $I_0 = 37.9$, $\theta_{fwhm} = 50.4°$

$$\text{Norm} = \left[\int_0^{2\pi} \int_0^{\pi} I(\theta) \sin\theta \, d\theta \, d\Phi \right]^{-1}. \tag{12.115}$$

For carbon nanofiber/polypropylene composite:

$$a_{11} = 0.354$$

$$a_{1111} = 0.22. \tag{12.116}$$

Advani and Tucker [28] gave the stiffness of the composite as:

$$C_{ijkl}^* = B_1^* (a_{ijkl}) + B_2^* (a_{ij}\delta_{kl} + a_{kl}\delta_{ij}) + B_3^* (a_{ik}\delta_{jl} + a_{il}\delta_{jk} + a_{jl}\delta_{ik} + a_{jk}\delta_{il})$$
$$+ B_4^* (\delta_{ij}\delta_{kl}) + B_5^* (\delta_{ik}\delta_{jl} + \delta_{il}\delta_{jk}). \tag{12.117}$$

Composite elastic modulus is replaced by the corresponding complex modulus according to the elastic–viscoelastic correspondence principle, and composite damping can be calculated as shown in Eq. (12.92). For the longitudinal direction, we get:

$$C_{1111}^* = 0.22B_1^* + 0.708B_2^* + 1.416B_3 + B_4^* + 2B_5^*$$
$$= E_{longitudinal} = E_c' + iE_c'', \tag{12.118}$$

where $B_1^*, B_2^*, B_3^*, B_4^*, B_5^*$ are the constants defined in terms of the stiffness matrix of the unidirectional aligned composite in complex form as:

$$B_1^* = C_{11}^* + C_{22}^* - 2C_{12}^* - 4C_{66}^*$$

$$B_2^* = C_{12}^* - C_{23}^*$$

$$B_3^* = C_{66}^* + \frac{1}{2}(C_{23}^* - C_{22}^*) \tag{12.119}$$

$$B_4^* = C_{23}^*$$

$$B_5^* = \frac{1}{2}\left(C_{22}^* - C_{23}^*\right).$$

Components of the stiffness matrix without taking into consideration the orientation tensors, C_{ij}^*, are calculated from the Halpin–Tsai equations for the unidirectional aligned composite in the complex form. By combining the self-consistent approach of Hill with the solutions of Hermans and making a few additional assumptions, Halpin and Tsai provide a simpler analytical form for predicting material properties of fiber composites. Halpin–Tsai equations need only one equation to find all the composite moduli and the longitudinal Poisson's ratio is simply found from the rule of mixtures. It should be noted that the Halpin–Tsai equations are partially empirical where one parameter ξ is found by fitting the equations to numerical results:

$$C_{11}^* = E_{11}^* = \frac{1 + 2\left(\frac{l}{d}\right)\eta_L^* V_f}{1 - \eta_L^* V_f} E_m^* \tag{12.120}$$

$$C_{22}^* = E_{22}^* = \frac{1 + 2\ \eta_T^* V_f}{1 - \eta_T^* V_f} E_m^* \tag{12.121}$$

$$C_{12}^* = C_{66}^* = E_{12}^* = E_{66}^* = \frac{1 + \eta_G^* V_f}{1 - \eta_G^* V_f} G_m^* \tag{12.122}$$

$$C_{23}^* = E_{23}^* = \frac{1 + \xi\eta_{TS}^* V_f}{1 - \eta_{TS}^* V_f} G_m^*, \tag{12.123}$$

where

$$\eta_L^* = \frac{\left[\frac{E_f^*}{E_m^*} - 1\right]}{\left[\frac{E_f^*}{E_m^*} + 2\left(\frac{l}{d}\right)\right]}, \quad \eta_T^* = \frac{\left[\frac{E_f^*}{E_m^*} - 1\right]}{\left[\frac{E_f^*}{E_m^*} + 2\right]}, \quad \eta_G^* = \frac{\left[\frac{G_f^*}{G_m^*} - 1\right]}{\left[\frac{G_f^*}{G_m^*} + 1\right]} \tag{12.24a,b,c}$$

$$\eta_{TS}^* = \frac{\left[\frac{G_f^*}{G_m *} - 1\right]}{\left[\frac{G_f^*}{G_m *} + \xi^*\right]}, \quad \xi^* = \frac{\frac{K_m^*}{G_m^*}}{\left[\frac{K_m^*}{G_m^*} + 2\right]}, \quad K_m^* = \frac{E_m^*}{\left[\ 3\,(1 - 2\nu_m)\right]}, \tag{12.25a,b,c}$$

where

V_f = fiber volume fraction

l/d = fiber aspect ratio

$E_f^* = E_f' + iE_f''$ = complex elastic modulus of the fiber

$E_m^* = E_m' + iE_m''$ = complex elastic modulus of the matrix

$G_f^* = G_f' + iG_f''$ = complex elastic shear modulus of the fiber

$G_m^* = G_m' + iG_m''$ = complex elastic shear modulus of the matrix.

TABLE 12.2
Dynamic Properties of Vapor-Grown Carbon Fibers and Polypropylene

Dynamic Properties	VGCF	Polypropylene
Storage modulus (E′), GPa	1000 (measured)	1.23 (measured)
Loss modulus (E″), GPa	1.5 (calculated)	0.0738 (calculated)
Loss factor (η)	0.0015 (from the literature)	0.06 (from the literature)
Fiber volume fraction (V_f)	0.03	

Dynamic properties of vapor-grown carbon fiber (VGCF) and polypropylene matrix as given by Gibson et al. [29] are used as input and shown in Table 12.2. With the above data, Eqs. (12.118–12.125) are solved using a mathematical tool such as MATLAB®. By writing a program in MATLAB, we can solve the equations and can find the effect of various parameters such as aspect ratio and fiber volume fraction on loss modulus, storage modulus, and damping loss factor.

12.3.7 Hashin–Shtrikman Model

A rather different approach to modeling stiffness is based on finding upper and lower bounds for the composite moduli. All bounding methods are based on assuming an approximate field for either the stress or the strain in the composite. The unknown field is then found through a variational principle, by minimizing or maximizing some functional of the stress and strain. Resulting composite stiffness is not exact, but it can be guaranteed to be either greater than or less than the actual stiffness, depending on the variational principle. This rigorous bounding property is the attraction of bounding methods. Historically, the Voigt and Reuss averages were the first models to be recognized as providing rigorous upper and lower bounds. To derive the Voigt model, Eq. (12.23), one assumes that the fiber and the matrix have the same uniform strain, and then minimizes the potential energy. Since the potential energy will have an absolute minimum when the entire composite is in equilibrium, the potential energy under the uniform strain assumption must be greater than or equal to the exact result, and the calculated stiffness will be an upper bound on the actual stiffness. The Reuss model, Eq. (12.24), is derived by assuming that the fiber and the matrix have the same uniform stress, and then maximizing the complementary energy. Since the complementary energy must be maximum at equilibrium, the model provides a lower bound on the composite stiffness.

Detailed derivations of these bounds are provided by Wu and McCullough [30]. Voigt and Reuss bounds provide isotropic results (provided the fiber and matrix are themselves isotropic), when in fact we expect aligned fiber composites to be highly anisotropic. More importantly, when the fiber and the matrix have substantially different stiffnesses, Voigt and Reuss bounds are quite far apart and provide little useful information about the actual composite stiffness. This latter point motivated Hashin and Shtrikman to develop a way to construct tighter bounds. Hashin and Shtrikman [31] developed an alternate variational principle for heterogeneous materials.

Their method introduces a reference material and bases the subsequent development on the differences between this reference material and the actual composite. Rather than requiring two variational principles, like the Voigt and Reuss bounds, their single variational principle gives both the upper and lower bounds by making appropriate choices of the reference material. For an upper bound, the reference material must be as stiff as, or stiffer than, any phase in the composite (fiber or matrix), and for a lower bound, the reference material must have a stiffness less than or equal to any phase. In most composites, the fiber is stiffer than the matrix, so choosing the fiber as the reference material gives an upper bound and choosing the matrix as the reference material gives a lower bound. If the matrix is stiffer than the fiber, the bounds are reversed. The resulting bounds are tighter than the Voigt and Reuss bounds, which can be obtained from the Hashin–Shtrikman theory by giving the reference material infinite or zero stiffness, respectively. Hashin and Shtrikman's original bounds apply to isotropic composites with isotropic constituents. Frequently, the bounds are regarded as applying to composites with spherical particles, though a fiber composite with 3-D random fiber orientation must also obey the bounds. Walpole [32] re-derived Hashin–Shtrikman bounds using classical energy principles and extended them to anisotropic materials. Walpole also derived results for infinitely long fibers and infinitely thin disks in both aligned and 3-D random orientations. Hashin–Shtrikman–Walpole bounds were extended to short-fiber composites by Willis [33] and by Wu and McCullough [30]. These workers introduced a two-point correlation function into the bounding scheme, allowing aligned ellipsoidal particles to be treated. Based on these extensions, explicit formulae for aligned ellipsoids were developed by Weng [34] and Eduljee et al. [35]. The general bounding formula, shown here in the format developed by Weng, gives the composite stiffness C as:

$$\mathbf{C} = \left[V_f \mathbf{C}^{\mathbf{f}} \cdot \mathbf{Q}^{\mathbf{f}} + V_m \mathbf{C}^{\mathbf{m}} \mathbf{Q}^{\mathbf{m}}\right]\left[V_f \mathbf{Q}^{\mathbf{f}} + V_m \mathbf{Q}^{\mathbf{m}}\right]^{-1}, \tag{12.126}$$

where the tensors $\mathbf{Q}^{\mathbf{f}}$ and $\mathbf{Q}^{\mathbf{m}}$ are defined as:

$$\mathbf{Q}^{\mathbf{f}} = \left[\mathbf{I} + \mathbf{E}^0 \mathbf{S}^0 \left(\mathbf{C}^{\mathbf{f}} - \mathbf{C}^0\right)\right]^{-1}$$

$$\mathbf{Q}^{\mathbf{m}} = \left[\mathbf{I} + \mathbf{E}^0 \mathbf{S}^0 \left(\mathbf{C}^{\mathbf{m}} - \mathbf{C}^0\right)\right]^{-1}. \tag{12.127}$$

Here $\mathbf{E}^0$ is Eshelby's tensor associated with the properties of the reference material, which has stiffness $\mathbf{C}^0$ and compliance $\mathbf{S}^0$. When the matrix is chosen as the reference material, Eq. (12.126) gives a strain concentrator of:

$$\hat{\mathbf{A}}^{\text{lower}} = \left[\mathbf{I} + \mathbf{E}^{\mathbf{m}} \mathbf{S}^{\mathbf{m}} \left(\mathbf{C}^{\mathbf{f}} - \mathbf{C}^{\mathbf{m}}\right)\right]^{-1}. \tag{12.128}$$

This result is labeled here as the lower bound, on the presumption that the fiber is stiffer than the matrix. The composite stiffness is found by substituting $\hat{\mathbf{A}}^{\text{lower}}$ in Eqs. (12.20–12.21). Eduljee et al. [35] argue that the lower bound provides the most accurate estimate of composite properties, and recommend it as a model. Note that this lower bound prediction is identical to the Mori–Tanaka model, Eq. (12.57). This correspondence lends theoretical support to the Mori–Tanaka approach, and guarantees that it will always obey the bounds.

12.3.8 Lielens Model

At fiber volume fractions close to unity, the matrix stiffness strongly influences the composite stiffness for the lower bound/Mori–Tanaka models, despite the tiny amount of it that is present. Packing considerations suggest that the only way to approach such high volume fractions is for the fiber phase to become continuous, and Lielens et al. [36] suggested that at very high fiber volume fractions, the composite stiffness should be much closer to the upper bound, or equivalent to the Mori–Tanaka prediction using the fiber as the continuous phase. This insight prompted Lielens and co-workers to propose a model that interpolates between the upper and lower bounds, such that the lower bound dominates at low volume fractions and the upper bound dominates at high volume fractions (again presuming the fiber is the stiffer phase). They performed this interpolation on the inverse of the strain concentration tensor $\hat{\mathbf{A}}$, producing the predictive equation:

$$\hat{\mathbf{A}}^{\text{lielens}} = \left\{(1-f)\left[\hat{\mathbf{A}}^{\text{lower}}\right]^{-1} + f\left[\hat{\mathbf{A}}^{\text{upper}}\right]^{-1}\right\}^{-1}. \tag{12.129}$$

The interpolating factor f depends on fiber volume fraction, and they propose

$$f = \frac{V_f + V_f^2}{2}. \tag{12.130}$$

This theory reproduces the lower bound and Mori–Tanaka results at low volume fractions, but is said to give improved results at reinforcement volume fractions in the 40%–60% range.

12.3.9 Self-Consistent Model

Another approach to account for finite fiber volume fraction is the self-consistent method. This approach is generally credited to Hill [37] and Budiansky [38], whose original work focused on spherical particles and continuous aligned fibers. The application to short-fiber composites was developed by Laws and McLaughlin [39] and by Chou et al. [40]. In self-consistent scheme, one finds the properties of a composite in which a single particle is embedded in an infinite matrix that has the average properties of the composite. For this reason, self-consistent models are also called embedding models. Building on Eshelby's result for an ellipsoidal particle, we can create a self-consistent version of Eq. (12.53) by replacing the matrix stiffness and compliance tensors by the corresponding properties of the composite. This gives the self-consistent strain concentration tensor as:

$$\mathbf{A}^{SC} = \left[\mathbf{I} + \mathbf{ES}\left(\mathbf{C}^f - \mathbf{C}\right)\right]^{-1}. \tag{12.131}$$

Of course, the properties C and S of the embedding "matrix" are initially unknown. When the reinforcing particle is a sphere or an infinite cylinder, the equations can be manipulated algebraically to find explicit expressions for the overall properties. For short fibers, this has not proved possible, but numerical solutions are easily obtained by an iterative scheme. One starts with an initial guess at the composite properties, evaluates E and then $\mathbf{A}^{SC}$ from Eq. (12.131), and by substituting the result into Eq. (12.21) to get an improved value for the composite stiffness. The procedure is repeated using this new value, and the iterations continue until the results for C converge. An additional, but less obvious, change is that Eshelby's tensor E depends on the "matrix" properties, which are now transversely isotropic. Expressions for Eshelby's tensor for an ellipsoid of revolution in a transversely isotropic matrix are given by Lin and Mura [41] and by Chou et al. [42]. With these expressions in hand, one can use Eq. (12.131) together with Eq. (12.21) to find the stiffness of the composite. This is the self-consistent approach used for short-fiber composites. A closely related approach, called "generalized self-consistent model," also uses an embedding approach. However, in these models, the embedded object comprises both fiber and matrix materials. When the composite has spherical reinforcing particles, the embedded object is a sphere of the reinforcement encased in a concentric spherical shell of matrix; this is in turn surrounded by an infinite body with the average composite properties. The generalized self-consistent model is sometimes referred to as a "double embedding" approach. For continuous fibers, the embedded object is a cylindrical fiber surrounded by a cylindrical shell of matrix. The first generalized self-consistent models were developed for spherical particles by Kerner [43] and for cylindrical fibers by Hermans [44]. While the generalized self-consistent model is widely regarded as superior to the original self-consistent approach, no such model has been developed for short fibers.

Based on the above discussion, some conclusions that can be inferred from the study of different models are listed below:

i. Cox model gives good approximation of average fiber axial stress and the total strain energy in the composite, yet this model is not applicable for low fiber volume fractions. This model gives good results for longer fibers but falls below for short fibers. Cox model treats the fiber as a slender body. The true axial displacements lag behind the beam theory predictions.
ii. Eshelby model is used when the volume fraction of fibers is very small. This model does not consider interfiber interaction and assumes that the strain is uniform in the fiber and highly nonuniform in the matrix. The stiffness predicted by this model increases linearly with fiber volume fraction.
iii. Mori–Tanaka model assumes that the fiber in a concentrated composite sees the average strain of the matrix; that is, the interaction between the fibers is taken into account by averaging the strain fields acting on a given fiber from all the other fibers. Most of other models reproduce the results of Mori–Tanaka. Hence, this model is widely used to predict stiffness and damping properties of short-fiber composites at different fiber volume fractions and aspect ratios of fiber.

iv. Chow model is equivalent to the Mori–Tanaka approach. Chow concluded that at high fiber volume fractions, the fiber strain would be the sum of the result given by Eshelby and the average strain in the matrix.
v. Modified Halpin–Tsai or Finegan model uses the elastic–viscoelastic correspondence principle. This model gives satisfactory results for short fibers but falls below for longer fibers. It contains no dependence on aspect ratio for E_{22}', G_{23}', and G_{12}'. This model is semi-empirical where one parameter $\xi = 2l/d$ is found by fitting the equations to numerical results.
vi. Hashin–Shtrikman model provides upper and lower bounds on the composite moduli. Choosing the fiber as reference material gives an upper bound, and choosing the matrix as reference material gives a lower bound on the composite moduli.
vii. Lielens model interpolates between the upper and lower bounds, such that lower bound dominates at low fiber volume fractions and upper bound dominates at high volume fractions. This theory reproduces the lower bound and Mori–Tanaka results at low volume fractions.
viii. In self-consistent model, we find the properties of the composite in which a single fiber is embedded in an infinite matrix that has the average properties of the composite. This model usually gives higher results as it starts with making an assumption of the composite properties.

12.4 FAST FOURIER TRANSFORM NUMERICAL HOMOGENIZATION METHODS

The properties of composites not only depend upon the constituent materials but also on their microstructures. Some paradigms are a diversity of biological materials which can assemble a few ingredients like carbon and minerals into a large number of structures featured with outstanding properties and multifunctionalities. The effective properties of such composites can be estimated by theories like Hashin–Shtrikman bounds and Budiansky bounds without the consideration of structural complexity. In order to design advanced materials, however, it is highly desirable to know their accurate properties after the microstructure is determined. Based on Eshelby's equivalent inclusion theory and combined with the average stress, some approaches such as self-consistent method, generalized self-consistent method, and Mori–Tanaka method have been proposed to predict the effective properties of composites. Initially, these methods are only applicable to composites with ellipsoid-shaped inclusions because this theory is based on an ellipsoid. With a step-by-step analysis, these methods were improved to be used for multiphase composite and graphene sheets dispersed in polymer nanocomposites [45,46].

For composites in more complex shapes, their representative volume element (unit cell) was simulated by finite element analysis with periodic boundary conditions. By averaging the local stress and the local strain within the unit cell, some mechanical properties can be obtained straightforward or in two steps. Virtual element method has also been developed to retrieve the effective properties for heterogeneous materials [47]. However, these finite element analysis-based methods lack a rigorous mathematic framework to explicitly illuminate the essential relationships between the effective properties and material layouts. In 1978, a homogenization theory was proposed to evaluate the effective properties of composites by assuming the microdisplacement of their unit cell is governed by a characteristic equation [48,49]. When the size of the unit cell tends to be infinitesimally small, the effective elastic tensor can be approximated by integrating the difference between test strains and micro strains. This method has been successfully used for porous materials, lattice, and fiber-reinforced composites. Because a well-defined formula is given, the sensitivity of the effective properties with respect to local density – namely, the ratio of solid material occupation in an element, can be derived. Such a sensitivity analysis has been used in structural topology optimization to systematically design advanced materials with specified properties or extreme properties on Hashin–Shtrikman bounds. Very recently, a novel homogenization method based on enriched micro-inertial continuum was explored for resonant meta-materials to analyze their negative stiffness and negative effective mass density [50]. Meanwhile, the effective medium model of periodic composites was analyzed by a multiscale nonlocal homogenization method [51].

To investigate the effective properties of linear and nonlinear periodic materials as well as their local response, a homogenization method based on fast Fourier transform (FFT) was developed by Moulinec and Suquet in 1994 [52]. Later, the authors extended this method to complex microstructures under nonlinear plastic deformation by imposing macroscopic stress [53]. Though this method works extremely well for composites with small contrasts in Young's modulus of constitutive materials, it converges slowly for porous materials with infinite ratio of stiffness [54]. To remedy such an insufficiency, an augmented Lagrangian method with the consideration of local strain potential was developed [55]. A conjugate gradients method [56,57] and a polarization FFT iterative scheme based on energy principle [58,59] were also established to solve this problem. To improve the efficiency of FFT-based homogenization, an accelerated scheme based on a contraction mapping and nested uniform grids was proposed [60]. Compared to asymptotic homogenization, the method of FFT-based homogenization avoids the time-consuming finite element analysis for calculating local displacement, but obtains the strain using the Green tensor operator in the frequency domain. Because of its versatility and high efficiency, FFT-based homogenization method has been extended to retrieve the effective physical properties used in rigid-plastics preformation [61], viscose-plastic preformation [62], crack propagation [63], electrical engineering [64,65], and optics [66].

In this section, we review the essential ideas and numerical implementation of FFT-based homogenization and asymptotic homogenization, and compare their efficiency and applicability. Numerical examples show that the former can be applicable to composites with an extremely large contrast of Young's modulus and eliminates inappropriate values (e.g., negative entries) in the effective elastic tensor. We propose an algorithm to calculate the local displacement of unit cell under test strains in principal directions for FFT-based homogenization. We find the boundaries of unit cell are wave-like even under test strains in principal directions, while they are straight in asymptotic homogenization. Because such deformation patterns reflect the periodic boundary conditions more reasonably, FFT-based homogenization is more likely to correctly retrieve the effective properties of composites with unsymmetrical microstructures.

12.4.1 FFT-Based Homogenization Method

The linear elastic composite considered in this section, without the loss of generality, is constructed by self-repeated unit cells. Under an external distributed force **f**, the stress $\boldsymbol{\sigma}$ in the unit cell is governed by the equation of elasticity:

$$div\,\sigma + \mathbf{f} = 0, \tag{12.132}$$

where the relation between the stress $\boldsymbol{\sigma}$ and strain $\boldsymbol{\varepsilon}$ is given as:

$$\sigma = C : \varepsilon = \left(C - C^0\right) : \varepsilon + C^0 : \varepsilon. \tag{12.133}$$

If the composite is assumed as a homogeneous material with an elastic tensor $\mathbf{C}^0$, at any arbitrary location occupied by solid material with an elastic tensor **C** or a void material, the material difference will result in the polarization stress tensor τ, given as:

$$\tau = \left(\mathbf{C} - \mathbf{C}^0\right) : \varepsilon. \tag{12.134}$$

Substituting Eq. (12.133) into Eq. (12.132) gives:

$$div\left(\tau + \mathbf{C}^0 : \varepsilon\right) + \mathbf{f} = 0. \tag{12.135}$$

Extending Eq. (12.135) and writing it in the Einstein tensor form, we get:

$$C^0_{ijkl} u_{k,lj} + f_i + \tau_{ij,j} = 0, \tag{12.136}$$

where the displacement **u** is subjected to periodic boundary condition. Based on the superposition law of a linear elastic problem, the displacement, namely, the solution to Eq. (12.136), is the sum of two components, given by:

$$\mathbf{u} = \mathbf{u}^0 + \mathbf{u}^1, \tag{12.137}$$

where $\mathbf{u}^0$ is the solution to:

$$C^0_{ijkl} u^0_{k,lj} + f_i = 0. \tag{12.138}$$

If deleting $\mathbf{f}_i$ and considering $\tau_{ij,j}$ as a load in Eq. (12.136), $\mathbf{u}^1$ represents the displacement induced by the polarization stress and can be determined by solving:

$$C^0_{ijkl} u_{k,lj} + \tau_{ij,j} = 0. \tag{12.139}$$

According to Green's function theory, the solution of Eq. (12.138) is:

$$u_j^0(\mathbf{x}) = \int_\Omega G_{ij}(\mathbf{x}-\mathbf{x}') f_i(\mathbf{x}') d\mathbf{x}', \tag{12.140}$$

where Green's function $G_{ij}(\mathbf{x}-\mathbf{x}')$ represents the displacement at point x in i direction under a unit force at point x′ in j direction. Mathematically, it is the solution to the equation of elasticity under load $\delta(x - x')\delta_{im}$, given by:

$$C^0_{imkl} G_{ij,\,lm}(\mathbf{x}-\mathbf{x}') + \delta(\mathbf{x}-\mathbf{x}')\delta_{kj} = 0, \tag{12.141}$$

where δ is the Kronecker function. In the composite, the Green's function only relies on the relative distance between two points. For simplicity, we set $\mathbf{x}' = (0,0,0)$ and obtain:

$$G_{ij}(\mathbf{x}-\mathbf{x}') = G_{ij}(\mathbf{x}). \tag{12.142}$$

Substituting Eq. (12.142) into Eq. (12.141), we get:

$$C^0_{imkl} G_{ij,lm}(\mathbf{x}) + \delta(\mathbf{x})\delta_{kj} = 0. \tag{12.143}$$

By applying Fourier transform on both sides of Eq. (12.141), we obtain:

$$\frac{1}{(2\pi)^3}\int_{-\infty}^{\infty}\left(C^0_{imkl}\ \hat{G}_{ij,\,lm}\ (\xi) + \delta_{kj}\right) e^{q\xi x}\, d\xi = 0. \tag{12.144}$$

Herein, $q = (-1)^{1/2}$ is the unit imaginary number and $\boldsymbol{\xi}$ denotes the frequency vector. By using the derivative rule, we get:

$$\frac{1}{(2\pi)^3}\int_{-\infty}^{\infty}\left(C^0_{imkl}\ (q\xi_1)(q\xi_m)\ \hat{G}_{ij}\ (\xi) + \delta_{kj}\right) e^{q\xi x}\, d\xi = 0. \tag{12.145}$$

Therefore, we obtain:

$$\left((\xi_1)(\xi_m) C^0_{imkl}\right)\ \hat{G}_{ij}\ (\xi) = \delta_{kj}. \tag{12.146}$$

Rearranging Eq. (12.146), we get:

$$\hat{G}_{ij}(\xi) = \left((\xi_k)(\xi_l) C^0_{ijkl}\right)^{-1}. \tag{12.147}$$

After obtaining the Green's function, the displacement $\mathbf{u}^0$ resulted from f_i can be easily calculated in accordance with Eq. (12.140). Similarly, $\mathbf{u}^1$ can be determined as:

$$u_j^1(\mathbf{x}) = \int_\Omega G_{ij}(\mathbf{x}-\mathbf{x}') \tau_{ij,j}(\mathbf{x}') d\mathbf{x}'. \tag{12.148}$$

Therefore, Eq. (12.137) becomes:

$$u_j(\mathbf{x}) = u_j^0(\mathbf{x}) - \int_\Omega h_{ijk}(\mathbf{x}-\mathbf{x}')\tau_{ik}(\mathbf{x}')d\mathbf{x}', \tag{12.149}$$

where h_{ijk} can be derived through integration by parts as:

$$h_{ijk}(x) = -\frac{1}{2}\left(G_{jk,\,i}(\mathbf{x}) + G_{ij,\,k}(\mathbf{x})\right). \tag{12.150}$$

With the consideration of strain field $\varepsilon_{ij} = 1/2(u_{i,\,j} + u_{j,\,i})$ in Eq. (12.149), we get the periodic Lippmann–Schwinger integral equation related to strain field as:

$$\varepsilon_{ij}(\mathbf{x}) = \varepsilon_{ij}^0(\mathbf{x}) - \int_\Omega \Gamma_{ijk}(\mathbf{x}-\mathbf{x}')\tau_{kl}(\mathbf{x}')d\mathbf{x}', \tag{12.151}$$

where Green tensor operator is

$$\Gamma_{ijkl}(\mathbf{x}) = \frac{1}{2}\left[\frac{\partial h_{ikl}}{\partial \mathbf{x}_j}(\mathbf{x}) + \frac{\partial h_{jkl}}{\partial \mathbf{x}_i}(\mathbf{x})\right]. \tag{12.152}$$

When the macroscopic strain field ε_0 is prescribed into the unit cell, without external force, Eq. (12.151) can be written into a terse form, given as:

$$\varepsilon(\mathbf{x}) = \varepsilon_0 - \Gamma(\mathbf{x}) * \tau(\mathbf{x}), \tag{12.153}$$

where "*" represents the convolution operator. Making use of the convolution theorem, after Fourier transform, Eq. (12.153) attains a form in frequency domain:

$$\hat{\varepsilon}(\xi) = -\hat{\Gamma}(\xi) : \hat{\tau}(\xi),\ \forall \xi \neq 0,\ \hat{\varepsilon}(0) = \varepsilon_0. \tag{12.154}$$

To calculate the effective elastic tensor, linearly independent unit test strains are prescribed in the unit cell. After obtaining the strain field from Eq. (12.154), the stress is calculated by Eq. (12.133). At last, the effective elastic tensor $\mathbf{C^H}$ can be naturally determined as:

$$\mathbf{C^H} = \int_\Omega \sigma d\mathbf{x} : \left(\int_\Omega \varepsilon d\mathbf{x}\right)^{-1}. \tag{12.155}$$

In the asymptotic homogenization method, the displacement has a similar expression to Eq. (12.137):

$$\mathbf{u} = \mathbf{u}^0(\mathbf{x},\mathbf{y}) + \gamma\mathbf{u}^1(\mathbf{x},\mathbf{y}) + o(\gamma^2), \tag{12.156}$$

where $\mathbf{y} = \mathbf{x}/\gamma$ denotes the position relation in macroscopic coordinate and microscopic coordinate with a constant $0 < \gamma \ll 1$. The first item in Eq. (12.156) represents the displacement in macroscale, similar to $\mathbf{u^0}$ in Eq. (12.137). The second item $\mathbf{u^1}$, namely, the characteristic microscale displacement, is the solution to the following characteristic equation:

$$\int_{\Omega} C_{ijmn}\, \varepsilon_{ij}\, \varepsilon_{mn}\left(u_i^{1(kl)}\right)d\Omega = \int_{\Omega} C_{ijkl}\, \varepsilon_{ij}\, \varepsilon_{mn}^{0(kl)}\, d\Omega, \tag{12.157}$$

where Ω is the domain occupied by the unit cell and the periodic admissible strain ε_{ij} is multiplied on both sides of Eq. (12.157) to obtain the weak form. In two dimensions, there are three linearly independent unit test strains $\varepsilon_{ij}^{0(kl)}$ given as $[1\ 0\ 0]^T$, $[0\ 1\ 0]^T$, and $[0\ 0\ 1]^T$, respectively.

In three dimensions, the number of test strains becomes six. After the characteristic displacements in Eq. (12.157) are solved using finite element analysis [21], the homogenized elasticity tensor is thereafter obtained as:

$$C_{ijkl}^{H} = \frac{1}{|\Omega|}\left(\int_{\Omega} C_{ijkl} - C_{ijmn}\frac{\partial u_m^{1(kl)}}{\partial y_n}\right)d\Omega, \tag{12.158}$$

where $|\Omega|$ denotes the volume of unit cell. The effective elasticity tensor can be rewritten in the form of strains as:

$$C_{ijkl}^{H} = \frac{1}{|\Omega|}\int_{\Omega} C_{mnrs}\left(\varepsilon_{mn}^{0(ij)} - \varepsilon_{mn}^{(ij)}\left(u_m^{1(ij)}\right)\right)\left(\varepsilon_{rs}^{0(kl)} - \varepsilon_{rs}^{(kl)}\left(u_m^{1(kl)}\right)\right)d\Omega, \tag{12.159}$$

where u_m is the characteristic displacement obtained from the characteristic displacement in terms of the strain–displacement relation.

12.4.2 Implementation of FFT-Based Homogenization Method

To obtain the strain in Eq. (12.153), we rearrange it by using Eqs. (12.134) and (12.154) and get:

$$(\mathbf{I} + \mathbf{B})\varepsilon = \varepsilon_0, \tag{12.160}$$

where **I** stands for a unit matrix and

$$\mathbf{B} = F^{-1}\left[\Gamma : F\left(\mathbf{C} - \mathbf{C}^0\right)\right], \tag{12.161}$$

with F and F^{-1} representing Fourier transform and inverse Fourier transform, respectively. In the real domain, the Green tensor operator can be derived as:

$$\Gamma_{ijkl} = -\frac{1}{4}\left[\frac{\partial^2 G_{il}}{\partial x_j \partial x_k}(\mathbf{x}) + \frac{\partial^2 G_{ik}}{\partial x_j \partial x_l}(\mathbf{x}) + \frac{\partial^2 G_{jl}}{\partial x_i \partial x_k}(\mathbf{x}) + \frac{\partial^2 G_{jk}}{\partial x_i \partial x_l}(\mathbf{x})\right]. \tag{12.162}$$

In the frequency domain, it becomes:

$$\hat{\Gamma}_{ijkl} = \frac{1}{4}\left[\hat{G}_{il}\ (\xi)\xi_j\xi_k + \hat{G}_{ik}\ (\xi)\xi_j\xi_l + \hat{G}_{jl}\ (\xi)\xi_i\xi_k + \hat{G}_{jk}\ (\xi)\xi_i\xi_l\right]. \tag{12.163}$$

Rewriting the elastic tensor of the isotropic reference material as:

$$C^0_{ijkl} = \lambda^0\delta_{ij}\delta_{kl} + \mu^0\left(\delta_{ik}\delta_{jl} + \delta_{il}\delta_{jk}\right), \tag{12.164}$$

where the Lamé parameters of reference material are represented as λ_0 and μ_0, respectively. Substituting Eq. (12.164) into Eq. (12.147), we get the Green's function in frequency domain as:

$$\hat{G}_{ij}\left(\xi\right) = \frac{1}{\mu^0}\frac{\delta_{ij}}{|\xi|^2} - \frac{\left(\lambda^0 + \mu^0\right)}{\mu^0\left(\lambda^0 + 2\mu^0\right)}\frac{\xi_i\xi_j}{|\xi|^4}. \tag{12.165}$$

Therefore, by substituting Eq. (12.165) into Eq. (12.162), the Green tensor operator in frequency domain can be explicitly expressed as:

$$\hat{\Gamma}_{ijkl} = \frac{\left(\delta_{ik}\xi_l\xi_j + \delta_{il}\xi_k\xi_j + \delta_{jk}\xi_l\xi_i + \delta_{lj}\xi_i\xi_k\right)}{4\mu^0|\xi|^2} - \frac{\left(\lambda^0 + \mu^0\right)}{\mu^0\left(\lambda^0 + 2\mu^0\right)}\frac{\xi_i\xi_j\xi_k\xi_l}{|\xi|^4}. \tag{12.166}$$

When we define $\mathbf{A} = \mathbf{I} + \mathbf{B}$, Eq. (12.160) can be rewritten as:

$$\mathbf{A}\varepsilon = \varepsilon_0, \tag{12.167}$$

where the inverse of matrix $\mathbf{A}$ can be obtained based on the Neumann expansion [67], given as:

$$\mathbf{A}^{-1} = \frac{1}{\mathbf{I} + \mathbf{B}} = \sum_{i=0}^{\infty}(-\mathbf{B})^i. \tag{12.168}$$

Therefore, an iterative procedure can be determined as:

$$\varepsilon^{(i+1)} = \left(-\mathbf{B}\right)\varepsilon^{(i)} + \varepsilon_0. \tag{12.169}$$

Inside the iterative scheme, the optimal rate of convergence is achieved by defining the reference material with specified Lamé parameters as follows:

$$\lambda^0 = \frac{\min\{\lambda\} + \max\{\lambda\}}{2}, \mu^0 = \frac{\min\{\mu\} + \max\{\mu\}}{2}. \tag{12.170}$$

The natural criteria for the iterative algorithms are based on the difference between $\varepsilon^{(i+1)}$ and $\varepsilon^{(i)}$, which can be expressed by:

$$\eta^i = \frac{\left\|\varepsilon^{(i+1)} - \varepsilon^{(i)}\right\|^2}{\left\|\varepsilon^0\right\|^2} \le \xi. \tag{12.171}$$

Although the standard FFT-based algorithm using the Neumann series iterative scheme is very simple, one drawback of the Neumann series expansion is that it suffers slow convergence when the contrast of material properties increases, which is undesirable for porous materials. But this linear algebra problem can be solved by the conjugate gradient method.

REFERENCES

1. Hashin Z. (1983), "Analysis of composite materials-a survey", *ASME, Journal of Applied Mechanics*, Vol. 50, pp. 481–505.
2. Hill R. (1963), "Elastic properties of reinforced solids: some theoretical principles", *Journal of Mechanics of Physics of Solids*, Vol. 11, pp. 357–372.
3. Cox H.L. (1952), "The elasticity and strength of paper and other fibrous materials", *British Journal of Applied Physics*, Vol. 3, pp. 72–79.
4. Kelly A. and Tyson W.R. (1965), "Tensile properties of fiber-reinforced metals: copper/tungsten and copper/molybdenum", *Journal of Mechanics and Physics of Solids*, Vol. 13(6), pp. 329–338.
5. Nairn J. (1997), "On the use of shear-lag methods for the analysis of stress transfer in unidirectional composites", *Mechanics of Materials*, Vol. 26(2), pp. 63–80.
6. Landis C.M., McMeeking R.M. (1999), "A shear-lag model for a broken fiber embedded in a composite with a ductile matrix", *Composite Science and Technology*, Vol. 59(3), pp. 447–457.
7. Wilczynski A.P. (1990), "A basic theory of reinforcement for unidirectional fibrous composites", *Composite Science and Technology*, Vol. 38, pp. 327–336.
8. Rosen B.W. (1964), "Tensile failure of fibrous composites", *AIAA Journal*, Vol. 2, pp. 1985–1991.
9. Carman G.P., Reifsnider K.L. (1992), "Micro-mechanics of short-fiber composites", *Composite Science and Technology*, Vol. 43, pp. 137–146.
10. Hashin Z., Rosen B.W. (1964), "The Elastic moduli of fiber-reinforced materials", *ASME Journal of Applied Mechanics*, Vol. 31, pp. 223–232.
11. Robinson I.M. and Robinson J.M. (1994), "The effect of fiber aspect ratio on the stiffness of discontinuous fiber-reinforced composites", *Composites*, Vol. 25, pp. 499–503.
12. Fukuda H. and Kawata K. (1974), "On Young's modulus of short fiber composites", *Fiber Science and Technology*, Vol. 7, pp. 207–222.
13. Eshelby J.D. (1957), "The determination of the elastic field of an ellipsoidal inclusion and related problems", *Proceedings of Royal Society A*, Vol. 241, pp. 376–396.
14. Mura T. (1982), *Micro-Mechanics of Defects in Solids*, The Hague, Martinus Nijho.
15. Russel W.B. (1973), "On the effective moduli of composite materials: effect of fiber length and geometry at dilute concentrations", *Journal of Applied Mathematics and Physics, (JAMP)*, Vol. 24, pp. 581–600.
16. Chow T.S. (1977), "Elastic moduli of filled polymers: the effect of particle shape", *Journal of Applied Physics*, Vol. 48, pp. 4072–4075.
17. Steif P.S., Hoysan S.F. (1987), "An energy method for calculating the stiffness of aligned short-fiber composites", *Mechanics of Materials*, Vol. 6, pp. 197–210.
18. Mori T. and Tanaka K. (1973), "Average stress in matrix and average elastic energy of materials with mis-fitting inclusions", *Acta Materialia*, Vol. 21, pp. 571–574.
19. Benveniste Y. (1987), "A new approach to the application of Mori-Tanaka's theory in composite materials", *Mechanics of Materials*, Vol. 6, pp. 147–157.
20. Wakashima K., Otsuka M., Umekawa S. (1974), "Thermal expansion of heterogeneous solids containing aligned ellipsoidal inclusions", *Journal of Composite Materials*, Vol. 8, pp. 391–404.

21. Taya M. and Mura T. (1981), "On stiffness and strength of an aligned short-fiber reinforced composite containing fiber-end cracks under uniaxial applied stress", *ASME Journal of Applied Mechanics*, Vol. 48, pp. 361–367.
22. Taya M. and Chou T.W. (1981), "On two kinds of ellipsoidal in-homogeneities in an infinite elastic body: an application to a hybrid composite", *International Journal of Solids and Structures*, Vol. 17, pp. 553–563.
23. Weng G.J. (1984), "Some elastic properties of reinforced solids, with special reference to isotropic ones containing spherical inclusions", *International Journal of Engineering Science*, Vol. 22(7), pp. 845–856.
24. Tandon G.P. and Weng G.J. (1984), "The effect of aspect ratio of inclusions on the elastic properties of uni-directionally aligned composites", *Polymer Composites*, Vol. 5, pp. 327–333.
25. Tucker C.L. and Liang E. (1999), "Stiffness predications for unidirectional short-fiber composites: review and evaluation", *Composite Science and Technology*, Vol. 59, pp. 655–671.
26. Chow T.S. (1978), "Effect of particle shape at finite concentration on the elastic modulus of filled polymers", *Journal of Polymer Science and Polymer Physics Edition*, Vol. 16, pp. 959–965.
27. Ferrari M. (1994), "Composite homogenization via the equivalent poly-inclusion approach", *Composite Engineering*, Vol. 4, pp. 37–45.
28. Advani S.G. and Tucker III C.L. (1987), "The use of tensors to describe and predict the fiber orientation in short fiber composites", *Journal of Rheology*, Vol. 31(8), p. 751.
29. Gibson R.F., Sun C.T., Wu J.K. (1985), "Prediction of material damping in randomly oriented short fiber polymer matrix composites", *Journal of Reinforced Plastics and Composites*, Vol. 4, pp. 262–272.
30. Wu C.T.D and McCullough R.L. (1977), "Constitutive relationships for heterogeneous materials", in *Developments in Composites Materials-1*, (G. S. Holister, ed.), Applied Science Publishers, London, pp. 118–186.
31. Hashin Z. and Shtrikman S. (1962), "On some variational principles in anisotropic and non homogeneous elasticity", *Journal of Mechanics and Physics of Solids*, Vol. 10, pp. 335–342.
32. Walpole L.J. (1966), "On bounds for the overall elastic moduli for inhomogeneous systems", *Journal of Mechanics and Physics of Solids*, Vol. 14, pp. 151–162.
33. Willis J.R. (1977), "Bounds and self-consistent estimates for the overall properties of anisotropic composites", *Journal of Mechanics of Physics of Solids*, Vol. 25, pp. 185–202.
34. Weng G.J. (1992), "Explicit evaluation of Willis' bounds with ellipsoidal inclusions", *Journal of Engineering Science*, Vol. 30, pp. 83–92.
35. Eduljee R.F., McCullough R.L., Gillespie J.W. (1994), "The influence of aggregated and dispersed textures on the elastic properties of dis-continuous fiber composites", *Composite Science and Technology*, Vol. 50, pp. 381–391.
36. Lielens G., Pirotte P., Couniot A., Dupret F., Keunings R. (1997), "Prediction of thermo-mechanical properties for compression-moulded composites", *Composites A*, Vol. 29, pp. 63–70.
37. Hill R. (1965), "A self-consistent mechanics of composite materials", *Journal of Mechanics of Physics of Solids*, Vol. 13, pp. 213–222.
38. Budiansky B. (1965), "On the elastic moduli of some heterogeneous materials", *Journal of Mechanics of Physics of Solids*, Vol. 13, pp. 223–227.
39. Laws N. and McLaughlin R. (1979), "The effect of fiber length on the overall moduli of composite materials", *Journal of Mechanics of Physics of Solids*, Vol. 27, pp. 1–13.
40. Chou T.W., Nomura S., Taya M. (1980), "A self-consistent approach to the elastic stiffness of short-fiber composites", *Journal of Composite Materials*, Vol. 14, pp. 178–188.

41. Lin S.C. and Mura T. (1973), "Elastic fields of inclusions in anisotropic media (II)", *Physica Statatus Solidi*, Vol. 15, pp. 281–285.
42. Chou T.W., Nomura S., Taya M. (1979), "A self-consistent approach to the elastic stiffness of short-fiber composites", in *Modern Developments in Composite Materials and Structures* (J. R. Vinson, ed.), ASME, New York, pp. 149–164.
43. Kerner E.H. (1956), "The elastic and thermo-elastic properties of composite media", *Proceedings of Physics Society B*, Vol. 69, pp. 808–813.
44. Hermans J.J. (1967), "The elastic properties of fiber reinforced materials when the fibers are aligned", *Proc Kon Ned Akad v Wetensch B*, Vol. 65, pp. 1–9.
45. Ji X., Cao Y., Feng X. (2010), "Micromechanics prediction of the effective elastic moduli of graphene sheet-reinforced polymer nanocomposites", *Modelling and Simulation in Materials Science and Engineering*, Vol. 18, p. 045005.
46. Feng X., Tian Z., Liu Y., Yu S. (2004), "Effective elastic and plastic properties of interpenetrating multiphase composites", *Applied Composite Materials*, Vol. 11, pp. 33–55.
47. Lo Cascio M., Milazzo A., Benedetti I. (2020), "Virtual element method for computational homogenization of composite and heterogeneous materials", *Composite Structures*, Vol. 232, p. 111523.
48. Bensoussan A., Lions J., Papanicolaou G. (1978), *Asymptotic Analysis for Periodic Structures*, North-Holland Publishers, Amsterdam.
49. Andrianov I., Bolshakov V., Danishevs'kyy V., Weichert D. (2007), "Asymptotic simulation of imperfect bonding in periodic fibre-reinforced composite materials under axial shear", *International Journal of Mechanical Sciences*, Vol. 49, pp. 1344–1354.
50. Sridhar A., Liu L., Kouznetsova V., Geers M. (2018), "Homogenized enriched continuum analysis of acoustic metamaterials with negative stiffness and double negative effects", *Journal of Mechanics and Physics of Solids*, Vol. 119, pp. 104–117.
51. Hu R. and Oskay C (2019), "Multiscale nonlocal effective medium model for in-plane elastic wave dispersion and attenuation in periodic composites", *Journal of Mechanics and Physics of Solids*, Vol. 124, pp. 220–243.
52. Moulinec H. and Suquet P. (1994), "A fast numerical method for computing the linear and nonlinear mechanical properties of composites", *Comptes rendus de l'Académie des Sciences*, Vol. 2, pp. 1417–1423.
53. Moulinec H. and Suquet P. (1998), "A numerical method for computing the overall response of nonlinear composites with complex microstructure", *Computer Methods in Applied Mechanics and Engineering*, Vol. 157, pp. 69–94.
54. Moulinec H., Suquet P., Milton G. (2018), "Convergence of iterative methods based on Neumann series for composite materials: theory and practice", *International Journal of Numerical Methods in Engineering*, Vol. 114, pp. 1103–1130.
55. Michel J., Moulinec H., Suquet P. (2001), "A computational scheme for linear and nonlinear composites with arbitrary phase contrast", *International Journal of Numerical Methods in Engineering*, Vol. 52, pp. 139–158.
56. Zeman J., Vond řejc J., Novák J., Marek I. (2010), "Accelerating a FFT-based solver for numerical homogenization of periodic media by conjugate gradients", *Journal of Computational Physics*, Vol. 229, pp. 8065–8071.
57. Gélébart L. and Mondon-Cancel R. (2013), "Non-linear extension of FFT-based methods accelerated by conjugate gradients to evaluate the mechanical behavior of composite materials", *Computational Materials Science*, Vol. 77, pp. 430–439.
58. Monchiet V. and Bonnet G. (2012), "A polarization-based FFT iterative scheme for computing the effective properties of elastic composites with arbitrary contrast", *International Journal of Numerical Methods in Engineering*, Vol. 89, pp. 1419–1436.
59. Brisard S. and Dormieux L. (2010), "FFT-based methods for the mechanics of composites: a general variational framework", *Computational Materials Science*, Vol. 49, pp. 663–671.

60. Eyre D. and Milton G. (1999), "A fast numerical scheme for computing the response of composites using grid refinement", *European Journal of Physics*, Vol. 6, pp. 41–47.
61. Bilger N., Auslender F., Bornert M., Moulinec H., Zaoui A. (2007), "Bounds and estimates for the effective yield surface of porous media with a uniform or a non-uniform distribution of voids", *European Journal of Mechanics - A/Solids*, Vol. 26, pp. 810–836.
62. Idiart M., Willot F., Pellegrini Y., Castaneda P. (2009), "Infinite-contrast periodic composites with strongly nonlinear behavior: effective-medium theory versus full-field simulations", *International Journal of Solids and Structures*, Vol. 46, pp. 3365–3382.
63. Vigliotti A. and Pasini D. (2012), "Linear multiscale analysis and finite element validation of stretching and bending dominated lattice materials", *Mechanics of Materials*, Vol. 46, pp. 57–68.
64. Willot F., Gillibert L., Jeulin D. (2013), "Microstructure-induced hotspots in the thermal and elastic responses of granular media", *International Journal of Solids and Structures*, Vol. 50, pp. 1699–1709.
65. Willot F. and Jeulin D. (2011), "Elastic and electrical behavior of some random multiscale highly contrasted composites", *International Journal of Multiscale Computational Engineering*, Vol. 9, pp. 305–326.
66. Azzimonti D., Willot F., Jeulin D. (2013), "Optical properties of deposit models for paints: fullfields FFT computations and representative volume element", *Journal of Modern Optics*, Vol. 60, pp. 519–528.
67. Michel J., Moulinec H., Suquet P. (1999), "Effective properties of composite materials with periodic microstructure: a computational approach", *Computer Methods in Applied Mechanics and Engineering*, Vol. 172, pp. 109–143.

13 Toughness of Composite Materials

13.1 BASICS

Fracture mechanics developed from the research work done by Inglis [1]. He demonstrated that the stresses near the crack tip in a material under load are much higher in comparison with anywhere in the bulk material (σ_∞). The stress at the crack tip, for a length c of the crack (or 2c for internal crack) and crack tip radius r, is written as given by Inglis [1] as:

$$\sigma = \sigma_\infty\left(1 + 2\sqrt{\frac{c}{r}}\right). \tag{13.1}$$

Thus, a circular hole ($c = r$) has a *stress concentration factor* of 3. While this is physically reasonable, the case of a sharp crack ($r \to 0$) presents difficulties, in that the stress concentration, according to Eq. (13.1), becomes very large. On this basis, most structures should fail, under low applied loads, at the fine surface scratches which are almost inevitably present. This was contrary to engineering experience at the time, since most artifacts, particularly metallic ones, were able to function even when small cracks and scratches were present. The problem was resolved by the pioneering work of Griffith [2], who pointed out that a crack cannot propagate unless the energy of the system is thereby decreased. The energy released when a crack advances comes from the associated release of stored elastic strain in the surrounding material (plus any work done by the loading system). If this is insufficient to counterbalance the energy absorbed in the material through the creation of new fracture surfaces and associated internal damage or deformation processes, then the crack cannot advance. In many materials, efficient mechanisms for internal energy absorption are stimulated by the high stresses at a crack tip, so that the energy balance for crack propagation is often unfavorable and they exhibit high toughness, that is, high resistance to fracture. This is particularly true for most metals since the dislocation motion which occurs is highly effective in this regard.

Griffith considered brittle materials, such as glasses, in which energy-absorbing processes are not readily stimulated and the only significant energy penalty of crack propagation comes from the new surface area of the crack faces. He showed that the change in the stored energy of a loaded plate of unit thickness, caused by the introduction of an interior crack of length $2c$, is given by:

$$U = -\frac{\pi\sigma^2 c^2}{E}, \tag{13.2}$$

where

$\sigma = \sigma_\infty$ is the applied stress and E is Young's modulus.

The other contribution to the overall energy change is that required to create the new surface area, which is positive and has a value of $4c\gamma$, where γ is the surface energy. The dependence of these two contributions on the length of the crack is shown in Figure 13.1. Cracks longer than a critical length (c^*) will grow spontaneously with a reduction in net energy. This critical length is found by differentiating the total energy with respect to crack length and setting the result equal to zero, leading to:

$$c^* = \frac{2\gamma E}{\pi\sigma^2}. \tag{13.3}$$

This approach was extended by Irwin [3] to encompass tougher materials. The surface energy term 2γ is supplemented by other contributions to the energy absorbed in the vicinity of an advancing crack tip. For a given applied stress and preexisting crack size, an expression can be obtained from Eq. (13.3) for the *energy release rate*, G, which has units of J m^{-2}.

$$G = \frac{\pi\sigma^2 c}{E}. \tag{13.4}$$

For fracture to occur, this must exceed a critical value, sometimes referred to as the *crack resistance*, R. This critical value represents the total energy absorbed, per unit of crack advance area, and is often termed as G_c, the *critical energy release rate*, or *fracture energy*. Values of G_c are fairly easy to obtain experimentally. For example, the work done in a tension or bending test is given by the area under a load–displacement plot and, provided this energy is all permanently absorbed in the specimen, the fracture energy is then found by simply dividing by the sectional area through which failure has occurred. The specimen is commonly pre-notched so as

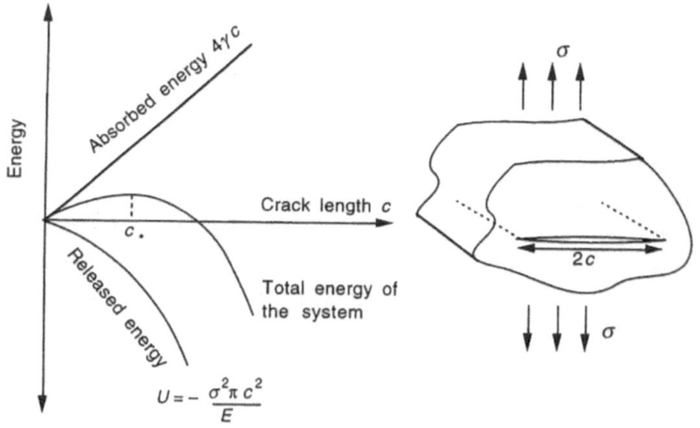

FIGURE 13.1 The contributions to the energy associated with a crack in a brittle material as a function of crack length.

to ensure that crack propagation occurs. Typical values of G_c are given in Table 13.1 for various materials. Tough (soft) metals have fracture energies of 100 kJ m^{-2} or more, whereas a brittle material, such as glass, can have a value as low as 0.01 kJ m^{-2}. Rearranging Eq. (13.4), the stress necessary to cause spontaneous fracture in a component with a preexisting crack of size c ($2c$ if internal) can be written as:

$$\sigma^* = \sqrt{\frac{G_c E}{\pi c}}. \tag{13.5}$$

This approach is particularly useful in practical terms, because attention is diverted from the complex problem of the precise nature of the stress field close to the tip of a crack to a more global approach involving macroscopic quantities which are measurable experimentally. However, there is still interest in the phenomena occurring locally near the crack tip. A useful link is provided between the energy and stress field approaches by the concept of a *stress intensity factor*, K. This parameter, which largely evolved from the work of Irwin in the 1950s, can be expressed as:

$$K = \sigma\sqrt{\pi c}. \tag{13.6}$$

TABLE 13.1
Typical Fracture Energy and Fracture Toughness Values for Various Materials

Material	Fracture Energy (G_c) (kJ m^{-2})	Fracture Toughness (K_c) (MPa m$^{1/2}$)
Polymers		
Epoxy resin	0.1–0.3	0.3–0.5
Nylon	2–4	3
Polypropylene	8	3
Metals		
Pure Al	100–1000	100–350
Al alloy	8–30	23–45
Mild steel	100	140
Ceramics		
Soda glass	0.01	0.7
SiC	0.05	3
Concrete	0.03	0.2
Natural Materials		
Woods (crack ⊥ grain)	8–20	11–13
Woods (crack // grain)	0.5–2	0.5–1
Bone	0.6–5	2–12
Composites		
Fiberglass	40–100	42–60
Al-based particulate metal matrix composites	2–10	15–30
SiC laminate (crack ⊥ layers)	5–8	45–55

It therefore encompasses the effects of both the applied load and the preexisting crack size, with the relative weighting that these two parameters have in determining the value of G, the energy release rate (see Eq. (13.4)). It characterizes the severity of the stress field around the crack tip. A critical value can be identified, corresponding to the case where the associated value of G reaches G_c:

$$K_c = \sigma^* \sqrt{\pi c} = \sqrt{EG_c}. \tag{13.7}$$

This *critical stress intensity factor* is often known as *fracture toughness.* Values are given in Table 13.1 for various materials. For tough materials, the fracture toughness can exceed 100 MPa$\sqrt{\text{m}}$, while a brittle material might typically have a value around 1 MPa$\sqrt{\text{m}}$. The usefulness of the stress intensity factor lies largely in the way it can be related to local crack tip features. For example, it can be shown that the size of the plastic zone ahead of the crack tip is related to the yield stress of the material by:

$$r_Y \approx \frac{1}{2\pi}\left(\frac{K}{\sigma_Y}\right)^2. \tag{13.8}$$

Similarly, the *crack opening displacement* (δ) can be expressed as:

$$\delta \approx \left(\frac{K^2}{\sigma_Y E}\right). \tag{13.9}$$

13.2 INTERFACIAL FRACTURE

Table 13.1 shows that a composite made from glass fibers and epoxy resin has a fracture energy comparable with those of metals (G_c ~ 50 kJ m^{-2}), even though the constituents are both brittle (G_c ~ 0.01–0.1 kJ m^{-2}). This high toughness of composites, which is very important in practical terms, is closely linked with interfacial effects. A first step in exploring this is to analyze the conditions under which interfacial debonding – that is, crack propagation along an interface between two different materials, occurs. For a given loading configuration, the propagation of a crack, along an interface between two constituents, gives rise to an energy release rate, G_i, in much the same way as for the case when the crack is in a homogeneous material. Also, there is a critical value, G_{ic}, and *interfacial fracture energy, G_i*, must reach for the crack to propagate.

Values of G_{ic} are not as readily available as G_c values for homogeneous materials. There are several reasons for this. First, the toughness of an interface is sensitive to the way in which the interface was manufactured, rather than being unique to the pair of constituents on either side. A second reason is slightly more complex. Interfacial cracks often propagate under *mixed mode* loading conditions. This is in contrast to a crack in a homogeneous material, which will always tend to advance in a direction such that the stress field at the crack tip is purely tensile (mode I). An interfacial crack, however, is constrained to follow a predetermined path. Depending on the loading configuration, the stress field at the crack tip may include a significant shear stress component acting on the plane of the interface (mode II). In general, the energy expended in debonding the interface is greater when there is a mode II component than for the case of pure mode I loading. This complicates the experimental measurement of G_{ic}. Not only can it be difficult to establish the exact stress field at the crack tip, but it may vary with position in the specimen, particularly for fiber–matrix interfaces. The situation is further complicated if any residual stresses are present.

The proportion of opening and shearing modes at the crack tip is often characterized by means of the *phase angle*, (ψ) (psi), which is defined in terms of the mode I and mode II stress intensity factors, as shown schematically in Figure 13.2:

$$\psi = \tan^{-1}\left(\frac{K_{\mathrm{II}}}{K_{\mathrm{I}}}\right). \tag{13.10}$$

The value of ψ is 0° for pure opening ($K_{\mathrm{II}} = 0$) and 90° for pure shear ($K_{\mathrm{I}} = 0$). The phase angle can be established for various loading arrangements, although often this calculation is not a simple one. Furthermore, ψ is likely to vary with position when the interface is nonplanar. The limited experimental data currently available suggest that the dependence of G_{ic} on ψ may be quite significant, depending on the type of system.

One of the main reasons for interest in interfacial toughness concerns *crack deflection*. For a composite to have a high toughness, a crack passing through the matrix must be repeatedly deflected at fiber–matrix interfaces, at least for materials based on polymer resins or ceramics. Early work by Cook and Gordon [4] on conditions for crack deflection was focused on the stress field ahead of a crack tip. It was pointed out that when a crack approaches a fiber in a composite loaded parallel to the fiber axis, there is a transverse stress ahead of the crack tip, tending to open up the interface and hence to "blunt" and deflect the crack from entering the fiber. Since the peak value of this transverse stress is always about 20% of the maximum axial stress, Cook and Gordon proposed that the interface debonds if its strength is less than about one-fifth that of the matrix. However, the "strength" of an interface is not well defined and is sensitive to the presence and distribution of flaws and to the method of loading. In view of the success of the Griffith treatment, it is clearly preferable to establish a criterion based on the energetics of crack propagation.

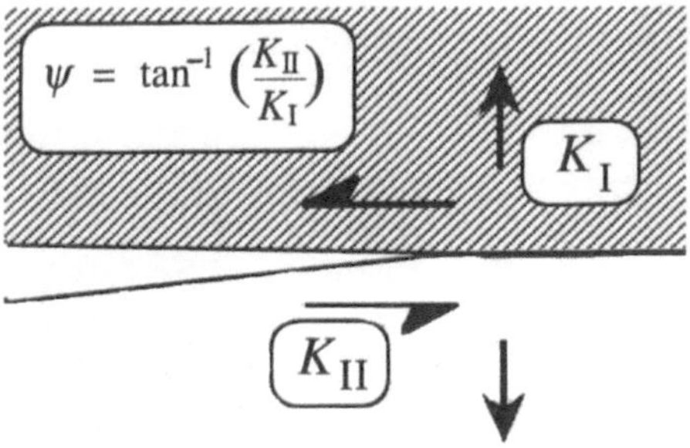

FIGURE 13.2 Stress field at an interfacial crack.

Energy-based crack deflection criteria have been proposed by Kendall [5] and by He and Hutchinson [6]. Kendall considered two blocks bonded together and loaded in tension parallel to the interface, one of the blocks having a crack approaching the interface, as shown in Figure 13.3. He estimated the applied loads under which a crack would penetrate the other block or deflect along the interface, assuming that the crack requiring the lower load would predominate. This produced the following criterion for deflection:

$$\frac{G_{ic}}{G_{fc}} \leq \left(\frac{h_m E_m + h_f E_f}{h_f E_f}\right)\left[\frac{1}{4\pi\left(1-\nu^2\right)}\right], \tag{13.11}$$

where

G_{ic} and G_{fc} are the fracture energies of interface and uncracked block (fiber),
h_m and h_f are the thicknesses of cracked (matrix) and uncracked (fiber) blocks,
E_m and E_f are the corresponding Young's moduli,
ν is the Poisson's ratio taken as equal in both constituents.

If h_m and h_f are taken as equal, corresponding approximately to a crack passing through the matrix between unbroken fibers in a typical composite, then this critical ratio is around 20% for $E_m \sim E_f$, falling to about 10% for $E_m \ll E_f$.

He and Hutchinson's analysis is based on considering whether the penetrating or deflecting crack gives a greater net release of energy. It is more complex than the Kendall's model, but also yields a critical fracture energy ratio of around 20% for the case of $E_m \sim E_f$. For $E_m \ll E_f$, however, an increase (rather than a decrease) is

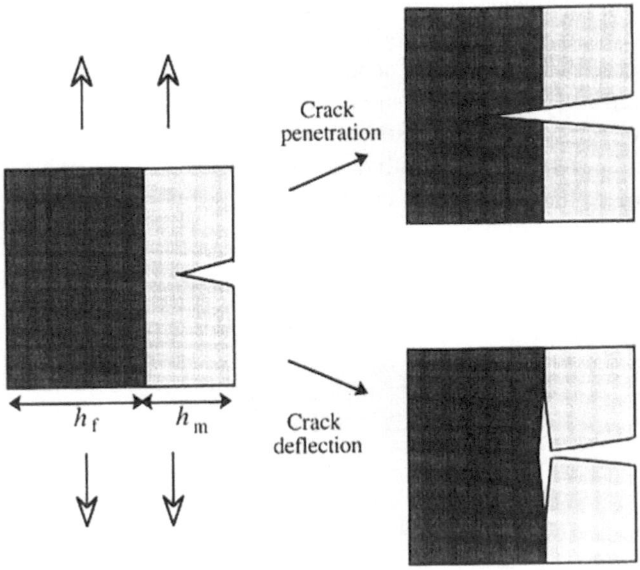

FIGURE 13.3 Crack deflection geometry at an interface.

predicted in this ratio. There are few experimental data available to validate either of these models. Kendall's experimental work with rubbers did seem consistent with his predictions, but few systematic measurements have been performed with systems of practical interest. It is, however, clear that the interfacial fracture energy must be appreciably lower than that of the reinforcement if matrix cracks are to be consistently deflected. Since (ceramic) fibers tend to have low fracture energies (Table 13.1), this means that interfaces of very low toughness are often required if crack deflection is essential. For ceramic matrix composites, in particular, the retention of low interfacial toughness, through processing stages which tend to promote sintering and chemical reactions, represents a major technological challenge.

13.3 WORK OF FRACTURE

A relatively high toughness is essential for most engineering materials. One of the advantages of composite materials is that there is often scope for the promotion of energy-absorbing mechanisms in the material. It is important to understand how these mechanisms are controlled.

13.3.1 Deformation of Matrix

Work of fracture data for some typical matrices is given in Table 13.1. Most metallic matrices have a high toughness, mainly as a result of the extensive dislocation movement which occurs near a crack. Polymers, particularly thermosets, and ceramics have low fracture energies. The extent of matrix deformation during composite fracture may differ appreciably from that in the same material when unreinforced. The main effect is one of increased *constraint*, so that the matrix is unable to deform freely because it is surrounded by stiff and strong fibers. This may be partly a result of *load transfer*, which reduces the magnitude of the matrix stresses. Of greater significance, however, is the tendency for *triaxial stress states* to be set up which inhibit plastic flow of the matrix. For example, when a region of matrix extends plastically, with associated lateral contraction, this contraction may be opposed by a surrounding rigid cluster of fibers. This sets up transverse tensile stresses which reduce the deviatoric (shape-changing) component of the matrix stress state. This in turn inhibits plastic flow, but may encourage cavitation and fracture.

When compared with unreinforced Al, this type of effect accounts for the lower fracture energy as given in Table 13.1 for the Al-based metal matrix composites (MMCs). This loss of toughness can be minimized by eliminating reinforcement clusters and other inhomogeneities such as pores, debonded interfaces, and cracked particles or fibers. Processing improvements for MMCs are currently aimed in this direction. It follows that a high interfacial strength is desirable for MMCs, and in most cases, this is quite readily achievable. Only for relatively low-toughness metals (e.g., zinc) reinforced with long fibers is there any interest in a low interfacial toughness, since in these cases toughening by fiber pullout is preferable to retaining as much of the toughness of the matrix as possible. The constraint effects outlined above also operate with nonmetallic matrices. In these cases, however, there is in any event limited scope for energy absorption in the matrix, and other mechanisms are often of greater significance. However, considerable improvements in the toughness of polymer matrix composites (PMCs) have been achieved by increasing the toughness of the matrix, which can alter the micromechanisms of damage and increase the associated energy absorption.

13.3.2 Fiber Fracture

Depending on the fiber architecture and loading configuration, component failure usually involves the fracture of fibers. The contribution this makes to the fracture energy of the material is small for most fibers. Typical fracture energies for fibers of glass, carbon, and SiC are only a few tens of J m^{-2}. Some polymeric fibers are not completely brittle and undergo appreciable plastic deformation. An example of this category is a Kevlar™ fiber. Most natural fibers are in a similar category, and fracture of the cellulose fibers makes a significant contribution to the fracture energy of wood (across the grain). In such cases, fiber fracture contributes up to a few kJ m^{-2} to the overall fracture energy. Metallic fibers can in principle make even larger contributions. Thus, even at low volume content, steel rods in reinforced concrete raise the toughness substantially, as well as enhancing the tensile strength. Nevertheless, for most composites, the fibers themselves make little or no direct contribution to the overall toughness.

13.3.3 Interfacial Debonding

Some interfacial debonding usually occurs during the fracture of a composite. If a crack is propagating normal to the direction of fiber alignment, debonding occurs provided the crack is deflected on reaching the interface. Debonding can also be stimulated under transverse or shear loading conditions. Provided the value of the interfacial fracture energy, G_{ic}, is known, an estimate can be made of the contribution from debonding to the overall fracture energy. The basis for such a calculation is shown in Figure 13.4 for aligned short fibers. A simple shear-lag approach is used. Provided the fiber aspect ratio, s (= L/r), is less than the critical value, s^* $(= \sigma_f^* / 2\tau_i^*)$, all of the fibers intersected by the crack debond and are subsequently pulled out of their sockets in the matrix. The work done when a single fiber undergoes interfacial debonding can be written as:

$$\Delta U = 2\pi r x_o G_{ic}, \tag{13.12}$$

where

x_o is the embedded fiber length on the side of the crack where the debonding occurs $(x_o \le L)$.

To obtain the local work of debonding for the composite, G_{cd}, this is summed over all of the fibers intersected by the crack. If there are N fibers per m^2, then there will be $(N\, dx_o/L)$ per m^2 with an embedded length between x_o and $(x_o + dx_o)$. The total work done in debonding is therefore given by:

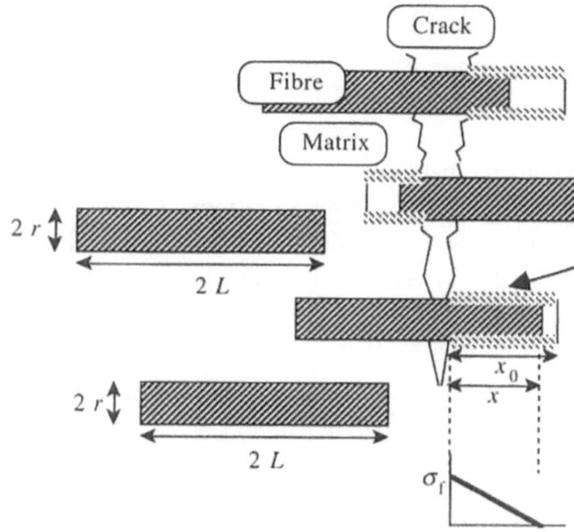

FIGURE 13.4 A crack passing through an aligned short-fiber composite depicting the interfacial debonding and fiber pullout.

$$G_{cd} = \int_0^L \frac{Ndx_o}{L} 2\pi r x_o G_{ic}. \tag{13.13}$$

The value of N is related to the fiber volume fraction, f, and the fiber radius, r, as:

$$N = \frac{f}{\pi r^2}. \tag{13.14}$$

Substituting this expression into Eq. (13.13) and integrating lead to:

$$G_{cd} = \frac{f 2\pi r G_{ic}}{\pi r^2 L}\frac{L^2}{2}. \tag{13.15a}$$

This gives:

$$G_{cd} = fsG_{ic}. \tag{13.15b}$$

Contributions from this mechanism are relatively small. For example, if s = 50, f = 0.5, and G_{ic} = 10 J m^{-2}, then G_{cd} = 0.25 kJ m^{-2}. If either the fiber aspect ratio or the interfacial fracture energy is much greater than these values ($s > s^*$ or $G_{ic} > G_{fc}$), then it is probable that the fibers will fracture in the crack plane and little or no debonding will occur. Slightly larger values are achievable in bending when debonding can propagate for long distances from the fracture plane without causing the reinforcement to break.

13.3.4 Frictional Sliding and Fiber Pullout

Potentially, the most significant source of fracture work for most fiber composites is interfacial frictional sliding. Depending on the interfacial roughness, contact pressure, and sliding distance, this process can absorb large quantities of energy. The case of most interest is pullout of fibers from their sockets in the matrix. Pullout aspect ratios can range up to several tens or even hundreds. The work done can be calculated using a similar approach to that in the previous section. Consider a fiber with a remaining embedded length of x being pulled out an increment of distance dx. The associated work is given by the product of the force acting on the fiber and the distance it moves.

$$dU = \left(2\pi r x \tau_i^*\right) dx, \tag{13.16}$$

where τ_i^* is the interfacial shear stress, taken here as constant along the length of the fiber. The work done in pulling this fiber out completely is therefore given by:

$$\Delta U = \int_0^{x_o} \left(2\pi r x \tau_i^*\right) dx = \pi r x_o^2 \tau_i^*. \tag{13.17}$$

The next step is a similar integration (over all of the fibers) to that used to obtain Eq. (13.13), leading to an expression for the pullout work of fracture, G_{cp}:

$$G_{cp} = \int_0^L \frac{N dx_o}{L} \pi r x_o^2 \tau_i^*. \tag{13.18}$$

Using Eq. (13.14):

$$G_{cp} = \frac{s^2 f r \tau_i^*}{3}. \tag{13.19}$$

This contribution to the overall fracture energy can be large. For example, taking $f = 0.5$, $s = 50$, $r = 10\,\mu\text{m}$, and $\tau_i^* = 20$ MPa gives a value of about 80 kJ m^{-2}. Since σ_f^* would typically be about 3 GPa, the critical aspect ratio, s^* $(= \sigma_f^* / 2\tau_i^*)$, for this value of τ_i^* would be about 75. Since this is greater than the actual aspect ratio, pullout is expected to occur (rather than fiber fracture), so the calculation should be valid. The pullout energy is greater when the fibers have a larger diameter, assuming that the fiber aspect ratio is the same. The potential for contribution to the toughness from fiber pullout is substantial. However, the mechanism can apparently only operate with relatively short fibers ($s < s^*$). Continuous fibers are expected to break in the crack plane since there will always be embedded lengths on either side of the crack plane which are long enough for the stress in the fiber to build up sufficiently to break it. (Interfacial debonding can still occur, perhaps over an appreciable distance, but it is clear from the previous section that this is unlikely to make a major contribution to the fracture energy. Using this argument, it appears that pullout should not occur in long-fiber composites. In fact, it is often highly significant in such materials. The reason for this is related to the variability in strength exhibited by most of the fibers. The effect of the variability in fiber strength, characterized by the Weibull modulus, m, is depicted schematically in Figure 13.5. Figure 13.5a shows how, in the case of a deterministic (single-valued) fiber strength ($m = \infty$), the probability, P_f, of the fiber breaking in the crack plane will become 100%, while remaining zero elsewhere. For a finite value of m, however (Figure 13.5b), not only are high values of P_f spread over an appreciable distance on either side of the crack plane, but it is now possible for the fiber to break almost anywhere. Prediction of the fracture energy becomes more difficult under these circumstances since account should be taken of these probabilities in calculating the contributions from different pullout lengths. In fact, strictly, the effect of this distribution of fracture probabilities should also be taken into account for the short-fiber case. The details of such treatments need not concern us here, but it may be noted that a low Weibull modulus always tends to raise the pullout work for long fibers, but is not necessarily beneficial for short ones.

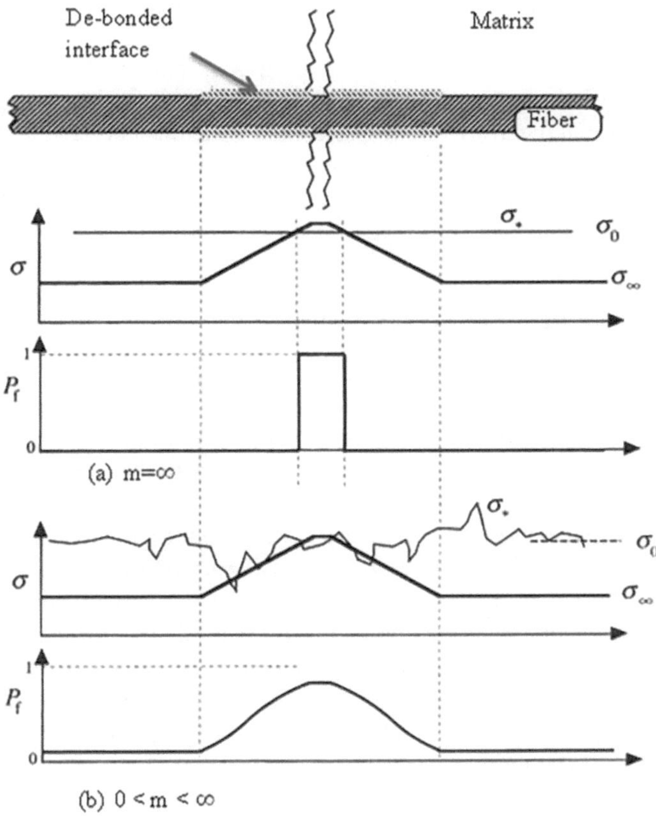

FIGURE 13.5 Stress distribution and probability of failure (P_f) along the length of a long-fiber bridging a matrix crack. (a) A fixed fiber strength ($m = \infty$) and (b) fiber strength varying along the fiber length (finite m).

13.3.5 Effect of Microstructure

There is considerable scope for controlling the fracture energy of composites by altering various features of the fiber length, orientation, and interfacial characteristics. The fracture energy of the unidirectional lamina falls off sharply as the angle between the crack plane and the fiber axis is reduced (loading angle increases). This is largely because fiber pullout becomes inhibited and fracture occurs parallel to the fiber axis. The cross-ply laminate and woven cloth-reinforced material exhibit more isotropic behavior. It is difficult to predict the dependence on loading angle in these cases since complex interactions occur between the different plies and fiber tows. However, fracture always involves a considerable degree of interfacial debonding, fiber fracture, and fiber pullout, so that the toughness is always relatively high. This is an important attribute of long-fiber laminates. The toughness of some metals can also be enhanced in this way. Good toughness can be obtained when the crack propagation direction lies in the plane or the laminate, provided the crack front is normal to this plane.

In particulate MMCs, the major concern is the retention of as much as possible of the inherent toughness of the matrix. As aging progresses, the tensile strength rises and then falls (over-aging). The fall in fracture energy during the initial rise in strength is expected on the grounds of reduced dislocation mobility. The failure of the toughness to rise again as over-aging occurs is more difficult to explain, but may be owing to precipitates forming preferentially at the interface, making it a preferred site for cavitation. Finally, evaluation of the energy absorbed can be particularly complex for laminates, for which contributions can come from interlaminar debonding as well as intra-ply processes.

13.4 SUBCRITICAL CRACK GROWTH

When the rate of energy release during crack propagation (driving force) is lower than the critical value, spontaneous fast fracture does not occur. Under some circumstances, however, an existing crack may advance slowly under this driving force. Since crack growth leads to an increased driving force (for the same applied load), this process is likely to lead to an accelerating rate of damage, culminating in conditions for fast fracture being satisfied. There are two common situations in which such subcritical crack growth tends to occur. First, if the applied load is fluctuating in some way, local conditions at the crack tip may be such that a small advance occurs during each cycle. Second, the penetration of a corrosive fluid to the crack tip region may lower the local toughness and allow crack advance at a rate determined by the fluid penetration kinetics or chemical interaction effects. In both cases, as indeed with fast fracture, the presence or absence of an initial flaw, which allows the process to initiate, is of central importance.

13.4.1 Fatigue

For metals, fatigue failure is an important topic which has been the subject of a detailed investigation. Analysis is commonly carried out in terms of the difference in stress intensity factor between the maximum and the minimum applied load (ΔK). This is because, while the maximum value, K_{max}, dictates when fast fracture occurs, the cyclic dissipation of energy is dependent on ΔK. It is, however, common to quote the *stress ratio*, R (= K_{min}/K_{max}), which enables the magnitudes of the K values to be established for a given ΔK. The resistance of a material to crack extension is given in terms of the crack growth rate per loading cycle (dc/dN). At intermediate ΔK, the crack growth rate usually conforms to the Paris–Erdogan relation (Paris and Erdogan [7]):

$$\frac{dc}{dN} = \beta(\Delta K)^n, \tag{13.20}$$

where β is a constant. Hence, a plot of crack growth rate (*m*/cycle) against ΔK, with log scales, gives a straight line in the Paris regime, with a gradient equal to n. At low stress intensities, there is a threshold, ΔK_{th} below which no crack growth occurs. The crack growth rate usually accelerates as the level for fast fracture, K_c, is approached. An alternative way of presenting fatigue data is in the form of S/N_f curves, showing the number of cycles to failure (N_f) as a function of the stress amplitude (S). Many materials show a rapid crack growth (low N_f) when the stress amplitude is high (a central portion of decreasing S with rising N_f), corresponding to the Paris regime, and a fatigue limit, which is a stress amplitude below which failure does not occur even after large numbers of cycles. This corresponds to a stress intensity below ΔK_{th}.

Some of the features of fatigue can be illustrated by examining how the presence of particulate reinforcement affects the behavior of a metal. This is shown schematically in Figure 13.6. Below ΔK_{th}, cracks are unable to grow at all. For Al/SiC_p MMCs, ΔK_{th} is typically around $2-4$ MPa$\sqrt{m}$, which is approximately twice that for unreinforced Al alloys ($\sim 1-2$ MPa$\sqrt{m}$). A number of explanations for this have been proposed (Christman and Suresh [8]), including crack deflection at interfaces and a reduction in slip-band formation due to the particles. This beneficial effect of

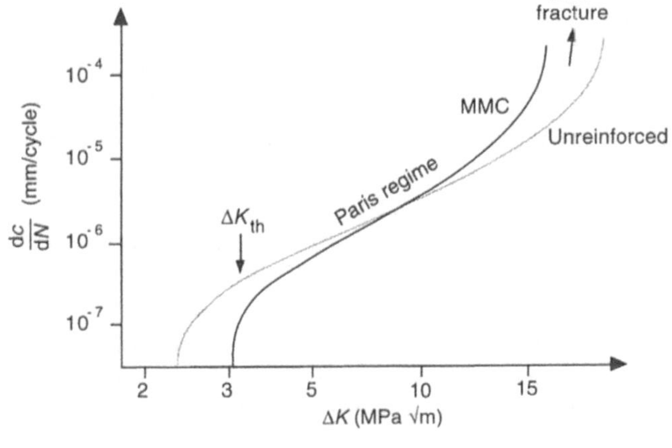

FIGURE 13.6 Fatigue crack growth rate as a function of applied stress intensity factor (ΔK) for a typical particle-reinforced MMC and the corresponding unreinforced alloy.

the reinforcement in inhibiting the onset of fatigue cracks is a useful feature of such MMCs. However, MMCs have a lower fracture toughness than unreinforced metals, mainly as a result of the constraint imposed on matrix plasticity. The Paris regime is usually short and the exponent n is often around 5–6, which is higher than those typical of unreinforced systems (~4). Fast fracture is initiated at lower ΔK values than for the unreinforced metal. In practical terms, this means that MMCs can offer improvements in fatigue performance compared with metals, providing that applied stress levels are low and/or flaw sizes are kept small. Control over processing of MMCs so as to eliminate inhomogeneities is important to ensure this.

While particulate reinforcement produces relatively minor modifications to the behavior of the matrix, the presence of long fibers has a more pronounced effect. This is particularly true for polymer composites, in which it is usually the type and orientation distribution of the fibers which is of most significance. The propagation of cracks through the matrix and along the interfaces usually dictates how fatigue progresses, but this is strongly influenced by how the fibers affect the stress distribution. The failure strain of the matrix is also important. A further difference from the particulate case lies in the distribution of damage. It is common in long-fiber composites for matrix and interfacial microcracks to form at many locations throughout the specimen. Fiber bridging across matrix cracks often occurs, reducing the stress intensity at the crack tip. In contrast to this, fatigue crack growth in monolithic and particle-reinforced materials usually involves a single dominant crack with a well-defined length. Composites reinforced with stiff fibers, such as boron and carbon, show excellent fatigue resistance, being able to withstand alternating loads of around 1 GPa for very large numbers of cycles. The fatigue performance of these materials is markedly superior to that of a typical aluminum alloy. With glass fibers, on the other hand, the lower stiffness of the fiber leads to reduced stress transfer, exposing the matrix to larger stresses and strains. This causes progressive damage at considerably lower applied loads than for the stiffer fibers.

While the axial fatigue resistance of long-fiber composites tends to be very good, particularly with high stiffness fibers, performance is usually inferior for laminates or under off-axis loading. The cross-ply and woven cloth laminates (for a glass-reinforced polymer) fail at appreciably lower loads than the unidirectional material and show little evidence of a fatigue limit stress value being identifiable. Damage to the transversely oriented regions starts at low applied loads, transferring extra load and eventually causing cracks to propagate into the axial regions. Nevertheless, the fatigue resistance of such materials compares quite well with that of many metals.

Broadly similar behavior is exhibited by MMC laminates. The best performance is shown by the unidirectional material, having the fibers parallel to the applied load. The performance of the other laminates can be rationalized by calculating the range of stress to which the 0° ply (parallel to the applied load) is subjected during loading. When this is plotted against the number of cycles to failure, the data for the different laminates fall on a common curve. This highlights the point that the laminate does not fail until the fibers in the 0° ply become fractured. The fatigue properties can, however, become badly degraded if matrix cracks are not deflected at the fiber–matrix interface. A further point worthy of note with respect to the fatigue of composites is that the behavior is often sensitive to the absolute values of the stresses being applied, rather than just the ΔK range. In particular, the introduction of compressive stresses usually reduces the resistance to fatigue. This is largely due to the axially aligned fibers having poor resistance to buckling. This results in damage to the fibers and the surrounding matrix, and also allows larger stresses to bear on neighboring, transversely oriented regions, accelerating their degradation. Aramid (e.g., Kevlar™) fibers are particularly prone to this effect since they have poor resistance to compression. The fatigue resistance of Kevlar™ composites becomes poor for negative R values. (It should, however, be noted that the effect is exaggerated by plotting the peak stress rather than the stress difference, which is larger for the lower R values.) Large diameter monofilaments, on the other hand, tend to be resistant to buckling, leading to improved performance at lower R values.

A final point concerns the frequency of cycling. For metals, this usually has little effect, but in polymer composites, a higher frequency often hastens failure. This is partly because of the viscoelastic response of polymers. Matrix damage is more likely if the strain is imposed rapidly, allowing no time for creep and stress relaxation. A second effect arises from the low thermal conductivity of polymers, particularly if the fibers are also poor conductors (glass, Kevlar™, etc.). Such composites tend to increase in temperature during fatigue loading, as a result of difficulties in dissipating the heat generated locally by damage and viscoelastic deformation. This is accentuated at high cycling frequencies. Since the strength of the matrix falls with increasing temperature, this tends to accelerate failure.

13.4.2 Stress Corrosion Cracking

Stress corrosion cracking is the term given to subcritical crack growth which occurs as a result of the effect of a corrosive fluid reaching the advancing crack tip. The micromechanisms responsible for this lowering of the local toughness of the material vary widely between different materials and environments. For example, in Al-based systems, the presence of moist or salt-laden air causes an acceleration of fatigue crack growth of between 5 and 10 times. A similar effect is observed for particulate MMCs (Crowe and Hasson [9]). For long-fiber reinforced aluminum, the fibers give rise to good retention of fatigue resistance under axial loading, whereas in the transverse direction, the fatigue resistance is affected by the corrosive environment to a degree directly related to the proportion of matrix occupied by the composite.

In the case of aluminum, stress corrosion cracking is usually due to the evolution of (atomic) hydrogen at the crack tip, which is promoted by the presence of various liquids. This tends to cause embrittlement by impeding dislocation motion. In MMCs, a further factor is introduced in the form of strong traps for hydrogen at the matrix–reinforcement interface, promoting cracking. For polymer composites, the behavior of the matrix is often quite sensitive to the presence of fluids. For example, the properties of many polymers are modified by the absorption of moisture. In many cases, this penetration occurs by diffusion through the matrix and is not confined to access along the crack. Various features of such absorption have been identified (Shen and Springer [10]). Commonly, water absorption raises the toughness and strain to failure of the matrix. Thus, the fatigue resistance of glass-reinforced epoxy can be raised by prior boiling in water (Harris [11]). However, water uptake also tends to promote interfacial debonding, impairing stiffness and strength, particularly under shear and transverse loading. Some fluids sharply degrade the strength of the fibers. They attack the fiber surface at the crack tip, reducing its strength considerably. Not only does this impair the strength of the composite, but the toughness and fatigue resistance are sharply reduced, since the fibers all break in the crack plane and pull-out work is virtually zero. The danger of such effects must always be borne in mind when glass-reinforced polymer composites are to be used under load in chemically aggressive environments, particularly mineral acids.

REFERENCES

1. Inglis C. (1913), "Stress in a plate due to the presence of cracks and sharp corners", *Transactions of the Royal Institution of Naval Architects*, Vol. 55, pp. 219–230.
2. Griffith A.A. (1920), "The phenomena of rupture and flow in solids", *Philosophical Transactions of the Royal Society A*, Vol. 221, pp. 163–197.
3. Irwin G.R. (1948), Fracture dynamics, in *Fracture of Metals*, ASM, Cleveland, pp. 152–169.
4. Cook J. and Gordon, J.E. (1964), "A mechanism for the control of crack propagation in all-brittle systems", *Proceedings of the Royal Society*, Vol. 282A, pp. 508–520.
5. Kendall K. (1975), "Transition between cohesive and interfacial failure in a laminate", *Proceedings of the Royal Society*, Vol. 344A, pp. 287–302.
6. He M.Y. and Hutchinson J.W. (1989), "Crack deflection at an interface between dissimilar and elastic materials", *International Journal of Solids and Structures*, Vol. 25, pp. 1053–1067.
7. Paris P. and Erdogan F. (1963), "A critical analysis of crack propagation laws", *Journal of Basic Engineering*, Vol. 85, pp. 528–534.
8. Christman T. and Suresh S. (1988), "Effects of SiC reinforcement and ageing treatment on fatigue crack growth in an Al-SiC composite", *Materials Science and Engineering*, Vol. 102A, pp. 211–216.
9. Crowe C.R. and Hasson D.F. (1982), "Corrosion fatigue of SiC/Al MMC in salt-laden moist air", in *Proceeding of ICMSA-6*, (R.C. Gifkins ed.), Pergamon, pp. 859–865.
10. Shen C.H. and Springer G.S. (1976), "Moisture absorption and desorption of composite materials", *Journal of Composite Materials*, Vol. 10, pp. 2–20.
11. Harris B. (1994), Fatigue-glass fibre reinforced plastics, in *Handbook of Polymer Fibre Composites,* (F.R. Jones ed.), Longman, London, pp. 309–316.

14 Interlaminar Stresses

The free edges in laminated composites cause an extra level of complexity. The stress state in the vicinity of the free edges is generally three-dimensional having nonzero through-thickness stresses. The through-thickness stresses include the interlaminar normal stress, σ_z, and two interlaminar shear stresses, τ_{yz} and τ_{zx} (Figure 14.1). These through-thickness stresses are called interlaminar stresses. It must be emphasized that while these stresses are called interlaminar, indicating that they act between layers, they are not limited to the interface between layers. Stress continuity requires that they vary in a continuous manner through the thickness of all layers of the laminate. Stress-free boundaries typically present the most severe interlaminar stresses. However, interlaminar stresses are also present in other structural configurations, including internal ply-drop in tapered laminates and geometric discontinuities such as the flange termination region of stiffened panels. We will confine our discussion here to the analysis of interlaminar stresses near free edges of finite-width coupons. Approximate elasticity solutions show that the interlaminar stresses are generally confined to a boundary-layer region adjacent to the free edge, and that the interlaminar components of stress can be singular at the intersection of a free edge with the interface between two distinct layers. Since the state of stress in this boundary-layer region is not a state of plane stress, the classical lamination theory is not valid in

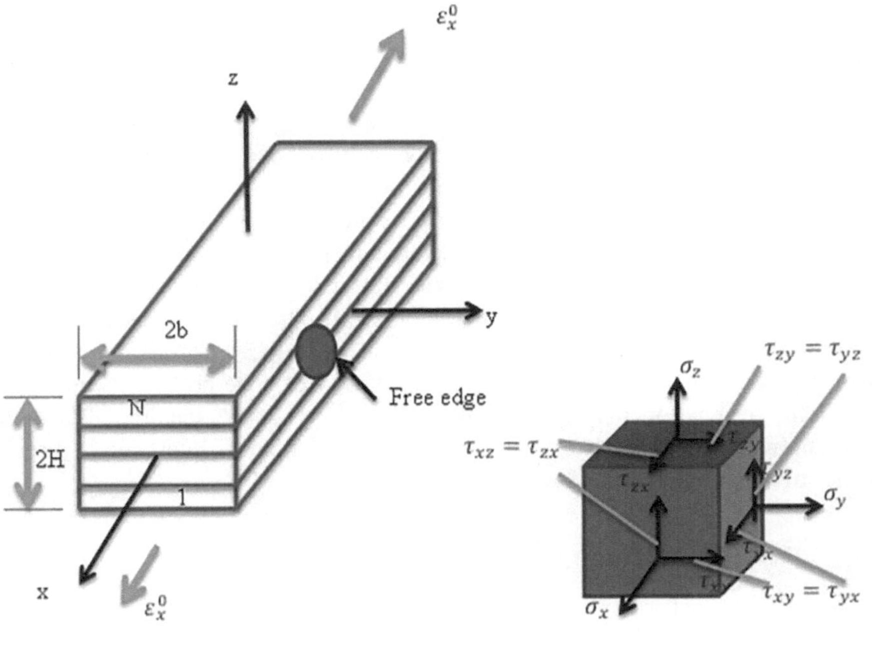

FIGURE 14.1 Finite-width laminated coupon under axial load.

this region. Away from the free edge (outside the boundary-layer region), the state of stress reverts to a planar state and lamination theory is again valid. Interlaminar stresses are not unique to anisotropic materials. They may be present in any nonhomogeneous material with free edges.

The interlaminar stresses are caused by the mismatch in material properties of bonded adjacent materials (layers) and gradients in in-plane components of stress [1]. For isotropic layers, the mismatch in properties is usually small and interlaminar stresses are typically ignored. In contrast to isotropic materials, orthotropic composite laminas, made of high-modulus fibers in a less stiff matrix, exhibit a very broad range of properties as a function of fiber orientation. As will be shown through equilibrium considerations, the mismatch in Poisson's ratios between layers requires the existence of interlaminar shear (τ_{yz}) and interlaminar normal stresses (σ_z). Further, off-axis laminas exhibit axial shear coupling as quantified by the coefficient of mutual influence. The potentially large mismatch in coefficients of mutual influence for polymeric matrix composites can lead to very large interlaminar shear stresses τ_{zx}. Other material property mismatches that can result in interlaminar stresses include the coefficients of thermal and hygroscopic expansion [2]. In real materials that exhibit inelastic response associated with matrix plasticity and damage (as contrasted with the idealized linear elastic materials under consideration here), the interlaminar stresses are not singular, but they do exhibit very large gradients near free edges. The large interlaminar stresses in the boundary layer are more severe for polymer matrix composites than for metal matrix composites. The polymer matrix composites are highly anisotropic, resulting in a greater mismatch of layer properties.

The interlaminar stresses in the boundary layer can be critical in structural applications because they can lead to delamination-type failures at loads well below those corresponding to in-plane failure. Interlaminar stresses may be present as a result of mechanical, thermal, or moisture loading; they should be considered whenever laminated composite materials are used in the design of a structure which has free edges, including the free edges around holes and cutouts.

Interlaminar stresses near free edges can be controlled to some extent through the choice of materials, fiber orientations, stacking sequence, layer thicknesses, and the use of functionally graded materials. However, when free edges are present, interlaminar stresses can be eliminated completely only through the use of homogeneous materials. Edge reinforcement techniques have been used to suppress the deleterious effects of interlaminar stresses, but at additional cost. These techniques do not eliminate the interlaminar stresses; they only provide a restraint against delamination. A thorough understanding of the mechanics of the free edge problem is indispensable to the designer of composite structures. With a basic understanding of the problem, laminates (and structures) can be designed to minimize interlaminar stresses while still meeting all other design requirements.

14.1 FINITE-WIDTH COUPON

Consider a symmetrically loaded, finite-width coupon of Figure 14.1 for studying the edge effects in composites. The N-ply laminate has total thickness $2H$ and finite width $2b$. For simplicity of exposition, we shall consider only laminates which are symmetric about the mid-plane and orthotropic ($A_{16} = A_{26} = 0$). Individual layers of the laminate are unidirectional (transversely isotropic) laminas of fibrous composites oriented at an angle θ_i, with respect to the global x axis of the coupon. It is assumed that all layers are perfectly bonded together and behave elastically under the application of load. Tensile load is applied at the ends of a long coupon, resulting in a constant, uniform axial strain ε_x^0 along the length of the coupon. Away from the ends, the states of stress and strain are independent of the axial coordinate x. The problem is quasi-three-dimensional in that all six components of stress and strain may be nonzero, but the stress analysis can be restricted to a generic two-dimensional cross section because of the x independence of stress and strain. In interior regions away from the edges, all layers are in a state of plane stress, and the intralaminar stresses σ_x, σ_y, and τ_{xy} can be determined from the classical lamination theory [3]. We note that z is taken positive upward in this development.

14.2 EQUILIBRIUM CONSIDERATIONS

Equilibrium considerations can be used to provide the basic justification for the existence of interlaminar stresses in a laminated composite with free edges. Consider the tensile loaded coupon of Figure 14.1. If we accept the fact (confirmed by elasticity theory) that a planar state of stress, independent of the axial coordinate, is recovered away from the free edge, then we can show that equilibrium of individual layers (or groups of layers) requires the existence of nonzero interlaminar stresses over some portion of interfacial planes. Integration of these interlaminar stresses over the region of application provides forces and moments that must be in equilibrium with the forces and moments calculated from the planar state of stress away from the edge. And since the state of stress away from the edges (i.e., along $y = 0$) is planar, these stresses can be calculated from lamination theory. These equilibrium considerations are developed in detail in the following paragraphs for unit length of a tensile coupon. A free-body diagram of a section cut from the unit length is shown in Figure 14.2. It is noted that the stresses σ_x, τ_{zx}, and τ_{zy} acting over the end surfaces $x = \pm L$ are self-equilibrating by virtue of the x independence of the stress state.

14.3 INTERLAMINAR F_{YZ} SHEAR FORCE

Figure 14.3 shows the partial free-body diagrams of a group of layers from the tensile coupon in Figure 14.1; the group includes the top free surface and free edge of the laminate. Shown in Figure 14.3a are the non-self-equilibrating y-component stresses acting on a generic cross-sectional plane above any surface $z = z^*$. Taking into consideration the x independence of stresses, y-force equilibrium per unit length reduces to:

$$\int_0^b \tau_{yz}\left(z^*\right)dy = -\int_{z^*}^{z_N} \sigma_y dz. \tag{14.1}$$

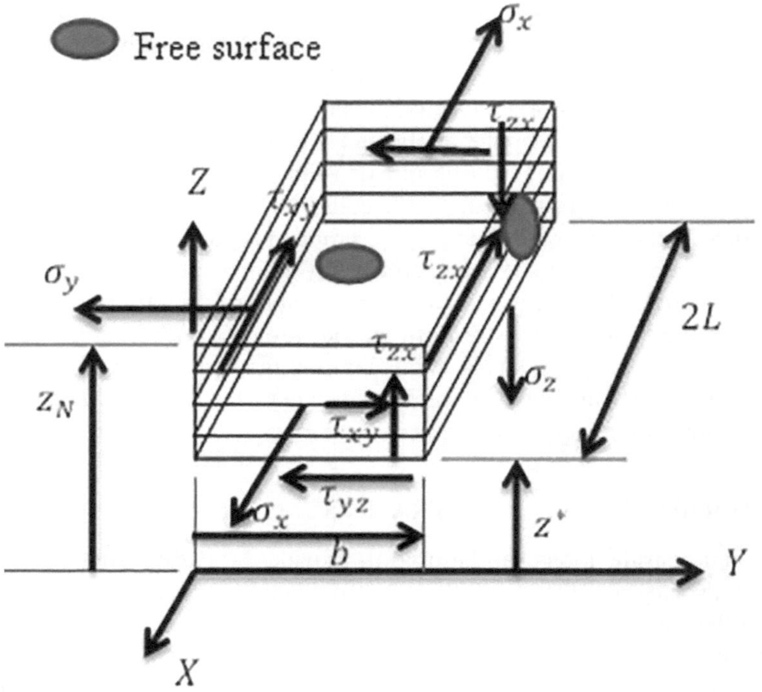

FIGURE 14.2 Free-body diagram of tensile coupon section.

The above equilibrium equation shows that the interlaminar shear stress τ_{yz} (z*) must be nonzero over some portion of the surface $0 \leq y \leq b$ at $z = z^*$ if the σ_y above this surface are not self-equilibrating. Since the integrals in Eq. (14.1) have the units of force per unit length, it is convenient to define the interlaminar shear force F_{yz} at any surface $z = z^*$ as:

$$F_{yz}\left(z^*\right) = \int_0^b \tau_{yz}\left(z^*\right)dy. \tag{14.2}$$

Combining Eqs. (14.1) and (14.2) gives the definition of the interlaminar shear force F_{yz} at any location $z = z^*$ in terms of the lamination theory stresses σ_y.

$$F_{yz}\left(z^*\right) = -\int_{z^*}^{z_N} \sigma_y dz. \tag{14.3}$$

More specifically, we can express the interlaminar shear force $F_{yz}^{(k)}$ at the kth interface (where $z^* = z_{k-1}$) in terms of in-plane stresses in the layers above the interface and the layer thicknesses t_j. The in-plane stresses above the interface are known (or determined) from lamination theory.

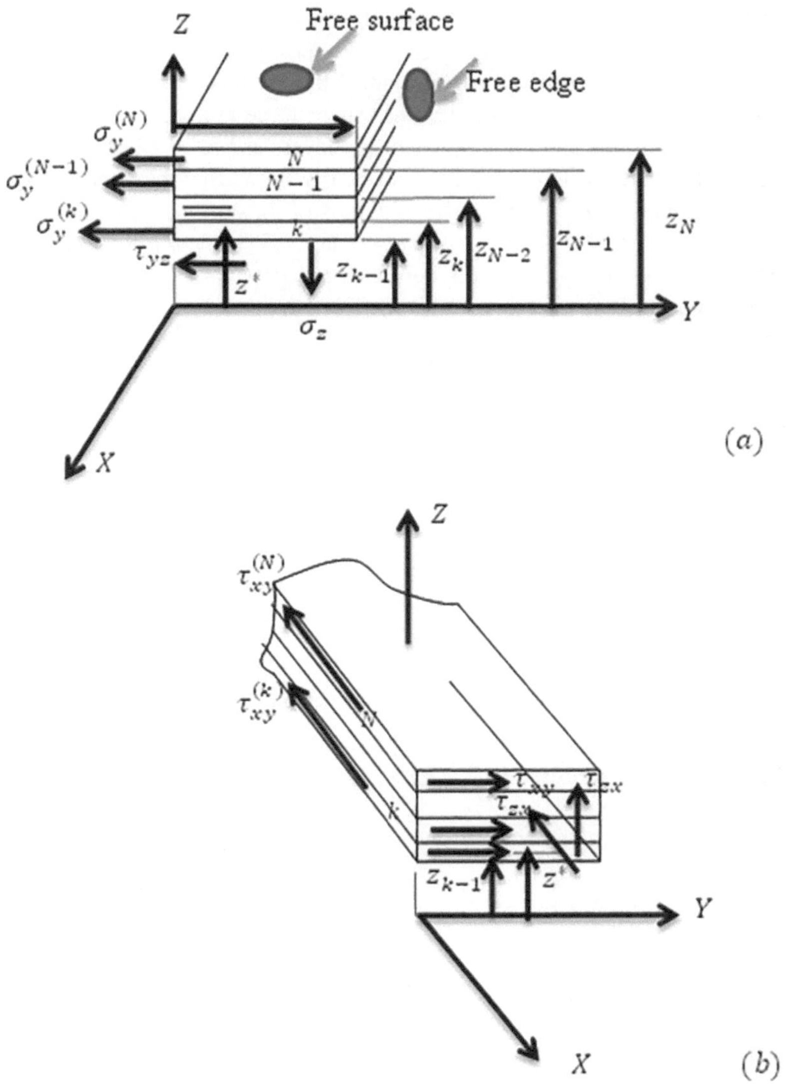

FIGURE 14.3 Partial free-body diagrams (a) showing σ_y, σ_z, and τ_{xy} (b) showing τ_{xy} and τ_{xz}.

14.3.1 Uniform Strain Loading

For loadings such that strain in the laminate is pure mid-plane strain (e.g., axial or uniform thermal loading of a symmetric laminate), the lamination theory stresses are constant through the thickness of each layer, hence:

$$F_{yz}^{(k)} = -\int_{z_{k-1}}^{z_N} \sigma_y dz = \sum_{j=k}^{N} \sigma_y^{(j)} t_j. \tag{14.4}$$

Appropriate modification of the thickness term of the $\underline{k}^{\text{th}}$ layer in Eq. (14.4) gives the interlaminar shear force at any location $z = z^*$ (not only the interface between layers) in terms of stresses determined from lamination theory. It is clear from Eq. (14.4) that for uniform strain loading, the interlaminar shear force at $z = z^*$ is only a function of the fiber orientations and thicknesses of the layers above z^* and not the stacking sequence of these layers. This equation also clearly shows that the interlaminar shear force will, in general, vary through the thickness of the laminate, that is, as z^* (or k) varies.

The free edge boundary condition requires that $\tau_{yz} = 0$ for all z at $y/b = 1.0$. Hence, the interlaminar shear stress τ_{yz} is zero at the free edge, may be nonzero near the free edge, but must return to zero in interior regions away from the free edge where a plane state of stress is recovered. Further, if the free-body diagram in Figure 14.3 is for the full laminate width rather than the half-width, equilibrium requires that:

$$\int_{-b}^{b} \tau_{yz}\left(z^*\right)dy = 0. \tag{14.5}$$

Thus, τ_{yz} must be an odd function of y.

14.3.2 Curvature Loading

For loadings that give rise to nonzero curvatures $\{k_x, k_y, k_{xy}\}$ with the mid-plane strains being zero, such as moment loading $M_x \neq 0$, the lamination theory stresses are linear in z, and the interlaminar shear force, F_{yz}, of Eq. (14.3) at an interface $z^* = z_{k-1}$ takes the form:

$$F_{yz}\left(z^*\right) = -\sum_{j=k}^{N}\left(\left(\bar{Q}_{12}^{(j)}k_x + \bar{Q}_{22}^{(j)}k_y + \bar{Q}_{26}^{(j)}k_{xy}\right)\frac{\left(z_j^2 - z_{j-1}^2\right)}{2}\right). \tag{14.6}$$

Appropriate modification of the term involving the z coordinates in Eq. (14.6) gives the interlaminar shear force at any location $z = z^*$ in terms of quantities known or determined from lamination theory.

14.4 INTERLAMINAR M_Z MOMENT

Moment equilibrium about the x axis through a point such as A for the partial free-body diagram of Figure 14.3a requires that the interlaminar normal stress σ_z be related to the laminate stresses σ_y through:

$$\int_{z^*}^{z_N} \sigma_y\left(z - z^*\right)dz = \int_{0}^{b} \sigma_z\left(z^*\right)y\,dy. \tag{14.7}$$

This moment equilibrium equation suggests a definition for the interlaminar moment M_z (the moment associated with the stress σ_z) at any surface $z = z^*$ in the form:

$$M_z\left(z^*\right) = \int_0^b \sigma_z\left(z^*\right) y dy. \tag{14.8}$$

Using the moment equilibrium Eq. (14.7), we can write the interlaminar moment M_z at any location z^* in terms of the stresses σ_y determined from the lamination theory as:

$$M_z\left(z^*\right) = \int_{z^*}^{z_N} \sigma_y\left(z - z^*\right) dz. \tag{14.9}$$

14.4.1 Uniform Strain Loading

Similar to the interlaminar shear force F_{yz}, the interlaminar moment $M_z^{(k)}$ at the kth interface (where $z^* = z_{k-1}$) for an axially loaded symmetric laminate can be expressed as a summation of constant layer stresses $\sigma_y^{(j)}$:

$$M_z^{(k)} = \int_{z_{k-1}}^{z_N} \sigma_y\left(z - z^*\right) dz = \sum_{j=k}^{N} \sigma_y^{(j)}\left(z_j - z_{j-1}\right)\left[\frac{\left(z_j + z_{j-1}\right)}{2} - z_{k-1}\right]. \tag{14.10}$$

It is clear from Eqs. (14.8) and (14.10) that the interlaminar moment is a function of the stacking sequence of the layers above the plane of interest. Also, the interlaminar moment will, in general, vary through the thickness of the laminate.

We have now established conditions for y-force equilibrium and moment equilibrium about an x axis of the generic free-body diagram of Figure 14.3. z-force equilibrium additionally requires that the total force associated with the distribution of σ_z along any plane $z = z^*$ be zero. Thus, the distribution of σ_z must be equivalent to a couple. The sign of this couple is a function of the through-thickness distribution of σ_y at $y = 0$. Hence, the interlaminar normal stresses vary through the thickness of the laminate and may be positive, negative, or zero at any point (y, z). Further, moment equilibrium of a free-body diagram of the full width shows that σ_z must be an even function of y.

14.4.2 Curvature Loading

For loadings that give rise to nonzero curvatures $\{k_x, k_y, k_{xy}\}$ with the mid-plane strains being zero, the linear z dependence of the stresses results in the following form of Eq. (14.9) at an interface $z^* = z_{k-1}$:

$$M_z\left(z^*\right) = -\sum_{j=k}^{N}\left(\left(\bar{Q}_{12}^{(j)} k_x + \bar{Q}_{22}^{(j)} k_y + \bar{Q}_{26}^{(j)} k_{xy}\right)\left(\frac{\left(z_j^3 - z_{j-1}^3\right)}{3} - z^* \frac{\left(z_j^2 - z_{j-1}^2\right)}{2}\right)\right). \tag{14.11}$$

14.5 INTERLAMINAR F_{ZX} SHEAR FORCE

Summation of forces in the x direction of the free-body diagram in Figure 14.3b requires that unbalanced stresses τ_{xy} along the laminate centerline ($y = 0$) be equilibrated by nonzero stresses τ_{zx} along the interface $z = z^*$. This equilibrium equation can be expressed as:

$$\int_0^b \tau_{zx}\left(z^*\right)dy = -\int_{z^*}^{z_N} \tau_{xy}dz. \tag{14.12}$$

This equation leads to the definition of the interlaminar shear force F_{zx} at any surface $z = z^*$:

$$F_{zx}\left(z^*\right) = \int_0^b \tau_{zx}\left(z^*\right)dy. \tag{14.13}$$

Combining Eqs. (14.12) and (14.13) gives the interlaminar force F_{zx} in terms of the shear stresses τ_{xy} determined from the lamination theory:

$$F_{zx}\left(z^*\right) = -\int_{z^*}^{z_N} \tau_{xy}dz. \tag{14.14}$$

14.5.1 Uniform Strain Loading

As for the previously defined interlaminar force and moment, $F_{zx}^{(k)}$ at the kth interface for uniform strain loading can be written in terms of the known (or determined) lamination theory stresses as:

$$F_{zx}^{(k)} = -\int_{z_{k-1}}^{z_N} \tau_{xy}dz = \sum_{j=k}^{N} \tau_{xy}^{(j)} t_j, \tag{14.15}$$

where t_j is the thickness of the jth layer and we have again used the fact that the lamination theory stresses are constant through the thickness of each layer for in-plane loading of a symmetric laminate. Appropriate modification of the thickness term in Eq. (14.15) gives the interlaminar shear force F_{zx} at any location $z = z^*$. From Eq. (14.12), nonzero interlaminar shear stresses τ_{zx} must exist over some region of the interface $z = z^*$ if the in-plane shear stresses τ_{xy} are not self-equilibrating above this surface. If the full width of the laminate is taken as the free-body diagram, we see that τ_{zx} must be an odd function of y as was τ_{yz}. However, unlike the case for τ_{yz}, there is no boundary condition on τ_{xz} at $y = b$. Hence, τ_{zx} is zero away from the edge where lamination theory is valid, is generally nonzero as the edge is approached, and is not limited in any manner at the free edge. As a final comment, we note that for a group of layers, with given fiber orientations above $z = z^*$, F_{zx} at $z = z^*$ is independent of the stacking sequence of the layers within the group above $z = z^*$.

14.5.2 Curvature Loading

For curvature loading, the interlaminar shear force F_{zx} at $z^* = z_{k-1}$ takes the form:

$$F_{zx}\left(z^*\right) = -\sum_{j=k}^{N}\left(\left(\bar{Q}_{16}^{(j)}k_x + \bar{Q}_{26}^{(j)}k_y + \bar{Q}_{66}^{(j)}k_{xy}\right)\frac{\left(z_j^2 - z_{j-1}^2\right)}{2}\right). \tag{14.16}$$

REFERENCES

1. Herakovich C.T. (1989), "Free edge effects in laminated composites." in *Handbook of Composites: Vol. 2, Structures and Design* (C.T. Herakovich and Y.M. Tarnopol'skii, eds.), Elsevier Science Publishing Co., New York, pp. 187–230.
2. Popov E.P. (1990), *Engineering Mechanics of Solids*, Prentice Hall, Englewood Cliffs, NJ.
3. Hyer M.W. (1998), *Stress Analysis of Fiber Reinforced Composite Materials*, McGraw-Hill, Singapore.

15 Laminated Plates

We want to be able to accurately describe the response at all points in a fiber-reinforced structure and examine the stresses, strains, and the issue of failure throughout the structure. We need to study the influence of the loading type, temperature change, laminate shape, laminate size, and boundary conditions, as well as fiber orientation, material properties, and lamination sequence, on the response. Maximum deflections, maximum stresses, loads to cause material failure, loads to cause buckling, and natural frequencies of vibration are some of the more important structural responses that depend on the details of the entire structure. To study these, it is necessary to develop the tools for determining how laminate response varies with x and y. Within the context of classical lamination theory, we now know how laminate response varies with z, but we must expand our thinking to variations in the other two coordinate directions. The development of the tools depends to a large degree on the issue being studied (do we want to know maximum deflections, or buckling loads, or natural frequencies?) as well as the type of the structure (is it a thin plate, a thick plate, a cylindrical shell, a conical shell, etc.?). In this chapter, we shall develop the tools necessary to study flat laminated plates and apply these tools to several problems that illustrate the unique response characteristics of fiber-reinforced structures in general, and plates in particular. Because the study of laminated plates can lead to several books in itself, this chapter will be limited to the study of the linear response of laminated plates. Thus, for example, buckling will not be addressed. Furthermore, the discussions of this chapter will be limited to rectangular plates. Our primary purpose here is to establish the principles of classical laminated plate theory and illustrate these principles with simple examples. Advanced topics can then be pursued from these basic principles. The next section develops the equations governing the behavior of plates.

15.1 GOVERNING EQUATIONS

There are several approaches to developing the equations that govern the behavior of plates [1–3]. It must be clear from the onset that the equations that govern the behavior must include the proper specification of the conditions at the boundary, as well as the specification of the conditions that govern the behavior away from the boundary, in the interior of the plate. The latter conditions are generally referred to as the governing differential equations, whereas the former are, naturally, referred to as boundary conditions. The most consistent method of deriving the governing conditions is through energy and variational principles. With this approach, we obtain the governing differential equations and the boundary conditions. This is the preferred approach, but unfortunately, not everybody feels comfortable with the principles of variational calculus. As a result, the alternative Newtonian approach, summing forces and moments, is often used to derive the governing differential equations.

The disadvantage of this approach is that there is no direct information regarding the boundary conditions. Nonetheless, we will use the Newtonian approach in this chapter; the differential equations of equilibrium will be derived by summing forces and moments on a differential element of laminate, and the boundary conditions will simply be stated.

Consider a plate, as in Figure 15.1, with length a in the x direction, width b in the y direction, and thickness H. The plate is made of a number of layers of fiber-reinforced material, as the inset shows, and the reference surface of the plate will be taken as the reference surface of the laminate. The plate is subjected to a known distributed force $q(x, y)$ acting in the $+z$ direction. Plate theory cannot distinguish whether this distributed force is pushing on the top surface or whether it is pulling on the bottom surface. The fact that the distributed force can vary with spatial location (x, y) is indicated by the functional dependence of q on x and y. In addition to the distributed force, the boundaries of the plate can be subjected to a variety of forces and moments. Figure 15.2 illustrates the possible forces and moments that can be applied on the edge $x = +a/2$, namely three forces and two moments. Note the nomenclature associated with these forces and moments. The superscript + denotes that the application is at $x = +a/2$, and the functional dependence on y indicates that these forces and moments can vary with distance along the edge. Figure 15.3 illustrates such a variation for the normal force resultant N_x^+. In Figure 15.2, four of the five resultants are familiar. These are the normal force resultant $N_x^+(y)$, the shear force resultant $N_{xy}^+(y)$, the bending moment resultant $M_x^+(y)$, and the twisting moment resultant $M_{xy}^+(y)$. The transverse shear force resultant $Q_x^+(y)$ has not previously appeared. This resultant is defined by:

$$Q_x^+(y) = \int_{z=-\frac{H}{2}}^{z=+\frac{H}{2}} \tau_{xz}\left(+\frac{a}{2}, y, z\right) dz \tag{15.1}$$

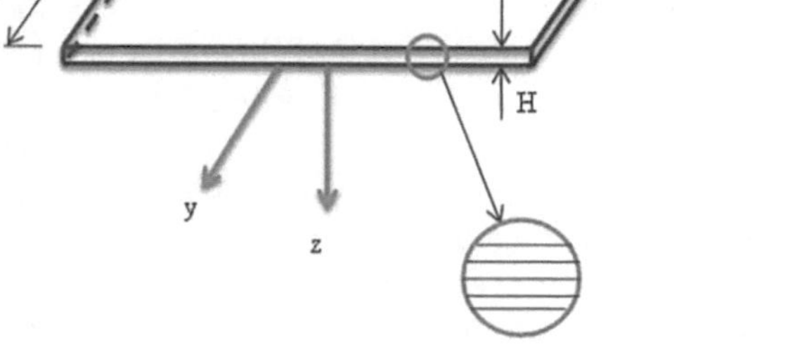

FIGURE 15.1 Geometry, loading, and nomenclature of a rectangular laminated plate.

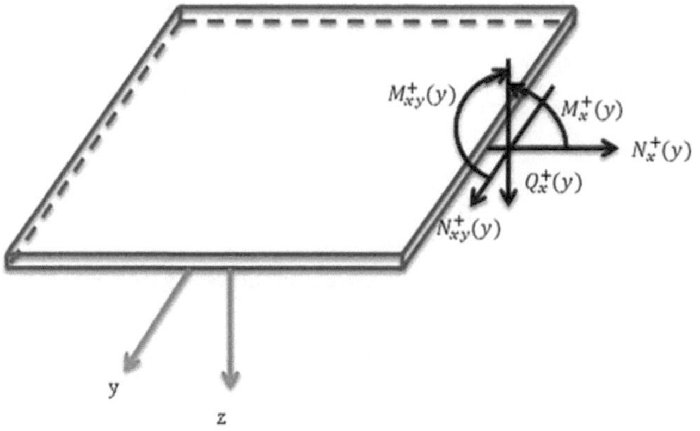

FIGURE 15.2 Stress resultants at the boundary $x = +a/2$.

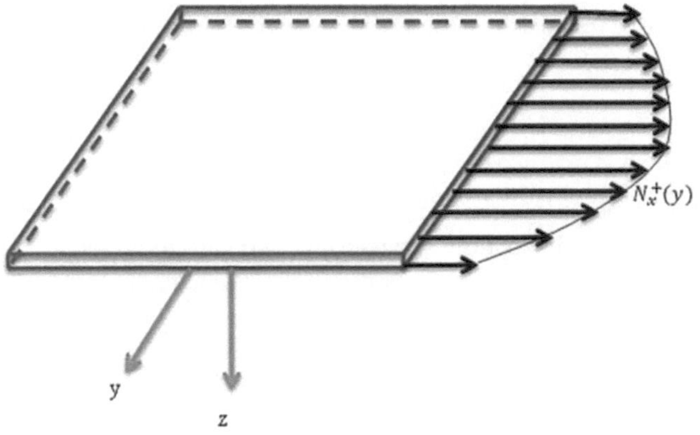

FIGURE 15.3 Variation of $N_x^+(y)$ with y.

One can immediately raise the issue that because of the plane-stress assumption, $\tau_{xz} = 0$, and thus, the transverse shear force resultant must be zero. Therefore, we should not be able to say anything about defining this resultant. As will be seen shortly, there must be transverse shear force resultants Q_x, and, shortly to be introduced, Q_y to maintain equilibrium in the z direction. Thus, there is an inconsistency in plate theory. The theory needs stresses τ_{xz} and τ_{yz} for equilibrium, yet the theory is based on the plane-stress assumption. The variational approach does not have to explicitly define Q_x and Q_y because it is not based on equilibrium principles.

Despite this inconsistency, the theory we are developing allows for the application of any or all of these five resultants on the edge $x = +a/2$. In particular, if the edge is free, as in a cantilevered plate, then all five resultants would be zero. On the edge $y = +b/2$ (Figure 15.4), five other force resultants can be applied. Again, note the superscript + signifying the location $y = +b/2$ and the possible variation of these resultants along the edge, in this case with the x-coordinate. The subscripts yx instead

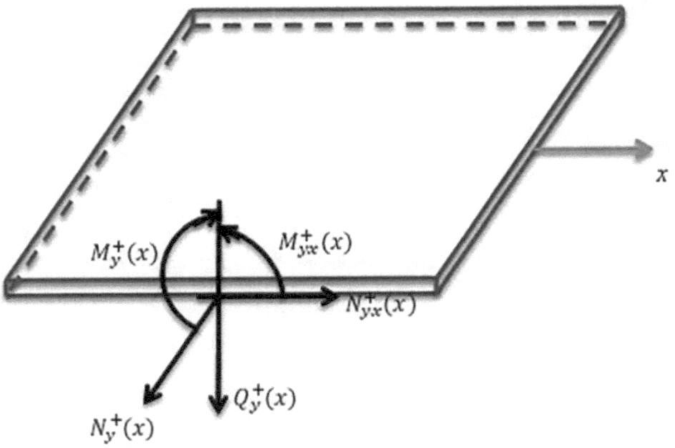

FIGURE 15.4 Stress resultants at the boundary $y = +b/2$.

of xy on N and M simply denote the difference between the edge $x = +a/2$ and the edge $y = +b/2$. The transverse shear force resultant $Q_y^+\ (x)$ is defined as:

$$Q_y^+\ (x) = \int_{z=-\frac{H}{2}}^{z=+\frac{H}{2}} \tau_{yz}\left(x, +\frac{b}{2}, z\right) dz \tag{15.2}$$

The forces and moments that can be applied on the other two boundaries are indicated in Figures 15.5 and 15.6. Note the superscript – indicating the location of the edges, $x = -a/2$ and $y = -b/2$. Note also the sense of the applied resultants, the sense being consistent with past notation. The sense of the Q's is consistent with the sense of τ_{xz} and τ_{yz}.

To derive the governing equilibrium equations, we must consider force and moment equilibrium of a differential element of laminated plate. Because equilibrium considers the sums of forces and moments, not the sums of force and moment *resultants*, the force and moment resultants must be multiplied by an appropriate length.

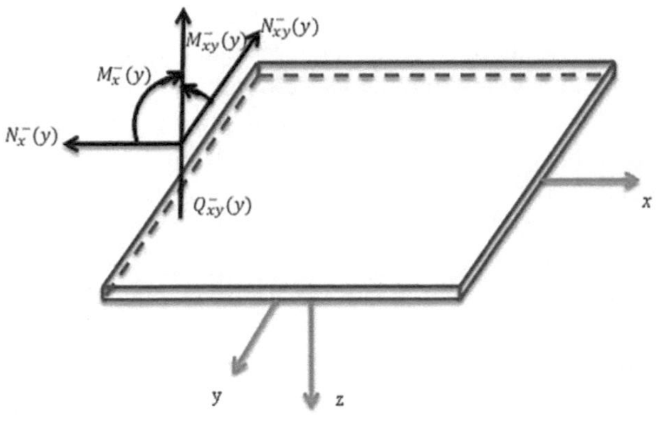

FIGURE 15.5 Stress resultants at the boundary $x = -a/2$.

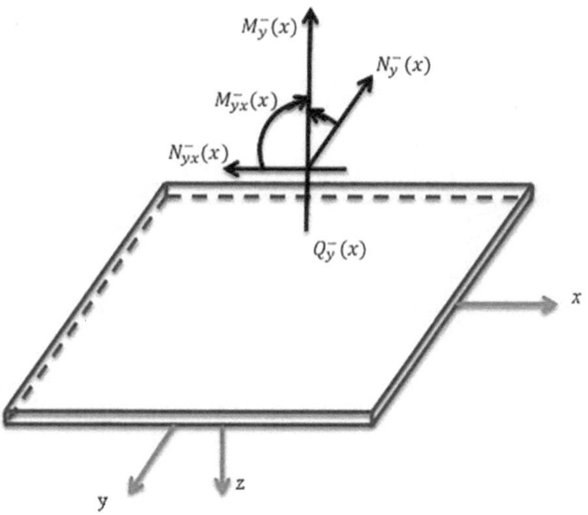

FIGURE 15.6 Stress resultants at the boundary $y = -b/2$.

Recall that because the units of N_x, for example, are N m^{-1}, N_x must be multiplied by a length to have the units of force. To derive the equilibrium equations, several figures of the differential element will be used, and each figure will depict a specific set of force or moment resultants. In reality, all force and moment resultants act simultaneously on the differential element, but to avoid cluttering the figure, sets of forces are illustrated separately.

Six equilibrium conditions must be considered. The sum of the forces in each of the three coordinate directions must be zero, and the sum of moments about each of the three coordinate axes must be zero. To sum forces in the x direction, consider Figure 15.7, which shows the differential element; the length in the x direction is Δx and the length in the y direction is Δy. The element is centered about the point (x, y). This differential element is cut from the interior of the plate, away from the edges, and includes the entire thickness of the laminate. Because the force and moment resultants vary with the x- and y-coordinates, and because the element is of differential size, we can use a Taylor series expansion to represent the resultants on the edges of the element in terms of the resultants at (x, y). In each case, only the first term of the Taylor series is retained.

In the figure, the force resultant on each edge is multiplied by the length of the edge to properly define a force. For convenience, only the forces affecting equilibrium in the x direction are shown and the force is assumed to act at mid-length along each edge. Summing forces in the $+x$ direction gives:

$$\left(N_x + \frac{\partial N_x}{\partial x}\left(\frac{\Delta x}{2}\right)\right)\Delta y + \left(N_{xy} + \frac{\partial N_{xy}}{\partial y}\left(\frac{\Delta y}{2}\right)\right)\Delta x - \left(N_x + \frac{\partial N_x}{\partial x}\left(-\frac{\Delta x}{2}\right)\right)\Delta y$$

$$-\left(N_{xy} + \frac{\partial N_{xy}}{\partial y}\left(-\frac{\Delta y}{2}\right)\right)\Delta x = 0 \qquad (15.3)$$

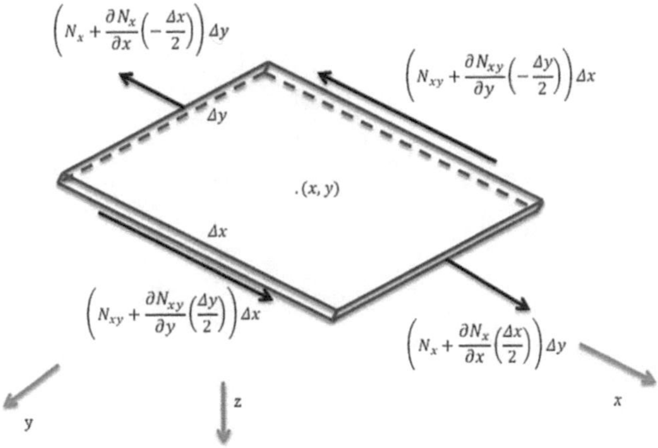

FIGURE 15.7 Summation of forces in the x direction on a differential element of plate.

The algebra leads to:

$$\frac{\partial N_x}{\partial x} + \frac{\partial N_{xy}}{\partial y} = 0 \tag{15.4}$$

This is one equilibrium equation. It is a partial differential equation, whose partial derivatives are gradients of the force resultants. The equation is an algebraic equation in the force resultant gradients, and the equation states that if N_x varies with x, then N_{xy} must vary with y. The second equilibrium equation can be derived by summing forces in the y direction, where the force resultants involved in this summation are illustrated in Figure 15.8. Repeating the steps that led to Eq. (15.4), we find that the second equilibrium equation is:

$$\frac{\partial N_{xy}}{\partial x} + \frac{\partial N_y}{\partial y} = 0 \tag{15.5}$$

To derive the third equilibrium equation, we must define the transverse shear force resultants in the interior of the plate; these definitions are similar to the definitions of the transverse shear force resultants at the edges of the plate (Eqs. (15.1) and (15.2)). Accordingly, the transverse shear force resultant Q_x is defined as:

$$Q_x = \int_{z=-\frac{H}{2}}^{z=+\frac{H}{2}} \tau_{xz}\, dz \tag{15.6}$$

while the transverse shear force resultant Q_y is defined as:

$$Q_y = \int_{z=-\frac{H}{2}}^{z=+\frac{H}{2}} \tau_{yz}\, dz \tag{15.7}$$

Both resultants vary with x and y.

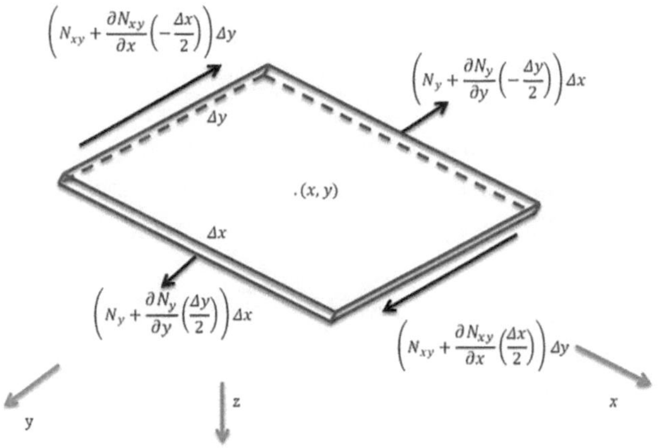

FIGURE 15.8 Summation of forces in the y direction on a differential element of plate.

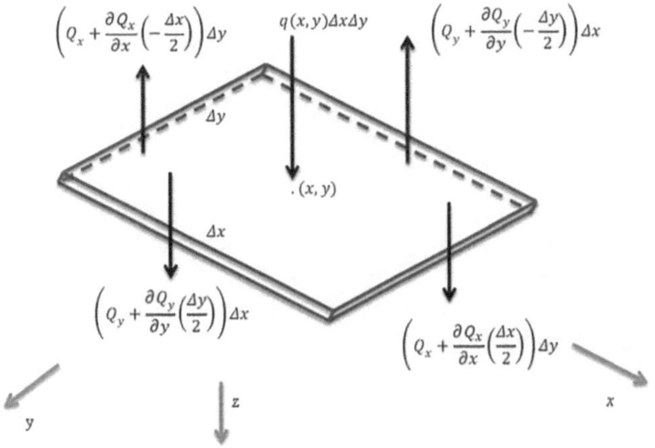

FIGURE 15.9 Summation of forces in the z direction on a differential element of plate.

Figure 15.9 shows the force resultants pertinent to summing forces in the z direction. The pertinent forces are these just-defined transverse shear force resultants, and the applied distributed force $q\ (x,\ y)$. To have the units of force, the distributed force must be multiplied by the differential area it acts on. The differential element is small enough that we can assume that the force due to the distributed load acts at point (x, y). Summing forces in the $+z$ (downward) direction results in:

$$\left(Q_x+\frac{\partial Q_x}{\partial x}\left(\frac{\Delta x}{2}\right)\right)\Delta y+\left(Q_y+\frac{\partial Q_y}{\partial y}\left(\frac{\Delta y}{2}\right)\right)\Delta x-\left(Q_x+\frac{\partial Q_x}{\partial x}\left(-\frac{\Delta x}{2}\right)\right)\Delta y$$
$$-\left(Q_y+\frac{\partial Q_y}{\partial y}\left(-\frac{\Delta y}{2}\right)\right)\Delta x=0 \tag{15.8}$$

The algebra leads to:

$$\frac{\partial Q_x}{\partial x}+\frac{\partial Q_y}{\partial y}+q=0 \tag{15.9}$$

We now must consider moment equilibrium; Figure 15.10 illustrates the force and moment resultants that produce a moment about the x-axis. Summing moments about the point (x, y) results in:

$$\begin{aligned}&\left(M_y+\frac{\partial M_y}{\partial y}\left(-\frac{\Delta y}{2}\right)\right)\Delta x+\left(M_{xy}+\frac{\partial M_{xy}}{\partial x}\left(-\frac{\Delta x}{2}\right)\right)\Delta y+\left(Q_y+\frac{\partial Q_y}{\partial y}\left(-\frac{\Delta y}{2}\right)\right)\Delta x\\&\times\left(\frac{\Delta y}{2}\right)+\left(Q_y+\frac{\partial Q_y}{\partial y}\left(\frac{\Delta y}{2}\right)\right)\Delta x\left(\frac{\Delta y}{2}\right)-\left(M_y+\frac{\partial M_y}{\partial y}\left(\frac{\Delta y}{2}\right)\right)\Delta x\\&-\left(M_{xy}+\frac{\partial M_{xy}}{\partial x}\left(\frac{\Delta x}{2}\right)\right)\Delta y\end{aligned} \tag{15.10}$$

Though the equivalent force of the distributed load, $q(x, y)\,\Delta x\Delta y$, acts vertically, it acts through point (x, y) and hence has no moment arm. Collecting terms, the above equation becomes:

$$\frac{\partial M_{xy}}{\partial x}+\frac{\partial M_y}{\partial y}=0 \tag{15.11}$$

This equation, rather than Eq. (15.7), could be considered the definition of the shear force resultant Q_y. With this definition, it is unnecessary to appeal to using stresses that have been defined as zero, in this case τ_{yz} to define a quantity that is absolutely necessary for equilibrium. Rather, Q_y is defined to be equal to derivatives of moment

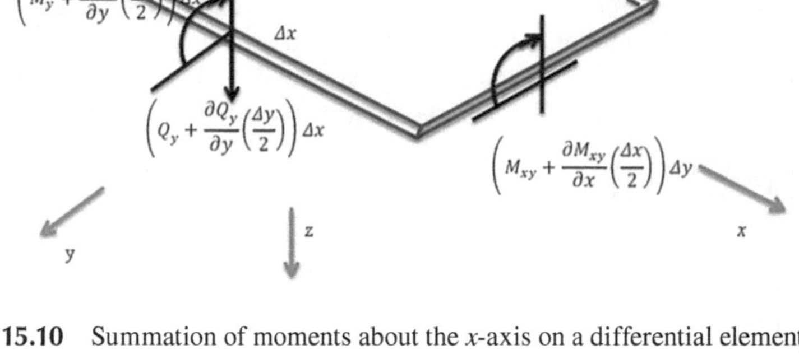

FIGURE 15.10 Summation of moments about the x-axis on a differential element of plate.

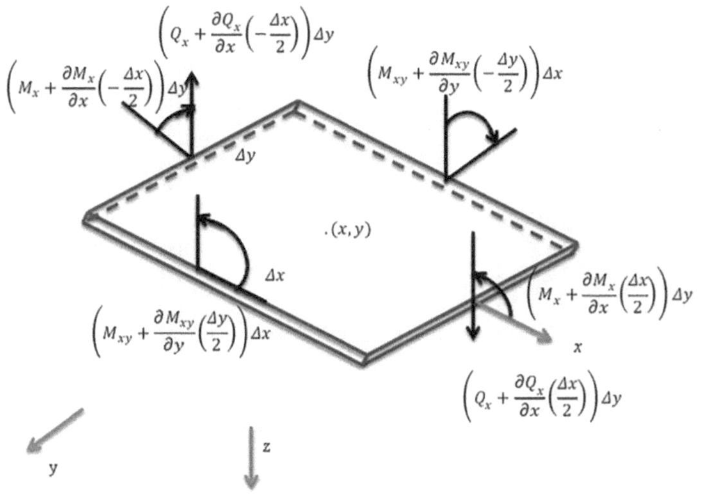

FIGURE 15.11 Summation of moments about the y-axis on a differential element of plate.

resultants M_x and M_{xy}. These moment resultants have been an integral part of classical lamination theory, and thus, we are familiar with them. As we can see, if it were not for Q_x and Q_y, there would be nothing to react to the applied load $q(x, y)$. Plate theory based on the plane-stress assumption indeed presents a dilemma; however, the theory is quite accurate. If we sum the moments about the y-axis (Figure 15.11), an equation involving Q_x is derived:

$$Q_x = \frac{\partial M_x}{\partial x} + \frac{\partial M_{xy}}{\partial y} \tag{15.12}$$

Finally, moments must be summed about the z-axis; Figure 15.12 illustrates a plan-form view of the important force resultants. Summing moments about point (x, y) results in:

$$-\left(N_{xy} + \frac{\partial N_{xy}}{\partial y}\left(-\frac{\Delta y}{2}\right)\right)\Delta x\left(\frac{\Delta y}{2}\right) - \left(N_{xy} + \frac{\partial N_{xy}}{\partial y}\left(\frac{\Delta y}{2}\right)\right)\Delta x\left(\frac{\Delta y}{2}\right)$$
$$+\left(N_{xy} + \frac{\partial N_{xy}}{\partial x}\left(-\frac{\Delta x}{2}\right)\right)\Delta y\left(\frac{\Delta x}{2}\right) + \left(N_{xy} + \frac{\partial N_{xy}}{\partial x}\left(\frac{\Delta x}{2}\right)\right)\Delta y\left(\frac{\Delta x}{2}\right) = 0 \tag{15.13}$$

Completing the algebra on the left-hand side leads to satisfaction of the equation identically. No new information is derived from this equation.

Generally, the equations defining Q_x and Q_y in terms of the moments, Eqs. (15.11) and (15.12), are used to eliminate these variables from the problem. If this is done, Eq. (15.9) becomes:

$$\frac{\partial^2 M_x}{\partial x^2} + 2\frac{\partial^2 M_{xy}}{\partial x \partial y} + \frac{\partial^2 M_y}{\partial y^2} + q = 0 \tag{15.14}$$

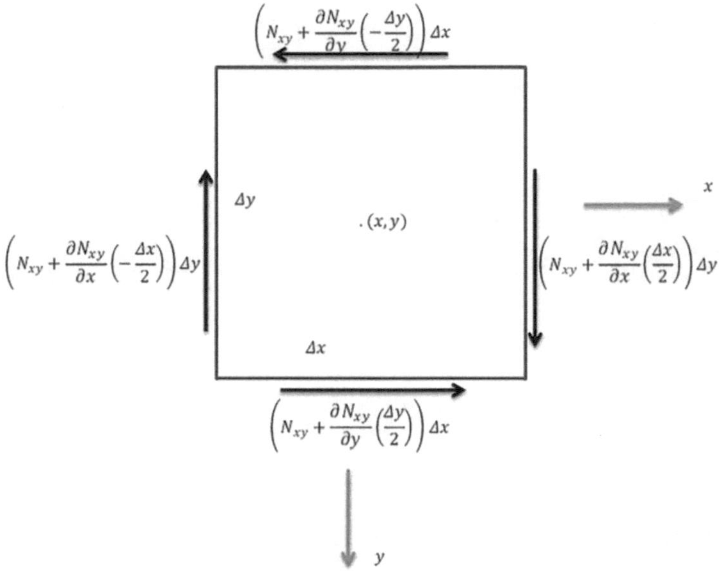

FIGURE 15.12 Summation of moments about the z-axis on a differential element of plate.

Thus, the equilibrium equations that govern the response of a laminated plate are:

$$\frac{\partial N_x}{\partial x}+\frac{\partial N_{xy}}{\partial y}=0$$

$$\frac{\partial N_{xy}}{\partial x}+\frac{\partial N_y}{\partial y}=0$$

$$\frac{\partial^2 M_x}{\partial x^2}+2\frac{\partial^2 M_{xy}}{\partial x\partial y}+\frac{\partial^2 M_y}{\partial y^2}+q=0 \tag{15.15}$$

Note that the issue of material properties has not entered the discussion. Material properties do not influence the equilibrium conditions; thus, the above equations are valid for any rectangular plate, independently of whether it is a single layer, a cross-ply laminate, a balanced laminate, an unsymmetric laminate, and so on.

The boundary conditions that must be satisfied along each edge are as follows:
Along the edge $x=+a/2$:

i. Either $N_x = N_x^+$ or u^0 must be specified
ii. Either $N_{xy} = N_{xy}^+$ or v^0 must be specified
iii. Either

$$\frac{\partial M_x}{\partial x}+2\frac{\partial M_{xy}}{\partial y}=Q_x^+ +\frac{\partial M_{xy}^+}{\partial y}$$

or w^0 must be specified (15.16a)

iv. Either $M_x = M_x^+$ or $\dfrac{\partial w^0}{\partial x}$ must be specified

Along the edge $x = -a/2$:

i. Either $N_x = N_x^-$ or u^0 must be specified
ii. Either $N_{xy} = N_{xy}^-$ or v^0 must be specified
iii. Either

$$\frac{\partial M_x}{\partial x} + 2\frac{\partial M_{xy}}{\partial y} = Q_x^- + \frac{\partial M_{xy}^-}{\partial y}$$

$$\text{or } w^0 \text{ must be specified} \tag{15.16b}$$

Either $M_x = M_x^-$ or $\dfrac{\partial w^0}{\partial x}$ must be specified

Along the edge $y = +b/2$:

i. Either $N_y = N_y^+$ or v^0 must be specified
ii. Either $N_{xy} = N_{yx}^+$ or u^0 must be specified
iii. Either

$$\frac{\partial M_y}{\partial y} + 2\frac{\partial M_{xy}}{\partial x} = Q_y^+ + \frac{\partial M_{yx}^+}{\partial x}$$

$$\text{or } w^0 \text{ must be specified} \tag{15.16c}$$

Either $M_y = M_y^+$ or $\dfrac{\partial w^0}{\partial y}$ must be specified

Along the edge $y = -b/2$:

i. Either $N_y = N_y^-$ or v^0 must be specified
ii. Either $N_{xy} = N_{yx}^-$ or u^0 must be specified
iii. Either

$$\frac{\partial M_y}{\partial y} + 2\frac{\partial M_{xy}}{\partial x} = Q_y^- + \frac{\partial M_{yx}^-}{\partial x}$$

$$\text{or } w^0 \text{ must be specified} \tag{15.16d}$$

Either $M_y = M_y^-$ or $\dfrac{\partial w^0}{\partial y}$ must be specified

These boundary conditions have important physical interpretations. At each edge, four conditions must hold. The four conditions are determined by satisfying one option from each of the four pairs of conditions, (i) through (iv). The first two conditions at each edge have rather simple interpretations. For example, along the edge at $x = +a/2$, condition (i) states that at that edge, either the normal force resultant in the plate, N_x, must equal the resultant applied normal to that edge, N_x^+, or the displacement normal to the edge must be given. Condition (ii) states that either the shear force resultant in the plate, N_{xy}, must equal the resultant applied tangent to the edge, N_{xy}^+, or the displacement tangential to that edge must be given. In contrast to the simplicity of the first two conditions, the interpretation of the third condition at each edge is not so obvious; however, it is a well-known and important condition. The condition states that the combination of moment resultant derivatives must equal the sum of the applied shear force resultant and the derivative of the applied twisting moment resultant; otherwise, the vertical displacement of the edge must be specified. In condition (iii), on each plate edge the derivative of the twisting moment resultant is taken with respect to the coordinate parallel with that edge. For example, at $x = +a/2$, the derivative of M_{xy}^+ is taken with respect to y, the coordinate parallel to that edge. Though this complicated boundary condition is a natural result of the variational process and is really nothing more than a result of the mathematical manipulation, the condition is not obvious when using the force equilibrium approach. Its physical interpretation is far-reaching. It basically implies that, for example, at $x = +a/2$, the edge of the plate does not respond to an applied transverse shear force resultant Q_x^+, and an applied twisting moment M_{xy}^+ separately if both are applied. Rather, the plate responds to the combination:

$$Q_x^+ + \frac{\partial M_{xy}^+}{\partial y} \tag{15.17}$$

In fact, because the derivative in the above has the units of force per unit length, the above combination can be considered an applied effective transverse shear force resultant at the edge of the plate at $x = +a/2$; that is:

$$Q_x^+ + \frac{\partial M_{xy}^+}{\partial y} = Q_{x\text{eff}}^+ \tag{15.18}$$

In the same vein, the quantity

$$\frac{\partial M_x}{\partial x} + 2\frac{\partial M_{xy}}{\partial y} \tag{15.19}$$

can also be considered as an effective transverse shear force resultant within the plate. Because the transverse shear force resultant within the plate is defined by Eq. (15.12) as:

$$Q_x = \frac{\partial M_x}{\partial x} + \frac{\partial M_{xy}}{\partial y} \tag{15.20}$$

the expression in Eq. (15.19) can be written:

$$Q_x + \frac{\partial M_{xy}}{\partial y} \tag{15.21}$$

or defining,

$$Q_{x\text{eff}} = Q_x + \frac{\partial M_{xy}}{\partial y} = \frac{\partial M_x}{\partial x} + 2\frac{\partial M_{xy}}{\partial y} \tag{15.22}$$

Boundary condition (iii) for the edge $x = +a/2$ can be stated as:

$$\text{Either } Q_{x\text{eff}} = Q^+_{x\text{eff}} \text{ or } w^0 \text{ must be specified} \tag{15.23}$$

meaning that at the edge either the effective transverse shear force resultant in the plate must be equal to the applied effective transverse shear force resultant, or the displacement w^0 must be specified. If the edge of the plate is free from forces or moments, then:

$$Q^+_x = M^+_{xy} = 0 \tag{15.24}$$

But rather than applying each condition independently, the boundary condition becomes:

$$\frac{\partial M_x}{\partial x} + 2\frac{\partial M_{xy}}{\partial y} = Q^+_x + \frac{\partial M^+_{xy}}{\partial y} \tag{15.25}$$

This is the famous *Kirchhoff free-edge condition.* Equation (15.22) defines what is referred to as the *Kirchhoff shear force resultant,* namely $Q_{x\text{eff}}$.

Condition (iv) is not so complicated. It states that either the bending moment resultant in the plate, M_x, must equal the bending moment applied to that edge, M^+_x, or the rotation (slope) of the edge must be given. For a clamped edge, for example, the slope would be specified to be zero and nothing could be stated regarding the moment. For a simply supported edge, the applied bending moment resultant M^+_x would be taken to be zero and nothing could be stated regarding the slope.

The boundary conditions on the other three edges follow similar interpretations. Within the plate, the effective transverse shear force resultant $Q_{y\text{eff}}$ can be defined as:

$$Q_{y\text{eff}} = Q_y + \frac{\partial M_{xy}}{\partial x} = \frac{\partial M_y}{\partial y} + 2\frac{\partial M_{xy}}{\partial x} \tag{15.26}$$

Likewise, the applied effective transverse shear force resultant at $y = +b/2$, for example, can be defined as:

$$Q^+_{y\text{eff}} = Q^+_y + \frac{\partial M^+_{xy}}{\partial x} \tag{15.27}$$

Thus, boundary condition (iii) at $y = +b/2$ can be alternatively stated as:

$$\text{Either } Q_{y\text{eff}} = Q^+_{y\text{eff}} \text{ or } w^0 \text{ must be specified} \tag{15.28}$$

The important conclusion from examining the boundary conditions is that *pairs* of variables must be considered on each boundary. Each pair always consists of what can be considered complementary response variables. One component of the pair involves a force or moment resultant, and the other component involves a displacement or rotation. In the study of plates, only the boundary conditions that conform to these conditions are legitimate. Imposing a boundary condition not covered by one of the four complementary pairs of variables results in an ill-posed problem. This will lead to erroneous conclusions if pursued, assuming, of course, you could even solve the problem. In addition to conditions along each edge, certain conditions must be specified at each comer. Comer conditions, which result because the circumferential boundary of the plate is not smooth (i.e., it has comers), will not be discussed here. However, each of the comer conditions requires that either a particular twisting moment within the plate equal the applied twisting moment at the comer, or the displacement of the plate comer must be specified. If the displacements of all four edges of the plate are specified, as with simply supported or clamped edges, the comer conditions are automatically satisfied. If two adjacent edges are free, then the comer condition must be satisfied by the twisting moments being zero. It might be added that for a simply supported or clamped edge, boundary condition (iii) is satisfied by specification of the displacement. This is opposed to a free edge, where condition (iii) must be satisfied by the Kirchhoff free-edge condition (i.e., the effective transverse shear force resultant is zero). The required satisfaction of the force portion of boundary condition (iii) and the moment portion of the comer condition poses a larger challenge in finding solutions to the governing equations than enforcing displacement conditions. In practice, fortunately, the edges of a plate are usually simply supported or clamped, so the comer conditions and boundary condition (iii) can easily be satisfied.

15.2 GOVERNING EQUATIONS (IN DISPLACEMENT FORM)

Because the displacements are the basic variables in the problem, all other responses being derivable from the displacements, it is meaningful to express the equilibrium equations and boundary conditions in terms of the displacements. In addition, we usually have more of a physical feel for the displacement response of a structure and so it is useful to write the equations in terms of these quantities. In Eq. (10.67), the stress resultants were expressed in terms of the reference surface strains and curvatures. In Eqs. (9.10–9.11), the reference surface strains and curvatures were defined in terms of the reference surface displacements. Accordingly, the stress resultants can be written in terms of the reference surface displacements as:

$$N_x = A_{11}\frac{\partial u^0}{\partial x} + A_{12}\frac{\partial v^0}{\partial y} + A_{16}\left(\frac{\partial v^0}{\partial x} + \frac{\partial u^0}{\partial y}\right) - B_{11}\left(\frac{\partial^2 w^0}{\partial x^2}\right) - B_{12}\left(\frac{\partial^2 w^0}{\partial y^2}\right) - 2B_{16}\frac{\partial^2 w^0}{\partial x \partial y}$$

$$N_y = A_{12}\frac{\partial u^0}{\partial x} + A_{22}\frac{\partial v^0}{\partial y} + A_{26}\left(\frac{\partial v^0}{\partial x} + \frac{\partial u^0}{\partial y}\right) - B_{12}\left(\frac{\partial^2 w^0}{\partial x^2}\right) - B_{22}\left(\frac{\partial^2 w^0}{\partial y^2}\right) - 2B_{26}\frac{\partial^2 w^0}{\partial x \partial y}$$

$$N_{xy} = A_{16}\frac{\partial u^0}{\partial x} + A_{26}\frac{\partial v^0}{\partial y} + A_{66}\left(\frac{\partial v^0}{\partial x} + \frac{\partial u^0}{\partial y}\right) - B_{16}\left(\frac{\partial^2 w^0}{\partial x^2}\right) - B_{26}\left(\frac{\partial^2 w^0}{\partial y^2}\right) - 2B_{66}\frac{\partial^2 w^0}{\partial x \partial y}$$

$$M_x = B_{11}\frac{\partial u^0}{\partial x} + B_{12}\frac{\partial v^0}{\partial y} + B_{16}\left(\frac{\partial v^0}{\partial x} + \frac{\partial u^0}{\partial y}\right) - D_{11}\left(\frac{\partial^2 w^0}{\partial x^2}\right) - D_{12}\left(\frac{\partial^2 w^0}{\partial y^2}\right) - 2D_{16}\frac{\partial^2 w^0}{\partial x \partial y}$$

$$M_y = B_{12}\frac{\partial u^0}{\partial x} + B_{22}\frac{\partial v^0}{\partial y} + B_{26}\left(\frac{\partial v^0}{\partial x} + \frac{\partial u^0}{\partial y}\right) - D_{12}\left(\frac{\partial^2 w^0}{\partial x^2}\right) - D_{22}\left(\frac{\partial^2 w^0}{\partial y^2}\right) - 2D_{26}\frac{\partial^2 w^0}{\partial x \partial y}$$

$$M_{xy} = B_{16}\frac{\partial u^0}{\partial x} + B_{26}\frac{\partial v^0}{\partial y} + B_{66}\left(\frac{\partial v^0}{\partial x} + \frac{\partial u^0}{\partial y}\right) - D_{16}\left(\frac{\partial^2 w^0}{\partial x^2}\right) - D_{26}\left(\frac{\partial^2 w^0}{\partial y^2}\right) - 2D_{66}\frac{\partial^2 w^0}{\partial x \partial y} \tag{15.29}$$

Substituting these expressions into the equilibrium equations (Eq. (15.15)) leads to three partial differential equations that govern the displacement response of a fiber-reinforced laminated plate, namely:

$$A_{11}\frac{\partial^2 u^0}{\partial x^2} + 2A_{16}\frac{\partial^2 u^0}{\partial x \partial y} + A_{66}\frac{\partial^2 u^0}{\partial y^2} + A_{16}\frac{\partial^2 v^0}{\partial x^2} + (A_{12} + A_{66})\frac{\partial^2 v^0}{\partial x \partial y} + A_{26}\frac{\partial^2 v^0}{\partial y^2} - B_{11}\frac{\partial^3 w^0}{\partial x^3} - 3B_{16}\frac{\partial^3 w^0}{\partial x^2 \partial y} - (B_{12} + 2B_{66})\frac{\partial^3 w^0}{\partial x \partial y^2} - B_{26}\frac{\partial^3 w^0}{\partial y^3} = 0 \tag{15.30a}$$

$$A_{16}\frac{\partial^2 u^0}{\partial x^2} + (A_{12} + A_{66})\frac{\partial^2 u^0}{\partial x \partial y} + A_{26}\frac{\partial^2 u^0}{\partial y^2} + A_{66}\frac{\partial^2 v^0}{\partial x^2} + 2A_{26}\frac{\partial^2 v^0}{\partial x \partial y} + A_{22}\frac{\partial^2 v^0}{\partial y^2} - B_{16}\frac{\partial^3 w^0}{\partial x^3} - (B_{12} + 2B_{66})\frac{\partial^3 w^0}{\partial x^2 \partial y} - 3B_{26}\frac{\partial^3 w^0}{\partial x \partial y^2} - B_{22}\frac{\partial^3 w^0}{\partial y^3} = 0 \tag{15.30b}$$

$$D_{11}\frac{\partial^4 w^0}{\partial x^4} + 4D_{16}\frac{\partial^4 w^0}{\partial x^3 \partial y} + 2(D_{12} + 2D_{66})\frac{\partial^4 w^0}{\partial x^2 \partial y^2} + 4D_{26}\frac{\partial^4 w^0}{\partial x \partial y^3} + D_{22}\frac{\partial^4 w^0}{\partial y^4} - B_{11}\frac{\partial^3 u^0}{\partial x^3} - 3B_{16}\frac{\partial^3 u^0}{\partial x^2 \partial y} - (B_{12} + 2B_{66})\frac{\partial^3 u^0}{\partial x \partial y^2} - B_{26}\frac{\partial^3 u^0}{\partial y^3} - B_{16}\frac{\partial^3 v^0}{\partial x^3} - (B_{12} + 2B_{66})\frac{\partial^3 v^0}{\partial x^2 \partial y} - 3B_{26}\frac{\partial^3 v^0}{\partial x \partial y^2} - B_{22}\frac{\partial^3 v^0}{\partial y^3} = q \tag{15.30c}$$

The four boundary conditions along each of the four edges can also be expressed in terms of the displacements. Substituting for the stress resultants N_x, N_{xy}, and so on, in terms of the displacements in Eq. (15.16), we find that the boundary conditions become:

Along the edge $x = +a/2$:

i. Either

$$\left\{A_{11}\frac{\partial u^0}{\partial x} + A_{12}\frac{\partial v^0}{\partial y} + A_{16}\left(\frac{\partial v^0}{\partial x} + \frac{\partial u^0}{\partial y}\right) - B_{11}\left(\frac{\partial^2 w^0}{\partial x^2}\right)\right.$$

$$\left.-B_{12}\left(\frac{\partial^2 w^0}{\partial y^2}\right) - 2B_{16}\frac{\partial^2 w^0}{\partial x \partial y}\right\} = N_x^+$$

or u^0 must be specified.

ii. Either

$$\left\{A_{16}\frac{\partial u^0}{\partial x} + A_{26}\frac{\partial v^0}{\partial y} + A_{66}\left(\frac{\partial v^0}{\partial x} + \frac{\partial u^0}{\partial y}\right) - B_{16}\left(\frac{\partial^2 w^0}{\partial x^2}\right)\right.$$

$$\left.- B_{26}\left(\frac{\partial^2 w^0}{\partial y^2}\right) - 2B_{66}\frac{\partial^2 w^0}{\partial x \partial y}\right\} = N_{xy}^+$$

or v^0 must be specified.

iii. Either

$$B_{11}\frac{\partial^2 u^0}{\partial x^2} + 2B_{66}\frac{\partial^2 u^0}{\partial y^2} + 3B_{16}\frac{\partial^2 u^0}{\partial x \partial y} + B_{16}\frac{\partial^2 v^0}{\partial x^2} + 2B_{26}\frac{\partial^2 v^0}{\partial y^2} + (B_{12} + 2B_{66})\frac{\partial^2 v^0}{\partial x \partial y}$$

$$- D_{11}\frac{\partial^3 w^0}{\partial x^3} - 2D_{26}\frac{\partial^3 w^0}{\partial y^3} - (D_{12} + 4D_{66})\frac{\partial^3 w^0}{\partial x \partial y^2} - 4D_{16}\frac{\partial^3 w^0}{\partial x^2 \partial y} = Q_x^+ + \frac{\partial M_{xy}^+}{\partial y}$$

or w^0 must be specified.

iv. Either

$$\left\{B_{11}\frac{\partial u^0}{\partial x} + B_{12}\frac{\partial v^0}{\partial y} + B_{16}\left(\frac{\partial v^0}{\partial x} + \frac{\partial u^0}{\partial y}\right) - D_{11}\left(\frac{\partial^2 w^0}{\partial x^2}\right)\right.$$

$$\left.- D_{12}\left(\frac{\partial^2 w^0}{\partial y^2}\right) - 2D_{16}\frac{\partial^2 w^0}{\partial x \partial y}\right\} = M_x^+$$

$$\text{or } \frac{\partial w^0}{\partial x} \text{ must be specified.} \tag{15.31a}$$

Along the edge $x = -a/2$:

i. Either

$$\left\{A_{11}\frac{\partial u^0}{\partial x}+A_{12}\frac{\partial v^0}{\partial y}+A_{16}\left(\frac{\partial v^0}{\partial x}+\frac{\partial u^0}{\partial y}\right)-B_{11}\left(\frac{\partial^2 w^0}{\partial x^2}\right)\right.$$

$$\left.-B_{12}\left(\frac{\partial^2 w^0}{\partial y^2}\right)-2B_{16}\frac{\partial^2 w^0}{\partial x\partial y}\right\}=N_x^-$$

or u^0 must be specified.

ii. Either

$$\left\{A_{16}\frac{\partial u^0}{\partial x}+A_{26}\frac{\partial v^0}{\partial y}+A_{66}\left(\frac{\partial v^0}{\partial x}+\frac{\partial u^0}{\partial y}\right)-B_{16}\left(\frac{\partial^2 w^0}{\partial x^2}\right)\right.$$

$$\left.-B_{26}\left(\frac{\partial^2 w^0}{\partial y^2}\right)-2B_{66}\frac{\partial^2 w^0}{\partial x\partial y}\right\}=N_{xy}^-$$

or v^0 must be specified.

iii. Either

$$B_{11}\frac{\partial^2 u^0}{\partial x^2}+2B_{66}\frac{\partial^2 u^0}{\partial y^2}+3B_{16}\frac{\partial^2 u^0}{\partial x\partial y}+B_{16}\frac{\partial^2 v^0}{\partial x^2}+2B_{26}\frac{\partial^2 v^0}{\partial y^2}+(B_{12}+2B_{66})\frac{\partial^2 v^0}{\partial x\partial y}$$

$$-D_{11}\frac{\partial^3 w^0}{\partial x^3}-2D_{26}\frac{\partial^3 w^0}{\partial y^3}-(D_{12}+4D_{66})\frac{\partial^3 w^0}{\partial x\partial y^2}-4D_{16}\frac{\partial^3 w^0}{\partial x^2\partial y}=Q_x^-+\frac{\partial M_{xy}^-}{\partial y}$$

or w^0 must be specified.

iv. Either

$$\left\{B_{11}\frac{\partial u^0}{\partial x}+B_{12}\frac{\partial v^0}{\partial y}+B_{16}\left(\frac{\partial v^0}{\partial x}+\frac{\partial u^0}{\partial y}\right)-D_{11}\left(\frac{\partial^2 w^0}{\partial x^2}\right)\right.$$

$$\left.-D_{12}\left(\frac{\partial^2 w^0}{\partial y^2}\right)-2D_{16}\frac{\partial^2 w^0}{\partial x\partial y}\right\}=M_x^-$$

$$\text{or } \frac{\partial w^0}{\partial x} \text{ must be specified.} \tag{15.31b}$$

Along the edge $y = +b/2$:

i. Either

$$\left\{A_{12}\frac{\partial u^0}{\partial x}+A_{22}\frac{\partial v^0}{\partial y}+A_{26}\left(\frac{\partial v^0}{\partial x}+\frac{\partial u^0}{\partial y}\right)-B_{12}\left(\frac{\partial^2 w^0}{\partial x^2}\right)\right.$$

$$\left.-B_{22}\left(\frac{\partial^2 w^0}{\partial y^2}\right)-2B_{26}\frac{\partial^2 w^0}{\partial x\partial y}\right\}=N_y^+$$

or v^0 must be specified.

ii. Either

$$\left\{A_{16}\frac{\partial u^0}{\partial x}+A_{26}\frac{\partial v^0}{\partial y}+A_{66}\left(\frac{\partial v^0}{\partial x}+\frac{\partial u^0}{\partial y}\right)-B_{16}\left(\frac{\partial^2 w^0}{\partial x^2}\right)\right.$$

$$\left.-B_{26}\left(\frac{\partial^2 w^0}{\partial y^2}\right)-2B_{66}\frac{\partial^2 w^0}{\partial x\partial y}\right\}=N_{yx}^{+}$$

or u^0 must be specified.

iii. Either

$$2B_{16}\frac{\partial^2 u^0}{\partial x^2}+B_{26}\frac{\partial^2 u^0}{\partial y^2}+(B_{12}+2B_{66})\frac{\partial^2 u^0}{\partial x\partial y}+2B_{66}\frac{\partial^2 v^0}{\partial x^2}+B_{22}\frac{\partial^2 v^0}{\partial y^2}+3B_{26}\frac{\partial^2 v^0}{\partial x\partial y}$$

$$-2D_{16}\frac{\partial^3 w^0}{\partial x^3}-D_{22}\frac{\partial^3 w^0}{\partial y^3}-(D_{12}+4D_{66})\frac{\partial^3 w^0}{\partial x^2\partial y}-4D_{26}\frac{\partial^3 w^0}{\partial x\partial y^2}=Q_y^{+}+\frac{\partial M_{yx}^{+}}{\partial x}$$

or w^0 must be specified.

iv. Either

$$\left\{B_{12}\frac{\partial u^0}{\partial x}+B_{22}\frac{\partial v^0}{\partial y}+B_{26}\left(\frac{\partial v^0}{\partial x}+\frac{\partial u^0}{\partial y}\right)-D_{12}\left(\frac{\partial^2 w^0}{\partial x^2}\right)\right.$$

$$\left.-D_{22}\left(\frac{\partial^2 w^0}{\partial y^2}\right)-2D_{26}\frac{\partial^2 w^0}{\partial x\partial y}\right\}=M_y^{+}$$

$$\text{or } \frac{\partial w^0}{\partial y} \text{ must be specified.} \tag{15.31c}$$

Along the edge $y = -b/2$:

i. Either

$$\left\{A_{12}\frac{\partial u^0}{\partial x}+A_{22}\frac{\partial v^0}{\partial y}+A_{26}\left(\frac{\partial v^0}{\partial x}+\frac{\partial u^0}{\partial y}\right)-B_{12}\left(\frac{\partial^2 w^0}{\partial x^2}\right)\right.$$

$$\left.-B_{22}\left(\frac{\partial^2 w^0}{\partial y^2}\right)-2B_{26}\frac{\partial^2 w^0}{\partial x\partial y}\right\}=N_y^{-}$$

or v^0 must be specified.

ii. Either

$$\left\{A_{16}\frac{\partial u^0}{\partial x}+A_{26}\frac{\partial v^0}{\partial y}+A_{66}\left(\frac{\partial v^0}{\partial x}+\frac{\partial u^0}{\partial y}\right)-B_{16}\left(\frac{\partial^2 w^0}{\partial x^2}\right)\right.$$

$$\left.-B_{26}\left(\frac{\partial^2 w^0}{\partial y^2}\right)-2B_{66}\frac{\partial^2 w^0}{\partial x\partial y}\right\}=N_{yx}^{-}$$

or u^0 must be specified.

iii. Either

$$2B_{16}\frac{\partial^2 u^0}{\partial x^2}+B_{26}\frac{\partial^2 u^0}{\partial y^2}+(B_{12}+2B_{66})\frac{\partial^2 u^0}{\partial x\partial y}+$$

$$2B_{66}\frac{\partial^2 v^0}{\partial x^2}+B_{22}\frac{\partial^2 v^0}{\partial y^2}+3B_{26}\frac{\partial^2 v^0}{\partial x\partial y}-2D_{16}\frac{\partial^3 w^0}{\partial x^3}-D_{22}\frac{\partial^3 w^0}{\partial y^3}$$

$$-(D_{12}+4D_{66})\frac{\partial^3 w^0}{\partial x^2\partial y}-4D_{26}\frac{\partial^3 w^0}{\partial x\partial y^2}=Q_y^-+\frac{\partial M_{yx}^-}{\partial x}$$

or w^0 must be specified.

iv. Either

$$\left\{B_{12}\frac{\partial u^0}{\partial x}+B_{22}\frac{\partial v^0}{\partial y}+B_{26}\left(\frac{\partial v^0}{\partial x}+\frac{\partial u^0}{\partial y}\right)-D_{12}\left(\frac{\partial^2 w^0}{\partial x^2}\right)\right.$$

$$\left.-D_{22}\left(\frac{\partial^2 w^0}{\partial y^2}\right)-2D_{26}\frac{\partial^2 w^0}{\partial x\partial y}\right\}=M_y^-$$

$$\text{or } \frac{\partial w^0}{\partial y} \text{ must be specified} \qquad (15.31\text{d})$$

The rather awesome display of algebra in Eqs. (13.30) and (13.31) is the proper formulation of the governing equilibrium equations and boundary conditions for the problem of determining the displacement response of a rectangular laminated fiber-reinforced plate subjected to loads on the plate edges, and/or normal loads acting on the top and/or bottom surfaces. All laminated plate problems that can be solved within the context of the assumptions of classical lamination theory are contained in the above formulation. Conversely, only within the above formulation can one solve for the response of a laminated plate that obeys the assumptions of classical lamination theory. In the next section, we shall discuss simplifications to the above equations which result from considering special plates (e.g., symmetrically laminated plates). Though both the governing differential equations and the boundary conditions will simplify, the basic concepts of having three governing differential equations for the three components of displacement, and four boundary conditions along each of the four edges, still are valid. In the sections following the simplifications, we will solve specific problems and obtain numerical results. These problems will illustrate the steps necessary for solving the governing equations and enforcing boundary conditions and will illustrate some important characteristics of the response of fiber-reinforced plates.

15.3 SIMPLIFICATION OF GOVERNING EQUATIONS

As we showed in Chapter 10, for specific classes of laminates some of the elastic coupling coefficients in the A, B, and D matrices vanish. The vanishing of these elastic coefficients can considerably simplify the differential equations and boundary conditions governing plate response. Some of the simplifications are minor, but some are quite significant.

15.3.1 Symmetric Laminates

By far, the most dramatic simplification to the governing equations and boundary conditions for a plate occurs when the plate is symmetrically laminated. For this situation, all B_{ij} terms are zero, and the equilibrium equations become:

$$A_{11}\frac{\partial^2 u^0}{\partial x^2}+2A_{16}\frac{\partial^2 u^0}{\partial x\partial y}+A_{66}\frac{\partial^2 u^0}{\partial y^2}+A_{16}\frac{\partial^2 v^0}{\partial x^2}+(A_{12}+A_{66})\frac{\partial^2 v^0}{\partial x\partial y}+A_{26}\frac{\partial^2 v^0}{\partial y^2}=0 \tag{15.32a}$$

$$A_{16}\frac{\partial^2 u^0}{\partial x^2}+(A_{12}+A_{66})\frac{\partial^2 u^0}{\partial x\partial y}+A_{26}\frac{\partial^2 u^0}{\partial y^2}+A_{66}\frac{\partial^2 v^0}{\partial x^2}+2A_{26}\frac{\partial^2 v^0}{\partial x\partial y}+A_{22}\frac{\partial^2 v^0}{\partial y^2}=0 \tag{15.32b}$$

$$D_{11}\frac{\partial^4 w^0}{\partial x^4}+4D_{16}\frac{\partial^4 w^0}{\partial x^3\partial y}+2(D_{12}+2D_{66})\frac{\partial^4 w^0}{\partial x^2\partial y^2}+4D_{26}\frac{\partial^4 w^0}{\partial x\partial y^3}+D_{22}\frac{\partial^4 w^0}{\partial y^4}=q \tag{15.32c}$$

The boundary conditions reduce to:

Along the edge $x = +a/2$:

i. Either

$$\left\{A_{11}\frac{\partial u^0}{\partial x}+A_{12}\frac{\partial v^0}{\partial y}+A_{16}\left(\frac{\partial v^0}{\partial x}+\frac{\partial u^0}{\partial y}\right)\right\}=N_x^+$$

or u^0 must be specified.

ii. Either

$$\left\{A_{16}\frac{\partial u^0}{\partial x}+A_{26}\frac{\partial v^0}{\partial y}+A_{66}\left(\frac{\partial v^0}{\partial x}+\frac{\partial u^0}{\partial y}\right)\right\}=N_{xy}^+$$

or v^0 must be specified.

iii. Either

$$-D_{11}\frac{\partial^3 w^0}{\partial x^3}-2D_{26}\frac{\partial^3 w^0}{\partial y^3}-(D_{12}+4D_{66})\frac{\partial^3 w^0}{\partial x\partial y^2}-4D_{16}\frac{\partial^3 w^0}{\partial x^2\partial y}=Q_x^+ +\frac{\partial M_{xy}^+}{\partial y}$$

or w^0 must be specified.

iv. Either

$$\left\{-D_{11}\left(\frac{\partial^2 w^0}{\partial x^2}\right)-D_{12}\left(\frac{\partial^2 w^0}{\partial y^2}\right)-2D_{16}\frac{\partial^2 w^0}{\partial x\partial y}\right\}=M_x^+$$

$$\text{or } \frac{\partial w^0}{\partial x} \text{ must be specified.} \tag{15.33a}$$

Along the edge $x = -a/2$:

i. Either

$$\left\{A_{11}\frac{\partial u^0}{\partial x} + A_{12}\frac{\partial v^0}{\partial y} + A_{16}\left(\frac{\partial v^0}{\partial x} + \frac{\partial u^0}{\partial y}\right)\right\} = N_x^-$$

or u^0 must be specified.

ii. Either

$$\left\{A_{16}\frac{\partial u^0}{\partial x} + A_{26}\frac{\partial v^0}{\partial y} + A_{66}\left(\frac{\partial v^0}{\partial x} + \frac{\partial u^0}{\partial y}\right)\right\} = N_{xy}^-$$

or v^0 must be specified.

iii. Either

$$-D_{11}\frac{\partial^3 w^0}{\partial x^3} - 2D_{26}\frac{\partial^3 w^0}{\partial y^3} - (D_{12} + 4D_{66})\frac{\partial^3 w^0}{\partial x \partial y^2} - 4D_{16}\frac{\partial^3 w^0}{\partial x^2 \partial y} = Q_x^- + \frac{\partial M_{xy}^-}{\partial y}$$

or w^0 must be specified.

iv. Either

$$\left\{-D_{11}\left(\frac{\partial^2 w^0}{\partial x^2}\right) - D_{12}\left(\frac{\partial^2 w^0}{\partial y^2}\right) - 2D_{16}\frac{\partial^2 w^0}{\partial x \partial y}\right\} = M_x^-$$

$$\text{or } \frac{\partial w^0}{\partial x} \text{ must be specified.} \tag{15.33b}$$

Along the edge $y = +b/2$:

i. Either

$$\left\{A_{12}\frac{\partial u^0}{\partial x} + A_{22}\frac{\partial v^0}{\partial y} + A_{26}\left(\frac{\partial v^0}{\partial x} + \frac{\partial u^0}{\partial y}\right)\right\} = N_y^+$$

or v^0 must be specified.

ii. Either

$$\left\{A_{16}\frac{\partial u^0}{\partial x} + A_{26}\frac{\partial v^0}{\partial y} + A_{66}\left(\frac{\partial v^0}{\partial x} + \frac{\partial u^0}{\partial y}\right)\right\} = N_{yx}^+$$

or u^0 must be specified.

iii. Either

$$-2D_{16}\frac{\partial^3 w^0}{\partial x^3}-D_{22}\frac{\partial^3 w^0}{\partial y^3}-(D_{12}+4D_{66})\frac{\partial^3 w^0}{\partial x^2\partial y}-4D_{26}\frac{\partial^3 w^0}{\partial x\partial y^2}=Q_y^+ +\frac{\partial M_{yx}^+}{\partial x}$$

or w^0 must be specified.

iv. Either

$$\left\{-D_{12}\left(\frac{\partial^2 w^0}{\partial x^2}\right)-D_{22}\left(\frac{\partial^2 w^0}{\partial y^2}\right)-2D_{26}\frac{\partial^2 w^0}{\partial x\partial y}\right\}=M_y^+$$

$$\text{or } \frac{\partial w^0}{\partial y} \text{ must be specified.} \tag{15.33c}$$

Along the edge $y = -b/2$:

i. Either

$$\left\{A_{12}\frac{\partial u^0}{\partial x}+A_{22}\frac{\partial v^0}{\partial y}+A_{26}\left(\frac{\partial v^0}{\partial x}+\frac{\partial u^0}{\partial y}\right)\right\}=N_y^-$$

or v^0 must be specified.

ii. Either

$$\left\{A_{16}\frac{\partial u^0}{\partial x}+A_{26}\frac{\partial v^0}{\partial y}+A_{66}\left(\frac{\partial v^0}{\partial x}+\frac{\partial u^0}{\partial y}\right)\right\}=N_{yx}^-$$

or u^0 must be specified.

iii. Either

$$-2D_{16}\frac{\partial^3 w^0}{\partial x^3}-D_{22}\frac{\partial^3 w^0}{\partial y^3}-(D_{12}+4D_{66})\frac{\partial^3 w^0}{\partial x^2\partial y}-4D_{26}\frac{\partial^3 w^0}{\partial x\partial y^2}=Q_y^- +\frac{\partial M_{yx}^-}{\partial x}$$

or w^0 must be specified.

iv. Either

$$\left\{-D_{12}\left(\frac{\partial^2 w^0}{\partial x^2}\right)-D_{22}\left(\frac{\partial^2 w^0}{\partial y^2}\right)-2D_{26}\frac{\partial^2 w^0}{\partial x\partial y}\right\}=M_y^-$$

$$\text{or } \frac{\partial w^0}{\partial y} \text{ must be specified.} \tag{15.33d}$$

As a result, for a symmetrically laminated plate, the determination of the out-of-plane response, and the strains and stresses that accompany it, is totally independent of the determination of the in-plane response. This is a major reduction in the level of effort if the out-of-plane response is the primary concern.

15.3.2 Symmetric Balanced Laminates

If, in addition to being symmetric, the laminate is also balanced, then both A_{16} and A_{26} in addition to B_{ij} are zero and the equations simplify further to become:

$$A_{11}\frac{\partial^2 u^0}{\partial x^2}+A_{66}\frac{\partial^2 u^0}{\partial y^2}+(A_{12}+A_{66})\frac{\partial^2 v^0}{\partial x\partial y}=0 \tag{15.34a}$$

$$(A_{12}+A_{66})\frac{\partial^2 u^0}{\partial x\partial y}+A_{66}\frac{\partial^2 v^0}{\partial x^2}+A_{22}\frac{\partial^2 v^0}{\partial y^2}=0 \tag{15.34b}$$

$$D_{11}\frac{\partial^4 w^0}{\partial x^4}+4D_{16}\frac{\partial^4 w^0}{\partial x^3\partial y}+2(D_{12}+2D_{66})\frac{\partial^4 w^0}{\partial x^2\partial y^2}+4D_{26}\frac{\partial^4 w^0}{\partial x\partial y^3}+D_{22}\frac{\partial^4 w^0}{\partial y^4}=q \tag{15.34c}$$

The boundary conditions reduce to:

Along the edge $x = +a/2$:

i. Either

$$\left\{A_{11}\frac{\partial u^0}{\partial x}+A_{12}\frac{\partial v^0}{\partial y}\right\}=N_x^+$$

or u^0 must be specified.

ii. Either

$$\left\{A_{66}\left(\frac{\partial v^0}{\partial x}+\frac{\partial u^0}{\partial y}\right)\right\}=N_{xy}^+$$

or v^0 must be specified.

iii. Either

$$-D_{11}\frac{\partial^3 w^0}{\partial x^3}-2D_{26}\frac{\partial^3 w^0}{\partial y^3}-(D_{12}+4D_{66})\frac{\partial^3 w^0}{\partial x\partial y^2}-4D_{16}\frac{\partial^3 w^0}{\partial x^2\partial y}=Q_x^+ +\frac{\partial M_{xy}^+}{\partial y}$$

or w^0 must be specified.

iv. Either

$$\left\{-D_{11}\left(\frac{\partial^2 w^0}{\partial x^2}\right)-D_{12}\left(\frac{\partial^2 w^0}{\partial y^2}\right)-2D_{16}\frac{\partial^2 w^0}{\partial x\partial y}\right\}=M_x^+$$

$$\text{or }\frac{\partial w^0}{\partial x}\text{ must be specified.} \tag{15.35a}$$

Along the edge $x = -a/2$:

i. Either

$$\left\{A_{11}\frac{\partial u^0}{\partial x}+A_{12}\frac{\partial v^0}{\partial y}\right\}=N_x^-$$

or u^0 must be specified.

ii. Either

$$\left\{A_{66}\left(\frac{\partial v^0}{\partial x}+\frac{\partial u^0}{\partial y}\right)\right\}=N_{xy}^-$$

or v^0 must be specified.

iii. Either

$$-D_{11}\frac{\partial^3 w^0}{\partial x^3}-2D_{26}\frac{\partial^3 w^0}{\partial y^3}-(D_{12}+4D_{66})\frac{\partial^3 w^0}{\partial x\partial y^2}-4D_{16}\frac{\partial^3 w^0}{\partial x^2\partial y}=Q_x^-+\frac{\partial M_{xy}^-}{\partial y}$$

or w^0 must be specified.

iv. Either

$$\left\{-D_{11}\left(\frac{\partial^2 w^0}{\partial x^2}\right)-D_{12}\left(\frac{\partial^2 w^0}{\partial y^2}\right)-2D_{16}\frac{\partial^2 w^0}{\partial x\partial y}\right\}=M_x^-$$

$$\text{or } \frac{\partial w^0}{\partial x} \text{ must be specified.} \tag{15.35b}$$

Along the edge $y = +b/2$:

i. Either

$$\left\{A_{12}\frac{\partial u^0}{\partial x}+A_{22}\frac{\partial v^0}{\partial y}\right\}=N_y^+$$

or v^0 must be specified.

ii. Either

$$\left\{A_{66}\left(\frac{\partial v^0}{\partial x}+\frac{\partial u^0}{\partial y}\right)\right\}=N_{yx}^+$$

or u^0 must be specified.

iii. Either

$$-2D_{16}\frac{\partial^3 w^0}{\partial x^3}-D_{22}\frac{\partial^3 w^0}{\partial y^3}-(D_{12}+4D_{66})\frac{\partial^3 w^0}{\partial x^2\partial y}-4D_{26}\frac{\partial^3 w^0}{\partial x\partial y^2}=Q_y^++\frac{\partial M_{yx}^+}{\partial x}$$

or w^0 must be specified.

iv. Either

$$\left\{-D_{12}\left(\frac{\partial^2 w^0}{\partial x^2}\right)-D_{22}\left(\frac{\partial^2 w^0}{\partial y^2}\right)-2D_{26}\frac{\partial^2 w^0}{\partial x\partial y}\right\}=M_y^+$$

$$\text{or } \frac{\partial w^0}{\partial y} \text{ must be specified.} \tag{15.35c}$$

Along the edge $y = -b/2$:

i. Either

$$\left\{ A_{12}\frac{\partial u^0}{\partial x} + A_{22}\frac{\partial v^0}{\partial y} \right\} = N_y^-$$

or v^0 must be specified.

ii. Either

$$\left\{ A_{66}\left(\frac{\partial v^0}{\partial x} + \frac{\partial u^0}{\partial y}\right) \right\} = N_{yx}^-$$

or u^0 must be specified.

iii. Either

$$-2D_{16}\frac{\partial^3 w^0}{\partial x^3} - D_{22}\frac{\partial^3 w^0}{\partial y^3} - (D_{12} + 4D_{66})\frac{\partial^3 w^0}{\partial x^2 \partial y} - 4D_{26}\frac{\partial^3 w^0}{\partial x \partial y^2} = Q_y^- + \frac{\partial M_{yx}^-}{\partial x}$$

or w^0 must be specified.

iv. Either

$$\left\{ -D_{12}\left(\frac{\partial^2 w^0}{\partial x^2}\right) - D_{22}\left(\frac{\partial^2 w^0}{\partial y^2}\right) - 2D_{26}\frac{\partial^2 w^0}{\partial x \partial y} \right\} = M_y^-$$

$$\text{or } \frac{\partial w^0}{\partial y} \text{ must be specified.} \tag{15.35d}$$

15.3.3 Symmetric Cross-Ply Laminates

Finally, if the plate is a symmetric cross-ply lamination, then in addition to having zero values for A_{16}, A_{26}, and B_{ij}, D_{16} and D_{26} are also zero, and the equilibrium equations reduce to:

$$A_{11}\frac{\partial^2 u^0}{\partial x^2} + A_{66}\frac{\partial^2 u^0}{\partial y^2} + (A_{12} + A_{66})\frac{\partial^2 v^0}{\partial x \partial y} = 0 \tag{15.36a}$$

$$(A_{12} + A_{66})\frac{\partial^2 u^0}{\partial x \partial y} + A_{66}\frac{\partial^2 v^0}{\partial x^2} + A_{22}\frac{\partial^2 v^0}{\partial y^2} = 0 \tag{15.36b}$$

$$D_{11}\frac{\partial^4 w^0}{\partial x^4} + 2(D_{12} + 2D_{66})\frac{\partial^4 w^0}{\partial x^2 \partial y^2} + D_{22}\frac{\partial^4 w^0}{\partial y^4} = q \tag{15.36c}$$

The boundary conditions reduce to:
Along the edge $x = +a/2$:

i. Either

$$\left\{ A_{11}\frac{\partial u^0}{\partial x} + A_{12}\frac{\partial v^0}{\partial y} \right\} = N_x^+$$

or u^0 must be specified.

ii. Either

$$\left\{A_{66}\left(\frac{\partial v^0}{\partial x}+\frac{\partial u^0}{\partial y}\right)\right\}=N_{xy}^{+}$$

or v^0 must be specified.

iii. Either

$$-D_{11}\frac{\partial^3 w^0}{\partial x^3}-2D_{26}\frac{\partial^3 w^0}{\partial y^3}-(D_{12}+4D_{66})\frac{\partial^3 w^0}{\partial x\partial y^2}=Q_x^{+}+\frac{\partial M_{xy}^{+}}{\partial y}$$

or w^0 must be specified.

iv. Either

$$\left\{-D_{11}\left(\frac{\partial^2 w^0}{\partial x^2}\right)-D_{12}\left(\frac{\partial^2 w^0}{\partial y^2}\right)\right\}=M_x^{+}$$

$$\text{or } \frac{\partial w^0}{\partial x} \text{ must be specified.} \tag{15.37a}$$

Along the edge $x=-a/2$:

i. Either

$$\left\{A_{11}\frac{\partial u^0}{\partial x}+A_{12}\frac{\partial v^0}{\partial y}\right\}=N_x^{-}$$

or u^0 must be specified.

ii. Either

$$\left\{A_{66}\left(\frac{\partial v^0}{\partial x}+\frac{\partial u^0}{\partial y}\right)\right\}=N_{xy}^{-}$$

or v^0 must be specified.

iii. Either

$$-D_{11}\frac{\partial^3 w^0}{\partial x^3}-(D_{12}+4D_{66})\frac{\partial^3 w^0}{\partial x\partial y^2}=Q_x^{-}+\frac{\partial M_{xy}^{-}}{\partial y}$$

or w^0 must be specified.

iv. Either

$$\left\{-D_{11}\left(\frac{\partial^2 w^0}{\partial x^2}\right)-D_{12}\left(\frac{\partial^2 w^0}{\partial y^2}\right)\right\}=M_x^{-}$$

$$\text{or } \frac{\partial w^0}{\partial x} \text{ must be specified.} \tag{15.37b}$$

Along the edge $y = +b/2$:

i. Either

$$\left\{A_{12}\frac{\partial u^0}{\partial x} + A_{22}\frac{\partial v^0}{\partial y}\right\} = N_y^+$$

or v^0 must be specified.

ii. Either

$$\left\{A_{66}\left(\frac{\partial v^0}{\partial x} + \frac{\partial u^0}{\partial y}\right)\right\} = N_{yx}^+$$

or u^0 must be specified.

iii. Either

$$-D_{22}\frac{\partial^3 w^0}{\partial y^3} - (D_{12} + 4D_{66})\frac{\partial^3 w^0}{\partial x^2 \partial y} = Q_y^+ + \frac{\partial M_{yx}^+}{\partial x}$$

or w^0 must be specified.

iv. Either

$$\left\{-D_{12}\left(\frac{\partial^2 w^0}{\partial x^2}\right) - D_{22}\left(\frac{\partial^2 w^0}{\partial y^2}\right)\right\} = M_y^+$$

$$\text{or } \frac{\partial w^0}{\partial y} \text{ must be specified.} \qquad (15.37c)$$

Along the edge $y = -b/2$:

i. Either

$$\left\{A_{12}\frac{\partial u^0}{\partial x} + A_{22}\frac{\partial v^0}{\partial y}\right\} = N_y^-$$

or v^0 must be specified.

ii. Either

$$\left\{A_{66}\left(\frac{\partial v^0}{\partial x} + \frac{\partial u^0}{\partial y}\right)\right\} = N_{yx}^-$$

or u^0 must be specified.

iii. Either

$$-D_{22}\frac{\partial^3 w^0}{\partial y^3} - (D_{12} + 4D_{66})\frac{\partial^3 w^0}{\partial x^2 \partial y} = Q_y^- + \frac{\partial M_{yx}^-}{\partial x}$$

or w^0 must be specified.

iv. Either

$$\left\{-D_{12}\left(\frac{\partial^2 w^0}{\partial x^2}\right)-D_{22}\left(\frac{\partial^2 w^0}{\partial y^2}\right)\right\}=M_y^-$$

$$\text{or } \frac{\partial w^0}{\partial y} \text{ must be specified.} \tag{15.37d}$$

REFERENCES

1. Herakovich C.T. (1989), "Free edge effects in laminated composites." in *Handbook of Composites:* vol. 2, *Structures and Design* (C. T. Herakovich and Y. M. Tarnopol'skii, eds.), Elsevier Science Publishing Co., New York.
2. Popov E.P. (1990), *Engineering Mechanics of Solids*, Prentice Hall, Englewood Cliffs, NJ.
3. Hyer M.W. (1998), *Stress Analysis of Fiber Reinforced Composite Materials*, McGraw-Hill, Singapore.

16 Viscoelastic and Dynamic Behavior of Composites

Composite structures are often subjected to dynamic loading caused by vibration or wave propagation. In addition, many composites exhibit time-dependent viscoelastic behavior under load; this is particularly true for composites having polymeric constituents. This chapter contains the basic information needed for the analysis of both viscoelastic and dynamic behavior of composites and their constituents.

The word "viscoelastic" has evolved as a way of describing materials that exhibit characteristics of both viscous fluids and elastic solids. Polymeric materials, which are known to be viscoelastic, may behave like fluids or solids, depending on the timescale and/or the temperature. For example, polycarbonate, a thermoplastic polymer, is a liquid during molding at processing temperatures, but is a glassy solid at service (ambient) temperatures. It will deform like a rubber at temperatures just above the glass transition temperature, T_g. At temperatures below T_g, however, it will deform just as much, and in the same way if the test time is long enough. We know that ideal Hookean elastic solids are capable of energy storage under load, but not energy dissipation, whereas ideal Newtonian fluids under non-hydrostatic stresses are capable of energy dissipation, but not energy storage. Viscoelastic materials, however, are capable of both storage and dissipation of energy under load. Another characteristic of viscoelastic materials is memory. Perfectly elastic solids are said to have only "simple memory" because they remember only the unstrained state and the current strains depend only on the current stresses. Viscoelastic materials have what is often referred to as "fading memory" because they remember the past in such a way that the current strains depend more strongly on the recent stress-time history than on the more distant stress-time history.

There are four important physical manifestations of viscoelastic behavior in structural materials, as illustrated by the various conditions of the uniaxially loaded viscoelastic rod in Figure 16.1. First, if the rod is subjected to a constant stress, the resulting strain will exhibit time-dependent "creep," as shown in Figure 16.1a. The time-dependent creep strains are superimposed on the initial elastic strains. Second, if the rod is subjected to a constant strain or displacement, the resulting stress will exhibit time-dependent "relaxation," as shown in Figure 16.1b. That is, the stress relaxes from the initial elastic stress. Third, if the bar is subjected to oscillatory loading, the resulting stress–strain curve will describe a "hysteresis loop," as shown in Figure 16.1c. The area enclosed by the hysteresis loop is a measure of the damping, or dissipation, of energy in the material. Fourth, if the bar is loaded at various strain rates, the stress–strain curves will exhibit a strain-rate dependence, as shown in Figure 16.1d. That is, the stress corresponding to a given strain depends on the rate of straining. An ideal elastic material exhibits none of the above characteristics.

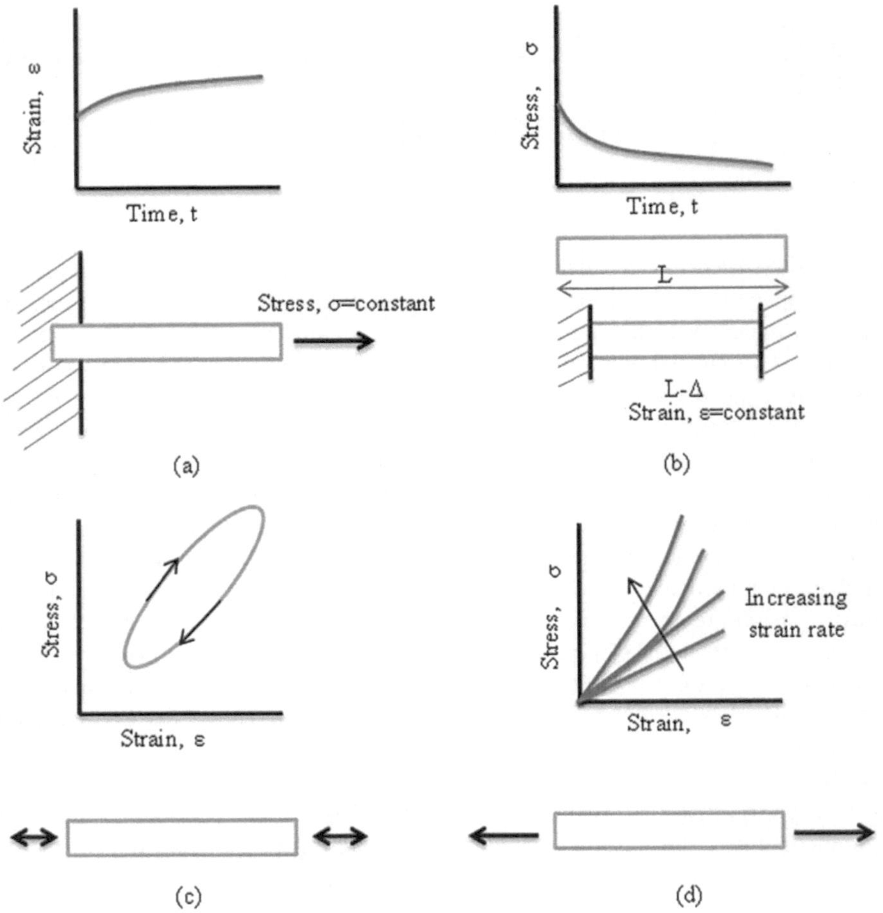

FIGURE 16.1 Viscoelastic behavior in structural materials: (a) creep under constant stress, (b) relaxation under constant strain, (c) hysteresis loop due to cyclic stress, and (d) strain-rate dependence of stress–strain curve.

All structural materials exhibit some degree of viscoelasticity, and the extent of such behavior often depends on environmental conditions such as temperature. For example, while a structural steel or aluminum may be essentially elastic at room temperature, viscoelastic effects become apparent at elevated temperatures approaching half the melting temperature. Polymeric materials are viscoelastic at room temperature, and the viscoelastic effects become stronger as the temperature approaches the glass transition temperature. Polymers with amorphous microstructures tend to be more viscoelastic than those with crystalline microstructures. On the other hand, crystalline microstructures consist of regular, ordered crystalline arrays of atoms. Some polymers have both amorphous and crystalline components in their microstructures, and some polymers are purely amorphous. In this chapter, we will be concerned with the development of stress–strain relationships for linear viscoelastic materials and composites made of those materials. These stress–strain relationships take on special forms for creep, relaxation, and sinusoidal oscillation.

Dynamic loading is usually categorized as being either impulsive or oscillatory. Dynamic response consists of either a propagating wave or a vibration, depending on the elapsed time and the relative magnitudes of the wavelength of the response and the characteristic structural dimension. Both types of excitation usually cause wave propagation initially. Wave propagation will continue if the response wavelength is much shorter than the characteristic structural dimension; otherwise, standing waves (i.e., vibrations) will be set up as the waves begin to reflect back from the boundaries. Wave propagation in composites may involve complex reflection and/or refraction effects at fiber/matrix interfaces or ply interfaces, complicating matters further. The dynamic response of composites may also be complicated by their anisotropic behavior. For example, the speed of a propagating wave in an isotropic material is independent of orientation, whereas the wave speed in an anisotropic composite depends on the direction of propagation. Anisotropic coupling effects often lead to complex waves or modes of vibration. For example, an isotropic beam that is subjected to an oscillatory bending moment will respond in pure flexural modes of vibration, but a non-symmetric laminate may respond in a coupled bending–twisting mode or some other complex mode. Composites generally have better damping than conventional metallic structural materials, especially if the composite has one or more polymeric constituents. It will be shown that the complex modulus notation and the elastic–viscoelastic correspondence principle from the viscoelasticity theory are particularly useful in the development of analytical models for predicting the damping behavior of composites.

16.1 VISCOELASTIC BEHAVIOR OF COMPOSITES

A linear elastic solid exhibits a linearity between stress and strain, and this linear relationship is independent of time. A linear viscoelastic solid also exhibits a linearity between stress and strain, but the linear relationship depends on the time history of the input. The mathematical criteria for linear viscoelastic behavior are similar to those for linear behavior of any system. Following the notation of Schapery [1], the criteria can be stated as follows:

Let the response R to an input I be written as $R = R\{I\}$, where $R\{I\}$ denotes that the current value of R is a function of the time history of the input I. For linear viscoelastic behavior, the response $R\{I\}$ must satisfy both of the following conditions:

i. Proportionality: i.e., $R\{cI\} = cR\{I\}$, where c is a constant
ii. Superposition: i.e., $R\{I_a + I_b\} = R\{I_a\} + R\{I_b\}$, where I_a and I_b may be the same or different time histories.

Any response not satisfying these criteria would be a nonlinear response. These criteria form the basis of the stress–strain relationship known as the Boltzmann superposition integral.

16.1.1 Boltzmann Superposition Integral

If the material is at a constant temperature and is "non-aging," then the response at any time t due to an input at time $t = \tau$ is a function of the input and the elapsed time $(t-\tau)$ only. Aging is a time-dependent change in the material which is different from viscoelastic creep or relaxation. Consider the one-dimensional isothermal loading of a non-aging, isotropic, homogeneous linear viscoelastic material by the stresses $\Delta\sigma_1$, $\Delta\sigma_2$, and $\Delta\sigma_3$ at times τ_1, τ_2, and τ_3, respectively, as shown in Figure 16.2. According to the Boltzmann superposition principle, the strain response is linearly proportional to the input stress, but the proportionality factor is a function of the elapsed time since the input stress. Thus, for the stress-time history in Figure 16.2, the total strain response at any time $t > \tau_3$ is given by:

$$\varepsilon(t) = \Delta\sigma_1 S(t-\tau_1) + \Delta\sigma_2 S(t-\tau_2) + \Delta\sigma_3 S(t-\tau_3) \tag{16.1}$$

where $S(t)$ is the creep compliance, which is zero for $t < 0$. For input stresses having arbitrary time histories, Eq. (16.1) can be generalized as the Boltzmann superposition integral, or hereditary law:

$$\varepsilon(t) = \int_{-\infty}^{t} S(t-\tau)\frac{d\sigma(\tau)}{d\tau}d\tau \tag{16.2}$$

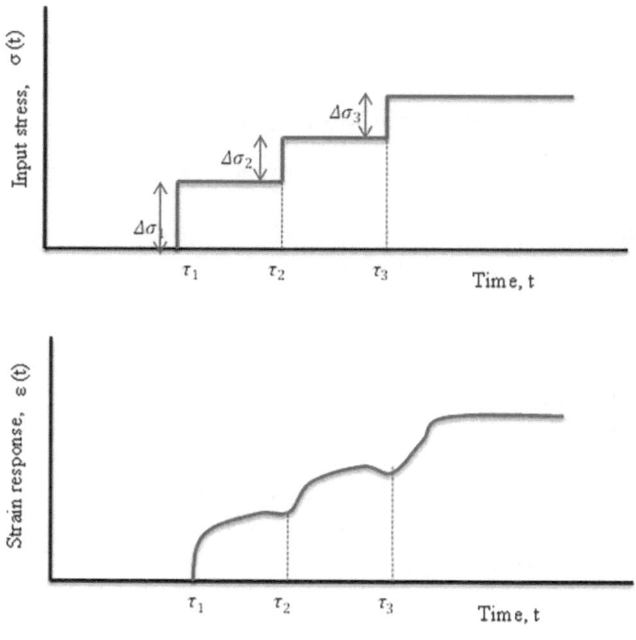

FIGURE 16.2 Boltzmann superposition principle.

Alternatively, the stress resulting from arbitrary strain inputs may be given by:

$$\sigma(t) = \int_{-\infty}^{t} C(t-\tau)\frac{d\varepsilon(\tau)}{d\tau}d\tau \tag{16.3}$$

where C(t) is the relaxation modulus, which is zero for $t<0$. Equation (16.2) can be extended to the more general case of a homogeneous, anisotropic, linear viscoelastic material with multiaxial inputs and responses by using the contracted notation and writing:

$$\varepsilon_i(t) = \int_{-\infty}^{t} S_{ij}(t-\tau)\frac{d\sigma_j(\tau)}{d\tau}d\tau \tag{16.4}$$

where
$i,j = 1,2,\ldots,6$
$S_{ij}(t)$ = creep compliances

For the specific case of the homogeneous, linear viscoelastic, specially orthotropic lamina in plane stress, Eq. (16.4) becomes:

$$\varepsilon_1(t) = \int_{-\infty}^{t} S_{11}(t-\tau)\frac{d\sigma_1(\tau)}{d\tau}d\tau + \int_{-\infty}^{t} S_{12}(t-\tau)\frac{d\sigma_2(\tau)}{d\tau}d\tau \tag{16.5}$$

$$\varepsilon_2(t) = \int_{-\infty}^{t} S_{12}(t-\tau)\frac{d\sigma_1(\tau)}{d\tau}d\tau + \int_{-\infty}^{t} S_{22}(t-\tau)\frac{d\sigma_2(\tau)}{d\tau}d\tau$$

$$\gamma_{12}(t) = \int_{-\infty}^{t} S_{66}(t-\tau)\frac{d\tau_{12}(\tau)}{d\tau}d\tau$$

Similarly, Eq. (16.3) can be generalized to the form:

$$\sigma_i(t) = \int_{-\infty}^{t} C_{ij}(t-\tau)\frac{d\varepsilon_j(\tau)}{d\tau}d\tau \tag{16.6}$$

where $C_{ij}(t)$ are the relaxation moduli. The creep compliances, $S_{ij}(t)$, for the viscoelastic material are analogous to the elastic compliances, S_{ij}, and the viscoelastic relaxation moduli, $C_{ij}(t)$, are analogous to the elastic stiffnesses, C_{ij}. Recall that in order to apply the stress–strain relationships at a point in a homogeneous material to the case of a heterogeneous composite, we can replace the stresses and strains at a point with the volume-averaged stresses and strains and also replace the elastic moduli of the heterogeneous composite by effective moduli of an equivalent homogeneous material. The criterion for the use of the effective modulus theory is that the scale of the inhomogeneity, d, has to be much smaller than the characteristic structural dimension, L, over which the averaging is done. Since this chapter also deals with dynamic behavior, it is appropriate to add another criterion related to dynamic effects. That is, the scale of the inhomogeneity, d, must also be much smaller than the characteristic wavelength, λ, of the dynamic stress distribution (Figure 16.3). Thus, the criteria for the use of the effective modulus theory in dynamic loading of viscoelastic composites are $d \ll L$ and $d \ll \lambda$. The second criterion only becomes important when dealing with the propagation of high-frequency waves that have very short wavelengths. On the other hand, the wavelengths associated with typical mechanical vibrations will almost always be sufficiently large so as to satisfy $d \ll \lambda$.

Thus, Eqs. (16.4)–(16.6) are valid for heterogeneous, anisotropic, linear viscoelastic composites if at an arbitrary time, t, we simply replace the stresses and strains at a point with the volume-averaged stresses and strains, replace the creep compliances with the effective creep compliances, and replace the relaxation moduli with the effective relaxation moduli. Thus, the effective creep compliance matrix for the specially orthotropic lamina in plane stress is given by:

$$S_{ij}(t) = \begin{bmatrix} S_{11}(t) & S_{12}(t) & 0 \\ S_{21}(t) & S_{22}(t) & 0 \\ 0 & 0 & S_{66}(t) \end{bmatrix} \tag{16.7}$$

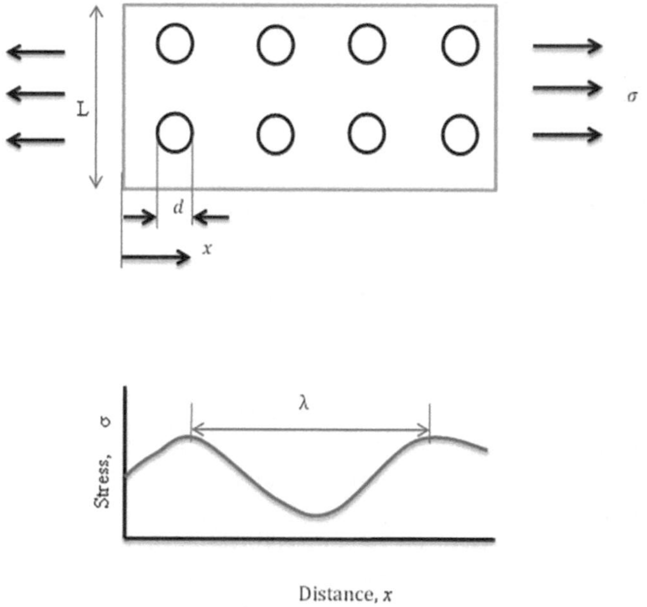

FIGURE 16.3 Critical dimensions used in effective modulus theory.

For the generally orthotropic lamina, we have:

$$\bar{S}_{ij}(t)=\begin{bmatrix} \bar{S}_{11}(t) & \bar{S}_{12}(t) & \bar{S}_{16}(t) \\ \bar{S}_{12}(t) & \bar{S}_{22}(t) & \bar{S}_{26}(t) \\ \bar{S}_{16}(t) & \bar{S}_{26}(t) & \bar{S}_{66}(t) \end{bmatrix} \tag{16.8}$$

where $\bar{S}_{ij}(t)$ are the transformed effective creep compliances. Halpin and Pagano [2] have shown that $\bar{S}_{ij}(t)$ are related to the $S_{ij}(t)$ by the transformations:

$$\bar{S}_{11}(t)=S_{11}(t)m^4+\left(2S_{12}(t)+S_{66}(t)\right)n^2m^2+S_{22}(t)n^4$$

$$\bar{S}_{12}(t)=\left(S_{11}(t)+S_{22}(t)-S_{66}(t)\right)n^2m^2+S_{12}(t)\left(n^4+m^4\right)$$

$$\bar{S}_{16}(t)=\left(2S_{11}(t)-2S_{12}(t)-S_{66}(t)\right)nm^3-\left(2S_{22}(t)-2S_{12}(t)-S_{66}(t)\right)mn^3$$

$$\bar{S}_{22}(t)=S_{11}(t)n^4+\left(2S_{12}(t)+S_{66}(t)\right)n^2m^2+S_{22}(t)m^4$$

$$\bar{S}_{26}(t)=\left(2S_{11}(t)-2S_{12}(t)-S_{66}(t)\right)mn^3-\left(2S_{22}(t)-2S_{12}(t)-S_{66}(t)\right)nm^3$$

$$\bar{S}_{66}(t)=2\left(2S_{11}(t)+2S_{22}(t)-4S_{12}(t)-S_{66}(t)\right)n^2m^2+S_{66}(t)\left(n^4+m^4\right) \tag{16.9}$$

where $m = \cos\theta$ and $n = \sin\theta$.

For the viscoelastic case, Schapery [1] has used thermodynamic arguments to show that if $S_{ij}(t)=S_{ji}(t)$ for the constituent materials, then the same is true for the composite. Halpin and Pagano [2] and others have presented experimental evidence that for transversely isotropic composites under plane stress, $S_{12}(t)=S_{21}(t)$. In both elastic and viscoelastic cases, further reductions in the number of independent moduli or compliances depend on material property symmetry and the coordinate system used.

16.1.2 Spring–Dashpot Models

Although the Boltzmann superposition integral is a valid mathematical expression of the stress–strain relationship for a linear viscoelastic material, it does not lend itself easily to the use of physical models that help us to understand viscoelastic behavior better. In this section, Laplace transforms will be used to convert the Boltzmann superposition integral to an ordinary differential equation involving time derivatives of stress and strain. Physical models for viscoelastic behavior can be easily interpreted by using differential equations.

The Laplace transform, $\mathcal{L}[f(t)]$ or $\bar{f}(s)$, of a function $f(t)$ is defined by:

$$\mathcal{L}\left[f(t)\right]=\bar{f}(s)=\int_0^\infty f(t)e^{-st}\,dt \tag{16.10}$$

where s is the Laplace parameter. For purposes of illustration, we now take the Laplace transform of the one-dimensional Boltzmann superposition integral given by Eq. (16.3). The Laplace transform of both sides of the equation is given by:

$$\mathcal{L}[\sigma(t)] = \bar{\sigma}(s) = \mathcal{L}\left[\int_{-\infty}^{t} C(t-\tau)\frac{d\varepsilon(\tau)}{d\tau}d\tau\right] \tag{16.11}$$

Noting that the right-hand side of Eq. (16.11) is in the form of a convolution integral, we can also write:

$$\bar{C}(s)\frac{d\bar{\varepsilon}(s)}{d\tau} = \mathcal{L}\left[\int_{-\infty}^{t} C(t-\tau)\frac{d\varepsilon(\tau)}{d\tau}d\tau\right] \tag{16.12}$$

Taking the inverse Laplace transform of Eq. (16.12):

$$\mathcal{L}^{-1}\left[\bar{C}(s)\frac{d\bar{\varepsilon}(s)}{d\tau}\right] = \left[\int_{-\infty}^{t} C(t-\tau)\frac{d\varepsilon(\tau)}{d\tau}d\tau\right] \tag{16.13}$$

Therefore, Eq. (16.11) can be written as:

$$\bar{\sigma}(s) = \mathcal{L}\left[\mathcal{L}^{-1}\left[\bar{C}(s)\frac{d\bar{\varepsilon}(s)}{d\tau}\right]\right] = \left[\bar{C}(s)\frac{d\bar{\varepsilon}(s)}{d\tau}\right] \tag{16.14}$$

Also,

$$\mathcal{L}\left[\frac{d\varepsilon(\tau)}{d\tau}\right] = \frac{d\bar{\varepsilon}(s)}{d\tau} = s\bar{\varepsilon}(s) - \varepsilon(0) \tag{16.15}$$

where $\varepsilon(0)$ is the initial strain. If we neglect the initial conditions, Eq. (16.14) becomes:

$$\bar{\sigma}(s) = s\bar{C}(s)\bar{\varepsilon}(s) \tag{16.16}$$

If we perform similar operations on Eq. (16.2), we find that:

$$\bar{\varepsilon}(s) = s\bar{S}(s)\bar{\sigma}(s) \tag{16.17}$$

This is another example of the correspondence between the equations for elastic and viscoelastic materials. Comparing Eqs. (16.16) and (16.17), the Laplace transform of the creep compliance and the Laplace transform of the relaxation modulus are related by:

$$\bar{S}(s) = \frac{1}{s^2\bar{C}(s)} \tag{16.18}$$

However, the corresponding time domain properties are not generally related by a simple inverse relationship. That is, in general:

$$S(t) \neq \frac{1}{C(t)} \tag{16.19}$$

Equation (16.17) can also be written as a ratio of two polynomials in the Laplace parameter s as follows:

$$\bar{\varepsilon}(s) = s\bar{S}(s)\bar{\sigma}(s) = \frac{Q(s)}{P(s)}\bar{\sigma}(s) \tag{16.20}$$

where

$$P(s) = a_0 + a_1 s + a_2 s^2 + \cdots + a_n s^n$$

$$Q(s) = b_0 + b_1 s + b_2 s^2 + \cdots + b_n s^n$$

Thus,

$$P(s)\bar{\varepsilon}(s) = Q(s)\bar{\sigma}(s) \tag{16.21}$$

But if we neglect the initial conditions, the Laplace transform of the nth derivative of a function $f(t)$ is:

$$\mathcal{L}\left[\frac{d^n f(t)}{dt^n}\right] = \left[s^n \bar{f}(s)\right] \tag{16.22}$$

Making use of Eq. (16.22) and taking the inverse Laplace transform of Eq. (16.21), we find that:

$$a_n \frac{d^n\varepsilon}{dt^n} + \cdots + a_2 \frac{d^2\varepsilon}{dt^2} + a_1 \frac{d\varepsilon}{dt} + a_0\varepsilon = b_0\sigma + b_1 \frac{d\sigma}{dt} + b_2 \frac{d^2\sigma}{dt^2} + \cdots b_n \frac{d^n\sigma}{dt^n} \tag{16.23}$$

Thus, linear viscoelastic behavior may also be described by an ordinary differential equation as well as by the Boltzmann superposition integral. As shown in Figures 16.4–16.6, useful physical models can be constructed from simple elements such as the elastic spring and the viscous dashpot, where the spring of modulus k is assumed to follow Hooke's law and the dashpot is assumed to be filled with a Newtonian fluid of viscosity μ. Thus, the stress–strain relationship for the elastic spring element is of the form $\varepsilon = \frac{\sigma}{k}$, whereas the corresponding equation for the viscous dashpot is $\frac{d\varepsilon}{dt} = \frac{\sigma}{\mu}$. The Maxwell model consists of a spring and a dashpot in series, as shown in Figure 16.4a.

The total strain across a model of unit length must equal the sum of the strains in the spring and the dashpot, so that:

$$\varepsilon = \varepsilon_1 + \varepsilon_2 \tag{16.24}$$

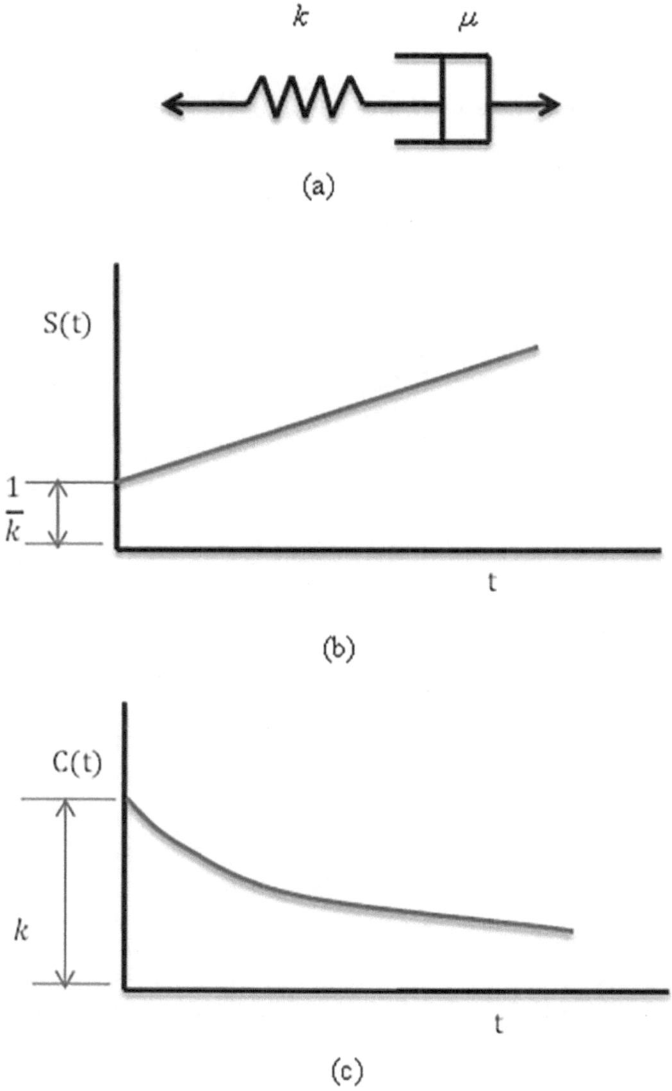

FIGURE 16.4 Maxwell model, with corresponding creep and relaxation curves: (a) spring–dashpot arrangement, (b) creep compliance vs. time, and (c) relaxation modulus vs. time.

and the strain rate across the model is then:

$$\frac{d\varepsilon}{dt} = \frac{d\varepsilon_1}{dt} + \frac{d\varepsilon_2}{dt} = \frac{1}{k}\frac{d\sigma}{dt} + \frac{\sigma}{\mu} \tag{16.25}$$

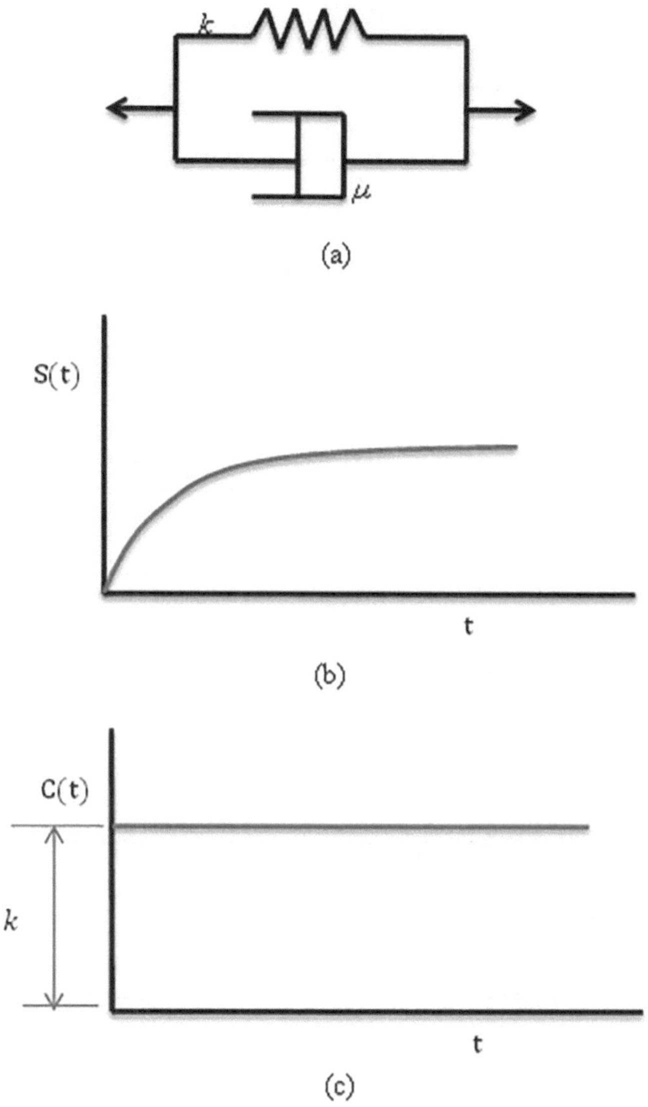

FIGURE 16.5 Kelvin–Voigt model, with corresponding creep and relaxation curves: (a) spring–dashpot arrangement, (b) creep compliance vs. time, and (c) relaxation modulus vs. time.

For creep at constant stress $\sigma = \sigma_0$, Eq. (16.25) reduces to:

$$\frac{d\varepsilon}{dt} = \frac{\sigma_0}{\mu} \tag{16.26}$$

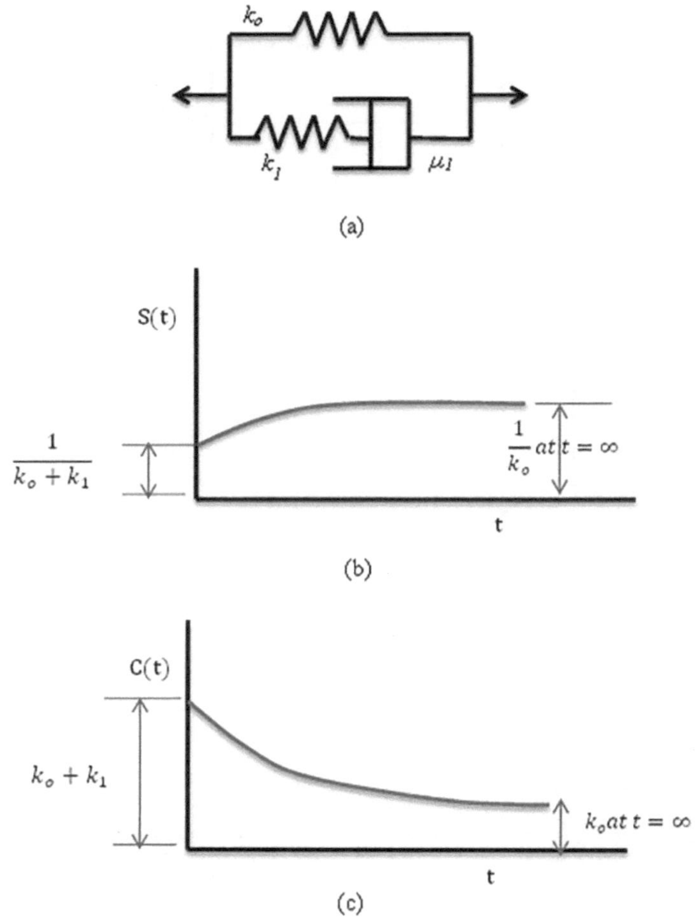

FIGURE 16.6 Standard linear solid, or Zener model, with corresponding creep and relaxation curves: (a) spring–dashpot arrangement, (b) creep compliance vs. time, and (c) relaxation modulus vs. time.

Integrating Eq. (16.26) once, we find that:

$$\varepsilon(t) = \frac{\sigma_0}{\mu} t + C_1 \tag{16.27}$$

where the constant of integration, C_1, is found from the initial condition:

$$\varepsilon(0) = C_1 = \frac{\sigma_0}{k}$$

Thus, the creep strain for the Maxwell model is given by:

$$\varepsilon(t)=\frac{\sigma_0}{\mu}t+\frac{\sigma_0}{k} \tag{16.28}$$

The corresponding creep compliance is given by:

$$S(t)=\frac{\varepsilon(t)}{\sigma_0}=\frac{t}{\mu}+\frac{1}{k} \tag{16.29}$$

A plot of the creep compliance vs. time according to Eq. (16.29) is shown in Figure 16.4b. The type of creep behavior that is actually observed in experiments is more like that shown in Figure 16.2, however. Thus, the Maxwell model does not adequately describe creep. For relaxation at constant strain $\varepsilon=\varepsilon_0$, the Maxwell model stress–strain relationship in Eq. (16.25) becomes:

$$0=\frac{1}{k}\frac{d\sigma}{dt}+\frac{\sigma}{\mu} \tag{16.30}$$

Integrating Eq. (16.30) once, we find that:

$$\ln\sigma=-\frac{k}{\mu}t+C_2 \tag{16.31}$$

where the constant of integration, C_2, is found from the initial condition:

$$\sigma(0)=\sigma_0$$

The resulting stress relaxation function is:

$$\sigma(t)=\sigma_0\, e^{-\frac{kt}{\mu}}=\sigma_0\, e^{-\frac{t}{\lambda}} \tag{16.32}$$

where $\lambda=\frac{\mu}{k}$ is the relaxation time, or the time required for the stress to relax to $1/e$, or 37% of its initial value. The relaxation time is therefore a measure of the internal timescale of the material. The corresponding relaxation modulus is:

$$C(t)=\frac{\sigma(t)}{\varepsilon_0}=\frac{\sigma_0}{\varepsilon_0}e^{-\frac{t}{\lambda}}=ke^{-\frac{t}{\lambda}} \tag{16.33}$$

Figure 16.4c shows the relaxation modulus vs. time from Eq. (16.33), which is in general agreement with the type of relaxation observed experimentally. Thus, the Maxwell model appears to describe adequately the relaxation phenomenon, but not the creep response. Figure 16.5a shows the Kelvin–Voigt model, which consists of a spring and a dashpot in parallel. Using the appropriate equations for a parallel arrangement and following a procedure similar to the one just outlined, it can be shown that the differential equation describing the behavior of the Kelvin–Voigt model is given by:

$$\sigma = k\varepsilon + \mu \frac{d\varepsilon}{dt} \tag{16.34}$$

It can also be shown that the creep compliance for the Kelvin–Voigt model is given by:

$$S(t) = \frac{1}{k}\left[1 - e^{-\frac{t}{\rho}}\right] \tag{16.35}$$

where $\rho = \frac{\mu}{k}$ is now referred to as the retardation time. Similarly, the relaxation modulus is given by:

$$C(t) = k \tag{16.36}$$

Equations (16.35) and (16.36) are plotted in Figure 16.5b and c, respectively. The creep compliance curve agrees with experimental observation, except that the initial elastic response is missing. On the other hand, the relaxation modulus has not been observed to be constant, as shown in Figure 16.5c. Thus, like the Maxwell model, the Kelvin–Voigt model does not adequately describe all features of experimentally observed creep and relaxation. One obvious way to improve the spring–dashpot model is to add more elements. One such improved model, shown in Figure 16.6a, is referred to as the standard linear solid, or Zener model. It can be shown that the differential equation for the Zener model is given by:

$$\sigma + \frac{\mu_1}{k_1}\frac{d\sigma}{dt} = k_0\varepsilon + \frac{\mu_1}{k_1}(k_0 + k_1)\frac{d\varepsilon}{dt} \tag{16.37}$$

where the parameters k_0, k_1, and μ_1 are defined in Figure 16.6a. The creep compliance for the Zener model is given by:

$$S(t) = \frac{1}{k_0}\left[1 - \frac{k_1}{(k_0 + k_1)}e^{-\frac{t}{\rho_1}}\right] \tag{16.38}$$

where

$$\rho_1 = \frac{\mu_1}{k_1 k_0}(k_0 + k_1) = \text{retardation time}$$

As shown in Figure 16.6b, the shape of the creep compliance curve from Eq. (16.38) matches the expected shape based on experimental observations. The relaxation modulus for the Zener model is given by:

$$C(t) = k_0 + k_1 e^{-\frac{t}{\lambda_1}} \tag{16.39}$$

where

$$\lambda_1 = \frac{\mu_1}{k_1} = \text{relaxation time}$$

Figure 16.6c shows the predicted relaxation modulus curve from Eq. (16.39), and again, the general shape of the curve appears to be similar to what is experimentally observed. Although the Zener model is the simplest spring–dashpot model that correctly describes all expected features of experimentally observed creep and relaxation behavior in linear viscoelastic materials, it still is not completely adequate. This remaining inadequacy is best described by plotting the relaxation modulus vs. the logarithm of time. Practically speaking, complete relaxation for the Zener model occurs in less than a decade in time, but relaxation for real polymers happens over a much longer timescale. For example, the glass-to-rubber transition, which is only one region of polymer viscoelastic behavior, takes about six to eight decades in time to complete. This extended relaxation period for polymers is due to the existence of a distribution of relaxation times. By using an improved Zener model such as the parallel arrangement shown in Figure 16.7, we can introduce such a distribution of relaxation times, A, which makes it possible to extend the range of relaxation to more realistic values. This form of the improved Zener model consists of n Maxwell elements in parallel with the elastic spring, k_0. It can be easily shown that the relaxation modulus for this improved Zener model is given by:

$$C(t) = k_0 + \sum_{i=1}^{n} k_i e^{-\frac{t}{\lambda_i}} \tag{16.40}$$

where

$$\lambda_i = \frac{\mu_i}{k_i} = \text{relaxation time for the } i\text{th Maxwell element.}$$

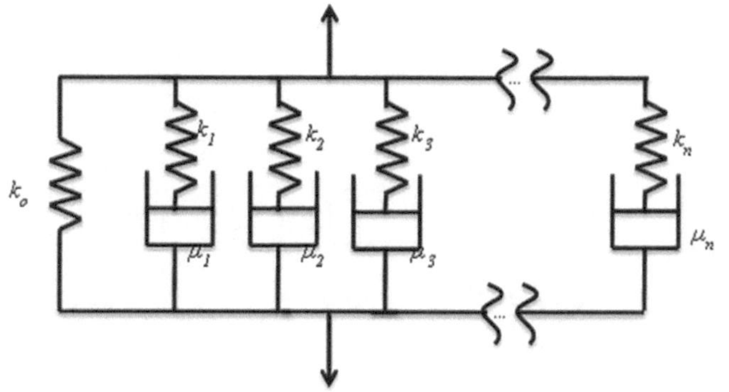

FIGURE 16.7 Improved Zener model, parallel arrangement.

For an infinite number of elements in the improved Zener model of Figure 16.7 and a continuous distribution of relaxation times, the relaxation modulus can be expressed as:

$$C(t) = k_0 + \int_0^{\infty} k(\lambda) e^{-\frac{t}{\lambda}} d\lambda \tag{16.41}$$

where $k(\lambda)$ is the distribution of relaxation times, or the relaxation spectrum.

By considering an alternative form of an improved Zener model consisting of a spring in series with n Kelvin–Voigt elements, as shown in Figure 16.8, it can be shown that the corresponding creep compliance expression is:

$$S(t) = \frac{1}{k_0} + \sum_{i=1}^{n} \frac{1}{k_i}\left[1 - e^{-\frac{t}{\rho_i}}\right] \tag{16.42}$$

where

$$\rho_i = \frac{\mu_i}{k_i} = \text{retardation time for the } i^{th} \text{ Kelvin} - \text{Voigt element.}$$

According to Schapery [1], if the elastic moduli are positive definite (i.e., always either positive or equal to zero), it can be shown using the thermodynamic theory that the generalized expressions corresponding to Eqs. (16.40) and (16.42) are, respectively:

$$C_{ij}(t) = \sum_{m=1}^{n} C_{ij}^{(m)} e^{-\frac{t}{\lambda_m}} + C_{ij} \tag{16.43}$$

$$S_{ij}(t) = \sum_{m=1}^{n} S_{ij}^{(m)}\left[1 - e^{-\frac{t}{\rho_m}}\right] + S_{ij} \tag{16.44}$$

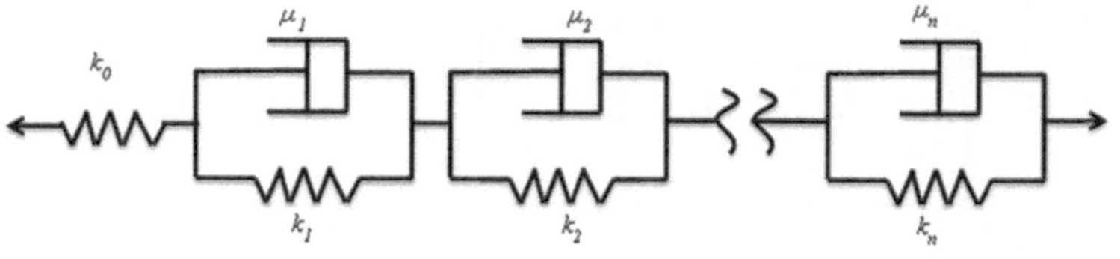

FIGURE 16.8 Improved Zener model, series arrangement.

where

$i, j = 1, 2, \ldots, 6$
C_{ij} and S_{ij} = elastic moduli and compliances, respectively
λ_m and ρ_m = relaxation and retardation times, respectively
$C_{ij}^{(m)}$ and $S_{ij}^{(m)}$ = coefficients corresponding to λ_m and ρ_m, respectively

As with the simple spring–dashpot models, the numerical values of the parameters on the right-hand side of Eqs. (16.43) and (16.44) must be determined experimentally.

16.1.3 Quasi-Elastic Approach

From the previous section, it should be clear that the generalized Boltzmann superposition integrals in Eqs. (16.4) and (16.6) can be Laplace-transformed to yield equations of the form:

$$\overline{\varepsilon_i}(s) = s\overline{S}_{ij}(s)\overline{\sigma_j}(s) \tag{16.45}$$

$$\overline{\sigma_i}(s) = s\overline{C}_{ij}(s)\overline{\varepsilon_j}(s) \tag{16.46}$$

These equations are of the same form as the corresponding elastic stress–strain relationships and are presumably easier to work with than the integral equations. In a practical analysis or design problem involving the use of these equations, however, the problem solution in the Laplace domain would then have to be inverse-transformed to get the desired time domain result, and this can present difficulties. Schapery [1] has presented several approximate methods for performing such inversions. If the input stresses or strains are constant, however, there is no need for inverse transforms and the time domain equations turn out to be very simple. Schapery refers to this as a "quasi-elastic analysis," and the equations used in such an analysis will be developed in the remainder of this section.

Consider a generalized creep problem with time-varying stresses $\sigma_j(t)$ given by:

$$\sigma_j(t) = \sigma_j' H(t) \tag{16.47}$$

where

$j = 1, 2, \ldots, 6$
σ_j' = constant stresses
$H(t)$ = unit step function or Heaviside function shown in Figure 16.9, defined as:

$$\lim_{\varepsilon \to 0} H(t) \begin{cases} = 0 & \text{for } t \le 0 \\ = \dfrac{t}{\varepsilon} & \text{for } 0 \le t \le \varepsilon \\ = 1 & \text{for } t \ge \varepsilon \end{cases} \tag{16.48}$$

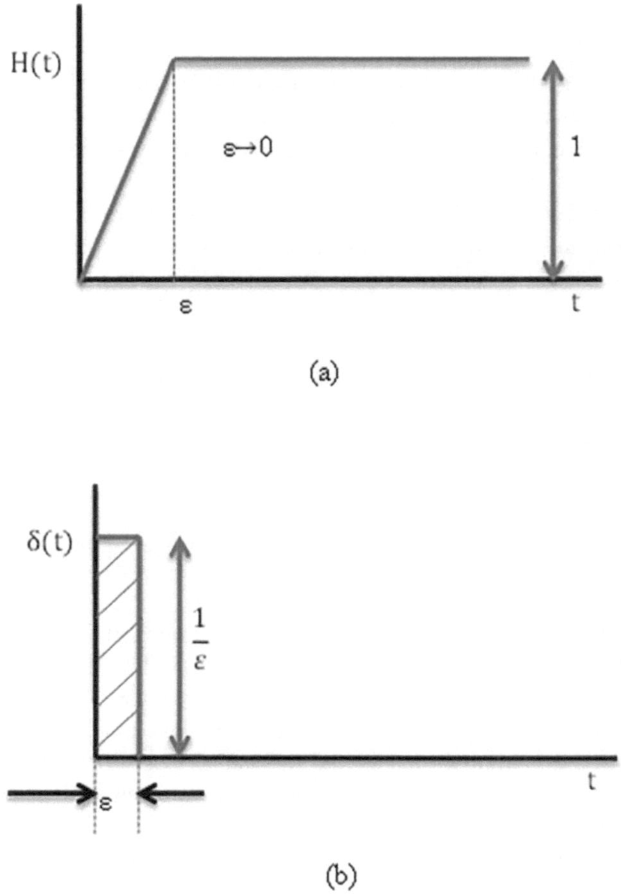

FIGURE 16.9 (a) Unit step function and (b) Dirac delta function.

The unit step function can be easily shifted along the time axis by an amount ξ by writing the function as $H(t-\xi)$. Substituting the stresses from Eq. (16.47) in the Boltzmann superposition integral, Eq. (16.4), we find that the resulting strains are given by:

$$\varepsilon_i(t) = \int_{-\infty}^{t} S_{ij}(t-\tau)\sigma'_j \frac{dH(\tau)}{d\tau} d\tau \tag{16.49}$$

From Eq. (16.48), the derivative of the step function must be:

$$\frac{dH(t)}{dt} = \delta(t) \begin{cases} = 0 & \text{for } t \le 0 \\ = \dfrac{1}{\varepsilon} & \text{for } 0 \le t \le \varepsilon \\ = 0 & \text{for } t \ge 0 \end{cases} \tag{16.50}$$

where the parameter ε can be made arbitrarily small, the derivative in Eq. (16.50) is taken before $\varepsilon \to 0$, and $\delta(t)$ is the Dirac delta function shown in Figure 16.9b. Thus, the integral in Eq. (16.49) can be written as:

$$\varepsilon_i(t) = \left\{ \int_{-\infty}^{t} S_{ij}(t-\tau)\delta(\tau)d\tau \right\} \sigma'_j \tag{16.51}$$

According to the properties of convolution integrals, we can also write:

$$\varepsilon_i(t) = \left\{ \int_{-\infty}^{t} S_{ij}(\tau)\delta(t-\tau)d\tau \right\} \sigma'_j \tag{16.52}$$

This integral can be broken down and rewritten as follows:

$$\varepsilon_i(t) = \left\{ \int_{-\infty}^{t-\varepsilon} (0)d\tau + \int_{t-\varepsilon}^{t} S_{ij}(t)\delta(t-\tau)d\tau \right\} \sigma'_j \tag{16.53}$$

where the $S_{ij}(\tau)$ evaluated over the interval $t-\varepsilon \le \tau \le t$ can be approximated as $S_{ij}(t)$ since ε is very small. The $S_{ij}(t)$ can now be moved outside the integral, leaving the integral of the Dirac delta function, which is defined as:

$$\int_{t-\varepsilon}^{t} \delta(t-\tau)d\tau = 1 \tag{16.54}$$

Therefore,

$$\varepsilon_i(t) = S_{ij}(t)\sigma'_j \tag{16.55}$$

Similarly, it can be shown that if the constant strain inputs:

$$\varepsilon_j(t) = \varepsilon'_j H(t) \tag{16.56}$$

are substituted in Eq. (16.6), the resulting stresses must be:

$$\sigma_i(t) = C_{ij}(t)\varepsilon'_j \tag{16.57}$$

Thus, the stress relaxation under constant strains can be found by replacing the elastic moduli, C_{ij}, in Hooke's law with the corresponding viscoelastic relaxation moduli, $C_{ij}(t)$. Equations (16.55) and (16.57) form the basis of the so-called "quasi-elastic analysis" and obviously eliminate the need for Laplace transform analysis in the stress–strain relationships. It should be emphasized again, however, that Eqs. (16.55) and (16.57) are only valid for constant or near-constant inputs.

16.1.4 Complex Modulus

In the previous section, it was shown that when the inputs are constant, the Boltzmann superposition integrals are reduced to simple algebraic equations that resemble the linear elastic Hooke's law. In this section, an analogous simplification will be demonstrated for the case of stresses or strains that vary sinusoidally with time. The results will make it much easier to analyze sinusoidal vibrations of viscoelastic composites. The general procedure here follows that presented by Fung [3]. Consider the case where the stresses vary sinusoidally with frequency (ω). Using the contracted notation and complex exponentials, such stresses can be written as:

$$\widetilde{\sigma_n}(t) = A_n e^{i\omega t} \tag{16.58}$$

where

$n = 1,2, \ldots, 6$
i = imaginary operator = $(-1)^{1/2}$
A_n = complex stress amplitudes

The superscript tilde (~) refers to a sinusoidally varying quantity.

Substituting Eq. (16.58) in Eq. (16.4), we find that the resulting sinusoidally varying strains are given by:

$$\widetilde{\varepsilon_m}(t) = \int_{-\infty}^{t} S_{mn}(t-\tau) i\omega A_n e^{i\omega t}\, d\tau \tag{16.59}$$

where

$m = 1,2, \ldots, 6.$

It is now convenient to define a new variable, $\xi = t - \tau$, so that:

$$\widetilde{\varepsilon_m}\,(t) = \int_{0}^{\infty} S_{mn}(\xi) e^{-i\omega\xi} i\omega A_n e^{i\omega t}\, d\xi \tag{16.60}$$

The terms not involving functions of ξ may be moved outside the integral, and since $S_{mn}(t)$ for $t < 0$, the lower limit on the integral can be changed to $-\infty$, so that:

$$\widetilde{\varepsilon_m}\,(t) = i\omega A_n e^{i\omega t} \int_{-\infty}^{\infty} S_{mn}(\xi) e^{-i\omega\xi}\, d\xi \tag{16.61}$$

The integral in Eq. (16.61) is just the Fourier transform of the creep compliances, $F\left[S_{mn}(\xi)\right]$, or $S_{mn}(\omega)$, which is written as:

$$F\left[S_{mn}(\xi)\right] = S_{mn}(\omega) = \int_{-\infty}^{\infty} S_{mn}(\xi) e^{-i\omega\xi}\, d\xi \tag{16.62}$$

Thus, the stress–strain relationship reduces to:

$$\widetilde{\varepsilon_m}(t) = i\omega S_{mn}(\omega) A_n e^{i\omega t} = i\omega S_{mn}(\omega)\widetilde{\sigma_n}(t) \tag{16.63}$$

In order to get this equation to resemble Hooke's law more closely, we simply define the frequency domain complex compliances as follows:

$$S^*_{mn}(\omega) = i\omega S_{mn}(\omega) \tag{16.64}$$

Hence, Eq. (16.63) gives:

$$\widetilde{\varepsilon_m}(t) = S^*_{mn}(\omega)\widetilde{\sigma_n}(t) \tag{16.65}$$

According to Eq. (16.64), the complex compliance, $S^*_{mn}(\omega)$, is equal to a factor of $i\omega$ times $S_{mn}(\omega)$, and $S_{mn}(\omega)$ is the Fourier transform of the creep compliance, $S_{mn}(t)$.

Alternatively, if we substitute sinusoidally varying strains in Eq. (16.5), we find that the sinusoidally varying stresses are:

$$\widetilde{\sigma_m}(t) = C^*_{mn}(\omega)\widetilde{\varepsilon_n}(t) \tag{16.66}$$

where the complex moduli are defined by:

$$C^*_{mn}(\omega) = i\omega C_{mn}(\omega) \tag{16.67}$$

And, $C_{mn}(\omega)$ are the Fourier transforms of the corresponding relaxation moduli. Alternatively, Eqs. (16.65) and (16.66) may be written in matrix form as:

$$\{\tilde{\varepsilon}(t)\} = [S^*(\omega)]\{\tilde{\sigma}(t)\} \tag{16.68}$$

$$\{\tilde{\sigma}(t)\} = [C^*(\omega)]\{\tilde{\varepsilon}(t)\} \tag{16.69}$$

where the complex compliance matrix and the complex modulus matrix must be related by $[S^*(\omega)] = [C^*(\omega)]^{-1}$.

Since the complex modulus is a complex variable, we can write it in terms of its real and imaginary parts as follows:

$$C^*_{mn}(\omega) = C'_{mn}(\omega) + iC''_{mn}(\omega) = C'_{mn}(\omega)[1 + i\eta_{mn}(\omega)] = \left|C^*_{mn}(\omega)\right| e^{-i\delta_{mn}(\omega)} \tag{16.70}$$

where

$C'_{mn}(\omega)$ = storage modulus
$C''_{mn}(\omega)$ = loss modulus
$\eta_{mn}(\omega)$ = loss factor $= \tan[\delta_{mn}(\omega)] = \dfrac{C''_{mn}(\omega)}{C'_{mn}(\omega)}$
$\delta_{mn}(\omega)$ = phase lag between $\widetilde{\sigma_m}(t)$ and $\widetilde{\varepsilon_n}(t)$

Thus, the real part of the complex modulus is associated with elastic energy storage, whereas the imaginary part is associated with energy dissipation, or damping. A physical interpretation of the one-dimensional forms of these equations may be given with the aid of the rotating vector diagram in Figure 16.10. The stress and strain vectors are both assumed to be rotating with angular velocity,ω, and the physical oscillation is generated by either the horizontal or vertical projection of the vectors. The complex exponential representations of the rotating stress and strain vectors in the diagram are:

$$\tilde{\sigma}(t) = \sigma\, e^{i(\omega t+\delta)}$$

$$\tilde{\varepsilon}(t) = \varepsilon\, e^{i\omega t} \tag{16.71}$$

so that the one-dimensional complex modulus is defined as:

$$C^*(\omega) = \frac{\tilde{\sigma}(t)}{\tilde{\varepsilon}(t)} = \frac{\sigma\, e^{i\delta}}{\varepsilon} = \frac{\sigma}{\varepsilon}(\cos\delta + \sin\delta)$$

$$= \frac{\sigma'}{\varepsilon} + i\frac{\sigma''}{\varepsilon} = C'(\omega) + iC''(\omega) = C'(\omega)\left[1 + i\eta(\omega)\right] \tag{16.72}$$

It is seen that the strain lags the stress by the phase angle, δ; the storage modulus, $C'(\omega)$, is the in-phase component of the stress, σ', divided by the strain, ε; the loss modulus, $C''(\omega)$, is the out-of-phase component of stress, σ'', divided by the strain, ε; and the loss factor, $\eta(\omega)$, is the tangent of the phase angle, δ. Experimental

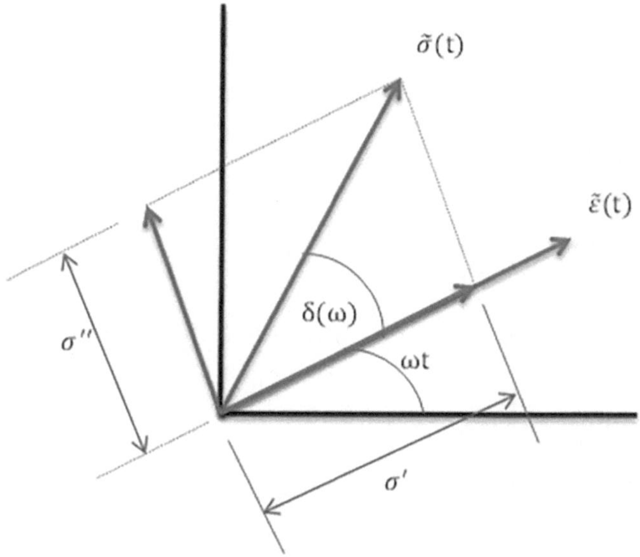

FIGURE 16.10 Rotating vector diagram for physical interpretation of the complex modulus.

determination of the complex modulus involves the measurement of the storage modulus, $C'(\omega)$, and the loss factor, $\eta(\omega)$, as a function of frequency, ω.

The inverse Fourier transform of the parameter $S_{mn}(\omega)$ is the creep compliance, $S_{mn}(t)$, as given by:

$$F^{-1}\left[S_{mn}(\omega)\right] = S_{mn}(t) = \frac{1}{2\pi}\int_{-\infty}^{\infty} S(\omega)e^{i\omega t}\,d\omega \tag{16.73}$$

where F^{-1} is the inverse Fourier transform operator. Equations (16.62) and (16.73) form the so-called Fourier transform pair, which makes it possible to transform back and forth between the time domain and the frequency domain. Since experimental frequency data are usually expressed in units of cycles per second, or Hz, it is convenient to define the frequency as $f = \frac{\omega}{2\pi}$ (Hz), so that the Fourier transform pair now becomes symmetric in form:

$$F\left[S_{mn}(t)\right] = S_{mn}(f) = \int_{-\infty}^{\infty} S_{mn}(t)e^{-i2\pi ft}\,dt \tag{16.74}$$

$$F^{-1}\left[S_{mn}(f)\right] = S_{mn}(t) = \int_{-\infty}^{\infty} S_{mn}(f)e^{i2\pi ft}\,df \tag{16.75}$$

16.1.5 Elastic–Viscoelastic Correspondence Principle

In most cases, the form of the stress–strain relationships for linear viscoelastic materials is the same as that for linear elastic materials. Such analogies between the equations for elastic and viscoelastic analysis have led to the formal recognition of an "elastic–viscoelastic correspondence principle." The correspondence principle for isotropic materials was apparently introduced by Lee [4], whereas the application to anisotropic materials was proposed by Biot [5]. The specific application of the correspondence principle to the viscoelastic analysis of anisotropic composites has been discussed in detail by Schapery [1,6] and Christensen [7]. If we have the necessary equations for a linear elastic solution to a problem, we simply make the corresponding substitutions in the equations to get the corresponding linear viscoelastic solution. One of the most important implications of the correspondence principle is that analytical models for predicting elastic properties of composites at both the micromechanical and the macromechanical levels can be easily converted for prediction of the corresponding viscoelastic properties. For example, the rule of mixtures for predicting the longitudinal modulus of a unidirectional composite can now be converted for viscoelastic relaxation problems by using the following equation:

$$E_1(t) = E_1^f(t)V^f + E^m(t)V^m \tag{16.76}$$

where

$E_1(t)$ = longitudinal relaxation modulus of composite.
$E_1^f(t)$ = longitudinal relaxation modulus of fiber

$E^m(t)$ = relaxation modulus of isotropic matrix
V^f = fiber volume fraction
V^m = matrix volume fraction

In most polymer matrix composites, the time dependency of the matrix material would be much more significant than that of the fiber, so the fiber modulus could be assumed to be elastic, and the time dependency of $E_1(t)$ would be governed by $E^m(t)$ alone. At the macromechanical level, equations such as the laminate force–deformation relationships can be converted to viscoelastic form using the correspondence principle. For example, the creep strains in a symmetric laminate under constant in-plane loading can be analyzed by employing the correspondence principle and a quasi-elastic analysis to give:

$$\begin{Bmatrix} \varepsilon_x^0(t) \\ \varepsilon_y^0(t) \\ \gamma_{xy}^0(t) \end{Bmatrix} = \begin{bmatrix} A'_{11}(t) & A'_{12}(t) & A'_{16}(t) \\ A'_{12}(t) & A'_{22}(t) & A'_{26}(t) \\ A'_{16}(t) & A'_{26}(t) & A'_{66}(t) \end{bmatrix} \begin{Bmatrix} N_x \\ N_y \\ N_{xy} \end{Bmatrix} \tag{16.77}$$

where

$A'_{ij}(t)$ = laminate creep compliances
N_x, N_y, N_{xy} = constant loads

When the correspondence principle is used for problems involving sinusoidally varying stresses and strains in viscoelastic composites, we must be particularly careful to make sure that the criteria for using the effective modulus theory are met. These restrictions are discussed in more detail, and applications of the correspondence principle to the prediction of complex moduli of particle and fiber composites are given in papers by Hashin [8,9]. For example [10], assuming that these criteria have been met, micromechanics equations such as Eq. (16.76) can be modified for the case of sinusoidal oscillations as:

$$E_1^*(\omega) = E_1^{f*}(\omega)V^f + E^{m*}(\omega)V^m \tag{16.78}$$

where

$E_1^*(\omega)$ = longitudinal complex modulus of composite.
$E_1^{f*}(\omega)$ = longitudinal complex modulus of fiber
$E^{m*}(\omega)$ = complex modulus of isotropic matrix

By setting the real parts of both sides of Eq. (16.78) equal, we find the composite longitudinal storage modulus to be:

$$E'_1(\omega) = E_1^{f'}(\omega)V^f + E^{m'}(\omega)V^m \tag{16.79}$$

where

$E_1'(\omega)$ = longitudinal storage modulus of composite.

$E_1^{f'}(\omega)$ = longitudinal storage modulus of fiber

$E^{m'}(\omega)$ = storage modulus of isotropic matrix

Similarly, by setting the imaginary parts of both sides of Eq. (16.78) equal, we find that the composite longitudinal loss modulus is:

$$E_1''(\omega) = E_1^{f''}(\omega)V^f + E^{m''}(\omega)V^m \tag{16.80}$$

where

$E_1''(\omega)$ = longitudinal loss modulus of composite

$E_1^{f''}(\omega)$ = longitudinal loss modulus of fiber

$E^{m''}(\omega)$ = loss modulus of isotropic matrix

The composite longitudinal loss factor is found by dividing Eq. (8.80) by Eq. (8.79):

$$\eta_1(\omega) = \frac{E_1''(\omega)}{E_1'(\omega)} = \frac{E_1^{f''}(\omega)V^f + E^{m''}(\omega)V^m}{E_1^{f'}(\omega)V^f + E^{m'}(\omega)V^m} \tag{16.81}$$

The complex forms of the other lamina properties can be determined in a similar manner. For oscillatory loading of symmetric viscoelastic laminates, Eq. (16.77) can be rewritten, so that the sinusoidally varying strains are related to the sinusoidally varying loads by:

$$\begin{Bmatrix} \tilde{\varepsilon}_x^0(t) \\ \tilde{\varepsilon}_y^0(t) \\ \tilde{\gamma}_{xy}^0(t) \end{Bmatrix} = \begin{bmatrix} A_{11}'^*(\omega) & A_{12}'^*(\omega) & A_{16}'^*(\omega) \\ A_{12}'^*(\omega) & A_{22}'^*(\omega) & A_{26}'^*(\omega) \\ A_{16}'^*(\omega) & A_{26}'^*(\omega) & A_{66}'^*(\omega) \end{bmatrix} \begin{Bmatrix} \tilde{N}_x(t) \\ \tilde{N}_y(t) \\ \tilde{N}_{xy}(t) \end{Bmatrix} \tag{16.82}$$

where $A_{ij}'^*(\omega)$ are the laminate complex extensional compliances.

16.2 DYNAMIC BEHAVIOR

The basic premise of all analyses presented in this section is that the criteria for valid use of the effective modulus theory have been met. That is, the scale of the inhomogeneity is assumed to be much smaller than the characteristic structural dimension and the characteristic wavelength of the dynamic stress distribution. Thus, all heterogeneous composite material properties are assumed to be effective properties of equivalent homogeneous materials. If the wavelength is not long in comparison with the scale of the inhomogeneity in the material, the wave shape is distorted as it travels through the material, and this is referred to as dispersion.

16.2.1 Longitudinal Wave Propagation

The longitudinal wave propagation and vibration in a homogeneous, isotropic, linear elastic bar (Figure 16.11) are governed by the one-dimensional wave equation:

$$\frac{\partial}{\partial x}\left(AE\frac{\partial u}{\partial x}\right)=\rho A\frac{\partial^2 u}{\partial t^2} \tag{16.83}$$

where

x = distance from the end of the bar
t = time
$u = u(x,t)$ is the longitudinal displacement of a cross section in bar at a distance x and time t
$A = A(x)$ is the cross-sectional area of the bar
ρ = mass density of the bar
$E = E(x)$ is the modulus of elasticity of the bar

It is assumed that the displacement $u(x,t)$ is uniform across a given cross section. Using effective modulus theory for a heterogeneous, specially orthotropic, linear elastic composite bar, we simply replace the properties ρ and E with the corresponding effective properties of an equivalent homogeneous material. The effective modulus E then depends on the orientation of fibers relative to the axis of the bar. For fibers oriented along the x direction, $E = E_1$; for fibers oriented along the transverse direction, $E = E_2$; and for a specially orthotropic laminate, we use the effective laminate engineering constant $E = E_x$. If the area and the modulus are not functions of position, Eq. (16.83) reduces to:

$$c^2\frac{\partial^2 u}{\partial x^2}=\frac{\partial^2 u}{\partial t^2} \tag{16.84}$$

where

$$c=\sqrt{\frac{E}{\rho}} = \text{wave speed}$$

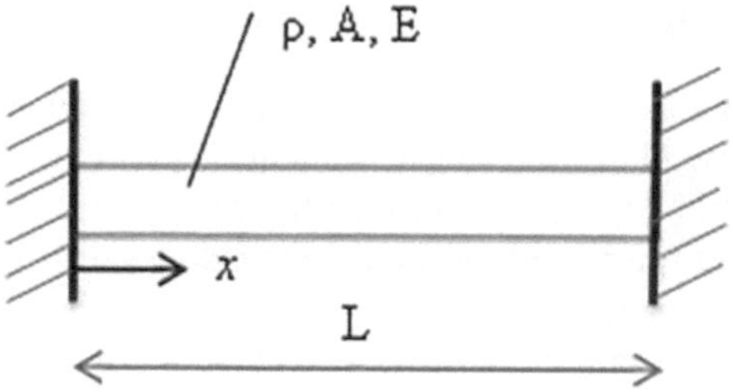

FIGURE 16.11 Bar of density ρ, cross-sectional area A, and length L.

The most common solutions to the one-dimensional wave equation are of the d'Alembert type or the separation-of-variables type. The d'Alembert solution is of the form:

$$u(x,t) = p(x+ct) + q(x-ct) \tag{16.85}$$

The function $p(x + ct)$ represents a wave traveling to the left with velocity c. That is, a point located at $\xi = x + ct$ moves to the left with velocity c if ξ is a constant, since $x = \xi - ct$. Similarly, $q(x - ct)$ represents a wave traveling to the right with velocity c. For a sine wave, we have:

$$u(x,t) = A \sin\frac{2\pi}{\lambda}(x+ct) + A \sin\frac{2\pi}{\lambda}(x+ct) \tag{16.86}$$

where

λ = wavelength

This wavelength must be greater than the scale of inhomogeneity, d, in order for the effective modulus theory to be valid. Alternatively, we can write Eq. (16.86) as:

$$u(x,t) = A \sin(2\pi kx + \omega t) + A \sin(2\pi kx - \omega t) \tag{16.87}$$

where

$k = \frac{1}{\lambda}$ = number of waves per unit distance

$\omega = \frac{2\pi c}{\lambda}$ = frequency of wave

Using trigonometric identities, we find that:

$$u(x,t) = 2A \sin 2\pi kx \cos\omega t \tag{16.88}$$

which represents a standing wave of profile $2A \sin 2\pi kx$, which oscillates with frequency ω. Generally, the combined wave motion in opposite directions is caused by reflections from the boundaries. Thus, wave propagation without reflection will not lead to a standing wave (or vibration). A separation-of-variables solution is found by letting:

$$u(x,t) = U(x)\, F(t) \tag{16.89}$$

where $U(x)$ is a function of x alone and $F(t)$ is a function of t alone. Substituting this solution in Eq. (16.84) and separating variables, we obtain:

$$c^2 \frac{1}{U}\frac{d^2u}{dx^2} = \frac{1}{F}\frac{d^2F}{dt^2} \tag{16.90}$$

The left-hand side of Eq. (16.90) is a function of x alone, and the right-hand side is a function of t alone; therefore, each side must be equal to a constant. If we let this constant be, say, $-\omega^2$, then Eq. (16.90) gives the two ordinary differential equations:

$$\frac{d^2F}{dt^2}+\omega^2F=0 \tag{16.91a}$$

$$\frac{d^2U}{dx^2}+\left(\frac{\omega}{c}\right)^2U=0 \tag{16.91b}$$

and the solutions to these equations are of the form:

$$F(t)=A_1\sin\omega t+B_1\cos\omega t \tag{16.92}$$

$$U(x)=A_2\sin\frac{\omega}{c}x+B_2\cos\frac{\omega}{c}x \tag{16.93}$$

where A_1 and B_1 depend on the initial conditions and A_2 and B_2 depend on the boundary conditions. For a bar that is fixed on both ends (Figure 16.11), the substitution of the boundary conditions $u(0, t) = u(L, t) = 0$ leads to the conclusion that $B_2 = 0$ and:

$$\sin\frac{\omega}{c}L=0 \tag{16.94}$$

Equation (16.94) is the eigenvalue equation, which has an infinite number of solutions, ω_n, such that:

$$\frac{\omega_n}{c}L=n\pi \tag{16.95}$$

where

n = mode number= 1,2,3, …, ∞
ω_n = eigenvalues or natural frequencies (rad s^{-1}) = $2\pi f_n$
f_n = natural frequencies (Hz)

Thus,

$$f_n=\frac{nc}{2L}=\frac{n}{2L}\left(\frac{E}{\rho}\right)^{1/2} \tag{16.96}$$

For the nth mode of vibration, the displacements are then:

$$u_n(x,t)=\left(A'\sin\omega_nt+B'\cos\omega_nt\right)\sin\frac{n\pi x}{L} \tag{16.97}$$

where

$A' = A_1A_2$ and $B' = B_1B_2$.

The mode shape for the nth mode is given by the eigenfunction:

$$U_n(x) = \sin \frac{n\pi x}{L} \tag{16.98}$$

And the general solution is the superposition of all modal responses:

$$u(x,t) = \sum_{n=1}^{\infty} (A' \sin \omega_n t + B' \cos \omega_n t) \sin \frac{n\pi x}{L} \tag{16.99}$$

Mode shapes, natural frequencies, and wavelengths for the first three modes of the fixed–fixed bar are shown in Figure 16.12. The most important point here is that as the mode number increases, the wavelength decreases and the use of effective modulus theory becomes more

questionable. In general, the wavelengths associated with typical mechanical vibration frequencies of structures in the audio frequency range will satisfy the criteria $d \ll \lambda$. Assuming that the criteria for the use of effective modulus theory have been met, the effective modulus of a specially orthotropic composite can be determined by measuring the longitudinal wave speed, c, in a specimen of density, ρ, and then solving for $E = c^2 \rho$. Alternatively, the nth mode natural frequency, f_n, can be measured in a vibration experiment, and the effective modulus can be found from an equation like Eq. (16.96).

16.2.2 Flexural Vibration

Transverse, or flexural, motion of a homogeneous, isotropic, linear elastic beam (Figure 16.13) without shear or rotary inertia effects is described by the well-known Bernoulli–Euler equation:

$$-\frac{\partial^2}{\partial x^2}\left(EI \frac{\partial^2 w}{\partial x^2}\right) = \rho A \frac{\partial^2 w}{\partial t^2} \tag{16.100}$$

where

I = moment of inertia of cross section about centroidal axis of beam
$w = w(x, t)$ is the transverse displacement of centroidal axis of beam

x, t, ρ, A, and E are as defined in Eq. (16.83). If the beam is such that EI is constant along the length, Eq. (16.100) reduces to:

$$\left(EI \frac{\partial^4 w}{\partial x^4}\right) + \rho A \frac{\partial^2 w}{\partial t^2} = 0 \tag{16.101}$$

Assuming that the criteria for the use of effective modulus theory have been met, these equations can be used for specially orthotropic composites or laminates without coupling if the modulus E is replaced by the effective flexural modulus, E_f.

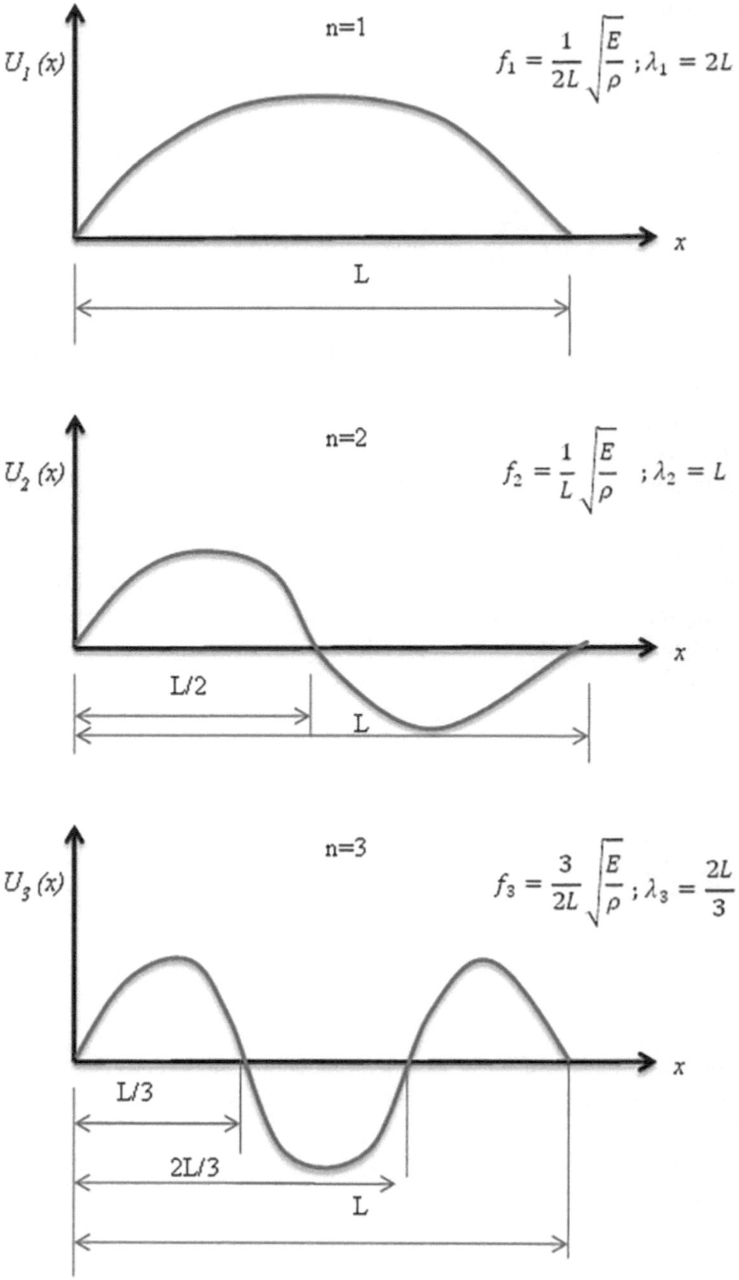

FIGURE 16.12 Mode shapes, natural frequencies, and wavelengths for the first three modes of longitudinal vibration of a bar with both ends fixed.

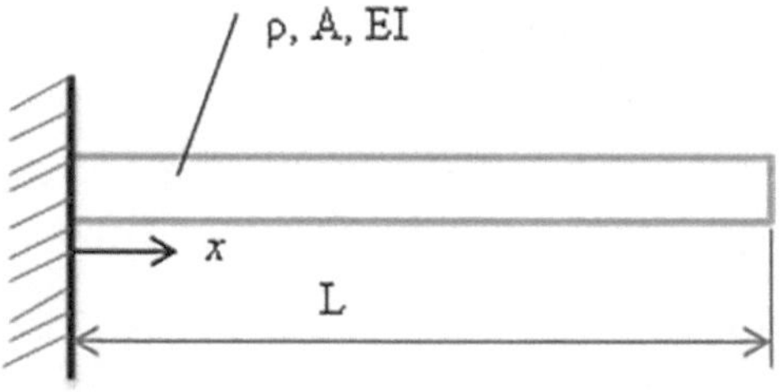

FIGURE 16.13 Cantilever beam for the Bernoulli–Euler beam theory.

As an example of a solution of the Bernoulli–Euler equation, consider a separation-of-variables solution for harmonic free vibration:

$$W(x,t) = W(x)e^{i\omega t} \tag{16.102}$$

where

ω = frequency
$W(x)$ = mode shape function

Substitution of this solution in Eq. (16.101) yields the equation:

$$\frac{d^4W(x)}{dx^4} - k^4W(x) = 0 \tag{16.103}$$

where

$$k = \left(\frac{\omega^2\rho A}{EI}\right)^{1/4}$$

The solution for Eq. (16.103) is of the form:

$$W(x) = C_1 \sin kx + C_2 \cos kx + C_3 \sinh kx + C_4 \cosh kx \tag{16.104}$$

where the constants C_1, C_2, C_3, and C_4 depend on the boundary conditions. For example, for a cantilever beam (Figure 16.13) the four boundary conditions yield the following relationships:

$$W(x) = 0 \text{ when } x = 0; \quad \text{therefore, } C_2 = -C_4$$

$$\frac{dW(x)}{dx} = 0 \text{ when } w = 0; \quad \text{therefore, } C_1 = -C_3$$

$$\frac{d^2W(x)}{dx^2} = 0 \text{ when } x = L;$$

Therefore,

$$C_1(\sin kL + \sinh kL) + C_2(\cos kL + \cosh kL) = 0$$

$$\frac{d^3W(x)}{dx^3} = 0 \text{ when } x = L;$$

Therefore,

$$C_1(\cos kL + \cosh kL) + C_2(\sin kL + \sinh kL) = 0$$

For nontrivial solutions C_1 and in the last two equations, the determinant of the coefficients must be equal to zero:

$$\cos kL \cosh kL + 1 = 0 \tag{16.105}$$

This is the eigenvalue equation for the cantilever beam, C_2 which has an infinite number of solutions, $k_n L$. The subscript n refers to the mode number. The eigenvalues for the first three modes are:

$$\begin{aligned} k_1L &= 1.875 \\ k_2L &= 4.694 \\ k_3L &= 7.855 \end{aligned} \tag{16.106}$$

Substituting the eigenvalues in the definition of k [Eq. (16.103)], rearranging, and using the relationship ($\omega = 2\pi f$), we have the frequency equation:

$$f_n = \frac{(k_nL)^2}{2\pi L^2}\left(\frac{EI}{\rho A}\right)^{1/2} \tag{16.107}$$

The mode shape function for the nth mode is then:

$$W_n(x) = C_2\left[\cos k_nx - \cosh k_nx + \sigma_n(\sin k_nx - \sinh k_nx)\right] \tag{16.108}$$

where

$$\sigma_n = \frac{\sin k_nL - \sinh k_nL}{\cos k_nL + \cosh k_nL}$$

The mode shapes and frequencies for the first three modes of the cantilever beam are shown in

Figure 16.14. The effect of increasing the mode number and the corresponding reduction in wavelength is again apparent. If transverse shear and rotary inertia effects are included in the derivation of the equation of motion for transverse vibration of a beam, the result is the well-known Timoshenko beam equation:

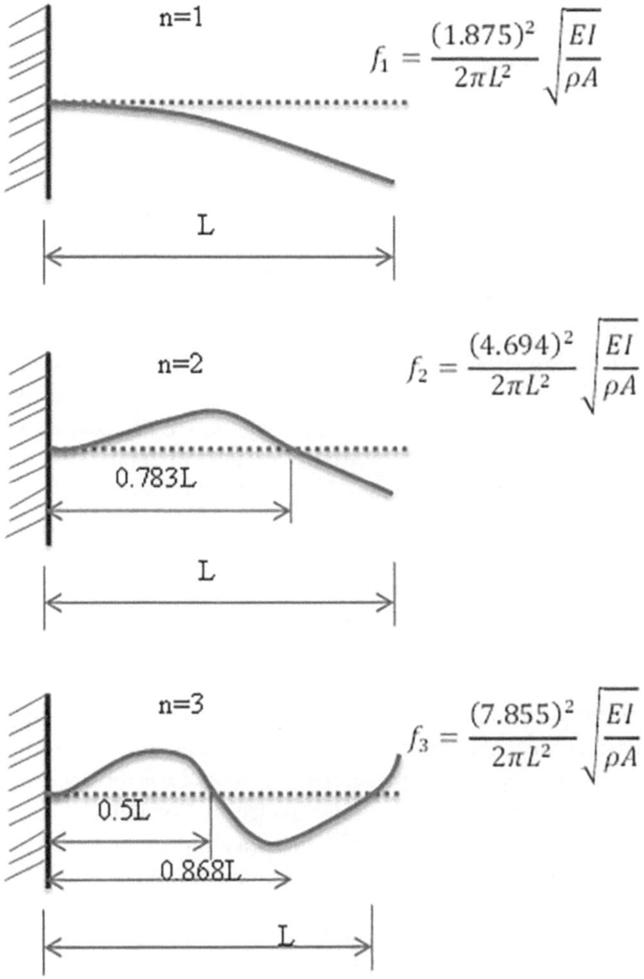

FIGURE 16.14 Mode shapes and natural frequencies for the first three modes of flexural vibration of the cantilever beam.

$$EI\frac{\partial^4 w}{\partial x^4} + \rho A\frac{\partial^2 w}{\partial t^2} + \frac{J\rho}{FG}\frac{\partial^4 w}{\partial t^4} - \left(J + \frac{EI\rho}{FG}\right)\frac{\partial^4 w}{\partial x^2 \partial t^2} = 0 \quad (16.109)$$

where

J = rotary inertia per unit length
F = shape factor for cross section
G = shear modulus

As in the previous section, the equations developed here can be used in dynamic mechanical testing to determine the effective moduli of a composite specimen.

The equations can also be converted to linear viscoelastic form by replacing the elastic moduli with the corresponding complex moduli, or by deriving the equation of motion from a viscoelastic stress–strain relationship.

16.2.3 Damping Analysis

Damping is simply the dissipation of energy during dynamic deformation. As structures and machines are pushed to higher and higher levels of precision and performance, and as the control of noise and vibration becomes more of a societal concern, it is becoming essential to take damping into account in the design process. In conventional metallic structures, it is commonly accepted that much of the damping comes from friction in structural joints or from add-on surface damping treatments because the damping in the metal itself is typically very low. On the other hand, polymer composites have generated increased interest in the development of highly damped, lightweight, structural composites because of their good damping characteristics and the inherent design flexibility that allows trade-offs between such properties as damping and stiffness. The purpose of this section is to give a brief overview of the analysis of linear viscoelastic damping in composites.

Viscoelastic behavior of fiber and/or matrix materials is not the only mechanism for structural damping in composite materials although it does appear to be the dominant mechanism in undamaged polymer composites vibrating at small amplitudes. Other damping mechanisms include thermoelastic damping due to cyclic heat flow, coulomb friction due to slip in un-bonded regions of the fiber/matrix interface, and energy dissipation at sites of cracks and/or delaminations. Thermoelastic damping is generally more important for metal composites than for polymer composites. Damping due to poor interface bonding, cracks, and/or delaminations cannot be relied upon in the design of structures, but the measurement of such damping may be the basis of a valuable nondestructive evaluation methodology.

In order to understand linear viscoelastic damping better, it is important to recognize the relationship between the timescale of the applied deformation and the internal timescale of the material. The timescale for cyclic deformation is determined by the oscillation frequency, ω. Recall that the relaxation times, λ_i, or retardation times, ρ_i, are measures of the internal timescale of the material. We will now use the Zener single-relaxation model to illustrate how damping depends on the relationship between these two timescales. For sinusoidal oscillation of the Zener single-relaxation model (Figure 16.6a), we can write:

$$\sigma = \sigma_0 e^{i\omega t} = \left(E' + iE''\right)\varepsilon \tag{16.110}$$

where

σ = stress
σ_0 = stress amplitude
ε = strain
ω = frequency

E' = storage modulus
E'' = loss modulus
i = imaginary operator = $(-1)^{1/2}$

Substituting Eq. (16.110) in the stress–strain relationship for the Zener model [Eq. (16.37)] and separating into real and imaginary parts, we find that:

$$E' = E'(\omega) = \frac{k_0 + (k_0 + k_1)\omega^2\lambda_1^2}{1+\omega^2\lambda_1^2} \tag{16.111}$$

$$E'' = E''(\omega) = \frac{\omega\lambda_1 k_1}{1+\omega^2\lambda_1^2} \tag{16.112}$$

$$\eta = \eta(\omega) = \frac{E''(\omega)}{E'(\omega)} = \frac{\omega\lambda_1 k_1}{k_0 + (k_0 + k_1)\omega^2\lambda_1^2} \tag{16.113}$$

where

$\lambda_1 = \dfrac{\mu_1}{k_1}$ = relaxation time from Eq. (16.39)

$\eta = \eta(\omega)$ = loss factor

The variations of E' and E'' with frequency, ω, are shown schematically in Figure 16.15. Note that when the frequency is the reciprocal of the relaxation time, $\omega = \dfrac{1}{\lambda_i}$, the loss modulus peaks and the storage modulus passes through a transition region. Such damping peaks in the frequency domain are often referred to as "relaxation peaks." The loss factor has a peak at a different frequency, not shown in Figure 16.15 because the relative position of that peak depends on the numerical

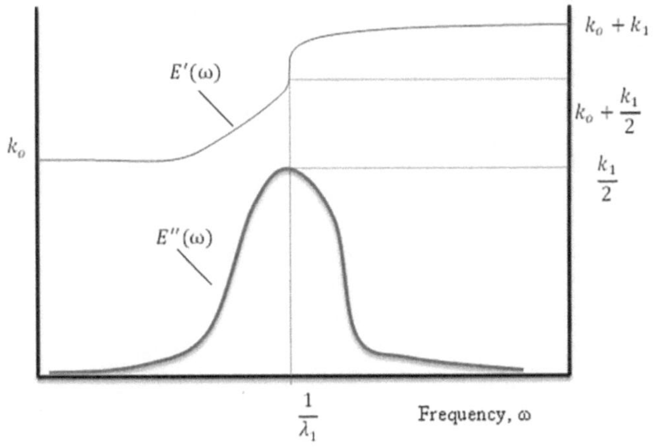

FIGURE 16.15 Variation of storage modulus, $E'(\omega)$, and loss modulus, $E''(\omega)$, with frequency for the Zener single-relaxation model.

values of the parameters. But the important point is that the dissipation of energy, whether characterized by the loss modulus or the loss factor, is maximized when the timescale of the deformation is the same as the internal timescale of the material. If the two timescales are substantially different, the energy dissipation is reduced. For example, notice in Figure 16.15 that $E'' \to 0$ as $\omega \to 0$ and as $\omega \to \infty$. This behavior is typical for viscoelastic materials.

If the damping mechanism is of the linear viscoelastic type, there are two basic approaches to the development of analytical models, both of which are based on the existence of experimental damping data for constituent materials. The two approaches are as follows:

i. The use of the elastic–viscoelastic correspondence principle in combination with elastic solutions from the mechanics of materials or the elasticity theory
ii. The use of a strain energy formulation that relates the total damping in the structure to the damping of each element and the fraction of the total strain energy stored in that element.

REFERENCES

1. Schapery R.A. (1974), Viscoelastic behavior and analysis of composite materials, in *Composite Materials Volume 2---Mechanics of Composite Materials*, (G. P. Sendeckyj, ed.), Academic Press, New York, pp. 85–168.
2. Halpin J.C. and Pagano N.J. (1968), "Observations on linear anisotropic viscoelasticity," *Journal of Composite Materials*, Vol. 2(1), pp. 68–80.
3. Fung Y.C. (1965), *Foundations of Solid Mechanics*, Prentice-Hall, Inc., Englewood Cliffs, NJ.
4. Lee E.H. (1955), "Stress analysis in viscoelastic bodies", *Quarterly of Applied Mathematics*, Vol. 13, pp. 183–190.
5. Biot M.A. (1958), "Linear Thermodynamics and the Mechanics of Solids", *Proceedings of the Third U.S. National Congress of Applied Mechanics*, pp. 1–18.
6. Schapery R.A. (1967), "Stress analysis of viscoelastic composite materials," *Journal of Composite Materials*, Vol. 1, pp. 228–267.
7. Christensen R.M. (1979), *Mechanics of Composite Materials*, John Wiley & Sons, New York.
8. Hashin Z. (1970), "Complex moduli of viscoelastic composites I---General theory and application to particulate composites," *International Journal of Solids and Structures*, Vol. 6, pp. 539–552.
9. Hashin Z. (1970), "Complex moduli of viscoelastic composites II---Fiber reinforced materials", *International Journal of Solids and Structures*, Vol. 6, pp. 797–807.
10. Gibson R.F. (2012), *Principles of Composite Material Mechanics*, CRC press, Boca Raton, FL, pp. 4, 387–424.

17 Mechanical Testing of Composites

The testing of materials has got immense importance because it gives the required data, that is, mechanical and other required properties for the designing and analysis of the structures for its safe, reliable, and cost-effective functioning. When one uses the data derived from tests, the following questions arise:

i. What tests should be carried out to give the required data?
ii. How precisely are these tests conducted and who guarantees it?
iii. What do the data actually mean?
iv. Are these data produced reliable?
v. Are data obtained from small test specimens meaningful when large structures are being designed?
vi. What will be the effect of operating environment?

These questions arise when one needs to establish the response of these materials to various types of loading such as tensile, compressive, or shear, for short-term or long-term duration, or cyclic. Further, the study of their behavior in the presence of high or low temperatures or other environments that might significantly modify their behavior is essential.

17.1 SOCIETIES FOR TESTING STANDARDS

There are few societies that develop the standards related to composites. They essentially provide the information and guidance necessary to design and fabricate end items from composite materials. Their primary purpose is the standardization of engineering data development methodologies related to testing, data reduction, and data reporting of property data for current and emerging composite materials. In the following, we briefly give the background of these societies.

a. *ASTM International*, formerly known as the American Society for Testing and Materials, was founded in 1898 by chemists and engineers from Pennsylvania Railroad, USA. At the time of its establishment, the organization was known as the American Section of the International Association for Testing and Materials. In 2011, the society became known as ASTM International. ASTM members deliver the test methods, specifications, guides, and practices that support industries and governments worldwide. ASTM International standards are developed in accordance with the guiding principles of the World Trade Organization for the development of international standards: coherence, consensus, development dimension, effectiveness, impartiality, openness, relevance, and transparency. The ASTM standards are also available in the volume form as *Composite Materials Handbook*.

b. *Composite Research Advisory Group* (CRAG) set about in the early 1980s to attempt to define what the best practice should be over a range of test methods. The CRAG recommendations were proposed to the British Standards Institution and subsequently had a considerable effect in the development of new international standards.
c. *Society of Automobile Engineers* (SAE) was formed in 1905. In the early 1900s, there were a lot of automobile companies worldwide, which needed to address their common design issues, patent protection, and the development of engineering standards. The development of standards for composite fabrication, testing, etc., is under progress.

17.2 OBJECTIVES OF MECHANICAL TESTING

The development of the mechanical testing of the materials depends upon other scientific factors. These factors help in better understanding and facilitate the progress in evaluating the various processes. These processes include the following:

i. Quality control of a process
ii. Quality assurance for the material developed and structure fabricated from thereof
iii. Better material selection
iv. Comparisons between available materials
v. Can be used as indicators in materials development programs
vi. Design analysis
vii. Predictions of performance under conditions other than test conditions
viii. Starting points in the formulation of new theories

It should be noted that these processes are dependent upon each other. However, if they are considered individually, then the data required can be different for the evaluation. For example, some tests are carried out as multipurpose tests using various processes. A conventional tensile test carried out under fixed conditions may serve quality control function, whereas the one carried out under varying factors such as temperature, strain rate, and humidity may provide information on load bearing capacity of the material. The properties evaluated for materials such as composite are very sensitive to various internal structure factors. However, these factors depend mainly upon the fabrication process or other factors. The internal structure factors that affect the properties are, in general, at the atomic or molecular level. These factors mostly affect the matrix and fiber–matrix interface structure. The mechanical properties of the fibrous composite depend on several factors of the composition. These factors are listed below again for the sake of completeness:

i. Properties of the fiber
ii. Surface character of the fiber
iii. Properties of the matrix material

iv. Properties of any other phase
v. Volume fraction of the second phase (and of any other phase)
vi. Spatial distribution and alignment of the second phase (including fabric weave)
vii. Nature of the interfaces

Another important factor is processing of the composites. There are many parameters that control the processing of composites that access the quality of adhesion between fiber and matrix, physical integrity, and the overall quality of the final structure.

In the case of composites, the spatial distribution and alignment of fibers are the most dominating factors that cause the variation of properties. The spatial distribution and alignment of the fibers can change during the same fabrication process. Thus, for a given fabrication process the property evaluated from the composite material may show a large variation.

17.3 EFFECT OF ANISOTROPY

The long-fiber composite exhibits the characteristics of inhomogeneity, anisotropy, and inelasticity. If the composite is viscoelastic, then the testing procedure demands much more things. However, we will not consider this fact here. We will consider the effect of anisotropy on mechanical testing. The following are the key points in the mechanical testing of long-fiber composites:

i. Generation of a uniform stress field in the critical reference volume
ii. Avoiding the "end-effects"
iii. Attainment of adequate loading levels without damage or failure near the loading points
iv. Appropriate specimen dimensions related to the scale of structural inhomogeneities
v. Tension–shear coupling

The first four considerations are similar to the testing of homogeneous isotropic materials. These considerations give rise to various constraints on specimen dimension, test configurations, and machine specifications. However, the fact of heterogeneity imposes more severe constraints and demands more considerations while testing. In the case of the composite, St. Venant's principle reflects a more stringent requirement. In anisotropic composites, the region of uniform stress is developed more gradually. It is shown that the decay length (λ) is of the following order:

$$\lambda = b\left(\frac{E_{11}}{G_{12}}\right)^{1/2} \tag{17.1}$$

where b is the maximum dimension of the cross section. In the case of rectangular strips subjected to end tractions,

$$\lambda \approx \frac{b}{2\pi}\left(\frac{E_{11}}{G_{12}}\right)^{1/2} \tag{17.2}$$

where λ is the distance over which a self-equilibrated stress applied at the ends decays to its end value of $1/e$. In the above expressions, the ratio $\frac{E_{11}}{G_{12}}$, that is, degree of orthotropy, is an important factor. For unidirectional composites, this ratio varies between 40 and 50, whereas for an isotropic material, this is about 3. Thus, the ratio of respective decay lengths is about 3.5:1.

Another problem with the composite testing arises when the loading directions do not coincide with the symmetry axes of the specimen. This situation gives rise to coupling between the normal and shear modes. This results in extraneous forces and deformations in the coupons. For example, a coupon in tension will exhibit the in-plane shear and a coupon in bending will exhibit the additional twisting. Further, if the laminate has layers oriented with respect to each other, then there will be a mismatch of interface deformation due to different degrees of tension–shear coupling of adjacent layers. This may lead to delamination. The severity of the mismatch of interface deformation depends upon various factors such as stacking sequence, test modes, degree of asymmetry, and end constraints. Thus, to summarize, the main practical consequences of anisotropy are as follows:

i. There will be severe end-effects, which extend in the direction of higher stiffness. Further, this is a function of both the specimen geometry and the anisotropy.
ii. A premature failure in grips or at other loading points may occur.
iii. A premature delamination at free edges or other unintended failure modes may take place. These failures emanate from the interactions between the macrostructure of the composite and system of external forces.
iv. There can be property imbalances of the lamina, for example, a tensile modulus (or strength) which is dominated by the properties of the fiber, and a shear modulus (or strength) which is dominated by the properties of the matrix.

17.4 NATURE AND QUALITY OF DATA

The quality of the mechanical properties derived from the mechanical testing depends upon various factors. These factors include the following:

i. Precision
ii. Accuracy
iii. Authenticity and repeatability
iv. Relevance to the test objective
v. Physical significance

The factors precision and accuracy can be attributed to statistical analysis. However, they cannot be separated if the data set is small. The remaining three factors cannot readily be quantified.

Usually, there will be scatter in the measured data. The scatter is attributed to the combined effect of the factors:

i. Precision with which the measurements are made
ii. Accuracy with which the measurements are made
iii. Variations in the structure of the test coupon in the set itself

The mean value and a measure of width of the distribution, such as standard deviation or range, are the two main statistical factors that are used to characterize the distribution of the values. Apart from their direct role as a measure of the variability in a set of data, the variance and the standard deviation, which is square root of variance, can be used to infer the following points:

i. Confidence limits for a set of data
ii. Reliability of apparent differences between sets of data
iii. Combined uncertainty of measurements when there are several sources of variability
iv. Separate variabilities when several factors have affected a set of data
v. Goodness of fit when a correlation between a dependent and an independent variable is derived.

17.5 SAMPLES AND SPECIMEN FOR TESTING

The samples from which the specimens are made for mechanical testing can be in the form of the following:

i. Pultrusions
ii. Filament wound tubes, and
iii. Flat sheets

The former two types are used because they represent the most important fabrication processes and because of their ease of fabrication process. As we know from our earlier studies, in pultrusion the fibers are aligned along the pultrusion axis. In the case of filament wound tubes, the fibers are aligned either spirally or circumferentially. However, the alignment can be optimally chosen in the case of other filament shapes. In these two fabrication processes, the degree of void content is less and there is better consolidation of structure. The advantage of using the tubular specimens is the ease with which specimen can be subjected to axial tension or compression, internal pressure, torsion, and multiaxial loading. However, the limitations with these specimens include the following: The cost of fabrication and testing is high, and fabrication may result in different microstructure and hence different equivalent properties. The flat sheets available for commercial use come in the following four categories:

i. Layers of unidirectional fibers aligned with reference to an axis
ii. Sheets of randomly oriented fibers in a plane
iii. Layers of woven fibers aligned with reference to an axis
iv. Sandwich structure

The flat sheets, depending upon the nature of alignment of the fibers, can result in various behaviors such as orthotropic, transversely isotropic, or even isotropic. When one uses the mechanical properties of the composite, it is essential to quote the volume fractions and spatial arrangement of fibers. The flat specimen is an obvious choice because of economic reason. The limitations with flat specimens can be the following:

i. Specific states of stress cannot be developed. For example, the state of pure shear is difficult to develop in such specimens.
ii. The axial compression is also a difficult issue due to buckling.
iii. Further, developing a combined state of stress in such specimens is also a difficult task.

17.6 MISCELLANEOUS ISSUES WITH TESTING

The other issues associated with the mechanical testing include the following:

i. Stress concentrations due to material discontinuities at free edges and ply drop-off regions, which result in early failures.
ii. In the case of compression testing, there is a susceptibility to buckling for thin specimens. This type of testing demands for additional fixtures.
iii. Flat specimens require special geometry for the purpose of gripping. For this reason, the flat specimens with end-tabs are favorably used.
iv. The composite is heterogeneous, and the volume fractions are the essential data required with the mechanical properties. Hence, additional tests are required to determine the volume fractions and void content, if any.
v. The mechanical properties determined are affected by moisture content. Hence, in some applications the amount of moisture present in the composite is required. Therefore, additional tests are required to determine the moisture content in the composite.
vi. Further, nondestructive evaluation (NDE) methods are required to assess the quality of the fabricated material and damage development during the loading.

The end-tabs are a special requirement in the case of flat specimens. Therefore, it needs additional information. The end-tabs are used almost universally to reduce the probability of failure initiating at the grips during a tensile test. End-tabs can also facilitate accurate alignment of the specimen in the test machine, provided that they are symmetrical and properly positioned on the specimen, but if they are deficient in these respects, they can cause misalignment and introduce stress concentrations.

17.7 PRIMARY PROPERTIES

There is a large set of properties of a long-fiber composite that one needs in design and analysis. However, it is generally agreed that the minimum requirements to assess the three main properties – modulus, strength, and ductility – are the parameters listed below:

i. Tensile modulus
ii. Compressive modulus (uniaxial)
iii. Flexural modulus
iv. Shear modulus (in plane)
v. Lateral contraction ratios
vi. Tensile strength
vii. Compressive strength (uniaxial)
viii. Flexural strength
ix. Apparent inter-laminar shear strength
x. Fracture toughness (various modes)

The requirement of property evaluation listed above is based upon the tests for an isotropic homogeneous sample. However, this minimum requirement is not sufficient to completely quantify the strength and stiffness tensors. Further, it neglects the viscoelastic behavior aspect of the composites. All the tests included in the minimum requirement of the property evaluation are not carried out by most of the industries working in composites. Their objective can be different and carry out some of the tests. For example, most of the fiber manufacturing companies give the properties of the composite which are fiber properties dominant. In such cases, the properties such as axial tensile and flexure are given more significance. However, the resin manufacturers tend to give more significance to compression and shear properties of the composite. In this case, these properties are dominated by matrix properties. The call for "open-hole" tests reflects reservations about the reliability of the theories of failure and about the relevance and relative paucity of the empirical evidence from conventional fracture toughness tests. The protagonists of such tests sometimes seem to be preoccupied with a search for authentic and/or definitive data which is perpetually frustrated by a preponderance of mixed-mode failures in their experiments.

The following are the primary engineering properties for preliminary selection of composite materials in a commercial aircraft industry:

i. Tensile strength at room temperature
ii. Uniaxial compression at room temperature
iii. Inter-laminar shear at room temperature
iv. Open-hole tension at room temperature
v. Open-hole compression at 93°C
vi. Hot/wet compression strength
vii. Edge-plate compression strength after impact at room temperature

The tests mentioned above are the essential tests in the *initial testing* phase. The quality assessment of the composites used in the specimens to be tested must be done prior to the testing whenever possible. If the composite used in the specimens is of low quality with defects, then the properties measured are spurious. It can mislead the design and analysis procedure and result in a premature and catastrophic failure of the structure. Hence, the quality assessment of composites before it is used in specimen or actual structure fabrication is essential.

The following quality assessment in composites is essential: quality of bond between fiber and matrix, voids, broken fibers, matrix cracks, and delaminations. In the following, we will see the various methods by which we can assess the quality of the composites.

17.7.1 Microscopy

The microscopy is one of the best methods that provide the firsthand information on the form of damage. The microscopy can provide the information such as the following:

i. Shape of the fibers
ii. Geometry and uniformity of the fiber spacing
iii. Presence of voids
iv. Regions rich or poor in matrix
v. Fiber alignment

The microscopy method has limitations like that it can give the information inside the composite. It can give the information on the surface as mentioned above. It cannot give the information like fiber–matrix bond, broken fibers, matrix cracks, and delaminations inside the composite.

17.7.2 Ultrasonic Inspection

Ultrasonic inspection is a nondestructive method of testing. Using this method, one can assess the quality of the composite. The ultrasonic testing method includes the propagation of mechanical waves through the object to be inspected. The mechanical waves propagated are in the range from 100kHz to 25MHz. Some of the waves propagated are reflected or transmitted at the other end. The intensity of the waves at the other end is measured by a receiving transducer.

There are two types of waves: longitudinal and transverse waves. In longitudinal waves, the direction of oscillation of atoms and the direction of propagation of the wave are along the same direction. In transverse waves, the direction of oscillation of atoms is perpendicular to the direction of propagation of the wave. The longitudinal waves propagate in all materials, whereas the transverse waves propagate only in solid materials. Further, due to the different type of oscillation, transverse waves travel at lower speeds. When the wave propagating in the material is intercepted by a defect and interfaces (like change from fiber to matrix material and vice versa or a foreign particle), the energy transmitted through the material also gets reduced due to the effect of reflection and attenuation. Thus, one can use both reflection and transmission forms of energy for ultrasonic inspection. The ultrasonic beam requires a transfer medium. In general, water is used as a transfer medium. This is a disadvantage of this method. Further, the use of water during the test process can lead to absorption of water by the composite.

In recent years, the new developments in the ultrasonic testing have made this process very sophisticated and attractive. One can get the complete map or intensity distribution corresponding to the discontinuity in the material. Such a map is called C-scan.

The detailed information on the ultrasonic inspection can be found in ASTM E114-90 for Pulse-Echo method, E214-68(91) for Reflection method, E317-93 for Pulse-Echo, and E494-91 for ultrasonic velocities.

17.7.3 X-Ray Inspection

The X-ray technique is a very useful technique. It uses the electromagnetic waves of extremely of short wavelength. These waves are capable of penetrating solid substances and are affected by discontinuities much as other waves. An X-ray opaque penetrant is introduced in the damaged area as a liquid solution or suspension so that it fills the cracks and delaminations, and makes them clearly visible on X-ray films as a dark region. Opaque dye penetrant such as tetrabromoethane (TBE) is used in these processes. It is cautioned that the frequent use of dye penetrant should be avoided. This is because the penetrant actually enhances the crack growth. Thus, under loading the frequent use of penetrant will increase the growth rate. Hence, this technique is treated as effectively destructive. The regions with lower density such as voids, defects, and cracks absorb less radiation. This results in a higher intensity of the radiation that reaches a photographic film or plate placed on the far side of the sample. The higher intensity causes the darkening of the film or the plate. Thus, the darker areas of the film indicate the outline of the low-density region. Standards related to this technique are E 94-93, E 142-92, and E 1316-94.

17.7.4 Thermography

This is one of the sophisticated techniques that are used in infrared thermography. The advantage of this technique is that it does not require any interruption for inspection. Thus, it is well suited for fatigue testing. This technique is based on the principles that the infrared thermography detects the heat generated from a source. In case of damage in the composite, there are two types of such heat sources. The first one is hysteresis evolving from resin and interface. The second source is heating due to friction between the cracks and delaminated interfaces. Thus, the area that appears hot on the thermographs is the area of damage. Once the area of damage is detected, one can zoom into it and get more details. The ASTM standard guide for nondestructive testing of polymer matrix composites used in aerospace applications is E2533-09.

17.8 PHYSICAL PROPERTIES

The physical properties of the composite play an important role in the measured mechanical properties. There is a direct dependence of mechanical properties on the physical properties. For example, the mechanical properties are directly dependent on fiber volume fractions. Here, we will consider the following physical property measurements.

17.8.1 Density

From our basic knowledge, the density of a material is defined as mass of the material per unit volume. A test method to determine the density of a material is detailed in ASTM standard D792-91. This method is used to determine the density of the composite and its constituents. The key points of the test procedure are explained in the following paragraphs.

The density of a material is determined using its weight in air, W_a, and in water, W_w. The densities of air (negligible) and water, ρ_w (= 0.9975 g cm^{-3} at 23°C), are taken as known parameters in this test. The volume of the specimen is determined from the difference between the weight of the material in air and the weight in water and using the known density of water. Then, the composite density ρ_c is given as:

$$\rho_c = \frac{W_a}{\dfrac{W_a - W_w}{\rho_w}} = \frac{\rho_w W_a}{W_a - W_w} \tag{17.3}$$

17.8.2 Fiber Volume Fraction

The fiber volume fraction is an important factor in composites as it governs the properties of a composite. The usual fiber volume fraction ranges from 30 to 65%. As we know, the lower end depends upon the significance of property contribution of the fibers, whereas the upper end depends upon the effective, defect-free packing. In the following, we will see some methods to determine the fiber volume fractions.

i. *First method*: In this method, the number of fibers is counted in several measured representative areas of a polished surface of the composite under magnification. Then, the diameter or the cross-sectional area of one or more fibers is measured. Then, the average fiber volume fraction is calculated as the percentage of area that is fiber. The advantage of this method is that it is simple and provides information about the type and uniformity of fiber spacing as well as indication of the void content. However, it should be noted that it is a crude method.
ii. *Second method*: In this method, the matrix material is digested or dissolved by putting a measured volume of composite in an acid bath. Then, the (dry) fibers remaining after digestion are weighed. Thus, knowing the density of the fibers, the volume of fibers and the fiber volume fraction can be determined. One should be careful to choose the liquid for digestion such that the fibers are not digested. Generally, hot nitric acid is used for carbon/epoxy. The ASTM standards used for digestion method are D3171-76 (1990) for polymeric composites and D3553-76 (1989) for metal matrix composites.
iii. *Third method*: In this method, one determines the density of the composites and then calculates the fiber volume fraction knowing the density of the fiber and the matrix. This method makes an assumption that the void content is negligible; however, it is not true for any composite. Hence, the results of this method may vary with the results of the earlier two methods.

17.8.3 Void Content

Unlike in other conventional materials, polymer and ceramic matrix materials have to be tested for one more physical property like void content. These composites have voids after fabrication. A composite with voids can affect the mechanical and thermal properties, strengths, and resistance to fatigue and corrosion. A composite with less than 1% void content is treated as a well-fabricated composite. Further, a composite with up to 7% void content is regarded as a poorly fabricated composite. The void content (V_c) is measured from experimental composite density (ρ_e) and theoretical composite density (ρ_t).

The void content in percent is simply the ratio of difference between experimental and theoretical densities to theoretical density. If one knows the densities of the constituents and resin content, then theoretical density can be found. The methods are described in ASTM standard D2534-91, which requires the use of ASTM standard D2584-68(1985) for determination of the resin content. The theoretical density of the composite is the weight of the composite per unit volume. The volume of the composite is the sum of the volume of the fiber and resin. The volume of the fiber and resin can be found from their weights and respective densities. Thus, the theoretical composite density is given by:

$$\rho_t = \frac{\text{Weight of composite}}{\text{Volume of composite}} \tag{17.4}$$

Using the volumes of fiber and resin, V_f and V_m:

$$\rho_t = \frac{W_c}{V_m + V_f} = \frac{W_c}{\frac{W_m}{\rho_m} + \frac{W_f}{\rho_f}} \tag{17.5}$$

where W_f is the weight of the fiber, W_m is the weight of the matrix, ρ_f is the density of the fiber, and ρ_m is the density of the matrix. The percentage void content is given as:

$$V_c(\%) = \frac{100(\rho_t - \rho_e)}{\rho_t} \tag{17.6}$$

17.8.4 Moisture Content

The composite materials when exposed to the environment or water absorb moisture. The absorption of the moisture results in the expansion. This also affects the degradation of the mechanical as well as thermal properties. The moisture content in a polymeric composite is given in terms of percent of moisture by weight. Hence, to measure the moisture content a sample is weighed at the ambient conditions. Then, the sample is dried and weighed again. The difference in these two weights per unit weight of the dry sample gives the weight change due to moisture content. Figure 17.1 shows a qualitative variation in weight change over time due to two different drying conditions for a polymer composite. From Figure 17.1, it can be seen that the drying

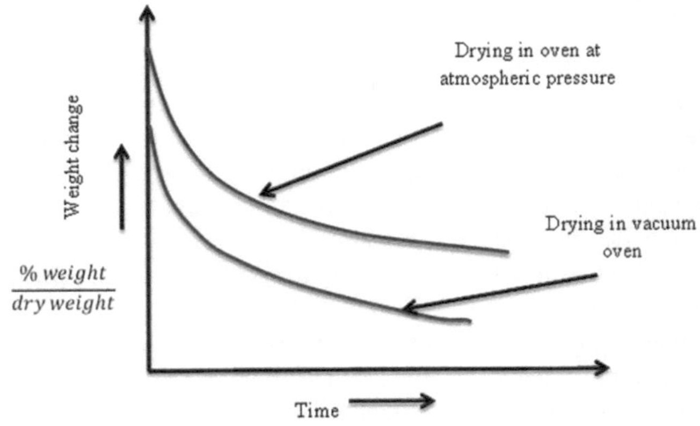

FIGURE 17.1 Effect of moisture on weight change.

in vacuum increases the desorption rate significantly. That is why the vacuum ovens are used in the laboratory for drying the specimens.

17.9 TENSILE AND COMPRESSIVE TESTING

The strain measurement along with stress measurement is an essential component of the measurements during the testing. The stress–strain variation depicts the behavior of the composite material under test. It gives the information about the linear and nonlinear behavior. This information is very essential when one is measuring the other properties like moduli, Poisson's ratios, and other engineering properties. When one is deriving the engineering properties from these behaviors, then the initial linear portion of the behavior must be used. This is because in this initial linear behavior it is assumed that there are no damages due to the loading and plasticity effects, if any, introduced. There are three commonly used methods to measure the strains. The strains can be measured using extensometers, strain gages, and optical methods.

i. *Extensometers*: They provide average strain over a finite length, typically of the order 1 in (25.4 mm). These are used primarily for measurement of axial strains. They can also be used to measure transverse strains.
ii. *Strain gages*: The strain gages can be uniaxial, bidirectional, and rosettes. They measure strain in one, two, and three directions, respectively.
iii. *Rosettes*: They provide a complete description of the average strain over the region of measurement through strain transformation equations.

The strain gages are available in a range of sizes with the smallest of the order 1.59 mm bidirection, and rosettes can be stacked or adjacent. The length of the gauge may be specified by the relevant standards, but should always be significantly shorter than the gauge length of the specimen. Composites can cause particular difficulties not encountered with metals. The issues that must be addressed are as follows:

i. High gauge resistances are desirable because high voltages (2–4 V) with low current can then be used; this improves hysteresis effects and zero-load stability.
ii. If possible, use gauges with lead wires attached, or solder wires to the gauge before installation; this should avoid soldering damage to the composite.
iii. Ideally, the pattern of the autoclave scrim cloth should be removed before gauge installation; this is particularly important if contact adhesives are used.
iv. Corrections may be necessary to gauge transverse sensitivity effects; errors of over 100% between actual and measured strains can be obtained.
v. Gauges must be precisely aligned; errors of 15% can result from a misalignment. There is no universally acceptable way of ensuring alignment. The scrim cloth pattern can be misleading. Sometimes C-scan after installation can be useful or checking with failure surfaces after fracture.
vi. Dummy gauges are the preferred method for temperature compensation but, again, precise alignment is needed. It is necessary to mount the dummy gauges on an "identical" piece of laminate, with the same orientation relative to the fibers as used for the active gauges.

17.9.1 Rosette Principle

The normal strains are measured along any three directions, θ_1, θ_2, and θ_3, and then, strain transformation equations are used to determine the global strains ε_{xx}, ε_{yy}, and γ_{xy}.

$$\varepsilon_1(\theta_1) = \cos^2\theta_1\varepsilon_x + \sin^2\theta_1\varepsilon_y + \sin\theta_1\cos\theta_1\frac{\gamma_{xy}}{2}$$

$$\varepsilon_2(\theta_2) = \sin^2\theta_2\varepsilon_y + \cos^2\theta_2\varepsilon_x + \sin\theta_2\cos\theta_2\frac{\gamma_{xy}}{2}$$

$$\varepsilon_3(\theta_3) = \sin^2\theta_3\varepsilon_y + \cos^2\theta_3\varepsilon_x + \sin\theta_3\cos\theta_3\frac{\gamma_{xy}}{2} \tag{17.7}$$

When the normal strains, ε_1, ε_2, and ε_3 are known, then Eq. (17.7) is a system of three equations with three global strains, ε_{xx}, ε_{yy}, and γ_{xy}, as the unknowns. It should be noted that when the rosettes are placed on both sides of flat specimen, it provides the most complete information as to axial, transverse, and shear strains. Further, it provides the information on the development of specimen curvature. Figure 17.2 shows the arrangement of strain gages in a rosette. One should pay attention while using the rosettes that the measurements are sensitive to temperature. Hence, appropriate gage should be used for the test. Further, the measurements are sensitive to fiber orientation of the composite and gage alignment. Therefore, it is very important to measure the fiber orientation and gage alignment accurately and must be used appropriately in the analysis of experimental results.

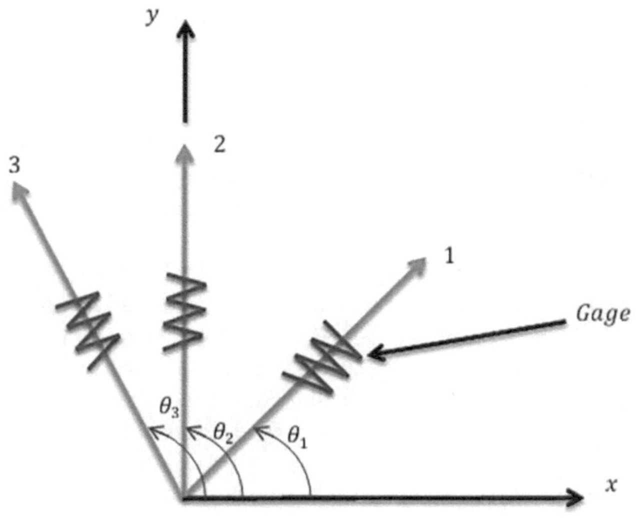

FIGURE 17.2 Strain gage rosette principle.

17.9.2 Tensile Test

The well-known purpose of the tensile testing is to measure the ultimate tensile strength and modulus of the composite. However, one can measure the axial Poisson's ratio with additional instrumentations. The standard specimen used for tensile testing of continuous fiber composites is a flat, straight-sided coupon. Flat coupons in ASTM standard D 3039/D 3039M-93 for 0° and 90° are [1] shown in Figure 17.3a and b, respectively. The specimen, as

mentioned above, is a flat rectangular coupon. The tabs are recommended for gripping the specimen. It protects the specimen from load being directly applied to the specimen causing the damage. Thus, the load is applied to the specimen through the grips. Further, it protects the outer fibers of the materials. The tabs can be fabricated from a variety of materials, including fiberglass, copper, aluminum, or the material and laminate being tested. When the tabs of the composite material are used, then according to ASTM specifications, the inner plies of the tabs should match with the outer plies of the composite. This avoids the unwanted shear stresses at the interface of the specimen and tabs. However, the recent versions of the ASTM standards allow the use of tabs with reinforcement at ±45°. Further, end-tabs can also facilitate accurate alignment of the specimen in the test machine, provided that they are symmetrical and properly positioned on the specimen. The tabs are pasted to the specimen firmly with adhesive. This specimen can provide data on the following:

i. The axial modulus, E_x
ii. In-plane and through thickness Poisson's ratio, γ_{xy} and γ_{xz}
iii. Tensile ultimate stress, σ_x^{ult}
iv. Tensile ultimate strain, ε_x^{ult}
v. Any nonlinear, inelastic response

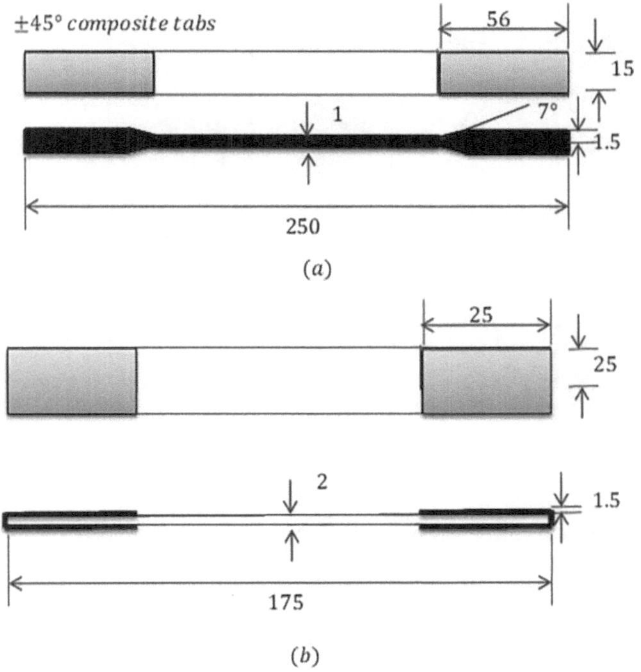

FIGURE 17.3 Composite tensile test specimens: (a) ASTM D 3039 for 0° and (b) ASTM D 3039 for 90°.

In general, the tensile tests are done on coupons with 0° laminas/laminate for the corresponding axial properties and coupons with 90° laminas/laminate for the corresponding transverse properties. The off-axis lamina specimen also provides data on coefficient of mutual influence and the in-plane shear response. The effective axial modulus and Poisson's ratio for orthotropic symmetric laminates with 0° and 90° laminas can be measured directly from a tensile test on a specimen of thickness t under axial force per unit length N_x as follows:

$$E_x = \frac{\sigma_{xx}}{\varepsilon_{xx}} = \frac{N_{xx}}{t\varepsilon_{xx}}$$

$$\nu_{xy} = -\frac{\varepsilon_{yy}}{\varepsilon_{xx}} \tag{17.8}$$

The tensile strength is defined as the average stress at failure. Thus, the tensile strength can be given using the maximum applied force per unit length N_x and thickness t as follows:

$$\bar{\sigma}_x^{ult} = \frac{N_x^{\max}}{t} \tag{17.9}$$

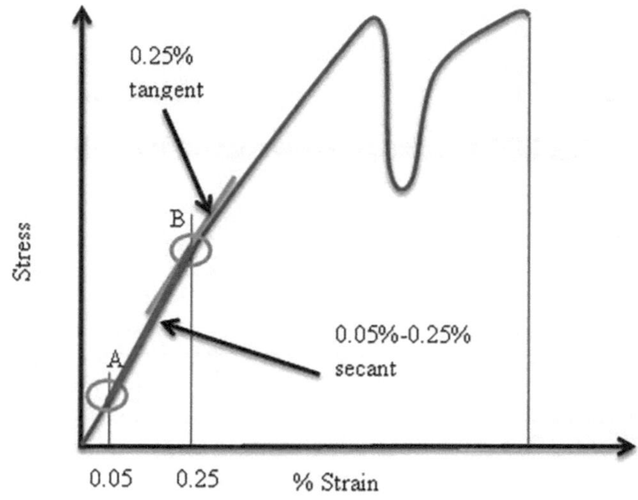

FIGURE 17.4 Typical tensile stress–strain curve.

It should be noted that due to progressive damage, the stiffness of the lamina or laminas/laminate changes causing the stress–strain curve to be nonlinear. The measurement of modulus in a tensile testing from a nonlinear loading curve can be done by three methods. In the first method, the modulus is taken as a tangent to the initial part of the curve. In the second method, a tangent is constructed at a specified strain level. For example, in Figure 17.4, the modulus is measured at 0.25% strain or 0.0025 strain (point B). In the third method, a secant is constructed between two points. For example, in Figure 17.4, a secant is constructed between points A and B. Typically, the strain values at these points are 0.0005 and 0.0025. In ASTM standards, the secant is called *chord*. The modulus measured by these methods is known as "initial tangent modulus," "B% modulus," and "A%–B% secant (chord) modulus," respectively.

17.9.3 Compression Test

Most of the structural members include the compression members. Such members can be loaded directly in compression or under a combination of flexural and compression loading. The axial stiffness of such members depends upon the cross-sectional area. Thus, it is proportional to the weight of the structure. One can alter the stiffness by changing the geometry of the cross section within limits. However, some of the composites have low compressive strength and this fact limits the full potential application of these composites. The compression testing of the composites is very challenging due to various reasons. The application of compressive load on the cross section can be done in three ways: directly applying the compressive load on the ends of a specimen, loading the edges in shear, and mixed shear and direct loading. These three ways of imposing the loads for compression testing are shown in Figure 17.5. During compression loading, the buckling of the specimen should be avoided. This demands a special requirement on the holding of the specimen for

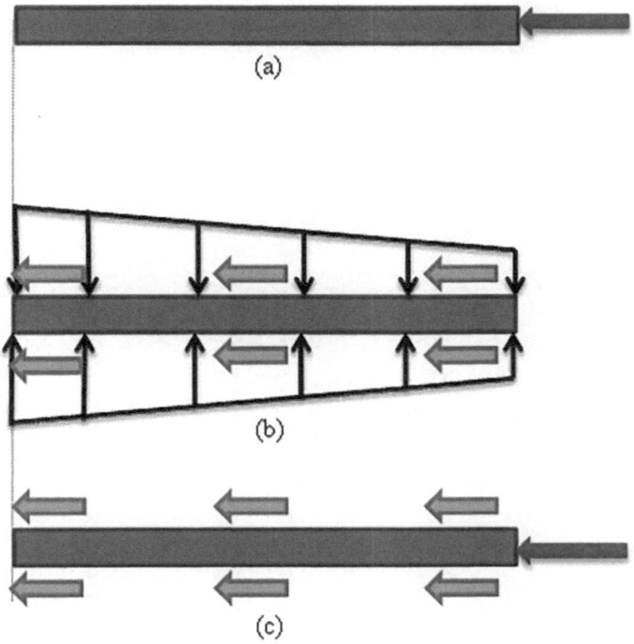

FIGURE 17.5 Load imposition methods for compression testing: (a) direct end loading, (b) shear loading, and (c) mixed shear and direct loading.

loading purpose. Further, it demands for special geometry of the specimen. These specimens are smaller in size as compared to the tensile testing specimens. A compression test specimen according to ASTM D695 (modified) standard [2,3] is shown in Figure 17.6.

17.10 SHEAR TESTING

Rail shear test is a very popular method used to measure in-plane shear properties. This method is extensively used in the aerospace industry. The shear loads are imposed on the edges of the laminate using specialized fixtures. There are two types of such fixtures: two-rail and three-rail fixtures. The ASTM D4255 [4–7] standard covers the specification for two- and three-rail specimens for both continuous and discontinuous (0° and 90° fiber alignment), symmetric laminates and randomly oriented fibrous laminates.

17.10.1 Two-Rail Shear Test

Figure 17.7 shows the two-rail shear test specimen geometry according to ASTM D4255 standard. It also shows the fixture for performing this test. The two-rail shear test fixture has two rigid parallel steel rails for loading purpose. The rails are aligned to the loading direction. Thus, it induces the shear load in the specimen that is bolted

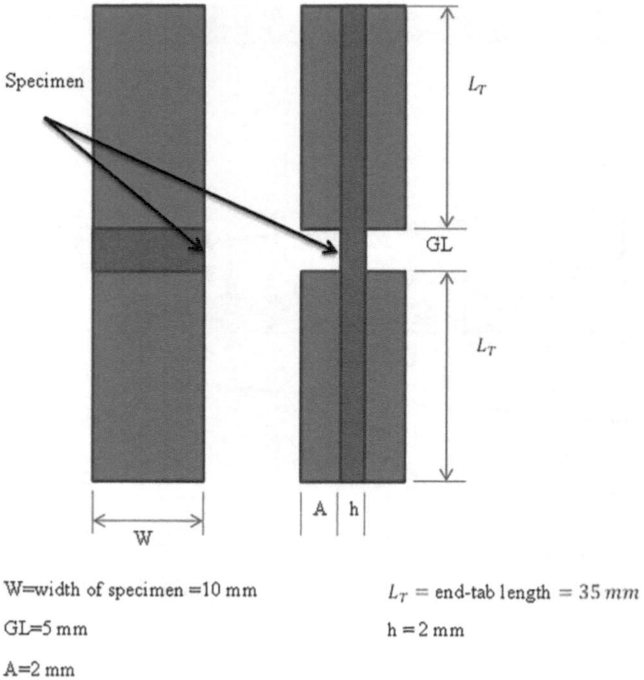

FIGURE 17.6 Composite compression test specimen according to ASTM D695 (modified) standard.

to these rails. A strain gage is bonded at 45° to the longitudinal axis of the specimen. The shear strength is obtained as:

$$\tau_{xy}^{ult} = \frac{P_{\max}}{Lh} \tag{17.10}$$

where

- P_{max} = ultimate failure load
- L = specimen length along the rails
- h = specimen thickness

The shear modulus is given as:

$$G_{xy} = \frac{\Delta P}{2Lh\Delta\varepsilon_{45}} \tag{17.11}$$

where

- ΔP = change in the applied load
- $\Delta\varepsilon_{45}$ = change in strain for +45° or −45° strain gage in the linear stress–strain regime.

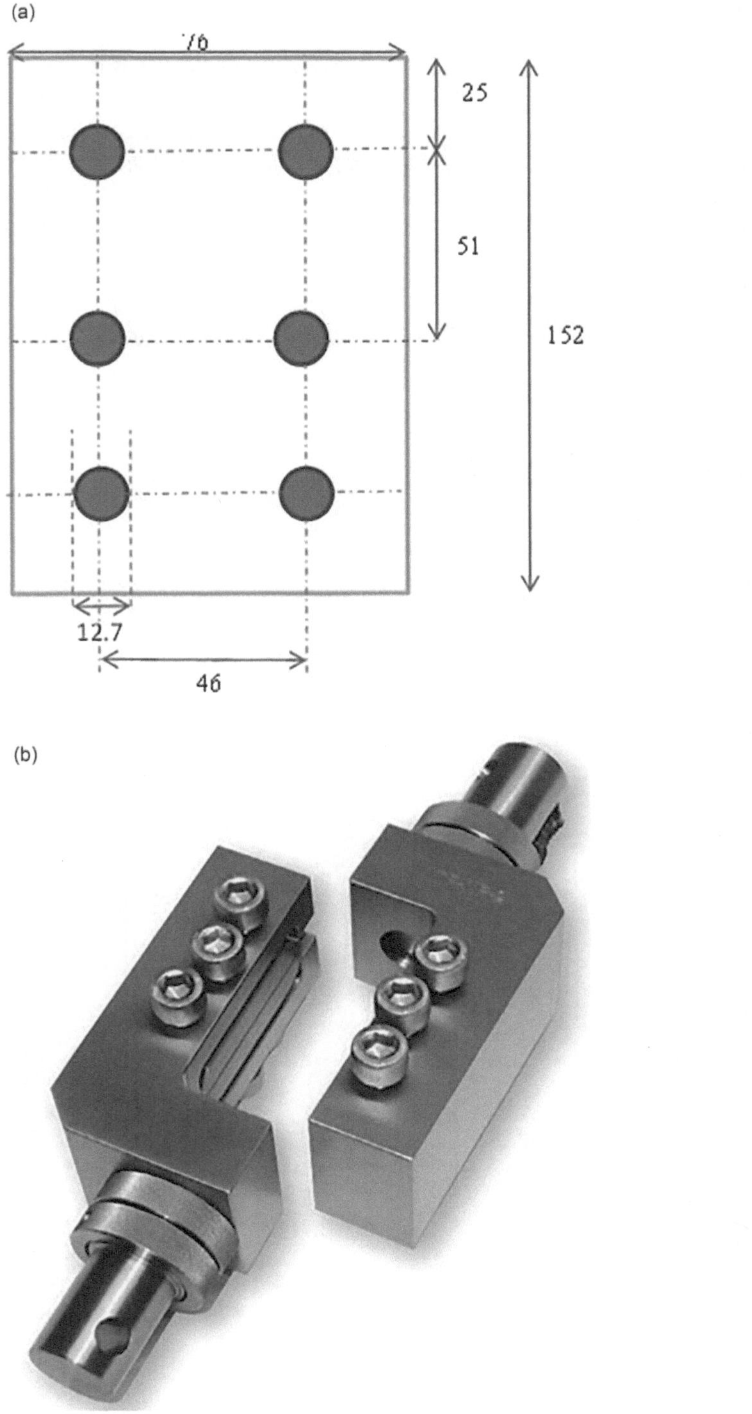

FIGURE 17.7 (a) Specimen for two-rail shear test and (b) fixture for holding the specimen.

17.10.2 Three-Rail Shear Test

The three-rail shear test is an improved version of the rail shear test. Using one more rail in two-rail shear test fixture, it can produce a closer approximation to pure shear. The fixture consists of three pairs of rails clamped to the test specimen. The outside pairs are attached to a base plate that rests on the test machine. Another (third middle) pair of rails is guided through a slot in the top of the base fixture. The middle pair is loaded in compression. The shear force in laminate is generated via friction between rail and specimen. The strain gages bond to the specimen at 45° to the specimen's longitudinal axis. The specimen geometry is shown in Figure 17.8.

The shear strength is obtained as:

$$\tau_{xy}^{ult} = \frac{P_{max}}{2Lh} \tag{17.12}$$

The shear modulus is given as:

$$G_{xy} = \frac{\Delta P}{4Lh\Delta\varepsilon_{45}} = \frac{\Delta\tau_{xy}}{\Delta\gamma_{xy}} \tag{17.13}$$

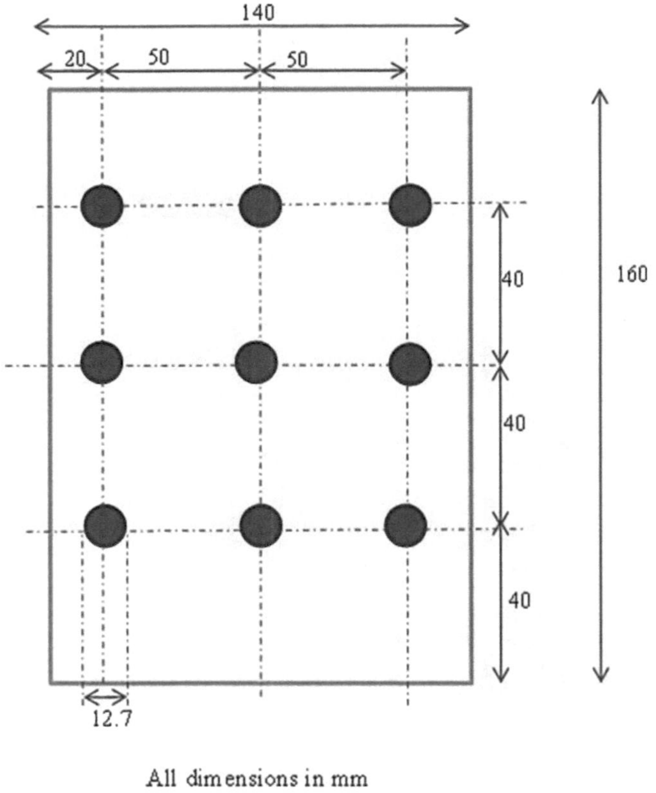

FIGURE 17.8 Specimen dimensions for three-rail shear test.

It should be noted that the holes in the specimen are slightly oversized than the bolts used for clamping. Further, the bolts are tightened in such a manner to ensure that there is no bearing contact between the bolt and specimen in the loading direction. It is recommended that each bolt is tightened with a 100 Nm torque.

REFERENCES

1. ASTM D3039/D3039M-08, Standard test method for tensile properties of polymer matrix composite materials. http://www.astm.org/Standards/D3039.htm.
2. ASTM D695, Standard test method for compressive properties of rigid plastics. http://www.astm.org/Standards/D695.htm.
3. ASTM D3410/D3410M, Test method for compressive properties of polymer matrix composite materials with unsupported gage section by shear loading. http://www.astm.org/Standards/D3410.htm.
4. ASTM D3518/D3518M, Test method for in-plane shear response of polymer matrix composite materials by tensile test of a 45 laminate. http://www.astm.org/Standards/D3518.htm.
5. ASTM D4255/D4255M, Test method for in-plane shear properties of polymer matrix composite materials by the rail shear method. http://www.astm.org/Standards/D4255.htm.
6. ASTM D7078/D7078M, Test method for shear properties of composite materials by v-notched rail shear method. http://www.astm.org/Standards/D7078.htm.
7. ASTM D3479/D3479M, Test method for tension-tension fatigue of polymer matrix composite materials. http://www.astm.org/Standards/D3479.htm.

Index